T0144486

ANCHORS IN THEORY AND PRACTICE
ANKER IN THEORIE UND PRAXIS

PROCEEDINGS OF THE INTERNATIONAL SYMPOSIUM ON ANCHORS IN THEORY
AND PRACTICE / SALZBURG / AUSTRIA / 9-10 OCTOBER 1995
BERICHTE DES INTERNATIONALEN SYMPOSIUMS ANKER IN THEORIE UND PRAXIS
SALZBURG / ÖSTERREICH / 9.-10. OKTOBER 1995

Anchors in Theory and Practice

Anker in Theorie und Praxis

Editor / Herausgeber
RICHARD WIDMANN
Österreichische Gesellschaft für Geomechanik

A.A.BALKEMA / ROTTERDAM / BROOKFIELD / 1995

The texts of the various papers in this volume were set individually by typists under the supervision of either each of the authors concerned or the editor.
Die Beiträge dieser Ausgabe wurden unter Aufsicht der einzelnen Autoren geschrieben.

Published by / Veröffentlicht durch
A.A. Balkema, P.O. Box 1675, 3000 BR Rotterdam, Netherlands (Fax: +31.10.413.5947)
A.A. Balkema Publishers, Old Post Road, Brookfield, VT 05036, USA (Fax: 802.276.3837)

ISBN 90 5410 577 1
© 1995 A.A. Balkema, Rotterdam
Printed in the Netherlands / Gedruckt in den Niederlanden

Anchors in Theory and Practice, Widmann (ed.) © 1995 Balkema, Rotterdam. ISBN 90 5410 577 1

Table of contents
Inhalt

2 *History cases*
Ausführungsbeispiele

3 Types of anchors
Ankertypen

4 Long-term behaviour
Langzeitverhalten

()

Preface

For decades anchors have increasingly been used as economic construction elements for temporary or permanent stabilization of slopes, tunnels and caverns in rock and soils, in mining, geotechnical engineering and for concrete structures. Various countries have developed standards for anchor engineering which are now being brought together in an EUROCODE.

Contributions to the conference on 'ANCHORS IN THEORY AND PRACTICE' are to provide an overview on the current state of engineering in this special subject which has often been covered only marginally in conferencing to date. The papers are focussing four major subjects that are of interest to designers, manufacturers, contractors and owners. The first halfday session is devoted to 'Load bearing capacity and effect of anchors'. Striking in this context are the different theoretical considerations to determine the load bearing capacity of dowels in concrete and anchors in rock or soil. When determining the load bearing capacity of usually short concrete dowels a sufficient strength over the bonding length is essentially assumed and the load bearing capacity is derived from the additional stresses in the surrounding concrete caused by the dowel. As far as anchors in rock or soil are concerned the anchor-length is determined by the weight of the retaining rock or soil mass providing safety for holding the load or prestressing force. The stresses along the bonding length are considered in detail. However, such anchors possible may cause effects on the change of stresses around the theoretical interface between the co-load-bearing rock or soil mass and its surrounding areas which will usually not be investigated – e.g. deformations or permeability. Also, the influence of bedding joints or rock discontinuities in general on the load bearing capacity is a subject that has rarely been investigated.

The mode of action of anchors, i.e. increased shear strength in the discontinuities or a certain doweling is covered in some papers, too.

'Case reports' are presented during the second halfday session. They cover the wide range of common anchor applications: slack or prestressed reinforcement of soils, retaining walls, temporary or permanent supports in mining, in the construction of roadways, tunnels and caverns. These contributions will emphasize the efficiency and the speed of anchoring applications.

The third halfday session deals with 'Anchor types'. The papers will reflect the efforts of manufacturers to gear their technical developments to the most various requirements of planners and contractors. Thus, new anchor materials will be discussed, e.g. glass-epoxy composites and improved mortars, epoxy resins or fast curing mortars. Reports will also cover tube anchors, multi anchor systems with different bonding lengths in the same borehole or self-drilling anchors.

The fourth halfday session will finally be devoted to the 'Long term behaviour and protection against corrosion'. Naturally these long term experiences may not always be representative for the present state of engineering, but they will certainly highlight some problem areas such as stress corrosion, brittle steels, creep etc., to be considered when using permanent anchors in future.

Finally the organizer wants to thank all those who have contributed to this meeting and hopes that it will provide valuable information on the current state and latest developments in anchor technology.

R. Widmann

Anker in Theorie und Praxis, Widmann (Herausgeber) © 1995 Balkema, Rotterdam. ISBN 90 5410 577 1

Vorwort

Anker werden seit Jahrzehnten als wirtschaftliches Konstruktionselement zur befristeten oder dauernden Erhöhung der Stabilität von Böschungen und Hohlräumen in Fels- und Lockergestein, im Bergbau wie im Tiefbau ebenso wie für Betonkonstruktionen in wachsendem Umfang eingesetzt. Dies hat auch in verschiedenen Ländern zu Normen in der Ankertechnik geführt, die nun in einem EUROCODE vereinheitlicht werden sollen.

Mit über 50 Beiträgen zur Tagung 'Anker in Theorie und Praxis' soll ein Überblick über den gegenwärtigen Stand der Technik auf diesem Spezialgebiet gebracht werden, das bisher auf Tagungen vergleichsweise wenig behandelt wird. Die Beiträge wurden zu vier Themen zusammengefaßt, die den Interessen der Projektanten, der Hersteller, der Ausführenden und der Erhalter Rechnung tragen.

Der erste Halbtag ist dem Thema 'Traglast und Wirkungsweise von Ankern' gewidmet. Auffallend erscheint die unterschiedliche theoretische Ermittlung der Traglast von Dübeln im Beton bzw. von Ankern in Boden oder Fels. Bei der Ermittlung der Traglast der meist kurzen Verankerungen im Beton wird vorwiegend die ausreichende Festigkeit der Haftstrecke vorausgesetzt und die Traglast aus den vom Anker im umgebenden Beton ausgelösten Zusatzspannungen abgeleitet. Bei Ankerungen in Boden oder Fels wird die Ankerlänge so bestimmt, daß durch das Gewicht des rückhaltenden Fels- oder Lockergesteinskörpers eine ausreichende Sicherheit für die Aufnahme der Traglast bzw. Vorspannkraft gewährleistet ist, im einzelnen werden mehr die Beanspruchungen der Haftstrecke betrachtet. Offen bleibt bei diesen Ankern die Untersuchung allfälliger Auswirkungen der Spannungsänderungen um die theoretische Grenzfläche zwischen dem mittragenden Boden- oder Felskörper und dem außerhalb liegenden Bereich, z.B. auf die Verformungen oder die Durchlässigkeit. Auch der Einfluß einer Schichtung oder Klüftung, allgemein der Trennflächen im Fels auf die Traglast der Anker wird nur vereinzelt untersucht.

Auch die Wirkungsweise von Ankern z.B. durch Erhöhung der Scherfestigkeit in den Trennflächen oder eine gewisse Verdübelung wird behandelt.

Im zweiten Halbtag wird über 'Ausführungsbeispiele' berichtet. Die Beiträge spiegeln die große Bandbreite der Einsatzmöglichkeiten von Ankern wider: schlaffe oder vorgespannte Bewehrung von Böden, Rückhaltung von Stützwänden, vorübergehende oder dauernde Sicherungen im Bergbau, Stollen-, Tunnel- und Kavernenbau. Ein Schwerpunkt dieser Beiträge ist die wirtschaftliche und rasche Ausführung der Ankerungen.

Im dritten Halbtag 'Ankertypen' werden in die Bemühungen der Hersteller aufgezeigt, mit der technologischen Entwicklung der Anker selbst den vielfältigen Anforderungen der Planer und Ausführenden zu entsprechen. So wird über neue Materialien für Anker, wie z.B. Glas – Epoxy – Verbundstoffe, und Mörtel, wie Kunstharze oder schnellbindende Zement-Mörtel, ebenso berichtet wie über Rohranker, Mehrfachanker mit unterschiedlicher Tiefe der Haftstrecke im gleichen Bohrloch oder selbstbohrende Anker.

Der vierte Halbtag schließlich ist dem Korrosionsschutz für Daueranker und den Langzeiterfahrungen gewidmet. Die Langzeiterfahrungen mögen naturgemäß nicht immer für den heutigen Stand der Technik repräsentativ sein; zweifellos aber geben sie Hinweise auf Problembereiche, die beim künftigen Einsatz von Dauerankern besondere Aufmerksamkeit erfordern, wie Spannungskorrosion, spröde Stähle, Kriechen usw.

Auch an dieser Stelle möchte der Veranstalter allen, die zum Gelingen dieser Tagung beigetragen haben, danken und der Hoffnung Ausdruck geben, einen Beitrag zur Information über den heutigen Stand der Ankertechnik und zur technologischen Weiterentwicklung geleistet zu haben.

R. Widmann

1 Bearing capacity and effect of anchors
Traglast und Wirkung von Ankern

Analytical model for hook anchor pull-out

Analytisches Modell für die Haftfestigkeit von Ankern mit Querplatte

Rune Brincker, Jens Peder Ulfkjær, Peter Adamsen, Lotte Langvad & Rune Toft
University of Aalborg, Denmark

ABSTRACT: A simple analytical model for the pull-out of a hook anchor is presented. The model is based on a simplified version of the fictitious crack model. It is assumed that the fracture process is the pull-off of a cone shaped concrete part, simplifying the problem by assuming pure rigid body motions allowing elastic deformations only in a layer between the pull-out cone and the concrete base. The derived model is in good agreement with experimental results, it predicts size effects and the model parameters found by calibration of the model on experimental data are in good agreement with what should be expected.

ZUSAMMENFASSUNG: Ein einfaches analytisches Modell für die Haftfestigkeit von Anker mit Querplatte ist behandelt. Das Modell basiert sich auf eine vereinfachte Fassung des fiktiven Riss modells. In diesem Modell wird angenommen, dass die Materialpunkte am Rissverlängerungspfad sich in einem der folgenden möglichen Stadien befinden: A) lineärelastisches Stadium, B) Riss stadium, wo das Material durch Kohäsionskräfte in der Riss prozesszone erweichert ist, und zuletzt C) ein Stadium mit keine Spannungstransmission. Im Bruchstadium ist der Riss prozess von einer erweicherten Relation beschrieben, die die Normalspannungen der Bruchfläche σ mit der Rissöffnung, w (Abstand zwischen die Bruchflächen) verbindet. Es wird angenommen, dass der Bruchform eine Entziehung eines kegelförmigen Betonkörpers entspricht. Das Problem wird vereinfacht durch die Annahme, dass reine Steifkörperbewegungen nur elastische Deformationen in einer Schicht zwischen den Haftfestigkeitskegel und das Betonbasis erlauben. Die steuernden Gleichungen werden dann mittels einfachen Gleichgewichtsbedingungen abgeleitet. Das abgeleitete Modell stimmt mit den experimentellen Resultaten gut überein. Es kann Gröss eneffekte voraussagen, und die Modellparameter, die durch kalibrierung des experimentellen Datenmodells mit experimentellen Daten gefunden ist, stimmen mit den Erwartungen gutüberein.

1 INTRODUCTION

Anchors are used in most reinforced concrete structures. It might be simple adhesive anchors, expansion anchors or hook anchors, figure 1. Usually, the simple adhesive anchor, figure 1.a is used where it is possible. It is simple and reliable. Further, since this anchor is usually designed in such a way that the load bearing capacity of the adhesive anchor relies on the shear resistance of the interface between the bar and the concrete, the failure process is ductile, and thus, as for all ductile failure problems, no or at least small size effects are observed. However, the simple adhesion anchor needs a relatively long embedment length to ensure enough load bearing capacity, and to ensure that the failure of the anchor will be pull-out of the anchor bar. If the space is limited and the embedment length

is reduced, there is a risk that the mode of failure will change from pull-out of the bar to pull-off of a concrete cone. In this case, the failure is more brittle, and the load-bearing capacity no longer depends on the shear resistance of the interface. In this case, the load-bearing capacity depends on the tensile strength and the fracture energy of the concrete material and of the size and the shape of the pulled-off concrete cone. The bigger the cone, the larger the load-bearing capacity, and, thus, it is natural to force the concrete cone to start as deeply as possible. This can be done by introducing an expansion part at the end of the anchor, figure 1.b, but the safest way of ensuring the cone to start at the end of the anchor is to provide the anchor with a "hook", usually shaped like an anchor plate at the end of thenchor bar, figure 1.c. As already mentioned, since the failure of the hook anchor mostly

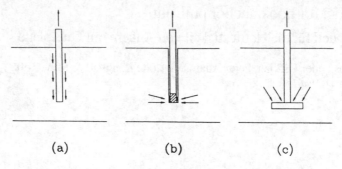

(a)   (b)   (c)

Figure 1. Different ways of transferring the load from the anchor bolt to the concrete, a) adhesive anchor, b) expansion anchor and c) hook anchor.

Figur 1: Verschiedene Methoden für Lasttransmission von Kopfbolzendübel zu Beton, a) Klebeanker, b) Expansionsanker und c) Anker mit Querplatte.

depends on the fracture mechanical properties of the concrete, the load-bearing capacity is expected to show a clear size effect.These size effects have been observed by several researchers, Bocca et al. 1990, and by Eligehausen and Savade ,1989. Their results indicate a strong size effect over embedment depth ranging from 50 mm to 500 mm. Eligehausen and Clausnitzer,1983, studied the behaviour of anchors by finite element models. Their investigation showed a clear influence of the type of model used. An ideal plastic model gave higher load-bearing capacities than a more brittle model using material softening.

Elfgren et al. ,1987, also studied the problem numerically using a fictitious crack model approach for the softening material. Their investigation indicates, that the shear stresses in the crack should be incorporated in the model. Also Rots,1990, investigated the problem numerically. He studied the influence of the number of radial cracks (cracking of the concrete cone) and used a smeared crack approach. His results indicate that the number of radial cracks tend to increase the ultimate load. Elfgren and Ohlson ,1990, studied the influence of tensile strength and fracture energy using a finite element analysis. As expected, their results indicate that the ultimate load and the ductility of the failure process increase with the fracture energy, and that increasing the tensile strength will increase the ultimate load and the brittleness of the failure process. Bocca et al. ,1990, made an analysis using the fictitious crack model in a finite element analysis using axi-symmetric elements and a re-meshing technique. They found good agreement with experimental results. Also Ozbolt and Eligehausen,1993, made a finite element analysis using axi-symmetric elements. They showed that cracking

starts at about 30 % of the ultimate load, and that the ultimate load is mainly determined by the fracture energy. Also, their results indicate a strong size effect on the ultimate load.

Tommaso et al. ,1993, investigated the influence of the shape of the crack opening relation. They found that a bi-linear softening curve predicts more realistic results than a single-linear curve. Similar results have been found by Urchida et al. ,1993.

2 BASIC ASUMPTIONS

In this section the basic assumptions of the simplified model describing anchorage pull-out using the fictitious crack approach is presented. The fictitious crack model is due to Hillerborg, 1977, but the basic idea is close to that of Dugdale, 1960, who used a similar approach assuming a constant yield stress in the fracture zone, and Barenblatt, 1962, who assumed a more general distribution of the stresses in the fracture zone. Usually the fictitious crack model concept is used in finite element programs using special no-volume elements, 1989, or using the smeared crack approach Rots 1989, or it might be formulated using sub-elements describing the elastic behaviour and introducing the softening only for the material in the pre-selected crack path, Petersson, 1982, Brincker and Dahl, 1989. However, these methods are complicated and time-consuming to use for design, and they do not provide simple analytical solutions indicating the degree of brittleness and indicating how strongly a certain problem might be influenced by size effects. Thus, it is desirable to have simple models that describe the basic fracture behaviour qualitatively correct in order to have simple tools especially for describing the brittleness of the failure process.

The intention of the models presented here is to formulate the simplest possible model that reflects the basic fracture mechanical behaviour. The model problem is illustrated in figure 2. The problem is assumed to be plane, i.e. the 3-dimensional problem is not considered, and thus radial crack are omitted from the analysis. Further, the crack path is assumed to be linear, the slope being described by the angle φ, and the deformation is assumed to be a rigid body motion as a rotation around the point where the crack path meets the surface of the concrete. The depth L is related to the radius of the cone by the equation $L = R$ tan(φ). The cone and the surrounding concrete is assumed to be perfectly rigid, all the elasticity being described by an elastic layer between the cone and the rest of the body. This simple approach has proved its value in modelling of the failure process for plain and reinforced beams, Ulfkjær et al. 1993a, 1993b, 1995.
In the distance r from the edge of the anchor stud the vertical deformation w is

$$w = u(1 - \frac{r}{R}) \qquad (1)$$

This deformation will cause vertical as well as horizontal stresses in the elastic layer, and horizontal as well as vertical reactions at the rotation point. However, in this simplified analysis, it will be assumed, that the geometry is chosen in such a way, that the vertical reactions at the rotation point can be neglected. Thus, considering only vertical stresses $\sigma = \sigma(r)$, the corresponding force is given by

$$F = \int_0^R 2\pi r \sigma(r) dr \qquad (2)$$

3 SINGLE-LINEAR SOFTENING

For the case of single-linear softening the physical relation of the layer is as shown in figure 3, i.e. the elastic part is linear and the softening part is linear. Here w is the the total deformation and, thus, it includes elastic as well as softening terms.

For any point in the crack path, as long as the stresses have not reached the ultimate stress σ_u, the response is linear, and no crack is present at that point. The deformation w_u where the softening starts is given by

$$w_u = \delta \frac{\sigma_u}{E} \qquad (3)$$

where E is Young's modulus of the concrete, and δ is the thickness of the elastic layer. eq. (3) defines the layer thickness δ.

The fracture energy is the area below the stress-deformation relation in figure 3, i.e. the fracture energy is

$$G_F = \frac{1}{2} w_c \sigma_u \qquad (4)$$

Using the introduced physical relation for the elastic layer the stress is given by

$$\sigma(r) = \begin{cases} w(r)\dfrac{\sigma_u}{w_u} & \text{for } w(r) \leq w_u \\[2mm] \sigma_u \left(1 - w(r) - \dfrac{w_u}{w_c} - w_u\right) & \text{for } w_u \leq w(r) < \\[2mm] 0 & \text{for } w_c \leq w(r) \end{cases} \qquad (5)$$

As it appears, this divides the fracture process into three phases. In phase I the deformation u has not reached the deformation w_u and thus, no fictitious

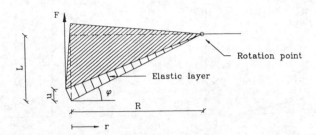

Figure 2. Geometry of simplified fictitious crack model.

Figur 2: Geometrie des vereinfachten fiktiven Riss modells.

5

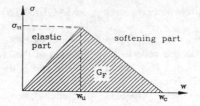

Figure 3. Stress-deformation relationship for the case of single-linear softening.

Figur 3: Spannungs-Deformationsverhältnis für monolineare Erweicherung.

crack is present. In phase II u is between w_u and w_c, i.e. a fictitious crack has developed. Finally, in phase III u has exceeded the critical crack opening w_c and a real crack has developed. The stress distributions for the three phases are illustrated in figure 4. Let c denote the length of the real crack, and let a denote the total length of the crack (real crack + fictitious crack). Now, using eq. (1) and (2) together with eq. (5) and carrying out the integrations, the following expression is obtained for the force

$$F = 2\pi\sigma_u\left(\frac{1}{2}(a^2-c^2)\left(1-\frac{u-w_u}{w_c-w_u}\right) + \right.$$
$$+ \frac{1}{3}(a^3-c^3)\frac{u}{R}(w_c-w_u)\right) +$$
$$+ 2\pi\frac{\sigma_u}{w_u}u\left(\frac{1}{6}R^2+a^2(\frac{a}{3}R-\frac{1}{2})\right) \tag{6}$$

where the crack parameters a and c are given by

$$a = \begin{cases} 0 & \text{for} \quad u<w_u \\ R(1-w_u/u) & \text{for} \quad u\geq w_u \end{cases} \tag{7}$$

$$c = \begin{cases} 0 & \text{for} \quad u<w_c \\ R(1-w_c/u) & \text{for} \quad u\geq w_c \end{cases} \tag{8}$$

The equations (6), (7) and (8) describe the pull-out of the concrete cone using the displacement u as the controlling parameter. To use the model the constitutive parameters σ_u, w_u and w_c most be known as well as the radius R of the cone at the concrete surface. Figure 5 shows a typical load-displacement curve simulated by the model using the values $R = 1000w_c$, $w_u = 2/17\ w_c$ and $G_F = 178\ \sigma_u/w_c$. The plot was made non-dimensional by dividing the force F by $\sigma_u R^2$ and the displacements u by w_c.

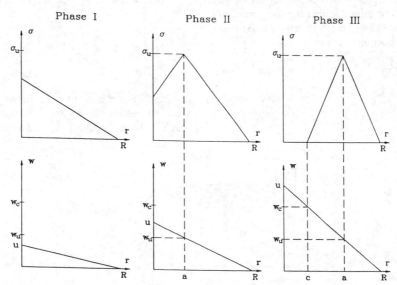

Figure 4. Stress distribution in the three phases.

Figur 4. Spannungfeld für die drei stadien

6

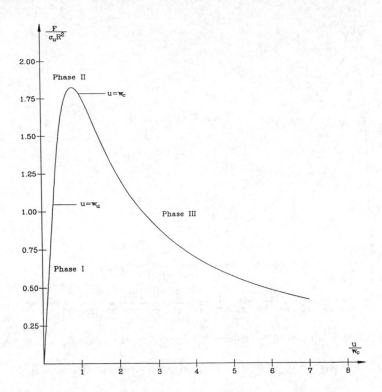

Figure 5. Relationship between force and deformation simulated by the model using single-linear softening.

Figur 5: Kraft-Deformationsverhältnis simuliert mittels monolinearer Erweicherung.

4 BRITTLENESS AND SIZE EFFECTS

The introduced analytical model provide a simple way of expressing the brittleness of the pull-out problem. The classical brittleness number $B = f_t^2 \, l/EG_F$ Bache , 1989, might be derived from the single-linear model of the fracture of a bar in uniaxial tension using the definition

$$B = 2\frac{w_u}{w_c} \qquad (9)$$

using this definition together with eqs. (3) and (4) yields the following expressions for the brittleness number for the pull-out problem

$$B = \frac{w_u \sigma_u}{G_F} = \frac{\delta \sigma_u^2}{EG_F} \qquad (10)$$

In this expression it would be natural to assume that the ultimate stress σ_u is proportional to the tensile strength f_t of the concrete. The thickness δ of the elastic layer might be estimated from the initial slope S of the relation between the force F and the

displacement u at the bottom plate. From the elastic regime of eq. (6) or (11) the relation is found as $\delta = \pi R^2 E/3S$. The shape of the pull-out relation depends on the brittleness number B. This effect is illustrated in figure 6 showing results for the single-linear case.

By introducing the brittleness number into equation (6), (7) and (8) and making all quantities non-dimensional the following simplfied equations are obtained

$$\frac{\mu}{2\pi\sigma_u L^2} = \frac{1}{2}(\alpha^2 - \gamma^2)\left(1 - \frac{\theta - 1}{\frac{2}{B} - 1}\right) + $$

$$\frac{1}{3}(\alpha^3 - \gamma^3)\frac{\theta}{\rho\left(\frac{2}{B} - 1\right)} + \qquad (11)$$

$$\theta\left(\frac{1}{6}\rho^2 + \alpha^2\left(\frac{\alpha}{\rho} - \frac{1}{2}\right)\right)$$

where the crack parameters α and γ are given by

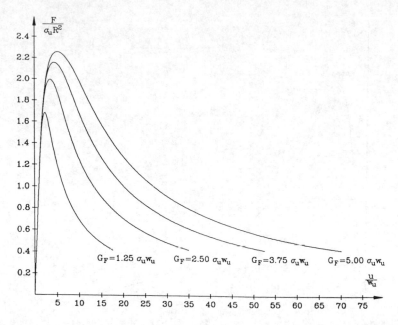

Figure 6. Influence of the brittlenes number B illustrated by varying the fracture energy G_F.

Figur 6: Einfluss von Sprödigkeitszahl B illustriert durch varierende Bruchenergie G_F.

$$\alpha = \begin{cases} 0 & \text{for } \theta < 1 \\ \rho\left(1 - - \frac{2}{\theta}\right) & \text{for } \theta \geq 1 \end{cases} \qquad (12)$$

$$\gamma = \begin{cases} 0 & \text{for } \theta < \frac{2}{B} \\ \rho\left(1 - - \frac{2}{B\theta}\right) & \text{for } \theta \geq \frac{2}{B} \end{cases} \qquad (13)$$

where $\theta = u/w_u$ is the controlling parameter and $\rho = R/L$ is a shape parameter. By writing the problem in non-dimensional form it ts seen that the problem is completely described by the brittleness number B and the shape parameter ρ simplifing the problem considerrably.

Size effects are studied by varying the size of the pull-out problem considering geometrically similar cases and comparing the ultimate load F_u normalised by $\sigma_u L^2$ corresponding to the normalization in equation (11). The result is shown in figure 7. The maximum size effect that can be predicted by the model is found using the stress distributions $\sigma = \sigma_u$ corresponding to ideal ductile behaviour (very small sizes) and $\sigma = \sigma_u (R-r)/R$ corresponding to brittle behaviour (large

sizes). Using these stress distributions in eq. (2) it is found that the maximum size effect predicted by the model is a factor of three - exactly the same as for a beam in bending, Ulfkjær et al., 1995. Note, that since the linear fracture mechanics is not incorporated in this model, no size effects are predicted when the size exceeds a certain level. This model predicts only the non-linear size effects, i.e. the size effects in the region where non-linear effects are dominating. In cases where the non-linear effects are not dominating, i.e. for very large specimens, using the results of the analysis carried out here might be misleading. However, the analysis indicates, figure 7, that for small embedment depths, for L ranging from 0 to about 50 mm, non-linear effects are dominating, and thus, the results predicted by the model should be representative.

5 EXPERIMENTS

In order to investigate the aplicability of the derived model a series of experiments with high strength concrete is performed.

5.1 Specimens

The specimens were rectangular with a depth of 150

8

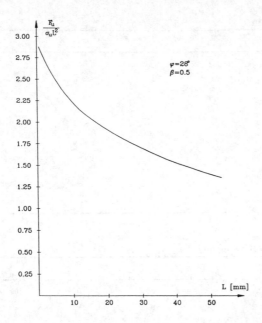

Figure 7. Model predicted size effects on the load-bearing capacity.

Figur 7: Modellierte Gröss eneffekte auf der Tragfähigkeit.

mm and base of either 1120 mm by 1120 mm or 620 mm by 620 mm dependent on the embedment depth of the anchor. The embedment depths were varied between 25 mm and 95 mm. In all 18 specimens were tested.

5.2 Concrete

A high strength concrete was used for the experiments. The composition of the concrete is shown in table 1. The cylinder compressive strength, the cylinder splitting strength and the modulus of elasticity were determined using standard methods on 100 mm by 200 mm cylinders. The fracture energy was determined on 840 mm by 100 mm by 100 mm beams in witch a notch was saw cut at the mid-section of the beam. The beams were subjected to three point bending. The physical parameters of the concrete are shown in table 2.

5.3 Casting

All the specimens were casted of the same batch, wich was delivered from a commercial concrete manufacturer. The anchor was embedded in the middle of the concrete matrix. The load was apllied at an anchor stud. The diameter of the anchor stud was 18 mm, the diameter of the anchor head was 34 mm and the thichkness of the head was 25 mm. On top of the anchor head an anchor extension with a diameter of 8 mm was mounted. The extension was so long so it would reach the opsite side of the speciemen. The anchor extension was used to measure the displacements of the anchor. The same dimension were used for all the anchors.

5.4 Test Set-up

The test set-up and geometri is shown in figure 8. The anchor is pulled out by downwards movement of the piston, which is linked to the anchor stud by an adaptor. At the top of the test specimen an LVDT (Linear Variable Differential Transformer) is situated. The LVDT meassures the vertical displacement of the bottom of the anchor head via the the extension mountet on the anchor head. Since the fitting for the LVDT is fixed at the concrete plate, deflection of the plate has no effect on the meassurements of the LVDT.

The above mentioned placing of the LVDT has been chosen because the displacement of the piston was not

Table 1. Composition of concrete.

Component	Amount [kg/m^3]
Cement	445
Silica	35
Gravel	881
Sand	822
Water	147
Peramin F (Plast.)	4.2
Pozz 80 (Super Plast)	0.7

Table 2. Physical parameters of the concrete

Property	Value
Modulus of Elasticity	39.6 GPa
Compressive strength	87.6 MPa
Splitting Strength	5.3 MPa
Fracture Energy	0.o93 N/mm

found suitable for estimating the movement of the anchor. This is due to friction between the anchor and the concrete, deformations of the test rig, elastic deformations of the test specimen and local crushing of the concrete. All thses phenomenon makes it impossible to correct for false deformations. In figure 9 is shown the difference between the piston deformation and the LVDT. Comparing the load displacement history of the piston and the LVDT a large difference is seen. This supports the importance of the method of the direct meassurements of the displacement of the anchor.

A servocontrolled material testing sysstem is used. Both the LVDT and the piston displacement isincluded in the feedback signal which controls the movements of the piston. The feedback signal is created by analog addition of the corresponding signals from the piston and the LVDT in the following proportion:

$$U_f = 0.9 U_{LVDT} + 0.1 U_{PISTON} \qquad (14)$$

where U_f is the feedback signal, U_{LVDT} is the signal from the LVDT and U_{PISTON} is the signal from the piston.

6 CALIBRATION AND EVALUATION

The model was calibrated on the 18 pull-out tests. The calibration was performed by inspecting the fit visually using a computer programme allowing for easy adjustment of all relevant parameters. Figure 10 shows the result of a typical calibration. As it appears, the fit is quite good over the entire measurement range.

The parameters were calibrated in the following way. First, the stress σ_u was chosen as a fixed value close to the measured tensile strength of the concrete. Then the initial slope was calibrated as explained in the preceding section. Then the peak load and peak deformation were calibrated by simultaneously changing the the radius R and the fracture energy G_F. A better estimate of the parameters can however be obtained by using an automized optimization algorithm as done for beams in three point bending in Ulfkjær and Brincker, 1992.

The results of the calibrations are shown in table 1. As it appears, two values are given for the radius, the value R for the final radius observed after the test, and the effective radius R' as it was estimated by calibration of the model, figure 11. As it appears, typically there is a factor 2-3 between the two values.

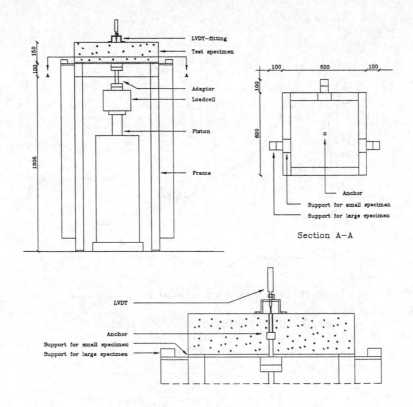

Section A–A

Figure 8. The test set-up

Figur 8. Die Versuchsaufstellung.

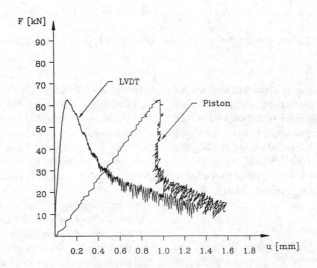

Figure 9. Force versus displacement. Displacement of piston and meassured by LVDT.

Figur 9. Kraft gegen Bewegen. Das Bewegen des Kolben und gemessen von LVDT.

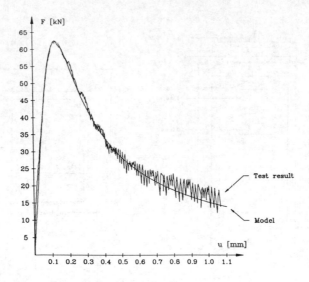

Figure 10. Calibration of model to test result.

Figur 10: Kalibrierung des Modells zu Testresultaten.

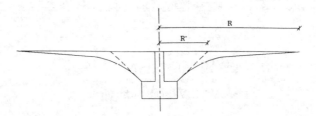

Figure 11. Radius R observed by test, and effective radius R' used in model.

Figur 11: Radius R gefunden im Test, und effektiver Radius R' benutzt im Modell.

The results do not necessarily represent any serious discrepancy between the model and reality. The cones that were pulled off during the test showed a curved crack path corresponding to large initial values of φ that were substantially decreased during the fracture process. Thus, since the model include only one value of φ, and since this value should be close to the initial value of φ observed during the test, relatively small values of the radius R should be expected when calibrating the model.

Further, the values of the fracture energy estimated by the model, the effective fracture energy G'_F, is substantially larger than usual fracture energies for concrete. Since an ordinary high-strength concrete was used, this effect must be due to the model. However, as before, this is to be expected considering the low values of the effective radius R'. Since the area under the force-displacement curve is approximately correct, the following relationship between the real and the effective parameters must hold $\pi R'^2 G'_F = \pi R^2 G_F$. If the values of the effective fracture energy is interpreted in this way, the results become close to the values of the fracture energy usually observed experimentally.

An examination of the estimated values for the effective radius R' and for the initial angle φ indicates that the problem is not geometrically independent of the size. Figure 12 shows the estimated values of φ as a function of the size. As it appears, the fracture angle does not seem to be constant. The results indicate a typical value of φ around 20-25 degrees for very small embedment depths, and a value of φ around 40 degrees for embedment length around 100 mm.

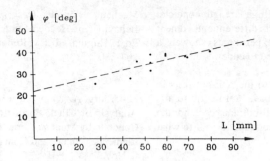

Figure 12. Estimated fracture angle φ as a function of the embedment depth.

Figur 12: Estimierter Bruchwinkel φ als Funktion der Haftstrecke.

Table 3. Model parameters estimated by calibration of the model

Name	L	R'	R	σ_b	G_F	φ
28 a	28	59	125	5.0	0.45	25.4
44 a	44	83	235	5.0	0.38	27.9
47 a	47	65	225	5.0	0.55	35.9
47 b	47	65	205	5.0	0.55	35.9
47 c	47	65	190	5.0	0.55	35.9
53 b	53	75	220	5.0	0.40	35.3
53 c	53	65	265	5.0	0.72	39.1
55 b	53	86	185	5.0	0.35	31.6
60 a	60	73	255	5.0	0.50	39.4
60 b	60	75	240	5.0	0.68	38.6
69 b	69	87	255	5.0	0.50	38.4
69 c	69	87	260	5.0	0.50	38.4
70 a	70	90	265	5.0	0.50	37.9
70 c	70	90	280	5.0	0.60	37.9
80 c	80	93	190	5.0	0.50	40.7
95 b	95	98	200	5.0	0.42	44.1

7 CONCLUSIONS

A simple model has been presented for the non-linear fracture mechanical problem of the pull-out of a concrete cone in a hook anchor failure test.

The model is formulated combining the fictitious crack model with very simple assumptions concerning the displacement field and the elastic response of the material around the crack path. Further, the solutions only correspond to an approximate satisfaction of the equilibrium equations.

To investigate the applicability of the model 18 pull

out tests were performed on a high strength concrete. In order to investigat size effects different embedment depths were studied. The experiments showed a significant size effect compared to codes.

The model was calibrated to the pull-out tests in different sizes. The model gave a fine fit to the experimentally measured pull-out curves, and the estimated model parameters correspond well to what should be expected.

8 REFERENCES

Bache, H., 1989: Brittleness/Ductility from a Deformation and Ductility points of View, in L. Elfgren (ed.) Fracture Mechanics of Concrete Structures, Chapman and Hall, pp. 202-207.

Barenblatt, G.I, 1962., in H.L. Dryden and T. Karman (eds.), pp. 56-131 : Advances in Applied Mechanics, Academic, Vol. 7.

Brincker, R and H. Dahl, 1989: On the Fictitious Crack Model of Concrete Fracture, Magazine of Concrete Research, Vol. 41, No. 147, pp. 79-86.

Bocca, P., A. Carpinteri and S. Valente, 1990: Fracture Mechanics Evaluation of Anchorage Bearing Capacity in Concrete}, Applications of Fracture Mechanics to Reinforced Concrete, Ed. A. Carpinteri, Elsevier Applied Science, pp. 231-265}

Dugdale, D.S, 1960.: Yielding of Steel Sheets Containing Slits, Journal Mech. Phys. Solids, Vol. 8, pp. 100-104.

Elfgren, L., U. Ohlson and K. Gylltoft, 1987: Anchor Bolts Analysed with Fracture Mechanics, Proc. of the International Conference on Concrete and Rock, Houston, Texas, June 17-19.

Elfgren, L., and U. Ohlson, 1990: Anchor Bolts Modelled with Fracture Mechanics, Applications of Fracture Mechanics to Reinforced Concrete, Ed. A. Carpinteri, Elsevier Applied Science, pp. 231-265.

Eligehausen, R. and Clausnitzer, 1983: Analytiches Modell zur Bescreibung des Tragverhaltens von Befestigungselementen, Report 4/1-83/3, Institut fü Werstoffe im Bauwesen, Universität Stuttgart.

Eligehausen, R. and G. Savade, 1989: A Fracture Mechanics based Description of the Pull-Out Behaviour of Headed Studs Embedded in Concrete, Ed. L. Elfgren, RILEM Report, Chapman and Hall, pp. 264-280.

Hillerborg, A., M. Modeer and P.E. Petersson, 1977: Analysis of Crack Formation and Crack Growth in Concrete by Means of Fracture Mechanics and Finite Elements, Cement and Concrete Research, Vol. 6, No. 6, Nov..

Hillerborg, A. and J.G. Rots, 1989: Crack Concepts and Numerical Modelling pp. 128-137 in L. Elfgren (ed.) Fracture Mechanics of Concrete Structures, Chapman and Hall.

Ozbolt, R. and Eligehausen, 1993: Fastening Elements in Concrete Structures, Fracture and Damage of Concrete and Rock, Ed. H.P. Rossmanith, Proc. of the 2nd International Conference on Fracture and Damage of Concrete and Rock, Vienna, Austria, E & FN Spon.

Petersson, P.E.: Crack Growth and Development of Fracture Zones in Concrete and Similar Materials, Report TVBM-1006, Division of Building Materials, Lund Institute of Technology, Sweden.

Rots, J.G, 1989.: Smeared Crack Approach, in L. Elfgren (ed.) Fracture Mechanics of Concrete Structures, Chapman and Hall, pp. 138-146.

Rots, J.G., 1990: Simulation of Bond and Anchorage: Usefulness of Softening Fracture Mechanics, Applications of Fracture Mechanics to Reinforced Concrete, Ed. A. Carpinteri, Elsevier Applied Science, pp. 231-265.

Tomasso, A.D., O. Manfrodi and G. Valente, 1993: Comparison of Finite Element Concrete Models Simulating Pull-Out Tests, Fracture and Damage of Concrete and Rock, Ed. H.P. Rossmanith, Proc. of the 2nd International Conference on Fracture and Damage of Concrete and Rock, Vienna, Austria, E & FN Spon.

Uchida, Y., K. Rokugo and W. Koyanagi, 1993: Numerical Analysis of Anchor Bolts Embedded in Concrete Plates, Fracture and Damage of Concrete and Rock, Ed. H.P. Rossmanith, Proc. of the 2nd International Conference on Fracture and Damage of Concrete and Rock, Vienna, Austria, E & FN Spon.

Ulfkjær, J.P. and Brincker R. 1992, Indirect Determination of the σ-w Relation of HSC through

Three-Point Bending, in Fracture and Damage of Concrete and Rock, FDCR-2 (Edited by H.P. Rossmanith), Vienna, Austria, E & FN Spon, London, pp.135-144.

Ulfkjær, J.P., O. Hededal, I. Kroon and R. Brincker, 1993a: Simple Applications of Fictitious Crack Model in Reinforced Concrete Beams, Proc. of the IUTAM Symposium on Fracture of Brittle Disordered Materials, Concrete, Rock and Ceramics, Queensland, Australia.

Ulfkjær, J.P., O. Hededal, I. Kroon and R. Brincker, 1993b: Simple Application of Fictitious Crack Model in Reinforced Concrete beams - Analysis and Experiment, Proc. of the JCI International Workshop on size Effects in concrete Structures, Sendai, Japan.

Ulfjkær, J.P., S. Krenk and R. Brincker, 1995: Analytical Model for Fictitious Crack Propagation in Concrete Beams, Journal of Engineering Mechanics, Vol. 121, No. 1.

Anchors in Theory and Practice, Widmann (ed.) © 1995 Balkema, Rotterdam. ISBN 90 5410 577 1

Observations on cone pull-out anchor behaviour in very weak rock

Untersuchungen über Verhalten von Ankern bei kegelförmigem Grundbruch in sehr weichem Gestein

T.G.Carter
Golder Associates, Toronto, Canada

Abstract Tests of anchors which induce full cone deformation and failure of the rock mass are rare. Most tests of adequately bonded anchors in rock, if taken to failure, fail either by tensile fracturing of the tendon usually through the threads, or by simply snapping at the collar of the hole. Some fail by pull-out at the anchor/grout interface as a result of shearing through the grout, particularly for poor grout mixes, or where inadequate bond length was provided. Some fail by shear at the rock/grout interface, but again usually because of poor hole cleaning, low quality grout, or wet hole conditions that have softened the borehole periphery. A few fail by pullout of the grout/rock bond, because of the low bond strength of the rock, but usually only because too short a bond length was provided, unless specifically testing shortened grout lengths simply as a means for checking achievable grout/rock bond.

This paper provides the results of some carefully controlled tests of anchors in weak sandstone where full-scale deformation and yielding of the rock mass was induced over a wide area around and including the collar zone of the anchor. Although these tests were performed over ten years ago, because these type of failure data is rare in the literature, it was considered that it would be useful that the basic data be made more generally available for reference by other practitioners.

Zusammenfassung Nur selten wird zur Untersuchung von Ankern ein völlig kegelförmiger Grundbruch und somit ein Versagen der Gebirgsmasse herbeigeführt. Die meisten Versuche an angemessenen, im Gestein eingebundenen Ankern, die zum Versagen gebracht werden, zeigen einen Zugbruch des Spanngliedes, im allgemeinen durch das Gewinde, oder einfach ein Abbrechen am Rand des Loches. Einige Brüche werden an der Anker/Injektionsgut-Kontaktfläche als Resultat einer Scherung im Injektionsgut beobachtet, besonders bei schlecht gemischtem Injektionsgut oder bei zu kurzer Länge der Einbindung. Andere Fehlverhalten treten an der Injektionsgut/Gestein-Kontaktfläche auf, wiederum als Resultat unsauberer Löcher, schlechten Injektionsgutes oder feuchter Randbedingungen, die das umgebende Gestein aufgeweicht haben. Außerdem gibt es als Fehlverhalten ein Herausziehen aus der Injektionsgut/Gestein-Einbindung wegen schlechtem Verbund im Gestein, welches allgemein ein Resultat zu kurzer Länge für die Einbindung ist, abgesehen von speziellen Tests, bei denen verkürzte Verbund-Längen benutzt werden, um den erreichbaren Injektionsgut/Gesteins-Verbund zu untersuchen.

In diesem Artikel werden die Resultate einiger sorgfältig kontrollierter Anker-Untersuchungen in weichem Sandstein präsentiert, bei denen eine völlige Verformung und ein Ausbruch der Gesteinsmasse über eine große Umgebungsfläche einschließlich des Ankerkopfes herbeigeführt wurde. Während die Versuche schon vor über 10 Jahren durchgeführt worden sind, berücksichtigt dieser Aufsatz, daß angesichts der Seltenheit vergleichbarer Untersuchungen die Basis-Daten anderen Praktikern als Referenz hilfreich sein könnten.

INTRODUCTION

As part of the investigations for the design of hold-down anchors for various of the foundation slabs for the 20 MW Annapolis Tidal Power Station in Nova Scotia, Canada (Figure 1) some detailed rock bolt pull-out tests were performed. In total nine anchors were installed and five tested to destruction. Of these, two were arranged for testing as full cone pull-out tests and one was tested to failure.

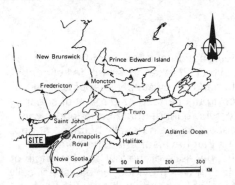

Figure 1: Site Plan

Abb. 1: Lageplan

As the rock mass at Annapolis had been recognized as weak, both from the initial site investigations and from the preliminary stages of construction, there was some concern that adequate anchorage would be available for required tie-down bolting to achieve the cyclical loading design requirements, which would occur because of tidal operating conditions (ref. Douma and Stewart,1981, Delory, 1986 or O'Kelly,1991). Two issues were specifically identified as needing field assessment:

a) a determination of the allowable bond stress that could be developed on the grout/rock interface and

b) a determination of whether the anchors could, because of the weak rock mass, realistically fail by cone pullout of the form shown in Figure 2.

Rock Conditions

The sandstone rock mass in which the permanent anchors were to be installed was known from surface exposures to be very friable and susceptible to degradation on exposure. Petrographically, the rock was only very weakly cemented with iron carbonate such that individual sand grains could

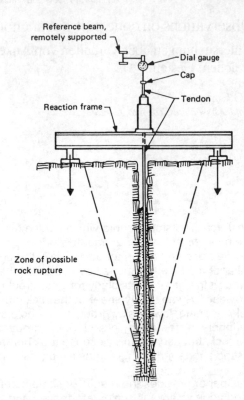

Figure 2: ISRM recommended arrangement for testing rock anchors and adjacent rock mass.

Abb. 2: ISRM-empfohlene Prüfanordnung für Anker und Fels

easily be brushed off from any piece of freshly drilled core. In most areas of the site the rock mass showed no evidence of significant structural fabric, apart from bedding. At outcrop, the beds, which ranged from 1 and 4 m in thickness, were noted to dip at 20-25° towards the north-northwest, dissected only by a few, ill-defined, widely spaced near-vertical joints.

TESTING PROGRAMME

Three series of tests were carried out just down-stream of the Tailrace (tidal basin) area of the Power Plant (Figure 3), with the bars, grout lengths and set-ups arranged to allow evaluation of all three likely modes of anchor failure, namely:

1. failure through the tendon steel
2. failure along the grout contact either
    a) at the grout/rock boundary or
    b) at the grout/anchor boundary, and
3. pull-out of rock cone around the anchor.

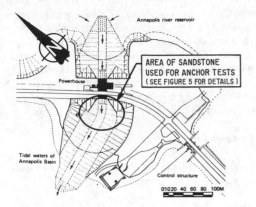

Figure 3 : Layout of Power Plant and Ancillary Structures showing test location area.

Abb. 3: Lage von Krafthaus und Nebenanlagen samt Prüfbereich

In this paper only the cone test is described in detail as the other tests were more routine. The results of all of the tests are however summarized on Table 1, with each test being conducted on 30 mm dia. 60 ksi deformed steel bars. These were grouted to various lengths into 1.5 m deep, 63 mm dia. holes. The grout used for the anchors was generally M-bed superflow with water:cement ratios set consistent with the manufacturer's recommendations. For the tests, in order to minimise any errors in calculating bond lengths based on bleed data, actual set bond

lengths, as listed in Table 1, were measured immediately prior to pull-testing. Also prior to pull-testing, each of the appropriate grout test cubes from the specific anchor to be tested, were crushed. Strengths were typically found to vary between 30 and 40 MPa.

Although the grout bond lengths for anchor tests 1 and 2 were arranged to allow evaluation of the grout/steel interface strength, while Tests No. 3 and 4 were more specifically designed to test the rock/grout interface; for comparison, in Table 1, the results have each been listed in terms of yield and failure stresses computed for both interfaces.

Of the four straight bar pull-out tests, anchor test No. 3 produced a clean failure zone on the rock/grout interface which suggested average yield and failure stress levels on the rock/grout contact of 0.6 N/mm^2 and 0.8 N/mm^2 respectively, values which are fairly low for sandstone, but of the same order as some of the weaker rock masses tested elsewhere, (ref. Table 4 - Littlejohn (1995)).

CONE TEST

The fifth test was specifically arranged to determine the yield characteristics of the sandstone rock mass under full cone pull-out conditions. Theoretical considerations for designs of many anchors assume 60° or 90° cones of rock to be mobilized at failure of the anchor (Figure 4). However, because most

Table 1 : Summary of Rock Anchor Pull test results

Test No.	Bond Length (mm)	Grout Strength[1,2] (MPa)	Mode of Failure	Load at Failure (tonnes)	Bond Stress at Grout/ Rock	Failure (N/mm^2) Grout/Bar	Yield Load[4] (tonnes)	Yield Stress[4] (N/mm^2) Grout/Rock
1	254	35.8	Grout/Bar	3.2	0.57	1.21	2.5	0.44
2	229	35.8	Grout/Bar	3.9	0.78	1.66	2.0	0.39
3	597	29.9	Grout/Rock	10.8	0.82	1.74	8.0	0.60
4	787	29.9	Grout/Rock?	26.1	1.50	1.94	16.0	0.90
Cone Pull	1575	40.8	Steel Bar	31.5	0.92[3]	1.92[3]	17.0 (21.0)[4]	0.48

Notes:
1. Grout strength determined from compression tests on 50 mm grout cubes cast at time of pouring and tested at time of pull testing.
2. Grout used was M-Bed Superflow manufactured by Sternsons Limited.
3. Tests 1 to 4 failed by pull-out along the contact indicated under the column headed "Mode of Failure". The Cone Pull Test failed through the steel of the anchor bar thus tabulated bond stresses are not strictly those at failure on the grout/rock or grout/bar contacts.
4. Values of yield of bar/grout/rock system tabulated based on interpretation of load/deformation graphs. Marked hysteresis in loading behaviour at 21 Tonnes in the cone test interpreted as yield of rock mass

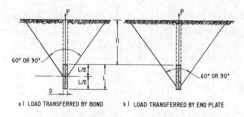

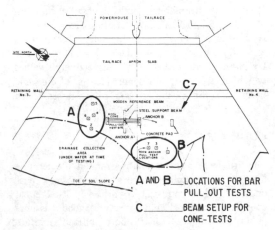

Figure 4: Geometry of Cone, assumed to be mobilised with failure in a homogeneous rock mass (from Littlejohn 1995)

Abb. 4: Abmessungen der Kegel, die durch Grundbruch in homogenen Fels entstehen (von Littlejohn 1995)

Figure 5 : Rock Anchor Test Locations

Abb. 5: Prüffeld der Felsanker

rocks are relatively strong, it is rare in practice that failure can be induced of a large rock cone to demonstrate whether such a mechanism is realistic or not. For this situation, however, with such a weak, generally structureless, sandstone, the test that is described in the following paragraphs provides some insight into rock mass behaviour at yield, and sheds some light on the cone failure concept.

Test Set-Up

Figure 5 shows the locations of each of the anchor tests with respect to the power plant.

All of the bar pull-out tests were carried out first, in order to establish key parameters for carrying out the cone test. Figures 6 and 7 show the general set up arranged for the cone test, with the cast concrete reaction blocks set well outside the zone of possible rock mass rupture. Each of these concrete reaction blocks were cast 7 days ahead of the testing period for the cone test. Each was 1.2 m long x 500 mm wide x 500 mm high, set 2.5 m radially away from the anchor position.

For each of the bar pull-out tests the anchors were installed with centralizers placed above the depth of the grouted zone in order to ensure that the bars were co-linear with the holes but that only the grout/bar contact influenced the test behaviour. For the cone pull-out test two centralizers were arranged, one near the lowest part of the anchor and one near the collar, and then the hole was fully grouted. Jacking was undertaken using a centre-

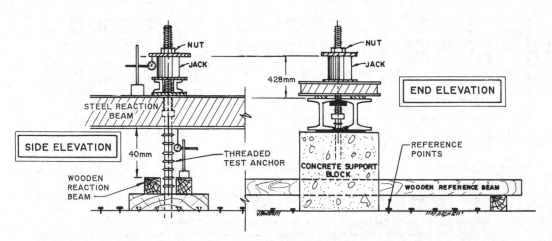

Figure 6: Detailed Arrangement of Reaction Beam and Support Blocks for Cone Test

Abb. 6: Detail der Anordnung von Pressenträgern und Widerlager für den Grundbruchversuch

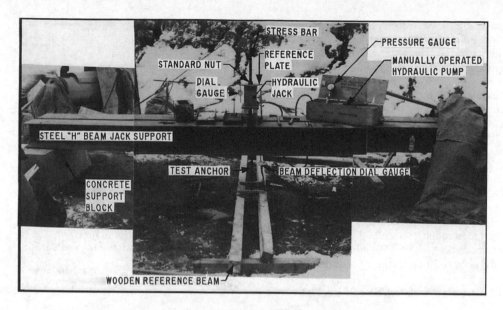

Figure 7 : Reaction Beam Set-up for Cone Pull-out Test

Abb. 7: Aufbau der Pressenträger für den Grundbruchversuch

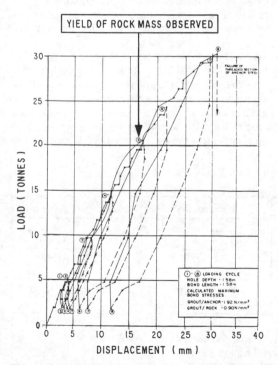

Figure 8: Load/Displacement Curves for Cone Test

Abb. 8: Last-Verformungskurven für den Grundbruchversuch

pull Enerpac 60 ton, custom calibrated, hydraulic jack. Loads were determined conventionally from calibrated pressure gauge readings of the hydraulic ram.

Prior to initiating the first loading cycle a nominal seating load was applied to remove slack in the system. This nominal tension was of the order of 1 tonne. Loads were then sequentially incremented in 1 tonne increments up to a maximum of 5 tonnes for the first load cycle (ref. Figure 8). Subsequent cycles were taken to 10, 15, 20, 21, 25 and 30 tonnes up to failure of the anchor. The anchor position and the behaviour of the rock mass adjacent to the anchor were monitored throughout each cycle by means of dial gauges mounted on a separate reference beam as shown on Figure 6.

Of the eight load displacement cycles undertaken during the testing, the last two cycles exhibited a somewhat softer response than the initial cycles, presumably as a result of deformation of the central zone of the rock mass adjacent to the anchor. As shown on Figure 8 the behaviour of the loading and unloading response up to 21 t, corresponding to the end of load cycle 5 was quite normal and typical of the results seen for the other bar pull-out tests. Beyond this point, however, the unloading cycles indicate marked hysteresis, interpreted as non-recoverable rock relaxation as the load was decreased at the end of each cycle, particularly at loads of less than 5 tonnes, corresponding approximately to the weight of a 90° cone of rock.

The behaviour of the rock mass to pull-out also changed somewhat post loading past 21 tonnes. Only a minor increase was noted in the deformation of the centre of the rock mass around the anchor. The maximum ground movement adjacent to the anchor up to this stage had reached 4 mm upward within a zone of about 20-50 cm from the anchor radiating outwards to about 2 mm upward movement recorded at 1 m from the bar, (ref. Figure 9). The test was completed to failure, which occurred through the steel of the anchor, after a total movement of 4.5 mm upward had been noted of the central 50 cm radius of the rock.

CONCLUSIONS & INTERPRETATION

While no disruptive failure of the rock mass developed in the test (as the bar broke prior to full pullout of the central zone around the anchor), the characteristic deformation behaviour shown on Figure 9 suggests that yield of the rock mass likely occurred during the 5th and 6th loading cycles (i.e. between 20 and 25 tonnes) although this is complicated by yield of the steel bar at about 18 t.

Although the deformation movements were small throughout the test, the trends are suggestive that some form of yield of a 1 m radius zone developed. Calculations of available capacity for the installed anchor (1.5 m long), if 21 t is taken as ultimate

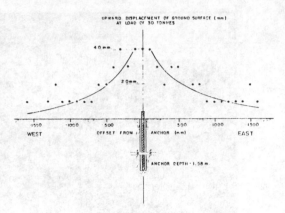

Figure 9: Characteristics of ground heave recorded around the anchor bar position

Abb. 9: Charakteristische Bodenhebungen rund um den Anker

would suggest safety factors of the order of 6 to 10 based on assuming a 90° cone of 0.9-1.1 m radius or 4 to 6 based on an assumption of the full depth of the anchor (again assuming an approximately 1 m radius disturbance zone - viz., a geometry approximately equivalent to a 60-70° cone).

These test results tend to corroborate the results of Saliman and Shaefer (1968) suggesting that considerable conservatism is inherent in the standard approach employed in conventional anchor design practise of assuming the "dead" weight of the rock cone around the anchor as the only resistance to pull-out. If, for the test reported in this paper, some

Table 2 : Back-calculated rock mass strengths based on Equivalent 90° Cone

Rock Mass Conditions	Homogeneous	Fissured, dry	Submerged, Fissured
Depth of Anchorage for 90° cone*	$\sqrt{(\dfrac{S_f T_w}{4.44\tau})}$	$\sqrt[3]{(\dfrac{3S_f T_w}{\gamma_s \pi \tan^2\phi})}$	$\sqrt[3]{(\dfrac{3S_f T_w}{(\gamma_s - \gamma_w)\pi \tan^2\phi})}$
Back Calculated, approximate Rock Strength Properties	$\tau \sim 20 - 60$ kPa	$\phi' \sim 55° - 70°$	$\phi' \sim 60° - 75°$

τ is the shear strength of the rock (kPa~ kN/m²)
S_f is the factor of safety against failure (assumed as 1 for calculations tabulated above,
          values of 2 to 3 are customary for current design practice);
ϕ' is the effective angle of friction across fractures in rock mass (degrees);
T_w is the working load on the anchorage (kN), assumed as failure load for above calculations;
γ_s is the unit weight of rock (kN.m³)
γ_w is the unit weight of water (kN.m³)
                    * Equations from Littlejohn, 1995 (after Hobst and Zajic, 1983)

consideration is given to the shear on the periphery of the presumed cone, as well as the dead weight of the rock cone, some interesting results are obtained. Specifically, if friction only on the cone periphery is assumed, by back-calculation, ϕ' for the rock ranges from 45 - 60°; the converse, assuming cohesion only, suggests c' to range between approximately 15 and 35 kPa; depending on the cone geometry, (60°/70° or 90°). These c', ϕ' values again suggest that the rock mass is more competent and of higher insitu strength than would currently be assumed in standard design approaches.

The same conservatism also appears evident in the range of back-calculated strengths derived from the formulas of Hobst and Zajic, 1983, (as listed within the central section of Table 2). While most rock anchor pull tests, are, by their very nature, site specific, such that absolute results are not widely applicable beyond the site, the value of such tests is immense where they demonstrate similarity of response in similar rock types. However, although nearly all of the tests that have been reported in the literature suggest that basing ultimate capacity on straight cone weight is conservative, not enough information is yet available to formulate an effective alternative. There is still a dearth of information where cone-failure has been induced in relatively strong rocks, and only a very limited number have been reported for soft rocks. Obviously, more site testing of cone failure mechanisms is essential if current design approaches are to be rationally improved.

ACKNOWLEDGEMENTS

Specific acknowledgements are due to the Nova Scotia Power Corporation for permission to publish the rock bolt pull-out data obtained during this test work, to Mr. Martin W. Hodgson who carried out the field testing program and to Dr. Steve McKinnon who participated in the detailed evaluation of the test behaviour.

REFERENCES

Delory, R.P. (1986). The Annapolis Tidal Generating Station Water for Energy, Brighton UK.

Douma, A and Stewart G.D. (1981). Annapolis Stratflo Turbine Will Demonstrate Bay of Fundy Tidal Power Concept, *Hydro Power*, pp. 1-8.

Hobst, L. and Zajic, J. (1983). *Anchoring in Rock and Soil*. Elsevier Publishing Comp., Amsterdam.

ISRM (1981): *Suggested Methods for Rockbolt Characterization, Testing, and Monitoring*, ISRM Commission on Testing Methods, E.T. Brown, Ed. (Pergamon, Oxford, 1981), 211 pp.

ISRM (1985): Suggested Methods for Rock Anchorage Testing "ISRM Commission on Testing Methods, *Int. J. Rock Mech. Min. Sci.*22 (2), 71-83.

Littlejohn, G.S. & D.A.Bruce. (1976) Rock Anchors:State of the Art. *Ground Engineering*, 8 (3), 35-32 (May 1975); 8 (4), 41-48 (July 1975); 8 (5), 34-35 (Sept. 1975); 8 (6), 36-45 (Nov. 1975), 9 (2), 20-29 (Mar. 1976); 9 (3), 55-60 (May 1976); 9 (4), 33-44 (July 1976).

Littlejohn, G.S. (1995). Rock Anchorages *ISRM News Journal*, pp. 18-38. Condensed from Chapter 15, Vol.4 *Comprehensive Rock Engineering* (1993).

O'Kelly, F. (1991). Harnessing the Ocean's Energy: Are We Ready for a Gift from the Sea? *Hydro Review*, Volume. X, #4 pp. 80-86.

Saliman, R and Shaefer, R. (1968) Anchored Foundations for Transmission Towers. *ASCE Annual, National Meeting on Structural Engineering.* Preprint 753, 28pp.

Anker in Theorie und Praxis, Widmann (Herausgeber) © 1995 Balkema, Rotterdam. ISBN 90 5410 577 1

Bemessung von Kopfbolzen und vergleichbaren Einlegeteilen

Design model for headed studs and comparable anchors

R. Eligehausen, J. Asmus & T. M. Sippel
Institut für Werkstoffe im Bauwesen, Universität Stuttgart, Germany

KURZFASSUNG: Als Verankerungsmittel dienende Kopfbolzen sind in Deutschland derzeit bis zu einer Verankerungstiefe $h_{ef} = 175$ mm in bewehrtem und unbewehrtem Normalbeton der Festigkeitsklasse $\geq$B25 bauaufsichtlich zugelassen [1, 2]. Die in diesen bauaufsichtlichen Zulassungen angegebenen Anwendungsbedingungen erschweren eine wirtschaftliche Anwendung der Kopfbolzen.

Ein neuer Bemessungsvorschlag für Kopfbolzenverankerungen [10] orientiert sich an der Bemessungsrichtlinie für Dübel [9], die auch in den Europäischen Richtlinien [13] Eingang gefunden hat. Hierbei erfolgt die Bemessung auf der Grundlage von Teilsicherheitsbeiwerten und die charakteristischen Lasten werden für alle Belastungsrichtungen und Versagensarten bestimmt. Die Richtigkeit der Bemessungsvorschläge wurde durch Nachrechnung der Ergebnisse von zahlreichen Versuchen überprüft.

In diesem Beitrag werden die Grundzüge dieses Bemessungsvorschlages insbesondere für Zugbeanspruchungen von Kopfbolzenverankerungen ohne Rückhängebewehrung erläutert.

ABSTRACT: In Germany at present, headed studs as anchor elements are permitted to be used in reinforced and unreinforced concrete of strength class $\geq$B25 with an embedment depth up to 175 mm according to current approvals [1,2]. The given application conditions in these approvals encomplicate an economic use of headed studs.

A new calculation model for headed studs, correspondending to both German and European current design guide-lines of fastenings, is proposed. The presented design model is based on partial safety factors and the characteristic loads are calculated for all loading directions and failure modes. The accuracy of the proposed equations has been checked by comparison with a large number of test results.

The essential features of this proposed calculation model, in particular for headed studs without hanger reinforcement under tensile load, are explained in this paper.

1 BEMESSUNG VON KOPFBOLZENVERANKERUNGEN OHNE RÜCKHÄNGEBEWEHRUNG

1.1 *Allgemeines*

Die Bemessung von Kopfbolzenverankerungen (Bild 1) soll analog der Bemessungsrichtlinie für Dübel nach dem Bemessungsverfahren A [9] erfolgen. Dies bedeutet:
• Die Ermittlung der Beanspruchung der einzelnen Kopfbolzen einer Verankerung erfolgt nach der Elastizitätstheorie. Die Ankerplatte muß ausreichend steif sein, damit die Voraussetzung der Elastizitätstheorie - Ebenbleiben der Ankerplatte - mit ausreichender Genauigkeit erfüllt wird.
• Die Bemessung der Kopfbolzenverankerungen erfolgt für die einzelnen Belastungsrichtungen und Versagensarten auf der Basis von Teilsicherheitsbeiwerten. Daher werden im folgenden Gleichungen zur Ermittlung der charakteristischen Tragfähigkeit und die zugehörigen Teilsicherheitsbeiwerte angegeben [10].

Zur Gewährleistung der ausreichenden Sicherheit des als Ankergrund dienenden Bauteils gelten die in [9], Abschnitt 6 angegebenen Regeln unvermindert.

Bei der Bemessung von Stahlbetonbauteilen wird i.a. von einer gerissenen Zugzone (Zustand II) ausgegangen, weil der Beton nur eine sehr geringe Zugfestigkeit besitzt, die zudem durch in der Berechnung nicht berücksichtigte Eigen- oder Zwangsspannungen ganz oder teilweise verbraucht werden kann. Die Erfahrung zeigt, daß die Rißbreiten bei überwiegender Beanspruchung unter quasi ständiger Last die als zulässig angesehenen Werte von w ≤ 0,3 bis 0,4 mm nicht überschreiten. Bei überwiegender Zwangsbeanspruchung können jedoch auch breitere Einzelrisse auftreten, wenn keine zusätzliche Bewehrung zur Beschränkung der Rißbreiten eingelegt wird. Für die Bemessung der Kopfbolzen wird generell von gerissenem Beton ausgegangen. Dadurch werden umfangreiche Nachweise über den Zustand des Betons (gerissen/ungerissen) vermieden und die Nachweise zur Gewährleistung einer ausreichenden Tragfähigkeit bei der Versagensart „Spalten des Betons" wesentlich erleichtert.

1.2 Teilsicherheitsbeiwerte

Kopfbolzenverankerungen weisen eine hohe Montagesicherheit auf, weil durch das industrielle Anschweißen der Kopfbolzen die vorgesehenen Werte für Verankerungstiefe und Achsabstand gewährleistet werden. Weiterhin sind die aufgestauchten Köpfe ausreichend groß, so daß das Versagen in der Regel durch Betonausbruch oder Stahlversagen erfolgt. Als Materialsicherheitsbeiwert wird $\gamma_{Mc} = 1,8$ angesetzt. Es sei jedoch darauf hingewiesen, daß der Beton im Bereich der Verankerungen gut verdichtet werden muß.

Bei Stahlversagen werden in [10] folgende Teilsicherheitsbeiwerte angegeben:
Stahlversagen Kopfbolzen:
zentrischer Zug: $\gamma_{Ms} = 1,50$
Querlasten: $\gamma_{Ms} = 1,25$
Stahlversagen Rückhängebewehrung:
$\gamma_{Mh} = 1,15$

1.3 Zugbeanspruchung

Die möglichen Versagensarten von zugbeanspruchten Kopfbolzen sind in Bild 2 dargestellt. Bei Verankerungen mit Kopfbolzen wird das Versagen in der Regel durch Betonausbruch hervorgerufen, Stahlbruch tritt nur gelegentlich und dann i.a. bei Verankerungen im hochfesten Beton auf. Herausziehen ist nur bei kleinen Aufstandsflächen (Schulterbreite) zu erwarten. Spalten des Betons kann durch anwendungstechnische Maßnahmen (Einhaltung der Mindestwerte für Achs- und Randabstände sowie Bauteilabmessungen) verhindert werden.

1.3.1 Stahlversagen

Die bei der Versagensart Stahlbruch zu erwartende charakteristische Höchstlast $N_{k,s}$ eines Kopfbolzens kann aus den Querschnittsabmessungen des Schaftes und der Mindeststahlzugfestigkeit ermittelt werden

$$N_{Rk,s} = A_s \cdot f_{uk} \tag{1}$$

mit A_s = Querschnittsfläche des Bolzens
f_{uk} = charakteristische Zugfestigkeit

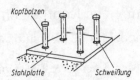

Bild 1: Ankerplatte mit angeschweißten Kopfbolzen
Fig. 1: Anchor plate with welded headed stud connectors

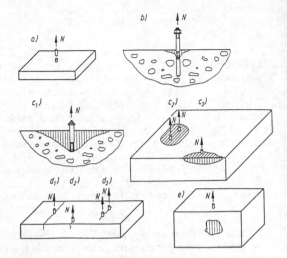

Bild 2: Versagensarten von Verankerungen bei Zugbeanspruchung
Fig. 2: Types of anchorage failure in tension

1.3.2 Betonausbruch

Bild 3 zeigt die in Versuchen gemessenen Bruchlasten in ungerissem Beton von Kopfbolzen mit großem Achs- und Randabstand in Abhängigkeit von der Verankerungstiefe bei gleicher Betondruckfestigkeit. Die Versuchsergebnisse (Mittelwerte) können durch die im Bild angegebene Gleichung angenähert werden. Die Bruchlast hängt danach nur von der Betonzugfestigkeit, die proportional zu $\sqrt{\beta_w}$ angenommen wird, und der Verankerungstiefe h_{ef} ab. Der Einfluß des Schaftdurchmessers ist vernachlässigbar gering. Das Verhältnis Rechnung zu Versuch beträgt ca. 1,0 bei einem Variationskoeffizient von V = 15%. Damit beträgt die charakteristische Betonausbruchlast $N_{k,c} \approx 0,75\, N_{u,m}$.

$$N_{u,m} = k \cdot \sqrt{\beta_w} \cdot h_{ef}^{1,5} \tag{2}$$

mit k = 15,5 für Kopfbolzen in ungerissenem Beton
β_w = Würfeldruckfestigkeit

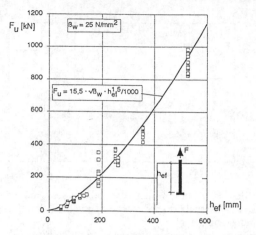

Bild 3: Betonausbruchlast von Kopfbolzen unter zentrischer Zugbelastung, nach /7/
Fig. 3: Concrete cone failure load of headed studs in tension

Die Ausbruchlasten von in Rissen verankerten Kopfbolzen, betragen bei der im Stahlbetonbau maximal zulässigen Rißbreite von 0,4 mm im Mittel das ca. 0,7fache der im ungerissenen Beton zu erwartenden Werte /7/ (siehe Bild 4).

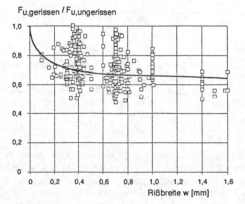

Bild 4: Einfluß von Rissen auf die Bruchlast von Kopfbolzen und Hinterschnittdübeln unter zentrischer Zubelastung
Fig. 4: Influence of crack width to the failure load of headed studs and undercut anchor

Eine Bruchlast nach Gleichung (2) wird nur erreicht, wenn eine ausreichend große Betonfläche pro Kopfbolzen zur Verfügung steht. Die zugehörigen Achs- und Randabstände betragen $s_{cr,N} = 2c_{cr,N} \geq 3h_{ef}$ (Bild 5 bis 8). Bei Unterschreitung dieser Abstände kann die charakteristische Betonausbruchlast nach dem CC-Verfahren (Gleichung (3)) berechnet werden. Das

CC-Verfahren liefert in der Regel die gleichen Bruchlasten wie das in [7] beschriebene κ-Verfahren. Mit dem CC-Verfahren werden die κ-Faktoren anschaulich berechnet.

$$N_{Rk,c} = \frac{A_{cN}}{A_{cN}^0} \cdot \psi_{s;N} \cdot N_{Rk,c}^0 \qquad (3)$$

mit $\quad N_{Rk,c}^0 = 8 \cdot \sqrt{\beta_{wN}} \cdot h_{ef}^{1,5} \qquad (3a)$

charakteristische Bruchlast einer Einzelbefestigung in gerissenem Beton

$A_{c,N}^0$ = Projektionsfläche einer Einzelverankerung mit großem Achs- und Randabstand auf der Bauteiloberfläche. Bei der Berechnung wird der Betonausbruchkörper als Pyramide mit der Höhe h_{ef} und der Länge der Basisseiten $l = s_{cr,N}$ idealisiert (siehe Bild 5a).

$\quad = s_{cr,N}^2$

$A_{c,N}$ = Vorhandene Projektionsfläche auf der Betonoberfläche. Bei der Berechnung ist der Ausbruchkörper eines Bolzens wie oben angegeben zu idealisieren und es ist der Einfluß von Bauteilrändern und benachbarten Bolzen zu beachten. Beispiele für die Berechnung von $A_{c,N}$ enthält Bild 5b.

$\psi_{s,N}$ = Einfluß der Störung des Spannungszustandes bei Verankerungen am Bauteilrand

$\quad = 0,7 + 0,3 \ c/c_{cr,N} \leq 1 \qquad (3b)$

β_{wN} = Festigkeitsklasse des Betons

Bei Verankerungen mit mehreren Randabständen (z.B. in der Bauteilecke) ist der kleinste Randabstand c in Gleichung (3b) anzusetzen. Durch zusätzliche Faktoren kann auch der Einfluß eines exzentrischen Lastangriffs sowie einer dichten Bewehrung (s<150 mm) berücksichtigt werden (siehe [10]).
In den Bildern 6 bis 8 sind Versuchsergebnisse und Rechnung gegenübergestellt.
Bei Verankerungen, bei denen der Abstand zu 3 oder 4 Rändern c ≤ c_{cr,N} beträgt, liefert Gleichung (3) auf der sicheren Seite liegende Ergebnisse. Genauere Ergebnisse erhält man mit den in [4] angegebenen Modifikationen.

1.3.3 *Lokaler Betonausbruch bei randnahen Verankerungen*

Bei randnahen Verankerungen (c ≤ 0,3 - 0,5 h_{ef}) tritt kein kegelförmiger Betonausbruch auf, sondern der Beton im Bereich des Kopfes bricht seitlich aus (vergl. Bild 9). Dieser örtliche Betonausbruch wird durch den quasihydrostatischen Druck im Bereich des Kopfes verursacht, der eine seitliche Abtriebskraft rechtwinklig zur Belastungsrichtung hervorruft (Bild 9). Überschreitet diese Kraft die Tragfähigkeit der zur Verfügung stehenden seitlichen Betonbruchfläche, kommt es zu einem lokalen Betonausbruch. Die Bruchlast ist unabhängig von der Verankerungstiefe. Sie steigt etwa proportional zum Randabstand an. Ein

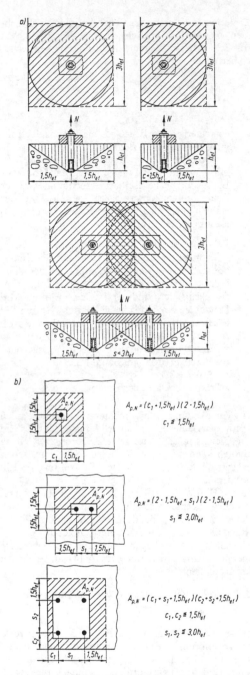

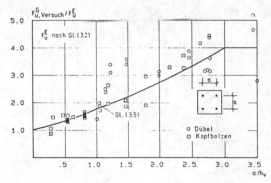

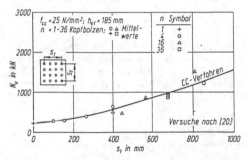

Bild 6: Einfluß des Achsabstandes auf die Betonausbruchlast bei zugbeanspruchten Vierfachbefestigungen
Fig. 6: Influence of spacing on the failure load of quadruple fastenings in tension

Bild 7: Einfluß des Achsabstandes auf die Betonausbruchlast mit bis zu 36 Kopfbolzen unter mittiger Zugbeanspruchung
Fig. 7: Influence of spacing on the failure load of multiple fastenings with up to 36 headed studs in tension

weiterer Einflußfaktor ist die Aufstandsfläche A_k. Dies kann dadurch erklärt werden, daß bei konstanter Last die Betonpressungen unter dem Kopf mit zunehmender Aufstandsfläche abnehmen und daher wegen des geringeren hydrostatischen Drucks das Verhältnis Abtriebskraft zu Zugkraft abnimmt.

Nach den Untersuchungen in [4] kann die charakteristische Bruchlast eines Einzelbolzens im gerissenen Beton bei dieser Versagensart nach Gleichung (4) berechnet werden.

$$N_{Rk,b}^0 = 7,5 \cdot c_1 \cdot \sqrt{A_k} \cdot \sqrt{\beta_{wN}} \tag{4}$$

mit

$N_{Rk,b}^0$ = Bruchlast eines Einzelbolzens bei lokalem Betonausbruch

$$A_k = \frac{\pi}{4} \cdot \left(d_k^2 - d_1^2 \right)$$

Bild 5: a: Einfluß des Rand- und Achsabstandes auf die Form des Betonausbruchkörpers bei zugbeanspruchten Einzel- und Zweifachbefestigungen
b: Projizierte Flächen für verschiedene zugbeanspruchte Verankerungen
Fig. 5: a: Influence of edge distance and spacing to the concrete cone for single and double fastenings
b: Projected areas for different anchorages in tension

28

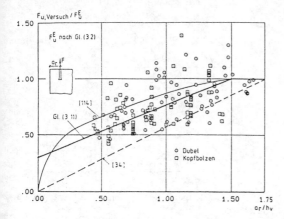

Bild 8: Einfluß des Randabstandes auf die Betonausbruchlast bei zentrischer Zugbeanspruchung

Fig. 8: Influence of edge distance on the failure load of fastenings in tension

d_1 = Bolzendurchmesser
d_k = Durchmesser des Bolzenkopfes
c_1 = Randabstand zur freien Kante
$ß_{wN}$ = Betonwürfeldruckfestigkeit

Berücksichtigt man die bei Kopfbolzen auftretenden Verhältnisse Kopfdurchmesser zu Schaftdurchmesser ergibt sich

$$N^0_{Rk,b} = 8,5 \cdot c_1 \cdot d_1 \cdot \sqrt{ß_{wN}} \qquad (5)$$

Gleichung (5) gilt für Einzelverankerungen nahe am Rand. Bei Anordnung mehrerer Bolzen mit geringem Achsabstand in Form einer Reihenverankerung parallel zum Rand ist eine Überschreitung der Betonausbruchkörper zu erwarten. Die Versuchsergebnisse zeigen, daß der Durchmesser der Ausbruchkegel relativ groß ist und im Mittel das 6fache des Randabstandes c_1 beträgt (vergl. [16]). Setzt man diesen Wert als kritischen Achsabstand an, bei dem keine Überschneidung der Ausbruchkörper stattfindet, fann die mittlere Bruchlast einer Reihenverankerung analog zum CC-Verfahren berechnet werden.

$$N_{Rk,b} = \frac{A_{c,Nb}}{A^0_{c,Nb}} \cdot \Psi_{ec,Nb} \cdot \Psi_{s,Nb} \cdot N^0_{Rk,b} \qquad (6)$$

mit $N^0_{Rk,b}$ nach Gleichung (5)

$A^0_{c,Nb}$ = projizierte Fläche eines Einzelbolzens auf der Seitenfläche des Betons. Dabei ist der Ausbruchkörper als Pyramide mit der Spitze in der Mitte des Bolzenkopfes, einer Höhe c1 und einer Länge der Basisseiten 6c1 anzunehmen.

$= 36 s_1^2$

$A_{c,Nb}$ = Vorhandene projizierte Fläche auf der Seitenfläche des Betons. Bei der Berechnung ist der Ausbruchkörper der Einzelbolzen wie oben angegeben zu idealisieren und es ist die Überschneidung der projizierten Flächen benachbarter Bolzen zu beachten.

$\Psi_{ec,Nb}$ = Einfluß der Exzentrizität der Zuglast der Reihenbefestigung

$$= \frac{1}{1 + 2e/(6c_1)} \qquad (6a)$$

e = innere Exzentrizität der gezogenen Bolzen

$\Psi_{s,Nb}$ = Störung des Spannungszustandes im Beton an der Bauteilecke

$$= 0,7 + 0,3 \cdot \frac{c_2}{3c_1} \leq 1 \qquad (6b)$$

c_1, c_2 = Randabstände

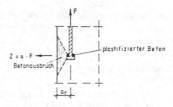

Bild 9: Örtlicher Betonausbruch eines randnahen Kopfbolzens, mechanisches Modell nach [7]

Fig. 9: Local Blow out of a headed stud close to a edge, mechanical model, according to [7]

1.3.4 *Herausziehen*

Herausziehen von Kopfbolzen tritt nicht auf, wenn die Pressung unter dem Kopf begrenzt wird. In [3] wurde die Versagensart "Herausziehen" eingehend untersucht. Danach ergibt sich ein kontinuierlicher Übergang von der Versagensart "Betonausbruch" zu "Herausziehen". Mit Anwachsen der Betonpressungen steigen die Betonverschiebungen an, wodurch die wirksame Verankerungstiefe und damit die Betonbruchlast absinkt. Zur Gewährleistung der mittleren Bruchlast nach Gleichung (2) dürfen die Pressungen unter dem Kopf im ungerissenen Beton je nach Verankerungstiefe im Mittel $p_u \sim 12\ ß_w$ (h_{ef} = 80 mm) bis $p_u \sim 15\ ß_w$ (h_{ef} = 200 mm) nicht überschreiten.

Im gerissenen Beton ist die o.g. kritische Pressung mit dem Faktor 0,7 zu multiplizieren. Berücksichtigt man die Streuung der Versuchsergebnisse, so beträgt die charakteristische Herausziehlast im gerissenen Beton

$$N_{Rk,p} = p_{u,k} \cdot A_k \qquad (7)$$

mit

29

$N_{Rk,p}$ = charakteristische Last bei der Versagensart Herausziehen

$p_{u,k}$ = charakteristische Pressung
~ 7,5 β_{wN}

A_k = Betonaufstandsfläche des Kopfbolzens

Zur Gewährleistung ausreichend niedriger Verschiebungen im Grenzzustand der Gebrauchstauglichkeit, insbesondere bei Lage der Bolzen in Rissen mit veränderlicher Breite (z.B. bei Be- und Entlastung des Bauteils) muß die charakteristische Pressung auf $p_{u,k}$ ~ 6 β_{wN} beschränkt werden.

1.3.5 Spalten

Bei Verankerungen in dünnen, schmalen Bauteilen aus unbewehrtem Beton kann nach [12] Spalten des Betons auftreten. Im vorliegenden Fall wird jedoch immer von bewehrtem Beton ausgegangen, so daß zwar Risse auftreten können, die Rissbreite jedoch durch eine entsprechend gewählte Bewehrung begrenzt wird. Somit wird die Bruchlast bei der Versagensart Spalten durch Risse im Beton nur unwesentlich beeinflußt.

1.4 Querzugbeanspruchung

Die möglichen Versagensarten von querbeanspruchten Kopfbolzen sind in Bild (10) dargestellt. Stahlbruch (Bild 10a) tritt bei großem Randabstand auf, wobei es kurz vor Erreichen der Höchstlast zu einem muschelförmigen Abplatzen des oberflächennahen Betons kommen kann. Diese Versagensart liefert die höchste Bruchlast. Desweiteren kann bei steifen Verankerungen mit kurzer Verankerungslänge einseitiger Betonausbruch auf der lastabgewandten Seite auftreten (Bild 10c).

Bei kleinen Randabständen kann die Betonkante ausbrechen (Bild 10b$_1$). Bei Gruppen kann sich ein gemeinsamer Ausbruchkörper bilden (Bild 10b$_2$) und bei Anordnung der Befestigung in einer Bauteilecke (Bild 10b$_3$), in einem dünnen (Bild 10b$_4$) oder schmalen (Bild 10b$_5$) Bauteil können sich die Bruchkörper nicht vollständig ausbilden. Die von einem Befestigungselement übertragbare Last ist in diesen Fällen geringer als die einer Einzelbefestigung nach Bild 8b$_1$. Bei steifen, nicht ausreichend tief verankerten Befestigungsmitteln kann es auch bei großem Randabstand zu einem Ausbrechen des Betons auf der lastabgewandten Seite kommen (Bild 10c).

Bei Betonkantenbruch hängt die Bruchlast maßgeblich von der Betonzugfestigkeit und dem Randabstand ab. Zusätzliche Einflußfaktoren sind die Steifigkeit des Befestigungsmittels, der Achsabstand und die Bauteildicke.

Das Betonbruchlast bei der Versagensart „Betonkantenbruch" kann ebenfalls nach dem CC-Verfahren berechnet werden. Die entsprechenden Gleichungen sind in [11] ausführlich dargestellt und erläutert.

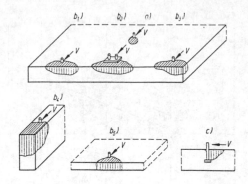

Bild 10: Versagensarten bei Querzugbeanspruchung

Fig. 10: Types of failure of anchorage loaded in shear

1.5 Schrägzugbeanspruchung

Das Tragverhalten von Verankerungen unter kombinierter Zug- und Querlast hängt vom Angriffswinkel der Last ab. In [6] werden Versuchsergebnisse mit Kopfbolzen- und Dübelverankerungen unter verschiedenen Beanspruchungsrichtungen ausgewertet. Danach kann die Höchstlast von Verankerungen durch folgende trilineare Interaktionsgleichung bestimmt werden

$$\left(\frac{N}{N_u}\right)^{1,5} + \left(\frac{V}{V_u}\right)^{1,5} = 1,0 \qquad (8)$$

N/N_u +	V/V_u =	1,2	(8a)
N/N_u	=	1,0	(8b)
V/V_u	=	1,0	(8c)

Dieser Vorschlag liefert für die Versagenskombination Stahlbruch/Stahlbruch auf der sicheren Seite liegende Ergebnisse. Für randnahe Verankerungen wird die tatsächliche Tragfähigkeit durch Gleichung (8) etwas überschätzt, jedoch ist die Überschätzung noch akzeptabel.

2 KOPFBOLZENVERANKERUNGEN MIT RÜCKHÄNGEBEWEHRUNG

Mit Hilfe einer speziellen Rückhängebewehrung kann die die Traglast von Verankerungen, bei denen ohne zusätzliche Bewehrung Betonversagen auftreten würde, deutlich verbessert werden [8]. Solche Verankerungen sind z.B. randnahe Verankerungen unter Querzugbeanspruchung in Richtung des freien Randes und Verankerungen unter Zugbeanspruchung mit einer relativ geringen Verankerungstiefe der Kopfbolzen.

Bemessungsgleichungen für die Berechnung der Traglasten unter Ansatz einer Rückhängebewehrung wurden inzwischen abgeleitet und sind in [15] ausführlich dargestellt und erläutert.

3 ZUSAMMENFASSUNG

Verankerungen mit Einlegeteilen sind in allen Bereichen des Bauwesens weit verbreitet. Im vorliegenden Beitrag werden Bemessungsgleichungen für Kopfbolzen vorgestellt und erläutert. Sie beruhen auf einem einfachen mechanischen Modell, dem sogenannten CC-Verfahren, und sind durch eine große Zahl von Versuchsergebnissen abgesichert.

Bisherige Bemessungsvorschläge beruhen auf zahlreichen Versuchen, die allerdings vorwiegend mit Dübeln und nur vereinzelt mit Kopfbolzen durchgeführt wurden.

Der neue Bemessungsvorschlag für Kopfbolzenverankerungen orientiert sich an der Bemessungsrichtlinie für Dübel, die auch in den Europäischen Richtlinien Eingang finden soll. Hiernach erfolgt die Bemessung auf der Grundlage von Teilsicherheitsbeiwerten und die charakteristischen Lasten werden für alle Belastungsrichtungen und Versagensarten bestimmt.

LITERATUR

[1]   Zulassungsbescheid Nr. Z-21.5-296 des Instituts für Bautechnik, Berlin, vom 14.08.1984 für "Verankerung von Stahlplatten mittels angeschweißter Kopfbolzen in Beton"

[2]   Änderungs-, Ergänzungs- und Verlängerungsbescheid des Instituts für Bautechnik, Berlin, vom 29.05.1989 für "Verankerung von Stahlplatten mittels angeschweißter Kopfbolzen in Beton"

[3]   *Furche, J.:* Zum Trag- und Verschiebungsverhalten von formschlüssigen Befestigungsmitteln bei zentrischem Zug, Dissertation an der Universität Stuttgart, 1993

[4]   *Furche, J.; Eligehausen, R.:* Lateral Blowout Failure of Headed Studs near a Free Edge. ACI-Special Publication 130 "Anchorages in Concrete - Design and Behaviour", Detroit, 1992

[5]   *Zhao, G.:* Tragverhalten von randfernen Kopfbolzenverankerungen bei Betonbruch, Dissertation an der Universität Stuttgart, 1993

[6]   *Zhao, G.; Eligehausen, R..:* Tragfähigkeit von Befestigungen unter kombinierter Zug- und Querlast. Bericht Nr. 10/17-92/2 des Instituts für Werkstoffe im Bauwesen der Universität Stuttgart, 1992, nicht veröffentlicht.

[7]   *Rehm, G.; Eligehausen, R.; Mallée, R.:* Befe-

stigungstechnik, Betonkalender, Vol. 2, Verlag W. Ernst & Sohn, Berlin, 1992

[8]   *Ramm, W.; Greiner, U.:* Verankerungen mit Kopfbolzen. Versuchsbericht der Universität Kaiserslautern, Fachgebiet Massivbau und Baukonstruktion, Oktober 1991, nicht veröffentlicht.

[9]   *Deutsches Institut für Bautechnik:* Bemessung von befestigungen mit Dübeln, Fassung Juni 1993

[10] Zulassungsentwurf für Verankerung von Stahlteilen mittels angeschweißter Kopfblzen in Beton, Entwurf 3/95

[11] *Fuchs, W.; Eligehausen, R.:* Das CC-Verfahren für die Berechnung der Betonausbruchlast von Verankerungen. Beton- und Stahlbetonbau 90 (1995), Heft 1,2,3

[12] *Furche, J.; Walser, P.:* Spalten von Betonbauteilen infolge Lasteinleitung durch formschlüssige Befestigungsmittel, Bericht-Nr. 16/1-91/6, Institut für Werkstoffe im Bauwesen, Universität Stuttgart, 1991, nicht veröffentlicht

[13] EOTA: Guideline for Anchors for Use in Concrete, Annex C: Design of Fastenings. Brüssel, Sept 1994.

[14] *Fuchs, W.; Eligehausen, R.; Breen, J.E.:* Concrete Capacity Design (CCD) Approach for Fastening to Concrete. ACI Structural Journal, V.92, No.1, Jan.-Feb.1995, S. 73 - 94.

[15] *Eligehausen, R.; Sippel, T.M.:* Bemessung von Kopfbolzen und Ankerschienen. Seminar „Befestigungstechnik im Bauwesen", Technische Akademie Esslingen, Feb. 1995

[16] *Zhao, G.; Eligehausen, R..:* Zusammenfassender Bericht über Versuche mit Kopfbolzenverankerungen. Bericht Nr. KRT-799/8-91/23 des Instituts für Werkstoffe im Bauwesen der Universität Stuttgart, 1991, nicht veröffentlicht.

Anchors in Theory and Practice, Widmann (ed.) © 1995 Balkema, Rotterdam. ISBN 90 5410 577 1

Mechanical behaviour of a single passive bolt: Theoretical approach

Theoretische Bestimmung des mechanischen Verhaltens eines passiven Felsankers

D. Hantz & C. Krocker
IRIGM-LGM, Grenoble University, France

ABSTRACT: 2D modelling has been used to determine the relation between the transverse force applied on a grouted hole and the induced displacement. An elastic-brittle-platic behaviour was assumed for rock and grout. An analytical solution has been established for an axisymmetrical loading, and a numerical method has been used for the case of an uniaxial loading. As the results for both types of loading are close, the analytical solution has been introduced in a 3D model to determine the normal reaction force on a bolt, induced by a transverse displacement. A numerical method is proposed to calculate the axial and transverse forces in a bolt submitted to axial and transverse displacements due to shearing of a joint. The method can be coupled with stress-strain methods used for simulation of discontinuous rock masses.

ZUSAMMENFASSUNG: Die Bestimmung der Reaktionskraft, welche durch das seitliche Eindrücken des Ankers in den Fels entsteht, erfolgt über ein zweidimensionales Modell der Theorie der plastischen Verformung eines zylindrischen Hohlraumes unter Innendruck sowie auf numerische Weise für einachsige Belastung. Es werden jeweils beide Materialien, Fels und Mörtel mit elastisch- sprödem Verhalten, berücksichtigt. Für typische Mörtel- und Gesteinsmaterialien, welche sich innerhalb einer getroffenen Klassifizierung befinden, werden bevorzugte Abfolgen von Typen der Rißausbreitung festgestellt. Die gute Übereinstimmung zwischen den Resultaten des symmetrischen und einachsigen Modells erlaubt die Anwendung der Theorie des expandierenden Zylinders in einem dreidimensionalen Modell zur Berechnung der Ankerkräfte. Diese basiert auf der Anbringung einer Kluftverschiebung und anschließender Interpolation zwischen Ankerkräften und der mit diesen ermittelten Geometrie des verformten Ankers. Der Anker ist zur Berechnung in Abschnitte, diese widerum in Querschnittselemente unterteilt. Zur Nachbildung des realen plastischen Verhaltens kommt die Arbeitslinie des gewählten Stahles aus einem Zugversuch zur Anwendung. Bei Änderung der Verformungsrichtung einzelner Elemente durch Einsatz vorgedehnter Stähle oder während der Ankerdeformation wird der Bauschinger-Effekt berücksichtigt. Ein Ausziehen des Ankers in Längsrichtung aus dem Verbund Mörtel-Fels in nicht durch Biegung plastifizierten Bereichen wird ebenfalls ermöglicht. Eine Kombination des Programmes mit Methoden zur Modellisierung des Verhaltens von Gesteinsklüften ist möglich.

1 INTRODUCTION

A passive bolt is a steel bar which is initially untensioned. In rock they are usually fully grouted; the term dowel is used too. In soil they may be grouted or not; here the term nail is used. In this paper fully grouted untensioned bolts in hard and weak rocks will be considered. The range of applications includes rock slopes and underground excavations.

Forces in passive bolts only develop when rock mass deformation occurs. They in turn influence the rock mass deformation. To account for the interaction between reinforcement and joint behaviours, stress-strain calculations appear to be the more convenient method for designing bolted rock masses. They are often used for underground excavations design. For slopes, limit equilibrium methods are more often used, with some simple assumptions about the rock mass deformation. These assumptions may be that deformation produces pure tension or pure shear in bolts, or that movement of a sliding mass has a given constant direction (Panet 1988, Pellet 1993).

Stress-strain calculations need constitutive equations for joints and reinforcement systems. Constitutive models for joints have been given by Amadei and Saeb (1990) or Souley and Homand (1991). Relations between the force vector in the bolt and the displacement vector of the joint have been proposed by Dight (1983), Lorig (1985), Holmberg (1991, 1992) and Pellet (1993). But they considered a bolt in an homogenous medium and the reaction of this medium was supposed normal and uniform along the bolt. Aydan et al. (1985) and Yasici and Kaiser (1992) considered a surrounding medium made up of grout and rock, but they were interested only in axial behaviour of the bolt. Schubert (1984) proposed a method to calculate the bolt forces, for the case of a non uniform distribution of the reaction along the bolt.

The purpose of this paper is to propose a relation between bolt forces and joint displacements, considering a two-phase medium and a non uniform distribution of the reaction of this medium along the bolt. Differentiating between rock and grout is necessary when these materials have very different mechanical properties, as for weak rocks and cement grouts.

The first part of the paper is a 2D analysis of the behaviour of a cylindrical grouted hole submitted to the dowel action of a bolt. The second part deals with the 3D problem of calculation of forces in a bolt submitted to joint displacements.

2 EXPANSION OF A GROUTED HOLE

For 3D equilibrium analysis of a dowel, the reaction supplied by the surrounding medium must be known. It depends on the transverse displacement of the bolt. Theoretical methods to calculate the reaction force per unit of length, P, are based on a 2D analysis. Their differences are in the distribution of stresses around the half-perimeter of the bolt section. Dight (1983) supposed that the ultimate stress on the surface of the bolt, σ_u, is normal and uniform on the half-perimeter. The ultimate reaction force, P_u, is then given by the expression:

$$P_u = \sigma_u D_1 \qquad (1)$$

where D_1 is the bolt diameter. σ_u was obtained from a theory proposed by Ladanyi (1966,1967) for expansion of a cylindrical cavity in an infinite homogenous elastic-brittle-plastic medium, with a uniform internal pressure and plane strain condition. First we will apply the same approach to a grouted hole. Then we will use a numerical method to deal with the case of a uniaxial load.

2.1 Grouted hole with axisymmetrical load

An internal pressure p_i is applied at a distance $r = R_1$ of the axis. The surrounding medium is made up with grout material from $r = R_1$ to $r = R_2$, and with rock for $r > R_2$ (figure 1). Grout and rock are assumed to behave as elastic-brittle-plastic materials, with peak strength given by the Fairhurst criterion (Fairhurst 1964) and residual strength given by the Mohr-Coulomb criterion. The ratio n, between uniaxial compressive and tensile strengths, is always greater than 5. It is assumed that stresses in the rock before application of p_i can be neglected. It is true at shallow depth.

As long as the internal pressure p_i is low, grout and rock behave elastically. Variations of radial and tangential stresses with the distance from the hole axis are shown on figure 2. When p_i reaches a critical value, a rupture occurs in grout or rock (which doesn't mean the failure of the reinforcement system). If it occurs in the rock, it will be with a tensile mode, according to the Fairhurst criterion and with the hypothesis $n > 5$. Radial cracks will appear in the rock at $r = R_2$. If rupture occurs in grout, the mode may be tensile or compressive. It will be tensile if the ratio between the shear modulus of the rock G_m and the shear modulus of the grout G_s is lower than a critical value:

$$G_m/G_s < (\lambda_s+G_s)[(m_s-1)^2+(R_2/R_1)^2(m_s^2-2m_s-1)]/ (\lambda_s+G_s)(m_s-1)^2-G_s(R_2/R_1)^2(m_s^2-2m_s-1) \qquad (2)$$

where λ_s is the second Lame's constant and m_s the Fairhurst's constant, the subscript s referring to grout.

$$m_s = (n_s + 1)^{1/2} \qquad (3)$$

The second member of (2) is always greater than 1. A compressive rupture is therefore possible only if G_m is greater than G_s. Assuming that R_2/R_1 is always higher than 1.4 and that υ_s is always smaller

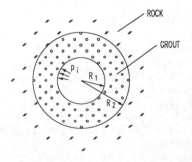

Figure 1. Model geometry - *Geometrie des Modells*

34

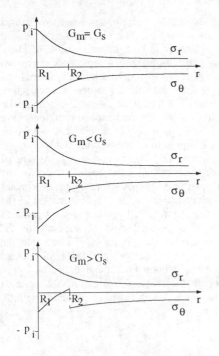

respectively in rock and grout. It does occur effectively in rock if $p_{im} < p_{is}$, and in grout if $p_{is} < p_{im}$. The first condition may be expressed as:

$$\sigma_{tm} / \sigma_{ts} < [2(1-\upsilon_s)G_m/G_s] /\{[1+(1-2\upsilon_s) G_m/G_s](R_2/R_1)^2 + (1- G_m/G_s)\} \qquad (4)$$

Table 1. Properties of typical materials - *Parameter typischer Modellmaterialien*

		Grout	Hard rock	Very weak rock
E	[MPa]	10000	52000	100
ν		0.25	0.25	0.25
σ_t	[MPa]	2.5	10	0.05
σ_c	[MPa]	50	200	1
ϕ_p	[°]	24.37	23.4	26
ϕ_r	[°]	30	30	30
c_p	[MPa]	16.12	66	0.31
c_r	[MPa]	0	0	0

The curves (σ_{tm}/σ_{ts}) versus (G_m/G_s) are shown on figure 3, for different values of $(R_2/R_1)^2$ and for $\upsilon_s = 0.25$.

Assuming typical properties for the grout material, a rock can be represented by one point on the diagram of the figure 3. To represent the most typical rocks, we have used the bivariate classification by Deere and Miller (Franklin & Dussault 1989). On a Deere and Miller diagram, most typical rocks (from hard to very weak) are located around the line defined by the following equation:

$$\log E = 1.18 \log \sigma_c + 2 \qquad (5)$$

Figure 2. Radial and tangential elastic stresses distribution along a radius in a two-phase-medium - *elastische Radial- und Umfangsspannungen über dem Radius in einem Fels-Mörtel-System*

than 0.33, G_m must be higher than 6 times G_s to have a compressive rupture. Therefore we will consider that the first ruptures in both materials are tensile.

Let p_{im} and p_{is} be the values of the internal pressure p_i, for which the rupture could occur

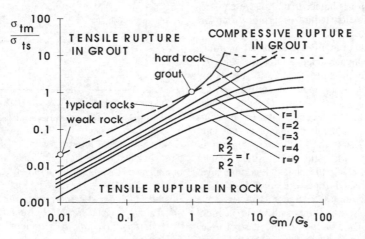

Figure 3. Type of initial rupture - *Typ der anfänglichen Rißausbreitung*

35

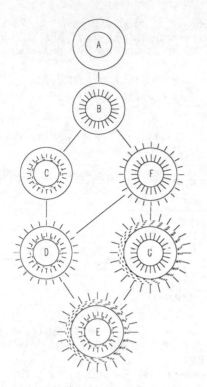

Figure 4. Occurrence of new modes of rupture - *Auftreten neuer Rißmoden*

where E is the elastic modulus and σ_c the compressive strength. Assuming $n = 20$, equation (5) is represented on figure 3 by the dashed line, which defines a typical series of rocks. Henceforth, rock material will be supposed to belong to this series. The mechanical parameters of the two extreme rocks and of the typical grout material are given in table 1. Assuming that R_1/R_2 is higher than 1.4, we can see on figure 3 that the first failure occurs in grout for rocks located on the dashed line.

After the first tensile rupture had occurred in grout, tension cracks propagate radially outward when p_i increases. The tangential stress reduces to zero in the fissured zone, which is then submitted to an increasing uniaxial compression in the radial direction. At this stage, a second rupture will occur either in grout in a compressive mode or in rock in a tensile one (figure 4). For the first case, compressive rupture will happen when the radial stress in the fissured zone will reach the uniaxial compressive strength, and a crushed zone will form inside the radially fissured one (case C in figure 4). Afterwards tensile and compressive ruptures will occur in rock (case D and case E). For the second case, tensile rupture will happen in rock when the tangential stress

will reach the tensile strength, and a radially fissured zone will form (case F). Afterwards two compressive ruptures will occur in both materials (G-E or D-E). Note that a sketch on figure 4 represents the modes of rupture which have occurred in each material and the straight lines indicate the order of occurrence. Although they are drawn on all sketches, the elastic zone and the radially fissured zone in grout may have disappeared. A more complex diagram would be necessary to represent all the possible states.

Relations between internal pressure p_i and radial internal displacement u_i have been derived for the different possible states. A dilatancy parameter has been introduced to calculate the deformation of the crushed zones. For typical rocks, the most common behaviour of a grouted hole corresponds to the succession ABFDE (see figure 4). For example, the curve p_i-u_i, for the hard rock defined in table 1 and with no dilatancy, is shown in the figure 5.

Applying these results to determine the reaction force acting on a dowel, the force per unit length, P, is given by:

$$P = p_i D_1 \qquad (6)$$

2.2 *Grouted hole with uniaxial load*

To test the reliability of the proposed analytical model, we have used a numerical method to calculate the displacements induced by a force P applied on a

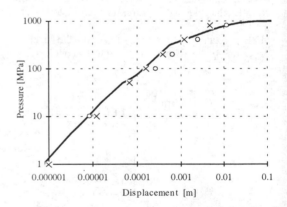

Figure 5. Internal pressure versus displacement for a hard rock with $R_2^2/R_1^2=2$. Line: analytical results, o: numerical results for axisymmetrical load, x: numerical results for uniaxial load
Innendruck über Verschiebung für hartes Gestein mit $R_2^2/R_1^2=2$. Linie: analytische Lösung, o: numerische Lösung für symmetrische Belastung, x: numerische Lösung für radiale Einzellast

dowel. This force results from a uniform pressure p_i, applied on a diameter normal to its direction, which is given by equation (6). A hybrid finite element/ boundary element program, PHASES, has been used (Hoek et al.). The mesh is shown on figure 6.

To check the numerical results, displacements induced by an axisymmetrical load have also been calculated with PHASES, for several load steps. A good agrement has been obtained between analytical and numerical results, as shown on figure 5 for the case of a hard rock.

On the same figure, numerical results for an uniaxial loading have been plotted. For a given displacement, the pressure obtained with a uniaxial loading is slightly over the one with axisymetrical loading, in particular the ultimate pressure. Using the analytical solution to determine the reaction force on a dowel is then safe. This feature applies to the rocks harder than the standard grout material.

For the very weak rock defined in table 1, both types of loading are compared on the figure 7. For small values of p_i, the displacements are very different. When an uniaxial load is applied, stresses are transfered to the rock which is strongly contracted in the direction of the load. On the other hand, for an axisymetrical load, stresses are supported mainly by grout as long as it is not completely fissured; the deformation of the rock is then smaller. When the radial fissures reach the interface, due to the increasing pressure, stresses are transfered to the rock and the displacement increases strongly up to a value slightly under the one for the uniaxial load (figure 7). This process occurs for the rocks weaker than the standard grout material.

For applying the analytical axisymetrical solution to calculate the reaction force acting on a dowel, it was necessary to modify it. For this, the internal pressure p_i has been transfered to the rock as if the grout was cut into two halves, normally to the direction of the real load. A pressure p_2, given by the following expression, has then been applied to the rock at $r = R_2$:

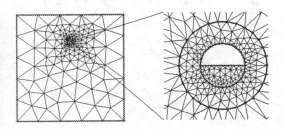

Figure 6. Mesh for uniaxial load - *Vernetzung für einachsige Belastung*

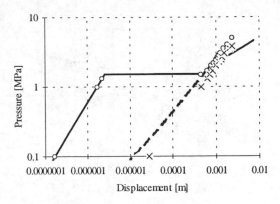

Figure 7. Internal pressure versus displacement for a very weak rock with $R_2^2/R_1^2=2$. Line: analytical results, dashed line: modified analytical results, o: numerical results for axisymmetrical load, x: numerical results for uniaxial load
Innendruck über Verschiebung für sehr weiches Gestein mit $R_2^2/R_1^2=2$. Linie: analytische Lösung, Strichlinie: modifizierte analytische Lösung, o: numerische Lösung für symmetrische Belastung, x: numerische Lösung für radiale Einzellast

$$p_2 = p_1 R_1 / R_2 \qquad (7)$$

After the grout has become completely fissured, the original analytical solution is used.

3 AXIAL AND TRANSVERSE FORCES IN A BOLT

3.1 *Hypothesis*

The figure 8 shows the geometry of a bolted joint with the initial opening of the joint y_i, the initial angle of the bolt θ_i, its diameter D_1, the hole diameter D_2 and the point of symmetry O.

As the curves pressure-displacement are defined by pairs of values (p_i, u_i) instead of an analytical function, computing the bolt is easier by a numerical method than by an analytical way. The principle of the method is to divide the bolt in slices and to calculate for each slice the forces and the moment transmitted to the next one (figure 9). The normal reaction force acting on a slice, P_s, is obtained from the equation (6):

$$P_s = p_i D_1 L_s \qquad (8)$$

Note that a force P_b may act on the opposite side, if this one is in contact with the other wall of the joint. Initially there is only a normal force P_s acting on the

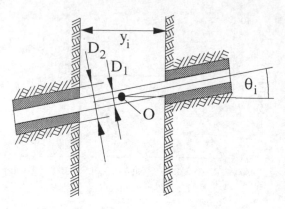

Figure 8. Geometry of a bolted joint - *Geometrie einer Ankerung*

bolt, but a shear force S_s is mobilized as soon as the bolt begins to bend. It is assumed that this force remains zero until a rupture occurs in grout, after which it is equal to the residual shear strength of the bolt-grout interface and the resulting force makes an angle ϕ with the normal, as on the figure 9.

3.2 *Numerical method*

For a displacement increment imposed at the point O, the following method is applied to obtain the axial and transverse forces, N_o and Q_o, acting in the bolt at this point (where the bending moment is zero).

For the first iteration, N_o et Q_o are set to the values obtained from the last displacement increment. For the current slice in the current iteration, the moment, axial and transverse forces applied to the next one are derived from equilibrium equations. If the slice is not between the joint walls, normal reaction force P_s is calculated by equation (8), in which p_i is given by

a curve like on the figure 5. Next the bending angle and the extension of the bolt are derived from the moment and the axial force. At this stage, it is necessary to assume a strain distribution in the cross-section, which must be divided into constant strain elements as shown in the figure 10. So the relative displacement between the ends of the slice is known, which leads to the displacement increment of the left end of the next slice.

For the latter the same method is applied to derive the extension, the bending angle and the next displacement increment. As the direction of this slice has changed, it is necessary to start again the calculations from the first slice.

This process stops when the location of the current slice reaches its initial position (at the beginning of the current increment). If this doesn't happen, Q_o was too low and a higher value is taken for the next iteration. If the angle θ of the slice meeting its initial position is greater than its initial value, Q_o was too high and a lower value is taken. In the same way the total extension of the bolt has to correspond with the axial displacement at point O. After some iterations Q_o and N_o are obtained, corresponding to the imposed displacement increment.

3.3 *Bauschinger effect*

The typical stress-strain response for an increasing loading of steel is shown on the figure 11. Strain-hardening is also shown for the case of unloading and reloading without changing the sign of stress (dotted line). If the stess sign changes, the strength is decreased instead of increased, as shown by the thin line on the figure 11. It is the Bauschinger effect.

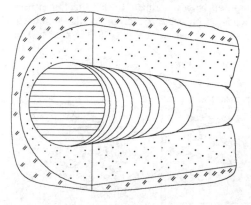

Figure 10. Division of the anchor in slices and their elements - *Unterteilung des Ankers in Scheiben und deren Oberflächenelemente*

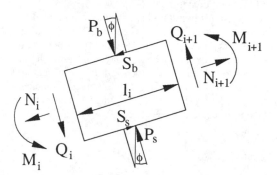

Figure 9. Forces and moments on a slice i - *Kräfte und Momente an einer Stabscheibe i*

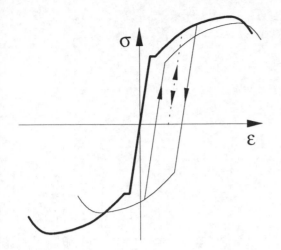

Figure 11. Stress-strain curve for a steel. Thick line: continuous loading; dotted line: strain hardening; thin line: strength decrease after the stress sign has changed - *Arbeitslinie eines Stahls. Dicke Linie: kontinuierliche Belastung; Strichlinie: Kaltverfestigung; dünne Linie: Festigkeitsabnahme nach Spannungsumkehr*

A bolt element may first be compressed because of bending and then be tensioned when the shear displacement of joint progresses. Moreover bars which have been predeformed by stretching are often used for reinforcement. For these reasons the Bauschinger effect is considered in the method of calculation of bending and axial strain of a slice.

4 CONCLUSION

The proposed method gives the axial and transverse forces in a bolt submitted to axial and transverse displacements applied in a joint. For a given direction of displacement, application of displacement increments allows to predict the complete force-displacement curves. The method can also be coupled with stress-strain methods used for simulation of discontinuous rock masses.

REFERENCES

Amadei, B. & S. Saeb 1990. Constitutive models of rock joints. *Proc. Int. Symp. Rock Joints*, Loen, Norway, pp. 581-594.

Aydan O., Ichikawa Y. & T. Kawamoto 1985. Load bearing capacity and stress distributions in/along rockbolts with inelastic behaviour of interfaces.

Proc. 5th Int. Conf. on Num. Meth. in Geom., Nagoya, pp. 1281-1292.

Dight P.M. 1983. Improvements to the stability of rock walls in open pit mines. *Ph. D. Thesis.* Monash University, Australia .

Fairhurst C. 1964. On the validity of the Brezilian test for brittle materials. *Int. J. Rock Mech. Min. Sci.* Vol.1, pp. 535-546.

Franklin J.A. & M.B. Dusseault 1989. Classification systems. In *Rock Engineering*, pp. 77-83. McGraw-Hill.

Hoek E., Carvalho J.L. & B.T. Corkum. PHASES. Rock Eng. Group, University of Toronto, Canada.

Holmberg M. 1991. The mechanical behaviour of untensioned grouted rock bolts. *Ph.D. Thesis,* Royal Inst. of Techn., Stockholm .

Holmberg M. 1992. The mechanical behaviour of a single grouted bolt. *Proc. Int. Symp. Rock Support in Mining and Underground Construction,* Sudbury, Canada, pp. 473-481. Balkema, Rotterdam.

Ladanyi B. 1966. Failure mechanism of a rock under a plate. *Proc. 1st ISRM Congr.*, Lisboa, Vol. 1, pp. 415-420. Lab. Nac. de Eng. Civil .

Ladanyi B. 1967. Expansion of cavities in brittle media. *Int. J. Rock Mech. Min. Sci.*, Vol. 4, pp. 301-328.

Lorig L. J. 1985. A simple numerical representation of fully bonded passive rock reinforcement for hard rock. *Computers and Geotechnics*, Vol. 1, pp. 79-97 .

Panet M. 1988. Renforcement des fondations et des talus à l'aide d'ancrages actifs et passifs. *Proc. 6th Int. Congr. Rock Mech.*, Montréal, Vol. 3, pp. 1569-1578. Balkema, Rotterdam .

Pellet F. 1993. Résistance et déformabilité des massifs rocheux stratifiés renforcés par ancrages passifs. *Thèse de Doctorat*, n° 1169. EPFL, Lausanne .

Schubert P. 1984. Das Tragvermögen des Mörterversetzten Ankers unter Aufgezwungener Kluftverschiebung. *Doctorate Thesis.* Montanuniversität, Loeben, Austria.

Souley M. et Homand F. 1991. Etude bibliographique des différentes lois de comportement des joints rocheux. *Rapport Lab. de Géomécanique*, Nancy .

Yasici S. & P.K. Kaiser 1992. Bond strength of grouted cable bolts. *Int. J. Rock Mech. Min. Sci.& Geomech. Abstr.*, Vol. 29, n° 3, pp. 279-292.

Anker in Theorie und Praxis, Widmann (Herausgeber) © 1995 Balkema, Rotterdam. ISBN 90 5410 577 1

Zusammenhang zwischen Tragverhalten und Rißbildung in der Haftstrecke eines Felsankers

Interaction between load capacity and fractures in the bonded length of the rock anchor

Peter Jirovec
Autobahndirektion Nordbayern, Nürnberg, Germany

Zusammenfassung Für vorgespannte Freispielanker wird die Krafteintragung entlang einer Haftstrecke in den umliegenden Fels vorgestellt. Um den Mechanismus des Tragverhaltens von Felsankern erklären zu können, muß auch die Haftstrecke so untersucht werden, daß Aufschlüsse über ihre mögliche Zerstörung erhalten werden. Die einzige Möglichkeit, eine solche Untersuchung des im Gebirge eingebauten Ankers durchzuführen, besteht darin, ihn mit einem größeren Durchmesser zu überbohren und aus dem Gebirge herauszuziehen. Nach dem Überbohren wurde die Haftstrecke mit einer Steinsäge durch Quer- bzw. Längsschnitte zerlegt.

Abstract The subject of this presentation is the interaction between prestressed anchors and the surrounding rock.
Particular attention was paid to the other length, that is critical to the reliable effectiveness of the anchor, in order to examine the bond length of the anchor it is necessary to drill it out in a larger diameter and to remove it from the rock. The anchors were set in 76 mm drill holes and they were drilled out with a 166 mm diameter bit, 8 mm lip and a casing pipe of 400 cm length.
After drilling the anchor shaft was cut up into sections and along its length with a stone saw. The extent of the fractures was particularly noticeable on the cut surfaces as they were still wet. Under this conditions the photographs were taken. The examinations were realized on the three following types of anchoring.
Type A: The single bar or mono-anchor, prestressing steel "Dywidag" of 26,5 mm diameter, was grouted in length of 110 cm (Fig. 1.1). Another anchor made of six parallel 7,5 mm wires was grouted in a length of 130 cm. The surface of the wires had 5 mm wide and 0,1 mm high ribs. It was put under 335 kN load. Fig. 1.2 to 1.4 illustrate the fracture pattern. In this anchor group the anchor consisting of smooth rods that are undulated at the rock end of the fixing shaft is included. Six wires of 6 mm diameter were grouted in a length of 200 cm. The undulated part was 40 cm long. The prestressing represented a load of 230 kN, see Fig. 1.5 and 1.6. The tension in the fissure was so large that even the surrounding rock was brocken apart. All of these anchors were prestressed up to 90 % of their load capicity.
Type B: For these examinations an anchor was produced of three strands with two spreads in the length of 120 cm. The spreads protruded 2,8 cm. Each strand had two fractures illustrated on Fig. 1.7 and 1.11.
Type C: No fractures were in evidence on anchor type C.
It is not possible to prevent cracks as they are an important factor in the load-bearing capacity of the anchoring. The negative affects of the cracks can be reduced, for example by encasing the anchor shaft in a plastic tube, Fig. 2.1 and 2.2.
The radical pressure produced by the prestressing of the anchor causes interaction between the anchor and the rock. The pressure is connected with the dilation and wedge effects of both anchor and rock face. Because of this effect the expansion of the anchor shaft is restricted at the point of friction, and the force is thereby transferred to a large extent in the bond area. Without these cracks in the bond length of the anchor the entire shaft would, above a certain load, slowly work itself loose and thus ruin the effectiveness of the anchor. The cracks are therefore to be acknowledged as a desirable property of the anchoring, and should be judged together with the anchor construction that causes them.

1 EINFÜHRUNG

Der vorliegende Beitrag befasst sich mit der Rißbildung in der Haftstrecke eines Ankers. Die Rißbildung an Felsankern war ein bislang ungeklärtes Problem. Der Hauptpunkt dafür liegt vermutlich darin, daß es sehr aufwendig ist, Felsanker „ausgraben".

Um den Mechanismus des Tragverhaltens von Felsankern erklären zu können, muß auch die Haftstrecke so untersucht werden, daß Aufschlüsse über ihre mögliche Zerstörung erhalten werden. Die einzige Möglichkeit, eine solche Untersuchung des im Gebirge eingebauten Ankers durchzuführen , besteht darin, ihn mit einem größeren Durchmesser zu überbohren und aus dem Gebirge herauszuziehen. Die verwendete Ausrüstung war geeignet, einen senkrecht in die Steinbruchsohle in ein Bohrloch von 76 mm Durchmesser eingebauten Anker wiederzugewinnen. Die Überbohrung geschah mit einer Bohrkrone mit einem Durchmesser von 166 mm, einer Lippenbreite von 8 mm und einer Futterrohrgesamtlänge von 400 cm. Nach dem Überbohren wurde die Haftstrecke mit einer Steinsäge durch Quer- bzw. Längsschnitte zerlegt. Auf diese Weise wurden drei Ankertypen untersucht.

Die Ursache der Rißbildung liegt in der unterschiedlichen Dehnung und Zugfestigkeit von Ankerstahl und Verpreßzementstein. Schon bei geringerer Anspannung ist die Dehnung des Stahles größer als jene, die eine Überschreitung der Zugfestigkeit des Zementsteines bewirkt. Die Folge davon ist eine erste Rißbildung. Bei Zuwachs der Dehnung entstehen weitere Risse. Da jede Rißbildung eine Volumenvergrößerung mit sich bringt, kommt es neben

der Dilatation auch noch zu einem Verkeilungseffekt im gerissenen Teil der Haftstrecke. Dadurch wird eine weitere Herauslösung des Ankers aus dem Mörtelbett verhindert. Wenn dieses nicht so wäre, würde sich das Zugglied entlang der ganzen Länge der Haftstrecke lösen. Die Länge des gerissenen Haftstreckenteiles ist im übrigen relativ klein.

Die Risse und deren geometrische Lage sind bei Felsankern von anderer Natur als die an Aluvial-Verpreßankern, ebenso unterscheidet sich das Tragverhalten beider Anker. Am Felsanker gibt es eine Zone zwischen Zugglied und Mörtel, in der er versagen kann. Beim Aluvial-Anker befindet sich eine derartige Zone dagegen zwischen dem Verpreßkörper und seiner Umgebung. An dieser Grenze bildet sich beim Aluvialanker eine dünne Scherfuge, deren Dicke zu etwa dem 10-fachen eines kennzeichnenden Korndurchmesser zu erwarten ist (Roscoe,1970).

Die Untersuchungen haben besonders ungünstige Fälle von Rißbildungen zum Gegenstand, da alle Anker bis zum 0,9- bzw. 0,84-fachen der Streckgrenze beansprucht wurden. Monoanker St 850/ 1050, Spanndrähte St 1500/ 1700. Wenn an einem Felsanker während seiner Einsatzdauer die Gebrauchslast nicht überschritten wird, werden die Risse keinesfalls so stark ausgeprägt sein wie in den hier vorliegenden Fällen.

Die Ausbreitung der Risse war besonders gut an der noch feuchten Schnittfläche erkennbar. In diesem Stadium wurden sie auch fotografiert.

1.1 Untersuchung am Ankertyp A

Der Einstab- oder Monoanker - ein Spannstahl „Dywidag" mit 26,5 mm Durchmesser - wurde auf

Abb. 1.1: Ausbildung der diskreten Scherfuge am „Dywidag" Felsanker

Fig. 1.1. Development of discrete shear joints on the "Dywidag" rock anchor

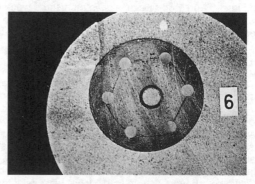

Abb.1.2: Charakteristisches Bild der tangentialen Risse

Fig. 1.2: Characteristic picture of the tangential cracks

Abb.1.3: Radiale und tangentiale Rißbildung

Fig. 1.3: Radial and tangential cracks

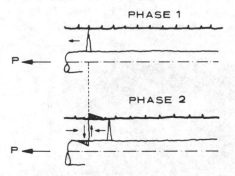

Abb.1.4: Schematische Darstellung des Verkeilungseffektes

Fig. 1.4: Schematic diagram of the wedge effect

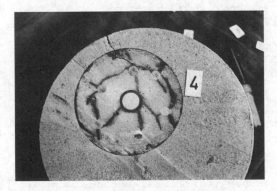

Abb.1.5: Die Fortpflanzung eines radialen Risses ins Gebirge

Fig. 1.5. The propagation of a radial crack into the rock

eine Länge von 110 cm verpreßt und bis 420 kN belastet. Der Unterschied im Tragverhalten von Fels- und Erdankern geht aus Abb. 1.1 hervor. Die ausgegrabenen Erdanker weisen radial verlaufende Risse auf, die offen sind und somit den Verpreßkörper zerteilen. Derselbe Einstab als Felsanker dagegen ist gekennzeichnet durch das Fehlen offener Risse und durch eine diskrete Scherfuge. An ihr wurde der Mörtel zwischen den Rippen abgeschert. Da die Freispielstrecke erst 9 cm hinter der Packermanschette endete, ist in diesem Bereich der Haftstreckenteil intakt geblieben.

Die Bildung einer diskreten Scherfuge ist für diesen Felsanker charakteristisch. Sie kann nur im Bereich der Stahlmantelfläche entstehen, denn an der Grenze Mörtel - Fels sind die dazu notwendigen Bedingungen nicht gegeben.

Ein Anker aus sechs parallel laufenden Drähten mit je 7,5 mm Durchmesser wurde auf einer Länge von 130 cm injeziert und bis 335 kN belastet. Die Oberfläche der Drähte war mit 5 mm breiten und 0,10 mm hohen Rippen versehen.

Einen Querschnitt durch diesen Anker, 10 cm vom Packer entfernt geschnitten, zeigt Abb. 1.2. In der Bohrlochmitte befinden sich der Injizierschlauch und um diesen herum die Drähte. Deutlich sind die tangentialen Risse, welche die einzelnen Drähte verbinden, zu sehen. Die Lage dieser Risse ist auf die Spaltzugkräfte der gleichmäßig gezogenen Drähte zurückzuführen. Dieser Versuch machte sowohl tangentiale Risse als auch radiale, normal zur Ankerachse stehende Risse sichtbar (Abb.1.3). In diesem Bereich befanden sich drei radiale Risse , d.h. die Haftstrecke dieses Ankers ist in vier Abschnitte einzuteilen. Der tangentiale Riß, parallel zur Ankerachse, endete rechts von dem dritten radialen Riß. Eine bessere Lage der radialen Risse im Längsschnitt könnte man sich nicht wünschen. Sie erzeugen optimal den oben erläuterten Verkeilungseffekt, welcher neben der Dilatation wesentlich zum Mechanismus des Tragverhaltens von Ankern beiträgt. In der Phase 1 (Abb.1.4) entsteht infolge der Überwindung der Mörtelzugfestigkeit ein Riß. Mit zunehmender Dehnung öffnet er sich, da durch ihn der Draht-Mörtel-Verbund überwunden wurde. Bei zunehmender Belastung wiederholt sich durch eine Verlagerung der Dehnung nach hinten, zum Bohrlochtiefsten hin, der beschriebene Vorgang. Da die Volumenvergrößerung nur begrenzt möglich ist, hat das Öffnen des neu entstehenden Risses das Schließen des vorherigen zur Folge - Phase 2. Die Schubspannungen entlang des Risses wirken einerseits auf die Bohrlochwandung, andererseits auf die Drathoberfläche. Da die hohen Dehnungen der Spannstähle schon innerhalb einer kurzen Strecke aufgefangen werden können, müssen Keilwirkung und Dilatation enorm dehnungsbehindernd wirken.

Zu diesem Ankertyp gehört auch der Anker aus glatten, am bergseitigen Ende der Haftstrecke gewellten Stäben. Für die Untersuchung der Rißausbreitung wurden sechs Drähte mit 6 mm Durchmesser auf einer Länge von 200 cm verpreßt. Der gewellte Bereich war 40 cm lang und der Anker bis 230 kN belastet. Die Bergseite des überbohrten Kernes, in dessen Mitte der Anker lag, zeigt Abb. 1.5. Links in der Abbildung ist das Ende der gewellten Drähte zu sehen. Die durch die Belastung aufgebrachten Spaltzugkräfte (vgl. auch Abb. 1.8) waren so groß, daß durch sie sogar das umliegende Gebirge aufgerissen wurde. Der Riß pflanzte sich

Drahtlage am Bohrlochende kann beim Einbau nicht abgeschätzt werden. Dies ist aber im Hinblick auf die Rißverhütungsmöglichkeit auch nicht notwendig.

1.2 Untersuchung am Ankertyp B

Für diese Untersuchungen wurde ein Anker aus drei Litzen hergestellt, welcher in der Haftstrecke von 120 cm Länge zwei Spreizungen aufwies. Die Einschnürung erfolgte durch Umwinden mit üblichem Montagedraht, die Spreizung mittels Kunststoffabstandhaltern. An der Spreizstelle betrug die Entfernung von der Litzenaußenkante zur Bohrlochwandung 1 cm, so daß die Stichhöhe der Spreizung 2,8 cm betrug. Dieser Anker wurde mit 350 kN belastet. Ein Querschnitt zwischen der ersten Einschnürung und Spreizung, also hinter der Packermanschette, ist in Abb. 1.7 dargestellt. Von jeder Litze ziehen sich

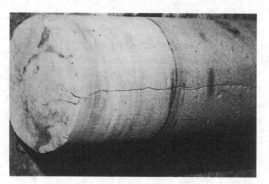

Abb.1.6: Rißbild im Bereich der gewellten Drähte
Fig. 1.6: Cracks in the vicinity of the undulated cables

Abb. 1.7: Rißbild am Litzenanker zwischen der Einschnürung und der Verspreizung
Fig. 1.7: Cracks on the strand anchor between the tie and the spread

über die Wellung auf einer Gesamtlänge von etwa 50 cm fort. Es ist wahrscheinlich an der Stelle aufgetreten, in welcher die in einem gewellten Draht enthaltene Ebene normal zur Bohrlochwandung stand. Wenn es möglich wäre, bei allen Drähten die Wellungsebene im Bohrloch parallel zur Bohrlochwandung zu placieren, würder es nur zur tangentialen Rißbildung kommen (Abb. 1.2). Ein Abbau der potentiellen Energie vom Draht zur Umgebung hin in Form der Rißbildung kann sowohl im Gebirge als auch in der Haftstrecke selbst stattfinden. Die gesamte Rißbildung (Abb. 1.6) im Querschnitt ist ein kompliziertes System von tangentialen und radialen Rissen. Von der drei an der Bohrlochwandung endenden Rissen konnte sich schließlich nur einer ins Gebirge fortpflanzen. Der Verlauf der Risse im Querschnitt hängt von der Lage der gewellten Drähte zueinander und zur Bohrlochwandung ab. Die

zwei Risse zur Bohrlochwandung hin, die durch die Spreizung hervorgerufen sind. Ein progressiver Gebirgsbruch ist oben links im Bild über der Litze zu sehen. Der Mechanismus der Kraftübertragung dieses Ankersystems kann (Abb. 1.8) folgenderweise erklärt werden: Die in einer Litze wirksame Reaktionskraft P_1 hat zwei Komponenten - eine axiale und eine radiale. Die axiale Komponente P_{l1} führt die einzutragende Kraft zum Bohrlochtiefsten hin, die radiale Komponente P_{r1}, die Spreizkraft, wirkt gegen das Gebirge und induziert die Risse. Im anderen Teil der „Zwiebel", zwischen den Punkten B und B' (Punkt B' liegt symetrisch zum Punkt B in der zweiten Zwiebelhälfte) induziert die Komponente P_{r2} Risse von den Litzen zur Ankermitte hin. Den

Beweis für die Richtigkeit dieser Hypothese liefert
Abb. 1.9, wo solche Risse beobachtet weren können.
Dieser Querschnitt lag zwischen den Punkten C und
B' der Abb. 1.8.

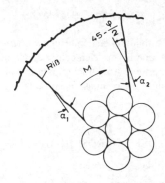

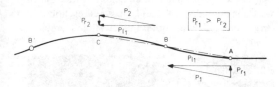

Abb.1.8: Schematische Darstellung der Kraftübertragung an einer „Zwiebel"
Fig. 1.8: Schematic diagram of transferring
force with an "onion"

An der Entstehung der beiden Risse, welche von
einer Litze nach außen abstrahlen, beteiligen sich
auch ein Moment M (Abb.1.10), das durch Zug der
verdrillten Litze zustande kommt. Aus diesem
Grund sind die Winkel α_1 und α_2 verschieden.

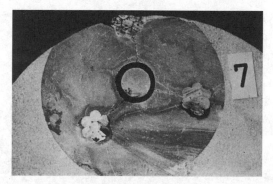

Abb.1.9: Rißbild am Litzenanker zwischen der Verspreizung und der Einschnürung
Fig. 1.9: Fracture on the strand anchor between
the spread and the tie

Die große Spannungskonzentration um die Litze
herum hatte den progressiven Gebirgsbruch zur Folge. Die Ergebnisse an einem ähnlich konstruierten
Anker, wo sechs Drähte, von etwa gleicher Fläche
wie diese drei Litzen, ähnlich belastet und untersucht
wurden, sind der Beweis dafür. An der selben Stelle
wie am Litzenanker wurde ein Querschnitt geöffnet
(Abb.1.11). Die radiale Komponente der in jeden
Draht einzutragenden Kraft wurde auf die Hälfte ihrer Größe je Draht reduziert. Deshalb ist es nur an
einer

Abb. 1.10: Schematische Darstellung des Einflusses der Litzenverdrillung an der Rißbildung
Fig. 1.10: Schematic diagram of the effect of the
strand drilling on the cracking of the rock

(Abb.1.1). Die radiale Komponente der in jeden
Draht einzutragenden Kraft wurde auf die Hälfte ihrer Größe je Draht reduziert. Deshalb ist es nur an
einer Stelle zur Fortpflanzung der Risse bis zur
Bohrlochwandung hin gekommen. Es bildeten sich
in erster Linie Risse zwischen den einzelnen Drähten, welche auf die Beschaffenheit der Drahtoberfläche zurückzuführen sind. Der kleine Riß am Inje-
zierschlauch ist mit großer Wahrscheinlichkeit der
End-

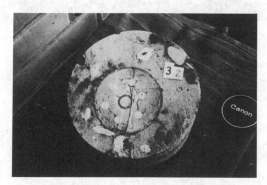

Abb.1.11: Einfluß der verminderten Spannungskonzentration auf die Rißbildung beim Ankertyp A.
Fig. 1.11: Effect of the reduced stress concentration
with anchor type A

bereich des in der anderen Zwiebelhälfte entstandenen Risses. Mit der Verminderung der Spannungskonzentration durch die Kraftverteilung auf mehrere
Drähte sind hier ähnliche Ergebnisse zu verzeichnen
wie bei der Messung der radialen Druckspannungen.

1.3 Untersuchung am Ankertyp C (Druckrohr)

Die festgestellte Rißbildung an den beiden vorherigen Ankertypen A und B wurde eindeutig durch die Belastung hervorgerufen. Am Druckrohranker wurden keine Belastungs- oder Entlastungsrisse festgestellt.

2 VERHÜTUNG DER KORROSION AN FELSANKERN

Die Mörtelumhüllung an Felsdaueranker kann nicht als allein ausreichender Schutz angesehen werden. Durch Risse, hauptsächlich durch die radial zur Bohrlochwandung verlaufende, ist die Gefahr einer

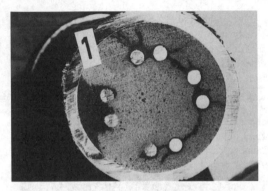

Abb. 2.1: Rißbildung im Wellbereich eines im Stahlrohr eingebauten Ankers

Fig. 2.1: Fractures in the vicinity of the undulation of an anchor set into a steel tube

möglichen Korrosion gegeben. Schon unter der Voraussetzung von Luftzufuhr in das den Anker umgebende Bergwasser ist Versagen zu befürchten. Noch mehr natürlich im Falle von Chloriden und anderen Chemikalien, welche speziell den Spannstahl angreifen. Ein Abreißen des Ankers in der Haftstrecke kann durch Vorhandensein von aggressiven Wässern erheblich beschleunigt werden.

Die Rißbildung in der Haftstrecke zu verhindern ist nicht möglich, da sie ein wesentlicher Faktor der Ankertragfähigkeit ist. Wohl aber läßt sich die schädliche Wirkung der Risse, zum Beispiel wie folgt, unterbinden. Es wird ein Kunstoffrohr über die Abstandhalter in der Haftstrecke aufgeschoben, welches bei vollständiger Verpressung den Einfluß der Korrosion verhindert.

In einem Querschnitt des in einem Stahlrohr ver-

preßten Ankers (Abb.1.12), dessen glatte Spanndrähte in der Haftstrecke gewellt wurden, kam es zur tangentialen und radialen Rißbildung. Dieser Anker war aus acht Drähten mit je 7,5 mm Durchmesser, und wurde bis zur Streckgrenze von 462 kN belastet. Bei Wiederholung dieses Versuches wurde über den Bereich der gewellten, glatten Drähte ein Kunstoffwellrohr mit 0,35 mm Wandstärke aufgeschoben. An derselben Stelle wie vorher im gewellten Bereich

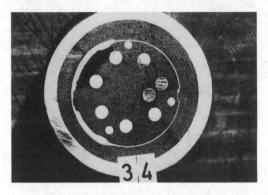

Abb. 2.2: Verhütung der Rißausbreitung mittels eines gewellten Kunststoffrohres

Fig. 2.2: Preventing cracks from spreading by an undulated plastic tube

wurde eine Querschnittprobe entnommen (Abb. 1.13). Es ist deutlich zu sehen, daß alle radialen Risse im Rohrinnern verblieben und ihre Ausbreitung im Rohr aufgehalten wurde. Nur ein radialer Riß (fünf-Uhr-Stellung) pflanzte sich auch außerhalb des Rohres fort, doch wurde das Rohr dadurch nicht zerrissen Die deutliche Unterbrechung des Risses hat für die Vermeidung der Korrosion elementare Bedeutung, da das Kunstoffrohr die Deformation ohne zu brechen mitmachte. Die scheinbar ungleichmäßige Wandstärke des Kunstoffrohres links im Bild wurde nur dadurch verursacht, daß das Rohr entlang eines Schenkels der Welle gesägt wurde. Die Quer- bzw. Längsschnitte der Haftstrecke wurden mittels Naßbohrung hergestellt.

3 SCHLUSSFOLGERUNGEN

Bei der Untersuchung der Rißbildung in der Haftstrecke der Felsanker ergab, daß ein direkter Zusammenhang zwischen der Ankerkonstruktion und jede Ankerkonstruktion ihr charakteristisches Rißbild aufweist.

Primär ist der Anker in der Haftstrecke durch radiale, zur Bohrlochwandung hin führende Risse gefährdet. Die tangentialen, zwischen den Drähten verlaufenden Risse bedeuten keine Korrosionsgefahr, wenn gewährleistet ist, daß sie weder den Pakker noch das Bohrlochende erreichen. Dies läßt sich erfüllen, indem die Freispielstrecke erst hinter dem Packer und das bergseitige Ankerende genügend weit vor dem Bohrlochtiefsten endet. Die nachfolgende Ergebnisse beweisen, daß die Ursache der Rißbildung in der Wirkungsweise der Felsanker zu finden ist.

- Durch Kernbohren gewonnene Kerne weisen weder radiale noch tangentiale Risse auf. Dies beweist eine intakte Bohrlochwandung vor dem Ankerbau

- Der in ein Stahlrohr eingebaute Anker erzeugte Risse, welche eindeutig allein auf die Funktionsweise der Anker zurückzuführen sind.

Es hätte sich vermuten lassen, daß die Rißbildung zum Teil von einer Entspannung des Gesteins oder des Mörtels nach dem Überbohren und durch den Transport der Kerne etc. entstanden sein könnten. Durch die Versuche ist nachgewiesen, daß die Rißbildung unabhängig von diesen Einflüssen erfolgt ist.

Radialer Druck, der durch die Vorspannung des Ankers hervorgerufen wird, erzeugt Wechselwirkung zwischen Anker und Gebirge. Dieser Druck wird durch die Keilwirkung noch gesteigert. Dadurch werden die Dehnungen des Ankergliedes im Reibungsbereich der Verpreßstrecke gebremst und aufgehalten und die zu übertragende Kraft zu einem esentlichen Teil im Haftungsbereich eingeleitet

Die Rißbildung muß also als ein wichtiger Bestandteil der Funktionsweise von Felsankern anerkannt werden. Gäbe es keine Rißbildung in der Haftstrecke des Felsankers, so würde sich das Zugglied oberhalb einer gewissen Laststufe nach und nach aus der ganzen Haftstreckenlänge lösen und so die Ankerwirkung zunichte machen. Die Rißbildung ist somit als eigentlich wünschenswerte Eigenschaft der Ankerung zu werten, deren Folgen nur zusammen mit der sie verursachenden Ankerkonstruktion zu beurteilen sind.

LITERATURVERZEICHNIS

Benz, G. 1975. Einpreßmörtel. Chem. Fabrik Grünau, Illertissen.

Eisenbiegler, W. 1975. Das Verbundverhalten druckbeanspruchter Betonrippenstähle im Beton. Dissertation, Universität Karlsruhe.

Jirovec, P. 1979. Untersuchung zum Tragverhalten von Ankern. Dissertation, Universität Karlsruhe.

Müller, L. 1963. Felsbau Band I. Ferdinand Enke Verlag, Stuttgart.

An analytical method improving the rock-support interaction analysis of ungrouted tensioned rockbolts

Eine analytische Methode, welche die Untersuchung der Anker durch die Gebirgskennlinienmethode verbessert

V. Labiouse
Laboratoire de Génie Civil, Université Catholique de Louvain, Louvain-la-Neuve, Belgium

ABSTRACT: As part of a doctoral thesis, an analytical method has been developed to improve the design of ungrouted tensioned rockbolts as support of tunnels in axisymmetric conditions. In this paper, the propounded improvements are first presented: 1) the reaction force transferred to the rock mass in the anchoring zone of the bolts, 2) the elastic recompression of the carrying ring surrounding the tunnel induced by the bolts preload, 3) the relative displacement of the bolts ends and its repercussion on the tension in the rockbolts. The principles of the analysis are then described and finally a numerical application is taken.

ZUSAMMENFASSUNG: Aufgrund der Feststellung, dass eine Einführung des Ankerbaus in der klassischen Gebirgskennlinienmethode nur begrenzt möglich ist, wurde eine neue analytische Methode erarbeitet. Diese Methode sowie zwei weitere komplementäre Studien sind Bestandteil einer Doktorarbeit an der Université Catholique de Louvain (Labiouse, 1993).

Die theoretischen Ausführungen basieren auf denselben Hypothesen wie die Gebirgskennlinienmethode: Axisymmetrie und ebener Verformungszustand. Der untersuchte Tunnel ist somit kreisförmig und in einem homogenen isotropen Felsmassiv gebohrt. Das ursprüngliche Spannungsfeld ist dadurch hydrostatisch.

Die Berechnung berücksichtigt die Merkmale der Anker (Länge, Querschnitt, Deformabilität), die longitudinalen und transversalen Abstände, der Plazierungsabstand zur Ortsbrust sowie die Vorspannung im Zeitpunkt des Einbaus. Diese neue Methode hebt sich vorallem durch drei Verbesserungen hervor: 1) So wird nicht nur die Auswirkung des Ankers auf die Tunnelwand berücksichtigt, sondern auch die Reaktionskraft, welche auf das Felsmassiv im Bereich der Verankerungszone übertragen wird. 2) Der Miteinbezug einer elastischen Kompression des verankerten Gewölbes unter Anwendung der ursprünglichen Vorspannung der Anker. 3) Die Untersuchung der Veränderung der Länge des Ankers nach dem Einbau anhand der relativen Verschiebung der beiden Enden und deren Auswirkung auf die Erhöhung der Zugspannung in der Ankerstange.

Die Einführung dieser Verbesserungen hat zur Folge, dass das Felsmassiv und der Ankerausbau nicht mehr wie dies bei der klassischen Methode der Fall ist, seperat behandelt werden können. Eine alternative Lösung wurde erarbeitet, welche darin besteht, dass dem Einfluss des Ankers auf die Konvergenzkurve des Felsmassivs Rechnung getreagen wird.

Von den mathematischen Ausführungen, welche mehr als 100 Seiten der Doktorarbeit ausmachen, werden hier nur die wichtigsten Lösungsprinzipien aufgeführt. Zu diesem Zweck werden die Entwicklung der Spannungen sowie der Verschiebungen des Gewölbes und des umgebenden Milieus während der Entspannung des Gebirges im Anschluss auf die Vorspannung der Anker folgt, aufgezeigt. Man stellt fest, dass das Gewölbe folgende Stadien durchläuft: 1) ein vollkommen elastisches Verhalten (kurz nach dem Vorspannen), 2) zwei Phasen teils elastisches teils plastisches Verhalten, und eventuell 3) ein vollkommen plastisches Verhalten (in Abhängigkeit des Ankereigenschaften). Die Reihenfolge dieser verschiedener Stadien zeigt sich ebenfalls auf der Konvergenzkurve des verankerten Gebirges.

Schliesslich unterstreicht eine numerische Anwendung den Einfluss der Eigenheiten der neuen Methode. Man stellt im Speziellen fest, dass die Analyse des systematischen Ankerausbaus durch die klassische Methode eine deutlich grössere Steifigkeit ergibt (75 %!) als jene welche durch die hier vorgeschlagenen Verbesserungen vorgenommen wurde.

Diese analytische Lösung hat zugleich den Vorteil einer kürzeren Rechnungszeit im Vergleich zu den herkömmlichen Methoden. Somit ermöglicht sie eine schnelle Erfassung des Einflusses der Verankerung und der mechanischen Eigenschaften des Felsmassivs, sowohl auf die Konvergenzen als auch auf die Zugspannung, welche im Anker bei erreichter Stabilität herscht.

1 INTRODUCTION

For a long time, the determination of the bolting support requirements for the stabilisation of underground excavations has been commonly based on in situ observations and on empirical approaches (rock mass classification systems). However, since a few years, this design procedure is completed by analytical and numerical evaluations which enable a better qualitative and quantitative analysis of the rockbolts parameters.

Within the framework of this integral design approach, this paper describes an analytical method which improves the usual rock-support interaction analysis (Hoek & Brown 1980, AFTES 1986) of ungrouted tensioned rockbolts.

After a recall of the basic assumptions, the three main improvements propounded in this novel theory are detailed and the definition of a new ground reaction curve is introduced. Then, the principles of the analysis are described, displaying the evolution of the stresses and displacement around the tunnel and showing the associated characteristic curve of the bolted rock mass. Finally, a numerical application is taken to underline the differences between the developed method and the usual theory.

2 BASIC ASSUMPTIONS

The developed method is based on the same fundamental assumptions as the usual rock-support interaction analysis :

1. The problem is axisymmetric; this means that the analysis applies to a circular gallery of radius r_i driven in a homogeneous, isotropic rock mass subjected to a hydrostatic in situ stress field (magnitude p_0). It implies as well to neglect the weight of the rock in the broken zone which develops around the excavation.

2. The length of the tunnel is such that the problem is in a plane strain situation. The three-dimensional response of the rock mass near the tunnel face has been reduced to this plane strain condition introducing a fictitious pressure on the

gallery wall (AFTES 1986, Corbetta 1990). This pressure depends on the unsupported span behind the excavation face and enables so to take into account the delay for support installation.

The rock mass behaviour has been idealised as elastic-perfectly-plastic with a Mohr-Coulomb strength criterion and a non-associated flow rule for the calculation of the post-failure strains.

The support is composed of a systematic pattern of ungrouted tensioned rockbolts (figure 1) characterised by :

s_t : the circumferential spacing
s_l : the longitudinal spacing
L_b : the free length of the shank
A_b : the area of the shank
E_b : the Young's modulus of the bolting system, taking into account the deformation characteristics of the anchor, washer plate and bolt head. It can be assessed from the load-extension curve obtained by means of a pull-out test.
T_{b0} : the pretension load of the bolts
T_{bf} : the ultimate strength of the bolting system determined from a pull-out test.

To keep the axisymmetry of the problem, it is assumed that the bolts spacings s_t and s_l are small enough to consider the tunnel behaviour as uniform and independent of the location of a single bolt. Then, the support system provides an uniform radial pressure p_p which is related to the load T_b in the bolts by :

$$p_p = \frac{T_b}{s_l \, s_t} \quad (1)$$

3 MAIN IMPROVEMENTS

The present approach effects three main improvements in the usual "convergence-confinement" analysis; it takes into consideration :

1. the reaction force transferred to the rock mass in the anchoring zone of the bolt,
2. the elastic recompression of the carrying ring surrounding the tunnel induced by the bolts preload,
3. the relative displacement of the bolts ends, and its repercussion on the tension in the rockbolts.

3.1 *Reaction force in the anchoring zone*

As the bolts are ungrouted (figure 2), the load T_b in the shank is transferred to the tunnel surface at the anchor head through a nut and an end plate; and to the rock mass in the anchoring zone through an expansion shell (mechanical anchor), a resin seal or a cement grout (chemical anchors).

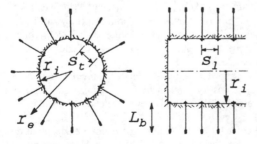

Fig. 1 :   Lie of the rockbolts
Ankeranordnung

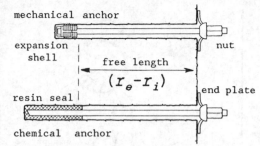

mechanical anchor

expansion shell

nut

free length

$$(r_e - r_i)$$

resin seal

end plate

chemical anchor

Fig. 2 :   Ungrouted tensioned rockbolts
Nicht injizierte vorgespannte Anker

In consequence, the effect of the ungrouted rockbolts must be simulated not only by an "action" pressure p_p applied on the tunnel surface (radius r_i), but also by a "reaction" pressure $p_p \cdot r_i/r_e$ on the inner face of the anchoring zone (radius r_e). This influence is schematically represented in figure 3.

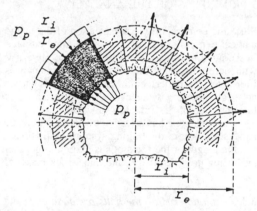

$p_p \dfrac{r_i}{r_e}$

p_p

r_i

r_e

Fig. 3 :   Support pressure p_p and reaction pressure $p_p \cdot r_i/r_e$
Ankerkräfte p_p und $p_p \cdot r_i/r_e$

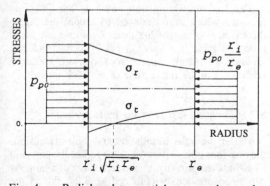

STRESSES

$p_{p0} \dfrac{r_i}{r_e}$

σ_r

p_{p0}

σ_t

0.

RADIUS

$r_i \quad \sqrt{r_i r_e} \qquad r_e$

Fig. 4 :   Radial and tangential stresses due to the bolts pressures
Radiale und tangentiale Spannungen aufgrund der Ankerkräfte

3.2   Elastic recompression due to the bolts preload

Figure 4 shows the stresses induced by the bolts preload T_{b0}, i.e. by the pressures $p_{p0} = T_{b0}/(s_1 \cdot s_t)$ and $p_{p0} \cdot r_i/r_e$ applied at the inner and outer faces of the bolted ring. The radial stress σ_r is bigger than the tangential one σ_t , and the latter takes negative values (i.e. tension) between r_i and $\sqrt{r_i r_e}$.

Now, this situation is just the opposite to the usual one encountered around underground excavations (where $\sigma_t > \sigma_r$ in both elastic and plastic zones) and consequently, the superposition of the preload to the initial situation before the bolts installation, will always be beneficial, owing to the reduction of the difference between the major and minor principal stresses.

More particularly, if a part of the carrying ring was characterised by a plastic behaviour before the bolts set-up, it will find temporarily an elastic behaviour again. Indeed, in a Mohr's diagram (Fig. 5), the circles initially tangent to the Coulomb's straight line before the bolts preload, become then detached owing to the recompression of the carrying ring.

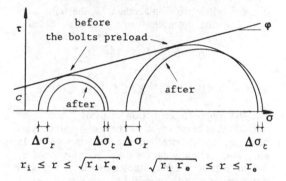

τ

before the bolts preload

φ

c

after

after

σ

$\Delta \sigma_r$ $\Delta \sigma_t \; \Delta \sigma_r$ $\Delta \sigma_t$

$r_i \leq r \leq \sqrt{r_i r_e}$ $\sqrt{r_i r_e} \leq r \leq r_e$

Fig. 5 :   Evolution of the Mohr's circles after the bolts preload
Veränderung des Mohrschen Kreises nach der Ankervorspannung

3.3   Tension increase in the rockbolts

After the support installation, the excavation restarts and the influence of the tunnel face reduces progressively (this effect is simulated by a gradual decrease of the fictitious pressure applied at the tunnel surface). The subsequent loosening of the rock mass induces an increase in the convergences. Figure 6 illustrates especially the relative displacement of the bolts ends and the elongation ΔL_b undergone by the shank :

$$\Delta L_b = (u_{ri} - u_{ri,pp0}) - (u_{re} - u_{re,pp0}) \qquad (2)$$

with $u_{ri,pp0}$ and $u_{re,pp0}$ the displacements after the bolts preloading.

This length variation ΔL_b has a repercussion on the tension T_h in the rockbolts :

$$T_b = T_{b0} + \Delta T_b = T_{b0} + \frac{\Delta L_b \, E_b \, A_b}{L_b} \qquad (3)$$

and consequently on the pressures p_p and $p_p . r_i/r_e$ generated at the inner and outer faces of the ring.

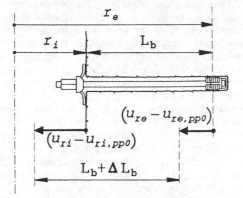

Fig. 6 : Relative displacement of the bolts ends and elongation of the shank. *Relative Verschiebung der Ankerenden und Verlängerung der Stange*

4 NEW GROUND CHARACTERISTIC LINE

The rock-support interaction analysis requires the calculation of two variables : the radial wall convergence u_{ri} and the internal support pressure p_i. In such a (u_{ri}; p_i) diagram, the equilibrium is found by the intersection of two characteristic lines : a ground response curve and a support reaction line. However this kind of presentation is only available when the rock mass and the support behave independently; e.g. a concrete or shotcrete lining and steel arches.

Now, as explained previously, the bolting system cannot be considered as an internal support since : 1) the bolts transfer their reaction into the rock mass; 2) their preload induces a recompression of the carrying ring; 3) their elongation depends on the rock mass convergences and has a repercussion on their tension.

For these reasons, an alternative solution has been explored : it consists to include the effect of the ungrouted bolts into the ground response curve. Such a curve C_{vb} is shown in figure 7 near the characteristic line of an unsupported rock mass C_v (usual presentation). A reaction line C_f for an internal support is also drawn. The equilibrium is reached when this line meets the curve of the bolts supported rock mass. Let's note that this kind of presentation has also been used by Stille & al (1989) and Holmberg (1991) for grouted bolts.

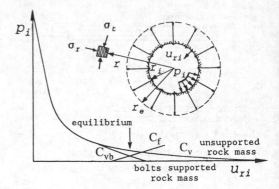

Fig. 7 : Ground reaction curve for a rock mass supported with rockbolts *Charakteristische Kurve eines verankerten Felsmassivs*

5 PRINCIPLES OF THE ANALYSIS

Since the comprehensive presentation of the mathematical expressions takes more than hundred pages in the author's doctoral thesis (Labiouse 1993), we will merely describe the principles of the analysis. Moreover, the undermentioned explanations will not cover all the cases which can be encountered in situ; but only one of the common situations :
- a broken zone has already developed around the excavation before the bolting system installation,
- the length of the rockbolts is chosen to get their anchoring point located in the intact rock mass.

5.1 *Situation before the bolts preloading*

The radial σ_r and tangential σ_t stresses as well as the radial displacement u_r occurring just before the bolts installation are represented in figure 8. The pressure p_{i0} applied on the gallery wall is the fictitious pressure which is introduced to model the three-dimensional effect of the face. As above-assumed, a broken zone ($r_i \leq r \leq r_{pe0}$) has already developed around the tunnel, but the rockbolts are anchored (radius r_e) in the intact zone of the rock mass.

5.2 *Situation just after the bolts preloading*

Just after the bolts installation and their pretension, the previous plastic part ($r_i \leq r \leq r_{pe0}$) of the carrying ring finds temporarily an elastic behaviour again. As explained previously (subheading 3.2), this is the consequence of the pressures p_{p0} and $p_{p0} . r_i/r_e$ applied at the inner and outer faces of the bolted ring. Figure 9 shows the stresses and displacements occurring at that moment.

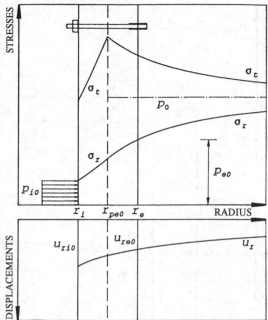

Fig. 8 :   Situation before the bolts preload
         Situation vor der Vorspannung der Anker

5.3   Wholly "elastic" behaviour of the bolted ring

The successive behaviours of the carrying ring and the rock mass occurring after the rockbolts pretension are studied decreasing progressively the fictitious pressure p_i from p_{i0} to zero. This way of proceeding enables to model the gradual advance of the tunnel face.

At first, the reinforced zone keeps an "elastic" behaviour (fig. 10); the quotation marks being used to bear in mind that a part of this ring ($r_i \le r \le r_{pe0}$) was characterised by a plastic behaviour before the favourable elastic recompression due to the bolts pretension.

The mathematical expressions which apply during this "elastic" phase can be obtained superposing the various "loading" stages undergone by the ring:

1. the initial elasto-plastic situation (pressures p_{i0} and p_{e0}),
2. the elastic recompression (pressures $p_p = p_{p0} + \Delta p_p$ and $p_p.r_i/r_e$) induced by the rockbolts preload as well as by the subsequent increase in tension owing to the relative displacement of their ends (subheading 3.3),
3. the elastic loosening due to the lowering of the internal ($p_{i0} \rightarrow p_i$) and external ($p_{e0} \rightarrow p_e$) pressures.

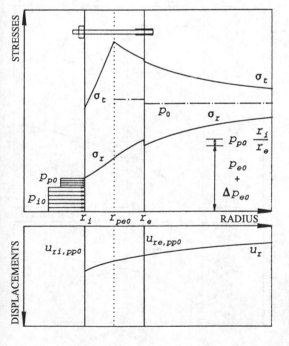

Fig. 9 :   Situation just after the bolts preload
         Situation kurz nach der Vorspannung

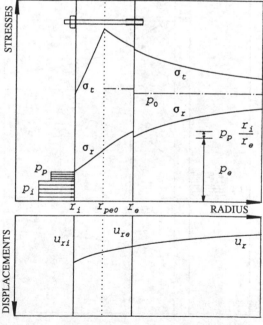

Fig. 10 : Wholly "elastic" behaviour of the ring
         Vollkommen "elastisches" Verhalten

5.4   Partly "elastic" - partly plastic behaviour

The wholly "elastic" behaviour of the carrying ring is only temporary. Indeed, one can note the speedy reappearance of a plastic zone at the inner face of the reinforced ring, and then its progression around the excavation.

At first (figure 11), the extent of this plastic zone (boundary radius r_{ie}; dashed line) remains smaller than the size it presented at the rockbolts installation (radius r_{pe0}; dotted line). At that moment, three zones with different behaviours coexist in the carrying ring:

$r_i \leq r \leq r_{ie}$ plastic zone

$r_{ie} \leq r \leq r_{pe0}$ zone with an "elastic" behaviour (which was previously characterised by a plastic behaviour before the bolts set-up)

$r_{pe0} \leq r \leq r_e$ elastic zone

Afterwards (fig. 12), the broken zone (boundary radius r_{ie}; dashed line) propagates beyond the position it occupied at the bolting system installation (radius r_{pe0}; dotted line); and consequently in this phase, the ring is only characterised by a plastic zone ($r_i \leq r \leq r_{ie}$) and an elastic zone ($r_{ie} \leq r \leq r_e$). Let's note that this progression of the broken zone into the carrying ring induces an important speeding up of the convergences and accordingly a more substantial increase in the bolts tension.

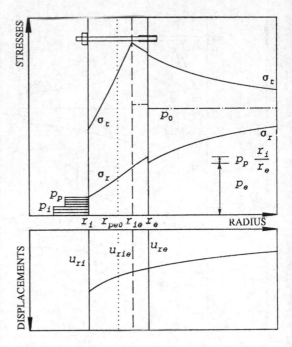

Fig. 12 :   Partly plastic behaviour - second stage
Teilweise plastisches Verhalten - Phase 2

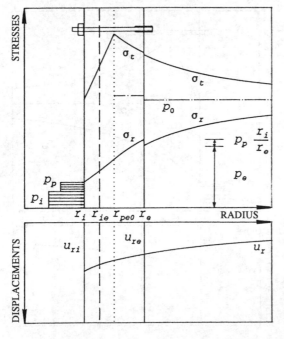

Fig. 11 :   Partly plastic behaviour - first stage
Teilweise plastisches Verhalten - Phase 1

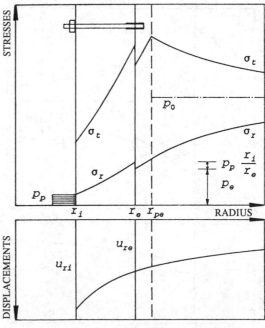

Fig. 13 :   Wholly plastic behaviour of the ring
Vollkommen plastisches Verhalten

5.5 Wholly plastic behaviour of the bolted ring

The partly "elastic"-partly plastic above-explained behaviour is apt to hold on till the rock-support equilibrium. However, depending on the mechanical properties of the rock mass as well as on the stiffness of the bolting system, it may sometimes happen that the carrying ring is completely plasticized before the equilibrium situation (fig. 13).

6 GROUND RESPONSE CURVE OF THE BOLTED ROCK MASS

The analytical developments (of which only the principles are explained in this paper) enable to relate the internal pressure p_i and the bolting support pressure p_p to the wall convergence u_{ri}. Accordingly, it is possible to represent :
- in an usual $(u_{ri}; p_i)$ diagram, the ground response curve for the rock mass supported with the ungrouted tensioned rockbolts,
- in a $(u_{ri}; p_p)$ graph, the tension increase in the bolts during the loosening of the rock mass.

Both curves $(u_{ri}; p_i)$ and $(u_{ri}; p_p)$ are shown in figure 14 for the case studied in the principles of the analysis; i.e. the bolts are anchored beyond the broken zone which has already developed around the excavation before their set-up. All the behaviours which have been described in the fifth point are represented, and the asterisks A to F point out the border-line situations :

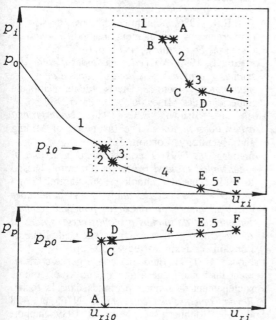

Fig. 14 : $(u_{ri}; p_i)$ and $(u_{ri}; p_p)$ curves
$(u_{ri}; p_i)$ *und* $(u_{ri}; p_p)$ *Kurven*

A : Situation at the rockbolts set-up
B : Situation just after the bolts preload
C : Border-line case corresponding to the reappearance of a plastic zone at the inner face of the reinforced ring
D : Intermediate situation between the two phases of the partly "elastic"-partly plastic behaviour (at that moment $r_{ie} = r_{pe0}$)
E : Border-line case when the plastic zone boundary reaches the outer face of the ring
F : Equilibrium situation.

Reducing gradually the internal pressure p_i to model the progressive advance of the tunnel face, the following situations (numbered from 1 to 5 in figure 14) are encountered in the reinforced ring :
1 : Partly elastic - partly plastic behaviour before the bolts installation
2 : Wholly "elastic" behaviour owing to the bolts pretension (subheading 5.3)
3 : Partly plastic behaviour - 1st stage (cfr. 5.4)
4 : Partly plastic behaviour - 2nd stage (cfr. 5.4)
5 : Wholly plastic behaviour (subheading 5.5).

Analysing the evolution of the convergences, it can be established that :
- their variation is very slow during the wholly "elastic" behaviour. At that moment, the gradient of the $(u_{ri}; p_i)$ curve is near the slope of the straight line which characterizes the elastic behaviour of the rock mass before the bolts installation,
- the reappearance of a plastic zone in the carrying ring induces a speeding up of the convergences, and this becomes more pronounced with the progression of this broken zone.

Since the tension increase in the rockbolts is related to the relative displacement of their ends, the above-mentioned establishments prove as well true for the increment of the bolting support pressure : slow during the "elastic" stage and then speedier as the plastic zone extends in the bolted ring.

7 NUMERICAL APPLICATION

A 10 m diameter gallery is driven at 400 m depth (initial in situ stress $p_0 = 10000$ kN/m^2) in a perfect elastic-plastic rock mass characterised by:
- Young's modulus $E = 10^6$ kN/m^2
- Poisson's ratio $\nu = 0.30$
- angle of friction $\varphi = 34\,°$
- cohesion $c = 750$ kN/m^2
- dilatancy parameter $\alpha = 1.2$

The bolting support is installed one metre behind the working face ($\rightarrow$ at that moment, the plastic zone has an extent of 1.2 m). Its specifications are :
- Young's modulus $E_b = 8\ 10^7$ kN/m^2
- diameter $\phi_b = 0.025$ m

- free length $\quad$ $L_b = 6$ m
- bolts spacings $\quad$ $s_l = s_t = 0.75$m
- pretension load $\quad$ $T_{b0} = 50$ kN
- yielding load $\quad$ $T_{by} = 245$ kN
- ultimate failure load $\quad$ $T_{bf} = 295$ kN

The stresses and displacements at equilibrium are represented in figure 15 for both methods: the usual one (dotted lines) and the novel one (solid lines). Although there are few differences in the figure, the numerical results allow to establish that :
- the tension in the bolts calculated by the new method (217 kN) is lower than the usual evaluation (245 kN). Moreover, in the latter the bolts reach their yielding load before the rock mass equilibrium,
- the convergences are larger in the new method,
- the bolting support stiffness is highly overestimated (75 % !) by the usual approach.

These differences can easily be explained by the assumptions of the calculation methods :
1. the usual one assumes that the bolt anchoring point doesn't converge ($u_{re} = u_{re,pp0}$), and consequently, the shank elongation becomes equal to the increment of convergence at the gallery surface: $\Delta L_b = u_{ri} - u_{ri,pp0}$. On the other hand, the present approach takes into account the convergence of the anchoring point and the relative displacement of the bolts ends : $\Delta L_b = (u_{ri} - u_{ri,pp0}) - (u_{re} - u_{re,pp0})$.
2. the usual method takes into consideration the action of the rockbolts on the excavation inner side, but neglects the reaction force transferred to the rock mass in the anchoring zone. On the other hand, the new approach presents a comprehensive study.

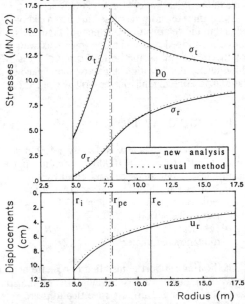

Fig. 15 : Stresses and displacements at equilibrium
Spannungen und Verschiebungen

These reflections emphasize that the stiffness of the bolting support evaluated by the new method is smaller than the one calculated by the usual theory.

8 CONCLUSION

The present paper has described an analytical method improving the calculation of ungrouted tensioned rockbolts supporting excavations under axisymmetric conditions. The main improvements in the usual theory have been explained, and the differences between both methods have been pointed out in a numerical application. In particular, it has been observed that a systematic bolting studied by the new approach is characterised by a much smaller stiffness than the one evaluated by the usual method.

Finally, let's point out the very fast computation of the developed method. Indeed, half a minute is enough to obtain the ground reaction curve of a rock mass supported with a bolting support, and to calculate the distributions of the stresses, strains and displacements around the excavation.

Consequently, such a method allows to perform quick parametrical studies; and so is helpful to understand the influence of the bolting characteristics (free length, delay of installation, pretension load, stiffness of the rockbolts) on the radial deformations and support stresses at equilibrium.

9 REFERENCES

A.F.T.E.S. 1986. Recommendations for use of convergence - confinement method. *Tunnels et Ouvrages Souterrains*, N°73, p.18-37.

Corbetta, F. 1990. *Nouvelles méthodes d'étude des tunnels profonds. Calculs analytiques et numériques*. Doctoral Thesis. Ecole Nationale Supérieure des Mines de Paris, Paris, 188 pp.

Hoek, E., Brown, E.T. 1980. *Underground excavations in Rock*. The Institution of Mining and Metallurgy, London, 525 pp.

Holmberg, M. 1991. *The mechanical behaviour of untensioned grouted rockbolts*. Doctoral Thesis. Royal Institute of Technology, Stockholm, 128 pp

Labiouse, V. 1993. *Etudes analytique et numériques du boulonnage à ancrage ponctuel comme soutènement de tunnels profonds creusés dans la roche*. Doctoral Thesis. Université Catholique de Louvain. Louvain-la-Neuve. 600 pp.

Labiouse, V. 1994. Etude par convergence-confinement du boulonnage à ancrage ponctuel comme soutènement de tunnels profonds dans la roche. *Revue Française de Géotechnique*, N°65, p.17-28

Stille, H., Holmberg, M., Nord, G. 1989. Support of weak rock with grouted bolts and shotcrete. *Int. Journal of Rock Mechanics and Mining Sciences*, Vol.26, N°1, p.99-113.

Numerical computation of 3D anchor bolts and application

Dreidimensionale Berechnungen für Ankerbolzen und deren Anwendung

X.Y. Lei
East China Jiaotong University, Nanchang, People's Republic of China

G. Swoboda
University of Innsbruck, Austria

A. Schulter
D2 Consult GmbH, Linz, Austria

ABSTRACT: The three dimensional anchor bolt element based on Aydan's theory is developed in this paper. In this model the deformation distributions of the bolt are assumed to be linear and constant in the axial and radial directions respectively. Theoretical solution of the axisymmetric body is used to describe the axial displacement distribution of the grout. The normal stresses transferred by the steel bar and the shear stresses due to different axial movements between the bar and the grout, as well as the transverse shear stresses from the lateral movements on the grout-rock border, are major stress components which are described in the element. The model is then used to simulate rockbolts in the analyses of pull-out tests. As an application example, an embankment with anchor bolts has been investigated and the results compared with a computation without rockbolts. The calculations illustrate the capabilities of the element.

ZUSAMMENFASSUNG: Die Entwicklung eines dreidimensionales Ankerlementes, basierend auf der Theorie von Aydan [1] wird in dieser Arbeit gezeigt. In diesem Modell wird angenommen, daß die Verformungsverteilung in axialer Richtung linear und in radialer Richtung konstant ist. Die theoretische analytische Lösung wird verwendet um die Längsverformung des Ankermörtels zu beschreiben. Die Normalspannungen des Ankers und die Schubspannungen auf Grund von unterschiedlichen Längsverformungen zwischen Anker und Ankermörtel, genauso wie die Querschubspannungen aus den Querverformungen an der Ankermörtel - Felsgrenze, sind wichtige Spannungskomponenten die in diesem Element berücksichtigt werden. Das heißt, daß in diesem Element auch näherungsweise der Dübeleffekt für Querverformungen berücksichtigt wird. Der sehr komplexe Weg eines eigenen Dübelelementes zur Simulation des Schubwiederstandes in Klüften, wie dies durch Marence und Swoboda [23] für zweidimensionale Elemente geschah, wurde damit vermieden. Der einfache Aufbau der Elementssteifigkeitsmatrix erlaubt einen sehr leichten Einbau in allgemeine Finite Elemente Programme. Durch die einfache Geometriebeschreibung die mit Hilfe von zwei bzw. vier Knoten erfolgt, ist auch die Generierung sehr komplexer dreidimensionaler Modelle durch Generierungsprogramme leicht möglich.

Mit derartigen Elementen können nun erstmalig auch die räumlichen Wirkungen von Ankern in numerischen Modellen untersucht werden. Insbesondere bei Baugruben kann die Tragwirkung jedoch nur wirklichkeitsnah erfaßt werden, wenn die räumliche Tragwirkung mit den zugehörigen Ankern berücksichtigt wird.

Das Ankermodell wird geeicht in einem Ausziehversuch. Dabei wird das Ausziehen eines Ankers aus einem unbelasteten Felsblock simuliert und die Resultate mit einer analytischen Lösung vergliechen. In einem Anwendungsbeispiel werden Teile einer Ankerwand untersucht. Die Resultate werden mit jenen ohne Ankerung gegenübergestellt. Diese zeigen die gute Anwendbarkeit des Ankerelementes.

1 INTRODUCTION

Finite element techniques are ideally suited to model rock masses reinforced with rockbolts because of the ease with which arbitrary loads, nonlinear material properties and special structural element types can be accommodated. Several different models have been proposed recently. The typical technique used in these proposed methods is to simulate the rockbolts as truss or beam elements [9][12]. These elements are not appropriate to describe rockbolts, especially grouted bolts. In these methods, the stiffness of the grout is not taken into consideration, and their main disadvantage is that it is not possible to consider different displacements in the bolt and on the surface of the borehole (a shear displacement which takes a part in the grout). To solve this problem some authors try to apply special interface elements between the rock mass and the bar, which is represented as a one dimensional element. These interface elements were springs in nodes [7][10][20][21], or special interface elements. Such elements are normally two or three dimensional elements with the characteristics of grouted material [15][17][18], or special joint elements such as the axisymmetrical joint element created by Ghamboussi et al. [6].

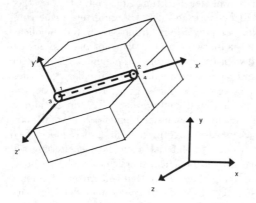

Figure 1: The three dimensional rockbolt element
Dreidimensionales Ankerelement

A further development of the rockbolt elements was the creation of an element which includes the stiffness of the bar and the interface. John and Van Dillen [8] gave a numerical representation for rockbolts in which the behaviour of the grout annulus as well as the axial and shear behaviour of the steel bar are incorporated. Aydan [3] developed a modified stiffness matrix for the bolt element with eight nodes, six nodes connected with the rock and two on the bar. For two dimensional cases he recommended an element type with six nodes. The stiffness of this element is the sum of the bar and the grout stiffness. This model describes accurately the true displacement field around the borehole, but it's disadvantage is the very complicated FE mesh around the bolt. Another type of bolt element was developed by Aydan et al. [1][2]. this element has four nodes, two representing the grout-rock border and two on the bar.

The explicit stiffness matrix of the bolt element based on Aydan's theory is developed here together with the analysis of an embankment reinforced with anchor bolts.

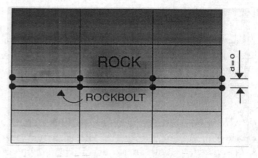

Figure 2: Four node bolt element and connection with finite element mesh
Vierknotiges Ankerelement und dessen Koppelung mit dem Finite Element Netz

2 STIFFNESS MATRIX OF THE BOLT ELEMENT

For different purposes, the bolt element can be classified with two [4] or a three dimensional models. The 3D model may be considered as a one dimensional bar passing through a cylindrical surface, to which elements representing surrounding material are attached. Such a concept is illustrated for the case of three dimensional space in Fig. 1. This element could be coupled to the regular finite element meshes with two nodes representing the grout-rock coupling. The two remaining nodes on the bar are connected with the rock through the stiffness of grout as shown in Fig.2. In the mesh generation, nodes on the bar and the grout-rock border have the same coordinates as the volume replaced by the rockbolt is insignificate while an original geometry will be used in the formulation of the stiffness matrix.

The global (x, y, z) and local (x', y', z') axes and the designation of the nodal points of the bolt element are illustrated in Fig.1. Let the global and local displacement vectors and force vectors be **a**, **a′**, **f** and **f′** respectively, where all terms correspond to absolute degrees of freedom in the global or local coordinate schemes. Also let :

$$\mathbf{a}'^e = \{u_1'\ v_1'\ w_1'\ u_2'\ v_2'\ w_2'\ u_3'\ v_3'\ w_3'\ u_4'\ v_4'\ w_4'\}^T$$

and

$$\mathbf{f}'^e = \{X_1'\ Y_1'\ Z_1'\ X_2'\ Y_2'\ Z_2'\ X_3'\ Y_3'\ Z_3'\ X_4'\ Y_4'\ Z_4'\}^T$$

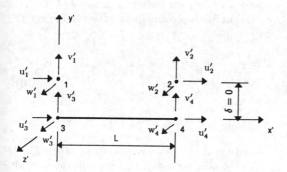

Figure 3: Bolt element as an axisymmetric body with possible node variables
Knotenparametern des rotationssymmetrischen Ankerelementes

where u', v' and w' refer to displacements in the x', y', z' directions, and X', Y' and Z' are forces in those directions. The subscripts 1 and 2 refer to the nodes on the outer surface connected with the surrounding rock of the annulus and 3 and 4 to the nodes on the bolt axis.

2.1 Elementary Assumptions

Some assumptions are made about the distribution of deformations and stresses within the element. In this study the deformation distributions of the bolt are assumed to be linear and constant in the axial and radial directions respectively. Theoretical solution of the axisymmetric body is used to describe the axial displacement distribution of the grout. The normal stresses transferred by the steel bar and the shear stresses due to different axial movements between the bar and the grout as well as the transverse shear stresses from the lateral movements on the grout-rock border are major stress components which are described in the element. Shear stresses which are the result of lateral bar movement are not taken into account in the stiffness matrix formulation and will be discussed later.

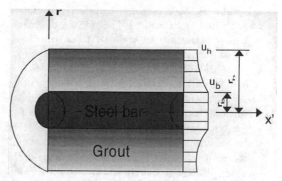

Figure 4: Axial displacement of the grout based on analytical solution
Analytische rotationssymmetrische Verformung des Ankermörtels

Based on assumptions about the displacement distributions, the displacements within the element $\mathbf{u}'^e$ could be expressed as a product of shape function **N** and element nodal displacement $\mathbf{a}'^e$

$$\mathbf{u}'^e = \mathbf{N}\mathbf{a}'^e \tag{1}$$

where

$$\mathbf{u}'^e = \{u', v', w'\}^T$$

in which u', v' and w' are displacements referring to local coordinate system at any point in the region of bar or grout.

For the steel bar

$$\mathbf{N} = \begin{bmatrix} N_1 & 0 & 0 & N_2 & 0 & 0 & 0 & 0 & 0 & 0 & 0 & 0 \\ 0 & N_1 & 0 & 0 & N_2 & 0 & 0 & 0 & 0 & 0 & 0 & 0 \\ 0 & 0 & N_1 & 0 & 0 & N_2 & 0 & 0 & 0 & 0 & 0 & 0 \end{bmatrix} \tag{2}$$

for the grout region

$$\mathbf{N} = \begin{bmatrix} 0 & 0 & 0 & 0 & 0 & 0 & N_1 & 0 & 0 & N_2 & 0 & 0 \\ 0 & 0 & 0 & 0 & 0 & 0 & 0 & N_1 & 0 & 0 & N_2 & 0 \\ 0 & 0 & 0 & 0 & 0 & 0 & 0 & 0 & N_1 & 0 & 0 & N_2 \end{bmatrix} \tag{3}$$

where:

$$N_1 = 1 - \xi \ , \quad N_2 = \xi \ , \quad 0 \le \xi \le 1 \quad (4)$$

2.2 Axial Displacement of the Grout Based on Analytical Solution

The grout annulus is assumed to be a cylindrical axisymmetric object, which transfer stresses between the bar and the rock. The transfer of stresses is mainly performed by a shear stress, which is a result of different displacements on the bar–grout and the grout–rock border. This assumption is proved in pull–out tests. It is possible to use a linear shape function to ascertain the distribution of axial displacements in the radial direction of grout, but a more accurate result can be achieved if the analytical solution for distribution is used.

Farmer [5] obtained the relationship between the shear displacement and the shear stress for grout, based on the assumption that the axial displacement is zero on the border of the grout and the rock. Aydan [3] extended this theory by considering the existence of any displacement on the interface, as shown in Fig.4.

Based on Aydan's assumption, the bolt can be considered as a one dimensional member and a force applied on the bar is transferred to the surrounding rock through shear stresses. If the forces are assumed to be transferred to the grout and the rock only by the shear stresses, the force equilibrium condition for an elementary slice of the grout annulus can be written as:

$$\Sigma X'_{grout} = 0 \quad \Rightarrow$$
$$2\pi(\tau'_{rx} + \Delta\tau'_{rx})(r + \Delta r)\Delta x' - 2\pi\tau'_{rx}r\Delta x' = 0 \quad (5)$$

Dividing the above expression by $2\pi r \Delta r \Delta x'$ and taking the limit yields, the governing equation is:

$$\lim_{\Delta r \to 0} \frac{\Delta\tau'_{rx}}{\Delta r} + \frac{\tau'_{rx}}{r} = 0 \quad \Rightarrow \quad \frac{d\tau'_{rx}}{dr} + \frac{\tau'_{rx}}{r} = 0 \quad (6)$$

where τ'_{rx} is a shear stress in the grout and r is a radial distance of the calculated elementary slice. Employing an elastic constitutive law for the shear stress τ'_{rx} and a strain γ'_g:

$$\tau'_{rx} = \tau'_g = G_g \, \gamma'_g = G_g \frac{du'}{dr} \quad (7)$$

where G_g is a shear modulus of grout, the differential equation (6) yields the following:

$$\frac{d^2u'}{dr^2} + \frac{1}{r}\frac{du'}{dr} = \frac{1}{r}\frac{d}{dr}\left(r\frac{du'}{dr}\right) = 0 \quad (8)$$

The general solution for the above differential equation is:

$$u' = C_1 \ln r + C_2 \quad (9)$$

where C_1 and C_2 are integration constants. With the use of boundary conditions given as:

$$u' = u'_h \quad at \quad r = r_h$$
$$u' = u'_b \quad at \quad r = r_b$$

where r_b and r_h are radii of the bolt and the hole, the axial displacement function in the radial direction is:

$$u' = \frac{1}{\ln(r_h/r_b)} [u'_h \ln r - u'_b \ln r + u'_b \ln r_h - u'_h \ln r_b] \quad (10)$$

2.3 Strain and Stress

According to the assumptions about stresses only four strains are taken into account-the normal strain in the bar ε'_b as a result of axial displacements in the bar, the shear strain in the grout γ'_g as a result of differential axial movements in the bar and on the grout-rock border, and the transverse shear stresses in the bar $\gamma'_{y'}$ and $\gamma'_{z'}$, results of different lateral movements of the grout-rock border. In a vector form this definition could be written as:

$$\boldsymbol{\varepsilon}'^e = \left\{ \begin{array}{c} \varepsilon'_b \\ \gamma'_g \\ \gamma'_{y'} \\ \gamma'_{z'} \end{array} \right\} = \left\{ \begin{array}{c} \partial u'/\partial x' \\ \partial u'/\partial r \\ \partial v'/\partial x' \\ \partial w'/\partial x' \end{array} \right\} \quad (11)$$

The derivation of the bolt strain components can be performed as follows:

$$\varepsilon'_b = \frac{\partial u'}{\partial x'} = -\frac{u'_3}{L} + \frac{u'_4}{L} \quad (12)$$

in which L is the length of the bolt element.

$$\gamma'_g = \frac{\partial u'}{\partial r} = \frac{u'_h - u'_b}{r \ln(r_h/r_b)} = R(u'_h - u'_b) \quad (13)$$

where:

$$R = \frac{1}{r\ln(r_h/r_b)}$$

$$\gamma'_{y'} = \frac{\partial v'}{\partial x'} = -\frac{v'_1}{L} + \frac{v'_2}{L} \tag{14}$$

$$\gamma'_{z'} = \frac{\partial w'}{\partial x'} = -\frac{w'_1}{L} + \frac{w'_2}{L} \tag{15}$$

The relationship between the bolt strain and the element nodal displacement vector $\mathbf{a}'^e$ gives:

$$\boldsymbol{\varepsilon}'^e = \mathbf{B}\mathbf{a}'^e \tag{16}$$

where:

$$\mathbf{B} = \begin{bmatrix} 0 & 0 & 0 & 0 & 0 & 0 \\ R\,N_1 & 0 & 0 & R\,N_2 & 0 & 0 \\ 0 & -1/L & 0 & 0 & 1/L & 0 \\ 0 & 0 & -1/L & 0 & 0 & 1/L \end{bmatrix}$$

$$\begin{bmatrix} -1/L & 0 & 0 & 1/L & 0 & 0 \\ -R\,N_1 & 0 & 0 & -R\,N_2 & 0 & 0 \\ 0 & 0 & 0 & 0 & 0 & 0 \\ 0 & 0 & 0 & 0 & 0 & 0 \end{bmatrix} \tag{17}$$

The components of stress corresponding to the strain $\boldsymbol{\varepsilon}'^e$ are given by

$$\boldsymbol{\sigma}'^e = \mathbf{D}\boldsymbol{\varepsilon}'^e \tag{18}$$

in which:

$$\boldsymbol{\sigma}'^e = \left\{ \begin{array}{c} \sigma'_b \\ \tau'_g \\ \tau'_{y'} \\ \tau'_{z'} \end{array} \right\} \tag{19}$$

$$\mathbf{D} = \begin{bmatrix} E_b & 0 & 0 & 0 \\ 0 & G_g & 0 & 0 \\ 0 & 0 & G_b & 0 \\ 0 & 0 & 0 & G_b \end{bmatrix} \tag{20}$$

where σ'_b is a normal stress in the bar, τ'_g a shear stress in the grout, $\tau'_{y'}$ and $\tau'_{z'}$ transverse shear stresses corresponding to the τ'_{dowel} in the y' and z' directions. E_b is an elasticity modulus of the bar, G_g a shear modulus of the grout, and G_b a shear modulus of the bar.

2.4 Element Stiffness Matrix

Based on the theorem of internal work, the stiffness matrix for the bolt element $\mathbf{K}'^e$ could be calculated as:

$$\mathbf{K}'^e = \int \mathbf{B}^T \mathbf{D} \mathbf{B} \; d(vol) \tag{21}$$

In this case the integral could be presented as:

$$\begin{aligned} \mathbf{K}'^e &= \int_0^L \int_0^{r_h} \int_0^{2\pi} \mathbf{B}^T \mathbf{D} \mathbf{B} \, d\theta\, r\, dr\, dx \\ &= 2\pi L \int_0^1 \int_0^{r_h} \mathbf{B}^T \mathbf{D} \mathbf{B} \, r\, dr\, d\xi \end{aligned} \tag{22}$$

Implementing the integral the explicit element stiffness matrix in the local coordinate system can be obtained as:

$$\mathbf{K}'^e = \begin{bmatrix} 2k_g & 0 & 0 & k_g & 0 & 0 \\ 0 & k_s & 0 & 0 & -k_s & 0 \\ 0 & 0 & k_s & 0 & 0 & -k_s \\ k_g & 0 & 0 & 2k_g & 0 & 0 \\ 0 & -k_s & 0 & 0 & k_s & 0 \\ 0 & 0 & -k_s & 0 & 0 & k_s \\ -2k_g & 0 & 0 & -k_g & 0 & 0 \\ 0 & 0 & 0 & 0 & 0 & 0 \\ 0 & 0 & 0 & 0 & 0 & 0 \\ -k_g & 0 & 0 & -2k_g & 0 & 0 \\ 0 & 0 & 0 & 0 & 0 & 0 \\ 0 & 0 & 0 & 0 & 0 & 0 \end{bmatrix}$$

$$\begin{bmatrix} -2k_g & 0 & 0 & -k_g & 0 & 0 \\ 0 & 0 & 0 & 0 & 0 & 0 \\ 0 & 0 & 0 & 0 & 0 & 0 \\ -k_g & 0 & 0 & -2k_g & 0 & 0 \\ 0 & 0 & 0 & 0 & 0 & 0 \\ 0 & 0 & 0 & 0 & 0 & 0 \\ k_b + 2k_g & 0 & 0 & k_g - k_b & 0 & 0 \\ 0 & 0 & 0 & 0 & 0 & 0 \\ 0 & 0 & 0 & 0 & 0 & 0 \\ k_g - k_b & 0 & 0 & k_b + 2k_g & 0 & 0 \\ 0 & 0 & 0 & 0 & 0 & 0 \\ 0 & 0 & 0 & 0 & 0 & 0 \end{bmatrix} \tag{23}$$

The stiffness matrix calculated in this way has four rows and columns of zeros as a *'dowel'* stiffness of the element is not included, which may create problems for the equation solver. This question is involved with the lateral displacements of the bar (v_3, w_3, v_4, w_4), which are not taken into account in the equation (11). In fact, the lateral displacements are important only when the bolt intersects a joint.

The singularity of the element stiffness matrix (23) can be eliminated by introducing the *'dowel'* stiffness k_d, which is equal to the shear stiffness of the crossed joint.

The modified stiffness matrix of the bolt element can be written as:

$$\mathbf{K'}^e =
\begin{bmatrix}
2k_g & 0 & 0 & k_g & 0 \\
0 & k_s + k_d & 0 & 0 & -k_s \\
0 & 0 & k_s + k_d & 0 & 0 \\
k_g & 0 & 0 & 2k_g & 0 \\
0 & -k_s & 0 & 0 & k_s + k_d \\
0 & 0 & -k_s & 0 & 0 \\
-2k_g & 0 & 0 & -k_g & 0 \\
0 & -k_d & 0 & 0 & 0 \\
0 & 0 & -k_d & 0 & 0 \\
-k_g & 0 & 0 & -2k_g & 0 \\
0 & 0 & 0 & 0 & -k_d \\
0 & 0 & 0 & 0 & 0
\end{bmatrix}$$

$$
\begin{bmatrix}
0 & -2k_g & 0 & 0 & -k_g & 0 & 0 \\
0 & 0 & -k_d & 0 & 0 & 0 & 0 \\
-k_s & 0 & 0 & -k_d & 0 & 0 & 0 \\
0 & -k_g & 0 & 0 & -2k_g & 0 & 0 \\
0 & 0 & 0 & 0 & 0 & -k_d & 0 \\
k_s + k_d & 0 & 0 & 0 & 0 & 0 & -k_d \\
0 & k_b + 2k_g & 0 & 0 & k_g - k_b & 0 & 0 \\
0 & 0 & k_d & 0 & 0 & 0 & 0 \\
0 & 0 & 0 & k_d & 0 & 0 & 0 \\
0 & k_g - k_b & 0 & 0 & k_b + 2k_g & 0 & 0 \\
0 & 0 & 0 & 0 & 0 & k_d & 0 \\
-k_d & 0 & 0 & 0 & 0 & 0 & k_d
\end{bmatrix}
$$

$$(24)$$

where:

$$k_b = \frac{E_b \pi r_b{}^2}{L}, \quad k_s = \frac{G_b \pi r_b{}^2}{L}, \quad k_g = \pi G_g \frac{L}{3 \ln(r_h/r_b)}$$

Before assembling into the global stiffness matrix, the local element stiffness matrix must be transformed to be consistent with the global coordinate scheme. This transformation is of the form:

$$\mathbf{K}^e = \boldsymbol{\alpha}^T \mathbf{K'}^e \boldsymbol{\alpha} \tag{25}$$

$$\boldsymbol{\alpha} =
\begin{bmatrix}
\mathbf{T} & 0 & 0 & 0 \\
0 & \mathbf{T} & 0 & 0 \\
0 & 0 & \mathbf{T} & 0 \\
0 & 0 & 0 & \mathbf{T}
\end{bmatrix} \tag{26}$$

in which $\mathbf{K}^e$ is a global stiffness matrix, $\boldsymbol{\alpha}$ a transformation matrix and $\mathbf{T}$ a matrix of direction cosines which are discussed in section 3. Equation (25) defines the rotation of the terms associated with the outside of the grout annulus, that is connected to nodes with absolute degrees of freedom. Similarly, the contribution to the global force vector is defined as:

$$\mathbf{f}^e = \boldsymbol{\alpha}^T \mathbf{f'}^e \tag{27}$$

The element stresses in the local coordinate system can be obtained from the element nodal displacement vector $\mathbf{a}^e$ in the global coordinate scheme:

$$\boldsymbol{\sigma'}^e = \mathbf{D} \boldsymbol{\varepsilon'}^e = \mathbf{DB} \boldsymbol{\alpha} \mathbf{a}^e \tag{28}$$

3 TRANSFORMATION MATRIX T

To obtain transformation matrix $\mathbf{T}$, define vector $\mathbf{x'}$ shown in Fig. 5:

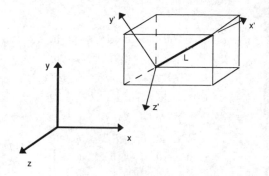

Figure 5: Local and global coordinate systems
Lokales globales Koordinatensystem

$$\mathbf{x'} = \alpha_{x'x}\mathbf{i} + \alpha_{x'y}\mathbf{j} + \alpha_{x'z}\mathbf{k} \tag{29}$$

in which:

$$\alpha_{x'x} = \frac{\Delta x}{L} \quad \alpha_{x'y} = \frac{\Delta y}{L} \quad \alpha_{x'z} = \frac{\Delta z}{L}$$

Then let

$$\mathbf{y'} = \mathbf{x'} \times \mathbf{i} = \alpha_{y'x}\mathbf{i} + \alpha_{y'y}\mathbf{j} + \alpha_{y'z}\mathbf{k} \tag{30}$$

in which:

$$\alpha_{y'x} = 0 \quad \alpha_{y'y} = \frac{s_y}{s} \quad \alpha_{y'z} = \frac{s_z}{s}$$

with:

$$s_y = \alpha_{x'z} \quad s_z = -\alpha_{x'y} \quad s = \sqrt{s_y^2 + s_z^2}$$

If $\mathbf{x'} \parallel \mathbf{i}$, we replace equation (30) with $\mathbf{y'} = \mathbf{x'} \times \mathbf{j}$. In this case:

$$\alpha_{y'x} = \frac{s_x}{s} \quad \alpha_{y'y} = 0 \quad \alpha_{y'z} = \frac{s_z}{s}$$

with:

$$s_x = -\alpha_{x'z} \quad s_z = \alpha_{x'x} \quad s = \sqrt{s_x^2 + s_z^2}$$

Finally defining:

$$\mathbf{z'} = \mathbf{x'} \times \mathbf{y'} = \alpha_{z'x}\mathbf{i} + \alpha_{z'y}\mathbf{j} + \alpha_{z'z}\mathbf{k} \qquad (31)$$

in which:

$$\alpha_{z'x} = \frac{t_x}{t} \quad \alpha_{z'y} = \frac{t_y}{t} \quad \alpha_{z'z} = \frac{t_z}{t}$$

with:

$$
\begin{aligned}
t_x &= \alpha_{x'y}\alpha_{y'z} - \alpha_{y'y}\alpha_{x'z} \\
t_y &= \alpha_{y'x}\alpha_{x'z} - \alpha_{x'x}\alpha_{y'z} \\
t_z &= \alpha_{x'x}\alpha_{y'y} - \alpha_{y'x}\alpha_{x'y} \\
t &= \sqrt{t_x^2 + t_y^2 + t_z^2}
\end{aligned}
$$

The transformation matrix $\mathbf{T}$ appears

$$\mathbf{T} = \begin{bmatrix} \alpha_{x'x} & \alpha_{x'y} & \alpha_{x'z} \\ \alpha_{y'x} & \alpha_{y'y} & \alpha_{y'z} \\ \alpha_{z'x} & \alpha_{z'y} & \alpha_{z'z} \end{bmatrix} \qquad (32)$$

4. EXAMPLE

For the illustration of the bolt element behaviovr a simple pull–out test is presented. The bolt is pulled out from an unloaded rock mass block. The rock mass is modelled as an elastic material. The computation is then compared with an analytical solution.

The system is calculated as a three dimensional problem with the geometry shown in Fig.6. The fixed supports on the bottom of the rock are selected so that they are the same as for the existing analytical solution. The bolts are placed in the middle of the rock mass.

The additional geometric parameters of the bolts and the elastic parameters of the rock mass, the grout and the bolt, are shown in Table 1.

The bolts are pulled out with a force of $P = 200, 0[kN]$ applied on top of the steel bar. The representative results of the numerical solution are shown in Fig.7 and Fig.8 which represent the stress σ_z distribution in the rock and the comparisons between the finite element method and analytical solution for the pull-out test [3]. No slip of the anchor is considered so the forces carried out by the bolts are quickly transffered into surrounding

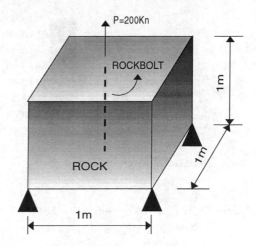

Figure 6: Pull-out test reinforced with bolts
Ausziehversuch eines Ankers

Table 1: Elastic material properties and dimensions

	Elasticity modulus [GN/m²]	Poisson's ratio [–]	Shear modulus [GN/m²]	Radius [m]
Bolt	210.00	0.30	–	0.0125
Grout	20.00	0.20	8.33	0.0325
Rock	15.00	0.20	6.25	–

rocks through the grout. The effects of the anchors only exist on the top region.

The results of Fig.8 also demonstrate that the linear force distribution in the bolt with few elements, indicates a satisfactory concurrence with the analytical solution.

5 EMBANKMENT ANALYSES

As an application example, an embankment with anchor bolts has been investigated here. The analyses contain five 13cm diameter anchor bolts passing through the centre of the embankment. The bolt is grouted along its entire length with concrete. The geometry and the arrangement of the rockbolts are pictured in Fig.9. The structure is fixed at the bottom of the rocks and only displacement y is allowed along the vertical boundary parallel to the z axis. Besides the initial stresses the uniform pressure $Q = 5000[kN/m^2]$, simulating traffic loads, is applied to the top of the embankment. The material properties of the study

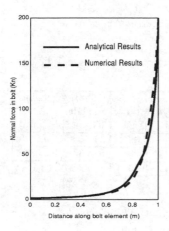

	0.12E+05
	0.97E+04
	0.70E+04
	0.44E+04
	0.17E+04
	-924.89
	-0.36E+04

Figure 7: Stress $\sigma_z[kN/m^2]$ distribution
Spannungsverteilung

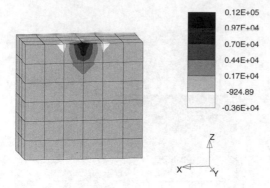

Figure 8: Axial normal force in the bolt compared with the analytical solution
Vergleich der Normalkräfte mit einer analytischen Lösung

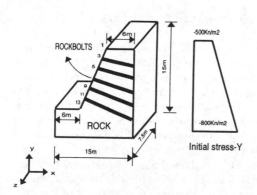

Figure 9: Geometry of embankment with bolts
Böschung mit Ankern

are listed in Table 2.

The results of the numerical computation and some comparisons with the analyses without rockbolts are demonstrated in Fig.10 and Fig.11. Figure.10 shows the horizontal displacement distribution in which the influence of the rockbolts is clear. A displacement reduction of up to 10% is observed in Fig.11. These reductions increase with the quantity of bolts employed.

Table 2: Material properties in an embankment analysis

	Elasticity modulus [GN/m²]	Poisson's ratio [–]	Shear modulus [GN/m²]	Radius [m]
Bolt	210.00	0.30	–	0.025
Grout	20.00	0.20	8.33	0.065
Rock	5.00	0.20	2.08	–

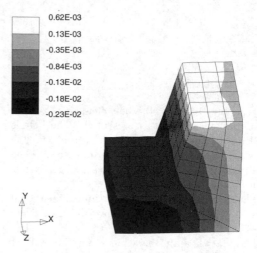

	0.62E-03
	0.13E-03
	-0.35E-03
	-0.84E-03
	-0.13E-02
	-0.18E-02
	-0.23E-02

Figure 10: Horizontal displacement distribution
Horizontalspannungsverteilung

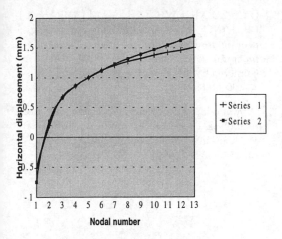

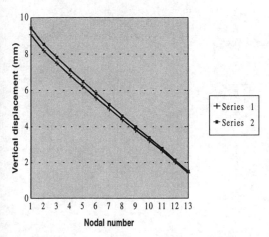

Figure 11: Horizontal and vertical displacements compared with results without bolts (Series 1: results without bolts; Series 2: current results)
        Horizontale und vertikale Verformungen, mit und ohne Anker

ACKNOWLEDGMENTS: The work reported herein was performed under contract of the Austrian National Bank, Project 4793, Rock anchors. This support is gratefully acknowledged.

REFERENCES

[1] Ö. AYDAN, T. KYOYA, Y. ICHIKAWA & T. KAWAMOTO: 'Anchorage perfomance and reinforcement effect of fully grouted rockbolts on rock excavations'. Proc. 6th Int. Cong. ISRM, Montreal, pp 757-760 (1987).

[2] Ö. AYDAN, T. KYOYA, Y. ICHIKAWA, T. KAWAMOTO, T. ITO & Y. SHIMIZU: 'Three-dimensional simulation of an advancing tunnel supported with forepoles, shotcrete, steel ribs and rockbolts'. Numerical Methods in Geomechanics, Swoboda(ed.), Innsbruck, Balkema, pp 1481-1486 (1988).

[3] Ö. AYDAN: The stabilisation of rock engineering structures by rockbolts. Thesis for the Degree of Doctor of Engineering, Department of Geotechnical Engineering, Nagoya University, Nagoya, Japan (1989).

[4] M. MARENCE, Numerical model for rockbolts, Dissertation, Innsbruck University, Chapter 3 (1992.)

[5] I.W. FARMER: 'Stress distribution along a resin grouted rock bolt'. Int. J. Rock Mech. Min. Sci. & Geomech. Abstr., Vol 12, pp 347-351 (1975).

[6] J. GHAMBOUSSI, E.L. WILSON & J. ISENBERG: 'Finite element for rock joints and interfaces'. Journal of the Soil Mechanics and Foundations Division, Proc. of ASCE, Vol 99, SM10, pp 833-848 (1973).

[7] G.W. HOLLINGSHEAD: ' Stress distribution in rock anchors'. Canadian Geotechnical Journal, Vol. 8, pp 588-592 (1971).

[8] C.M.ST. JOHN & D.E. VAN DILLEN: 'Rockbolts: A new numerical representation and its application in tunnel design'. Proc. of the 24th U.S. Symposium on Rock Mechanics, pp 13-25 (1983).

[9] F. LAABMAYR & G. SWOBODA: 'The importance of shotcrete as support element of the NATM'. Eng. Foundation Conf. Proc. of the 2nd Shotcret Conf., St. Anton, pp 65-79 (1978).

[10] M. KEUSER & G. MEHLHORN: 'Finite element models for bond problems'. Journal of Structural Engineering, Proc. of ASCE, Vol 113, No.10, pp 2160-2173 (1987).

[11] K. KLÖPPEL & M. YAMADA: 'Fließpolyester des Rechtek– und I–Querschnittes unter der Wirkung von Biegemoment, Normalkraft und Querkraft'. Stahlbau 27, pp 284-290 (1958).

[12] L. MÜLLER–SALZBURG: *Der Felsbau*. Ferdinand Enke Verlag, Stuttgart, (1978).

[13] G. NAMMUR, Jr & A. E. NAAMAN: *'Bond stress model for fiber reinforced concrete based on bond stress–slip relationship'*. ACI Journal, Jan–Feb 1989, Title no. 86-M6, pp 45-57.

[14] G.C. NAYAK & O.C. ZIENKIEWICZ: *'Elasto-plastic stress analysis, a generalization for various constitutive relations including strain softening'*. International Journal for Numerical Methods in Engineering, Vol 5, pp. 113-135 (1972).

[15] R.N. NITZSCHE & C.J. HAAS: *'Installation inducted stresses for grouted roof bolts'*. Int. J. Rock Mech. Min. Sci. & Geomech. Abstr., Vol 13, pp 17-24 (1976).

[16] D.R.J. OWEN & E. HINTON: *Finite Elements in Plasticity*. Pineridge Press, Swansea, United Kingdom, (1980).

[17] H.J. SIRIWARDANE: *'Numerical analysis of anchors in soil - Application brief'*. International Journal for Numerical and Analytical Methods in Geomechanics, Vol 13, pp. 427-433 (1989).

[18] K. SPANG: *Beitrag zur rechnerischen Berücksichtigung vollvermörtelter Anker bei der Sicherung von Felsbauwerken in geschichtetem oder geklüftetem Gebrige*. Ecole Polytechnique Fedarale de Lausanne, These No 740, (1988).

[19] G. SWOBODA: *Programmsystem FINAL - Finite element analyses program for linear and nonlinear structures*. Version 6.7, Print University Innsbruck, (1994).

[20] D. WULLSCHLÄGER & O. NATAU: *'Studies of the composite system of rock mass and non-prestressed grouted rock bolts'*. Proc. of the Int. Symp. on Rock Bolting, Abisko, Sweden, pp. 75-85 (1983).

[21] L. P. YAP & A. A. ROGER: *' A study of the behavior of vertical rock anchors using the finite element method'*. Int. J. Rock Mech. Min. Sci. & Geomech. Abstr., Vol 21, No. 2, pp 47-61 (1984).

[22] O.C. ZIENKIEWICZ, S. VALLIAPPAN & I.P. KING: *'Elasto-plastic solutions of engineering problems, 'Initial stress' Finite Element approach'*. International Journal for Numerical Methods in Engineering, Vol 1, pp. 75-100 (1969).

[23] G.Swoboda, M. Marence, *FEM modelling of rock bolts*, Proc. 7th Int. Conf. on Numerical Methods in Geomechanics, Cairns, Australien, 1515-1529 (1991).

Finite-Elemente Berechnung des Tragverhaltens verschiedener Dübeltypen für schwere Lasten

Finite-element-analysis of the loadbearing behavior of different heavy-duty anchors

R. Mattner, V. Raadschelders & Ch. Dietrich
HILTI AG, Konzern-Forschung, Schaan, Fürstentum Liechtenstein

ZUSAMMENFASSUNG: Lokale Lasteinleitungsprobleme in Betonuntergrund mit auftretenden Zug- und Querkräften von bis zu 80 kN können mit verschiedenen Schwerlastankern gelöst werden. Für die Entwicklung neuer sowie für die Weiterentwicklung bestehender Produkte stellt die Kenntnis der Spannungs- und Rissverteilung im Untergrund während des Setzvorganges sowie unter externer Last eine entscheidende Grundlage dar. Bei der Firma HILTI AG wurde ein Berechnungsprogramm auf Basis der Finite-Elemente-Methode entwickelt, das es ermöglicht, Aussagen über die Spannungen und über die Rissverteilung im Untergrundmaterial zu machen. In drei Beispielen werden diese Fähigkeiten des Programmes demonstriert. Die Ergebnisse zeigen besonders bei der Identifizierung der Versagensursache und beim charakteristischen Setz- und Auszugsverhalten sehr gute Übereinstimmung mit Versuchsresultaten.

ABSTRACT: Fastening elements such as anchors are widely used for transferring loads into concrete structural components. For the development of new anchors and the improvement of existing ones, the knowledge of the stress distribution and crack formation in the base material during the setting procedure and the pull-out procedure are of significant importance.

A numerical analysis based on the finite elemente methodology is performed to study the behavior of various anchors set in a predrilled hole in concrete. The FE-program developed by HILTI AG for the analysis of anchors provides very useful information on the stresses as well as the distribution of cracks in the concrete base material. This is shown in numerical investigations for three different types of heavy-duty fastenings. The examples include a heavy-duty anchor HSL, an undercut anchor HUC and an adhesive anchor HVA. These three examples represent the working principles: friction (HSL), keying (HUC) and bonding (HVA). The results cover the behavior of the fastening systems under axial loads. Although a number of interesting quantities are available, the investigation concentrates on the stresses in the concrete base material at the end of the setting phase and the crack distribution after the maximum load has been reached. With the crack distribution the different failure modes can be identified. These are break out for the HSL and the HUC. An anchor pull-out in case of the HVA occurs. As the results obtained in the numerical analysis agree quite well with experimental data, the FE program developed by HILTI AG proved to be a suitable tool for the improvement of existing anchors as well as for the development of new ones.

1. EINLEITUNG

Die Firma HILTI AG bietet im Standardprogramm Anker für Verankerungstiefen bis 220 mm an. Dabei werden im ungerissenen Beton bei einer Druckfestigkeit $f_{cc} = 20$ N/mm^2 charakteristische Lasten bis zu 163 kN und empfohlene Lasten bis 80 kN angegeben (HUC). Mit diesen Produkten wird ein weites Anwendungsspektrum in den Bereichen Kraftwerks- und Anlagenbau, schwere Maschinenbefestigungen sowie bei Anwendungen mit geringen Rand- und Achsabständen abgedeckt.

Um die physikalischen Vorgänge während des Setzvorganges und unter Einwirkung einer äusseren Last besser zu verstehen, werden heute bei der Firma HILTI AG Finite-Elemente-Berechnungsprogramme eingesetzt. Diese Programme dienen den Entwicklungsingenieuren als Werkzeug bei der Weiterentwicklung bestehender Produkte sowie bei der Entwicklung von Neuprodukten.

Bei Schwerlastankern ist aufgrund der hohen Versagenslasten der Aufwand für Versuche beträchtlich. Deshalb besteht besonders bei diesen Ankern durch konsequenten Einsatz von Simulationstools ein Einsparungspotential bei den Entwicklungskosten.

2. VERANKERUNGSMECHANISMEN

Im Folgenden wird ein kurzer Überblick über Verankerungsmechanismen und Versagensarten bei Verankerungen in Beton gegeben. Ausführliche Informationen darüber sind in der Literatur [1,2] zu finden.

Verankerungen in einem Bohrloch lassen sich auf die in Bild 1 dargestellten Wirkprinzipien Reibschluss, Formschluss und Stoffschluss zurückführen. Viele Verankerungen erfolgen in der Praxis durch eine Kombination dieser Wirkprinzipien.

Weiters können Verankerungen nach dem Verhalten bei Belastung über die Maximallast hinaus unterteilt werden. Die dabei unterschiedenen Versagensarten sind in Bild 2 dargestellt. Bei zentrischer Zugbeanspruchung einzelner Verankerungen treten in erster Linie die Versagensarten Ausbruch (1), Herausziehen (2) und Versagen des Ankers (3) auf. Darüber hinaus kann bei zu geringem Randabstand Kantenbruch (4) oder Spalten des Bauteils (5) auftreten.

3. NUMERISCHE SIMULATION

Aufgrund der spezifischen Anforderungen in der Befestigungstechnik, entwickelte die Firma HILTI AG ein Programm für die axialsymmetrische Berechnung von Dübelbefestigungen. Um Informationen über die Belastung des Befestigungselementes

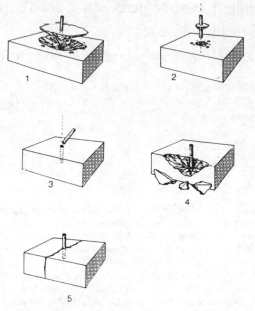

Bild 2: Versagensarten [3]
Fig. 2: Failure modes [3]

und des Untergrundes zu erhalten, erweist sich die in weiten Teilen der Konstruktionstechnik eingesetzte Methode der Finiten Elemente als geeignet. Die folgenden Beispiele geben eine Übersicht über wesentliche Fähigkeiten dieser Software.

Als Beispiele wurden drei Anker aus der aktuellen Produktepalette gewählt (vgl. Bild 3):

HSL M24: Schwerlastanker
Verankerungstiefe h_{ef} = 155 mm
max. Befestigungshöhe t_{fix} = 60 mm
empfohlene Last F_{30} = 45.5 kN

HUC M20: Hinterschnittanker
Verankerungstiefe h_{ef} = 220 mm
max. Befestigungshöhe t_{fix} = 90 mm
empfohlene Last F_{30} = 80.1 kN

HVA M24: Verbundanker
Verankerungstiefe h_{ef} = 210 mm
max. Befestigungshöhe t_{fix} = 54 mm
empfohlene Last F_{30} = 43.0 kN

Die drei Anker lassen sich, ihren wesentlichen Wirkprinzipien entsprechend, den oben erwähnten Kategorien Reibschluss (HSL), Formschluss (HUC) und Stoffschluss (HVA) zuordnen.

Bei den Ankern HSL und HVA wird das Bohrloch mittels einer Hammerbohrmaschine und einem Hartmetallschneidenbohrer erzeugt. Für den Hinter-

Reibschluss

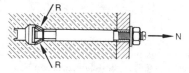

Formschluss

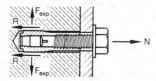

Stoffschluss

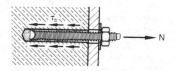

Bild 1: Wirkprinzipien [3]
Fig. 1: Working principles [3]

Anwendung	Dübeltyp
Säulen, Maschinen, Geräte, Krananlagen, schwere Maschinenbefestigungen	HSL Schwerlastanker 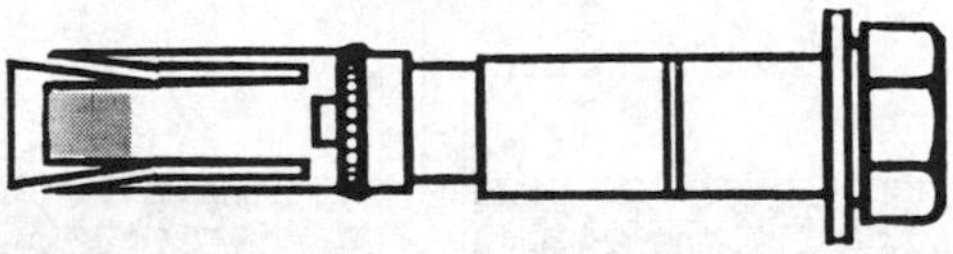
Kraftswerk- und Anlagenbau, Krananlagen, schwere Maschinenbefestigungen	HUC Hinterschnittanker 
Befestigungen mit geringen Achs- und Randabständen, hohe Verankerungswerte	HVA Verbundanker 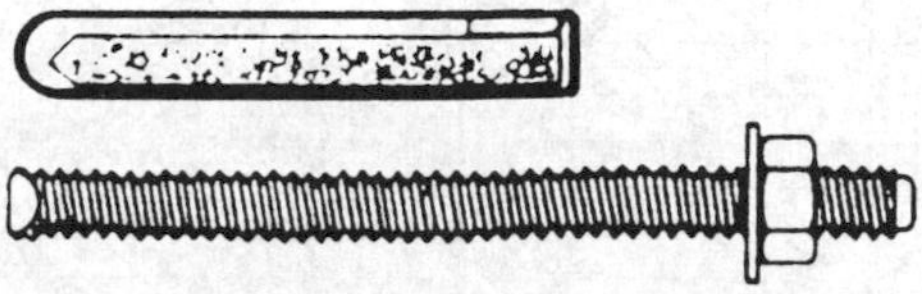

Bild 3: HILTI-Produkte für den Schwerlastbereich als Beispiele für den Einsatz der numerischen Simulation
Fig. 3: HILTI-products for heavy-duty fastenings chosen as examples for the numerical analysis

schnittanker HUC wird in einem ersten Arbeitsgang das Bohrloch mit einem Diamantbohrer und einer Diamantbohrkrone hergestellt. Diesem Arbeitsgang folgt das Erstellen des Hinterschnittes mit einem Hinterschnittgerät und einem Schleifkörper.

Durch Aufbringen eines Anzugsmomentes wird im Befestigungselement eine Vorspannkraft erzeugt. Dadurch erhöht sich die Steifigkeit der Verbindung und somit wird das Last-Verschiebungsverhalten im Bereich der Gebrauchslast bei statischer und dynamischer Belastung positiv beeinflusst.

Für detaillierte Informationen über die beschriebenen Produkte wird auf [4,5] verwiesen.

Randbedingungen

Als Untergrundmaterial wird Beton der Festigkeitsklasse C20 vorgegeben. Die Werkstoffe der einzelnen Ankerkomponenten werden den technischen Zeichnungen entsprechend definiert. In einem ersten Schritt wird der Anker gesetzt, d.h. eine axiale Vorspannung gemäss dem empfohlenen Anzugsmoment aufgebracht. Anschliessend wird eine äussere axiale Zugbelastung bis zum Versagen aufgebracht. Der Betonkörper wird dabei an der Oberfläche in ausreichendem Abstand vom Bohrloch festgehalten. Die Grösse des Betonuntergrundes wurde so gewählt, dass eine Beeinflussung der Rechenergebnisse durch die Randbedingungen ausgeschlossen werden kann.

Ergebnisse

Die dargestellten Ergebnisse behandeln die Zugspannungen und die Risse im Untergrundmaterial. Dabei wird auf die Belastung bei aufgebrachtem Anzugsmoment aber ohne äussere Last und dem Zustand nach Überschreiten der maximalen Last eingegangen. Aus der Vielzahl an Resultaten aus der FE-Berechnung werden hier diese Ergebnisse behandelt, da sie Auskunft über die primäre Versagensursache der Verankerung geben bzw. am besten dazu geeignet sind diese zu dokumentieren.

HSL M24

Das Finite-Elemente-Netz ist in Bild 4 dargestellt. Bild 5 zeigt die Last-Verschiebungskurve. Ein Vergleich mit einer gemessenen Last-Verschiebungskurve (Bild 6) von Ankern dieses Typs zeigt eine gute Übereinstimmung der Berechnungsresultate mit Versuchsergebnissen. Dies gilt sowohl in bezug auf das charakteristische Tragverhalten des Ankers, als auch was die Grösse der Auszugslast betrifft.

Die Darstellung in Bild 7 gibt Auskunft über die Zugspannungen im Untergrundmaterial bei aufgebrachtem Anzugsmoment.

Bei diesem Belastungszustand (vgl. Punkt 1 in

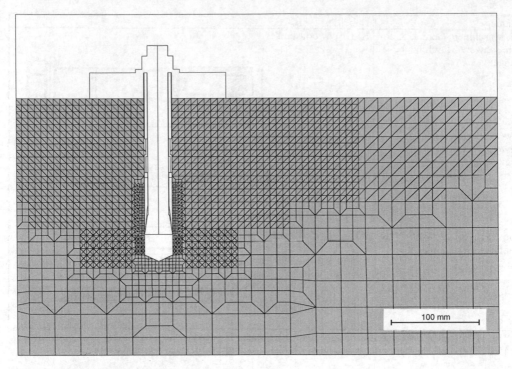

Bild 4: Finite-Elemente-Netz eines HSL M24
Fig. 4: Finite element mesh for a HSL M24

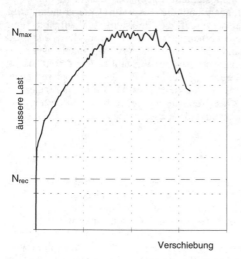

Bild 5: Berechnete Last-Verschiebungskurve
Fig. 5: Force-displacement curve (calculation)

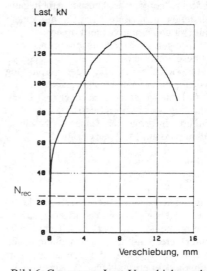

Bild 6: Gemessene Last-Verschiebungskurve [3]
Fig. 6: Force-displacement curve (experiment) [3]

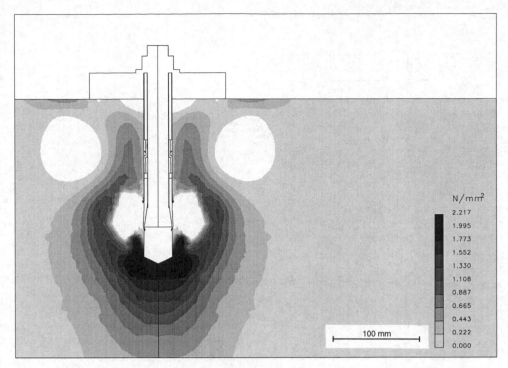

N/mm²
2.217
1.995
1.773
1.552
1.330
1.108
0.887
0.665
0.443
0.222
0.000

100 mm

Bild 7:   HSL M24 - Zugspannungen im Untergrundmaterial bei aufgebrachtem Anzugsmoment
Fig. 7:   HSL M24 - Tensile stresses in the base material at the end of the setting phase

der Last-Verschiebungskurve) ist keine äussere Last wirksam. Die dunklen Bereiche in der Darstellung stellen hohe Zugspannungen dar, die weissen Bereiche kennzeichnen Druckzonen. Bei dieser Belastung befinden sich die grössten Zugspannungen aufgrund der hohen Spreizkräfte am Grunde des Bohrloches. Im Bereich der Krafteinleitung in den Beton bildet sich eine Zone mit Druckspannungen.

Als Versagensursache bei externer Belastung der Verbindung ergibt sich Betonversagen. Die Darstellung der Risse im Untergrundmaterial nach Überschreiten der Maximallast (Bild 8) verdeutlicht dies. Die Breite der die Risse kennzeichnenden Linien ist proportional zur Grösse der Risse. Die Graustufen dokumentieren die Radialrisse.

HUC M20

Bild 9 zeigt die Geometrie des Finite-Elemente-Netzes eines HUC M20. Die errechnete Last-Verschiebungskurve ist in Bild 10 dargestellt.

Im gesetzten Zustand (empfohlenes Vorspannmoment nach [4]) wird das Untergrundmaterial wie in Bild 11 gezeigt belastet. Hier zeigt sich, dass nur ein kleiner Bereich in der Nähe der

Krafteinleitung auf Zug belastet wird. Dies ist darauf zurückzuführen, dass im Vergleich zum HSL eine geringe radiale Kraft in den Untergrund eingeleitet wird, da die Kraftübertragung zwischen Befestigungselement und Untergrund in erster Linie durch Formschluss erfolgt.

Das Rissbild (Bild 12) zeigt in diesem Fall den Versagenstyp Betonausbruch. Das Untergrundmaterial ist folglich das schwächste Glied der Verankerung. Auch dieses Ergebnis wird durch Versuche bestätigt (vgl. dazu charakteristische Lasten in [4]).

HVA M24

Als drittes Beispiel wurde ein Verbundanker berechnet. Das der Berechnung zugrunde gelegte Finite-Elemente-Netz zeigt Bild 13. Die errechnete Last-Verschiebungskurve in Bild 14 gibt Auskunft über das Tragverhalten des Ankers.

Im vorgespannten Zustand (Bild 15) zeigt sich eine schmale Zugzone entlang der Bohrlochwand. Die Form des Bereiches mit Zugbelastung unterscheidet sich dabei erheblich von den beiden zuvor vorgestellten Ankertypen. Die Tatsache, dass bei diesem Anker die Kraftübertragung zwischen Be-

71

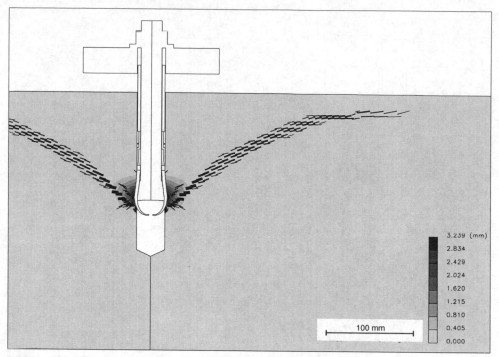

Bild 8: HSL M24 - Rissverteilung im Beton nach Überschreiten der Maximallast
Fig. 8: HSL M24 - Crack distribution in the base material after reaching the maximum load

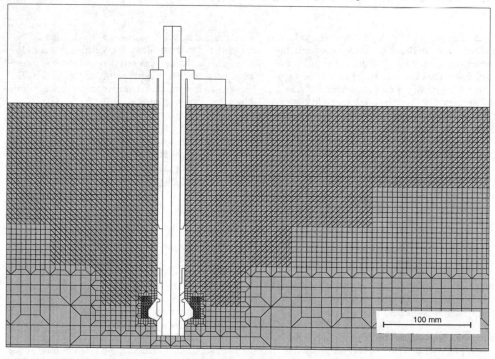

Bild 9: Finite-Elemente-Netz eines HUC M20
Fig. 9: Finite element mesh for a HUC M20

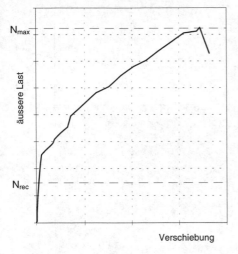

Bild 10: Berechnete Last-Verschiebungskurve
Fig. 10: Force-displacement curve (calculation)

festigungselement vom Verbund zwischen Mörtel und Beton bestimmt wird, bedingt, dass praktisch keine radialen Spreizkräfte auftreten.

Auch das Rissbild (Bild 16) und die Versagensursache weichen erheblich von den zuvor gezeigten Ankertypen ab. Es bildet sich an der Oberfläche ein kleiner Ausbruchkegel. Der restliche Teil des Ankers wird aufgrund des Versagens des Verbundes zwischen Mörtel und Beton aus dem Bohrloch ausgezogen.

Vergleich der drei Ankertypen

Der Hinterschnittanker HUC besitzt wegen des gros-sen Hinterschnittes wesentliche Vorteile bei Verankerungen in der Zugzone. Im Vergleich zum Schwerlastanker HSL ermöglicht er aufgrund der geringeren radialen Belastung des Untergrundes geringere Achs- und Randabstände. Ebenso eignet sich der Verbundanker HVA besonders für Verankerungen mit geringen Rand- und Achsabständen.

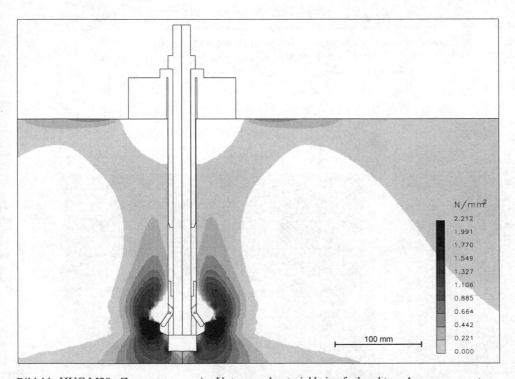

Bild 11: HUC M20 - Zugspannungen im Untergrundmaterial bei aufgebrachtem Anzugsmoment
Fig. 11: HUC M20 - Tensile stresses in the base material at the end of the setting phase

73

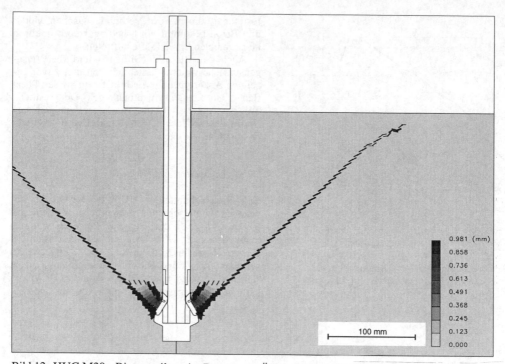

Bild 12: HUC M20 - Rissverteilung im Beton nach Überschreiten der Maximallast
Fig. 12: HUC M20 - Crack distribution in the base material after reaching the maximum load

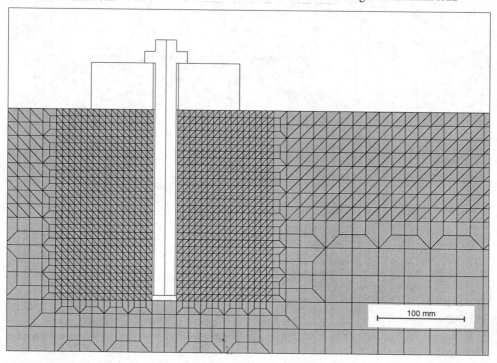

Bild 13: Finite-Elemente-Netz eines HVA M24
Fig. 13: Finite element mesh for a HVA M24

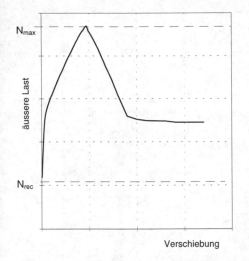

Bild 14: Berechnete Last-Verschiebungskurve
Fig. 14: Force-displacement curve (calculation)

4. SCHLUSSFOLGERUNGEN

Das bei der Firma HILTI AG entwickelte Finite-Elemente-Programm zur Berechnung von Dübelbefestigungen erweist sich als wichtiges Werkzeug bei der Entwicklung von Ankern für schwere Lasten. Die Berechnung von Spannungen und Rissen im Untergrundmaterial Beton ergeben dabei eine sehr gute Übereinstimmung mit Versuchsergebnissen.

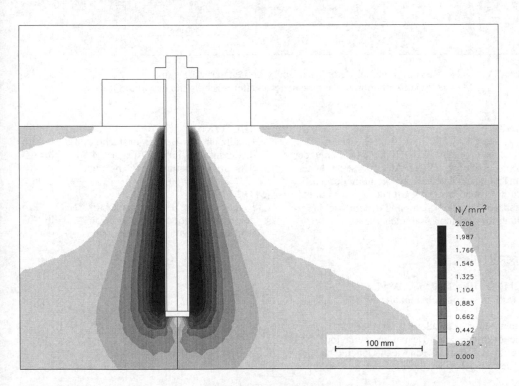

Bild 15: HVA M24 - Zugspannungen im Untergrundmaterial bei aufgebrachtem Anzugsmoment
Fig. 15: HVA M24 - Tensile stresses in the base material at the end of the setting phase

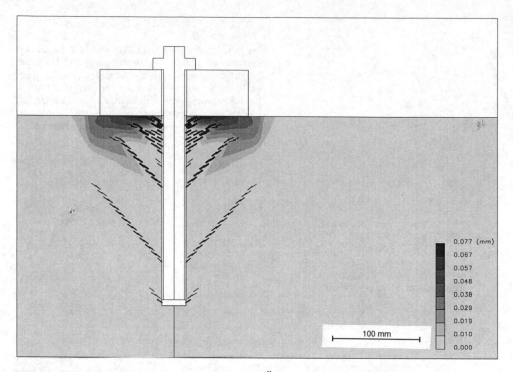

0.077 (mm)
0.067
0.057
0.048
0.038
0.029
0.019
0.010
0.000

|← 100 mm →|

Bild 16: HVA M24 - Rissverteilung im Beton nach Überschreiten der Maximallast
Fig. 16: HVA M24 - Crack distribution in the base material after reaching the maximum load

5. DANK

Wir danken Prof. Dr. W.J. Huppmann, Leiter der Konzern-Forschung, HILTI AG, für seine Unterstützung und sein Interesse an der numerischen Simulation. Besonderer Dank gilt auch allen Mitarbeitern der Konzern-Forschung und der Entwicklungsabteilung für die gute Zusammenarbeit.

6. LITERATUR

[1] Rehm G., Eligehausen R., Mallee R.
    Befestigunstechnik, Betonkalender 1992 Teil II

[2] Senkiw G.A., Lancelot H.B.
    Anchor in concrete - Design and behaviour, ACI Special Publication, American Concrete Institute, Detroit, MI, USA, 1992

[3] HILTI AG
    Handbuch der Befestigunstechnik, B1 - Anker- und Dübeltechnologie, Ausgabe 7/93, Hilti AG, Schaan, Fürstentum Liechtenstein

[4] HILTI AG
    Handbuch der Befestigunstechnik, B3.1 - Produkte Informationen, Ausgabe 4/94, Hilti AG, Schaan, Fürstentum Liechtenstein

[5] HILTI AG
    Handbuch der Befestigunstechnik, B3.2 - Produkte Informationen, Ausgabe 4/94, Hilti AG, Schaan, Fürstentum Liechtenstein

Anchors in Theory and Practice, Widmann (ed.) © 1995 Balkema, Rotterdam. ISBN 90 5410 577 1

Analysis of grouted soil anchors

Untersuchung injizierter Bodenanker

Jósef Mecsi
ME-SZI Engineering Ltd, Hungary

Abstract:
The paper presents a method for determining the load bearing capacity of grouted soil anchors and the force-elongation diagram. Computation reckons with changes in complex soil behaviour and changes in the stress/strain condition along with the imbedding technology. By introducing dimensionless coefficients a simple graphical method serves for the determination of approximate pull-out resistance of grouted soil anchors based on the expanded cylindrical cavity theory elaborated by the author. Anchor elongation may be calculated from anchor force distribution which will be determined from the activation process of the pull-out resistance due the relative axial displacement between the soil and the anchor body.

Zusammenfassung:
Der Artikel analysiert die Zusammenhänge der Tragfähigkeit injektierter Anker. Die Einbindelänge des Ankers wird in allgemeinen Zementmörtel unter höherem Druck injiziert. Während der Injektion erhöht das Wasser des Mörters den Porenwasserdruck, und der Festanteil die effektive Spannung. Durch die Injektion füllt der größere Teil des Festanteiles die Poren und vergrößert dadurch den Durchmesser des Ankers, der kleinere Teil (etwa 2-10%) befestigt den Anker dauernd und effektiv im Boden durch dessen volumetrische Ausdehnung. Durch diesen Effekt wird die Konzentration und die Größe der Druckspannung in der Umgebung des Ankers gesteigert.
Abbildung 2. analysiert den Spannungszustand des Bodens vor der Injektion und stellt die Bestimmung der durchschnittlichen Druckspannung (σ_{om}) dar, die auf die Ankerachse wirkt. Abbildung 4. zeigt die Verteilung der gesteigerten Spannungen und die Bestimmung der Grenze des verdichteten Bodens (Zusammenhänge ohne Dimensionen). Abbildung 5. gibt den Zusammenhang zwischen der Spannungsveränderung und der Volumenvergrößerung des Ankers. Der Ausreisswiderstand des Ankers wird nach Coulomb bestimmt, in Betracht genommen die durch die Injektion veränderte Spannungsverteilung. Die Ausreissfläche des Ankers ist annäherungsweise ein Zylindermantel. Der Ausreisswiderstand ist das Minimum des Produktwertes der Scherfestigkeit und der Scherfläche. Der Ausreissradius ist in der unmittelbaren Nähe des Ankerkörpers.
Maßgebend ist die Zugdehnung der Stahlbewehrung. Eine axiale Bewegung von Δ_{ult} ist zwischen dm Ankerkörper und dem Boden nötig um den Ausreisswiderstand und die Scherfestigkeit zu mobilisieren. Während dieses Prozesses kann die Einbindelänge des Ankers in zwei Teile gegliedert werden:

a.) mobilisierte Scherfestigkeitzone (l_o) $\Delta \geq \Delta_{ult}$

b.) Scherspannungszone $(l_{fixed}-l_o)$ $\Delta < \Delta_{ult}$

Die Kraftverteilung in der Scherspannungszone kann mit den Zusammenhängen der

Elastizitätstheorie berechnet werden. Die Dehnung des Ankers entspricht der Ankerkraft-verteilung, ihre Größe kann durch ein Diagramm geschätzt werden.
Der Artikel gibt auch die Bestimmung der nötigen Einbindelänge.
Der spezifische Ausreisswiderstand (t_{ult}) und die Länge der akutellen Scherfestigkeits-zone (l_o) können aus dem Diagramm abgelesen werden, das während der Verspannung des Ankers bestimmt wurde. Auf Grund praktischer Erfahrungen stellt der Artikel die wichtigsten Faktoren dar, die den Ausreisswiderstand beeinflussen.

1.INTRODUCTION

Several technologies are known for making grouted soil anchorages. Essential steps of the process in our investigation are:

- making a borehole of determined length and inclination (α) in the soil;
- replacing boring mud by cement mortar;
- inserting the anchor into the bore hole;
- after hardening of the cement mortar, intermittent grouting of the anchor;
- occasionally regrouting the fixed lengths of the anchor;
- prestressing and fixing of the anchorage.

The anchor length is divided longitudinally in two distinct, cross-sectionally different lengths. Since the anchorage force can not be transferred directly to the soil close to the supported structure, the anchor has to be put into a sleeve pipe along the first length of anchorage. This section is the freely displacing length of the anchor (l_{free}). Fixing the anchor into the soil occurs along the fixed length (l_{fixed}) Figure 1.
Backward overrun of the cement mortar is prevented by the so-called stop bag, which for the aspect of stress pattern has to considered as part of the fixed length.
When an anchor is made, first the bore hole is grouted by the boring mud which then will be extruded by the injected cement mortar, left in place for hardening. Anchorage forces develop in reality when the partly hardened injected cement mortar is broken into segments by an increased pressure of further mortar injection which stresses the perimeter

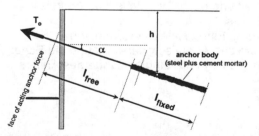

Fig. 1. Ground anchorage nomenclature
Abb. 1. Bezeichnung der Bodenanker

of the bore hole and fixes the anchor body permanently and effectively into the soil. (See 3.3). Specific pull-out resistance is unterstood as the resistance of 1 m of length of the anchor.

$$t_{ult} = \tau_{ult} \cdot \underbrace{2 \cdot r_k \cdot \pi \cdot 1}_{A_k} \qquad (1)$$

where τ_{ult}- ultimate shear strength of the anchor
$A_k = 2 \cdot r_k \cdot \pi \cdot 1$- is the shell surface area of the pulled-out anchor body.

Specific pull-out resistance depends on how the anchor was made and is influenced by several factors. Much depends on anchor grouting conditions, on compression and shear strength parameters of the surrounding soil, but less on the geometrical position of the anchor. Informative data of specific pull-out resistances for different soil types are shown in Table 1. under average grouting conditions.

Table.1.

Soil type	Specific pull-out resistance [kN/m]	Displacement activating the shear strength
Gravel	140-500	1-5 mm
Sand	100-300	1-5 mm
Fine sand	60-200	3-8 mm
Silt	50-150	3-14 mm
Clay	25-150	5-25 mm

Activation of the specific pull-out resistance requires a relative axial displacement between the pull-out anchor body and the soil. Informative values of displacements which activate the shear strengths in different soils are presented in Table 1.

2. DESIGN OF THE ANCHORS

Anchor design has to begin with the stability analysis of the anchorage system as a whole. Soil stratification, analysis of the stress pattern of the supported structure, and stability analysis underlie determination of the angle of inclination (α) and the freely displacing length of the anchor.

Suggested procedure of anchor design comprises the following steps:

- Determination of the needed and effective steel cross section of the anchor, number of high-strength tendons or cables.
- Computation of the specific pull-out resistance of the anchor (t_{ult}).
  - ¤ Analysis of the initial stress state in the soil, for determining the boundary conditions of the calculation.
  - ¤ Determining the soil stresses due to grouting.
  - ¤ Determining the specific pull-out resistance of the anchor.
- Analysis of the shear strength activation process for the determination of anchor force distribution.
- Determination of the anchor tension/strain diagram.

3. THE SPECIFIC PULL-OUT RESISTANCE

3.1. Basic assumptions:

We want to know the limiting value of resistance that can be exercised by an axially pulled cylinder after its expansion in the form as injection.

The analysis relies on the cylindrical expansion theory in the soil based upon the concepts of Mecsi (1982, 1991, 1993, 1994, 1995).

Determination of soil stresses due to grouting

The expansion is assumed to occur under conditions of plane strain and axial symmetry in a medium. The soil strength parameters (ϕ',c) and the non-linear deformation modulus (which is described by two parameters E_o and the power index a) have been introduced for the determination of structural changes in the soil.

In the initial condition state compressive stresses act in the soil, due to geostatic pressure.

Along the fixed length of anchor, *volume of solid sceleton of the grout, and the grouting pressure* much exceeding the geostatic pressure, *produce the stress concentration around the anchor body*.

Determination of the axial load bearing capacity

The stress condition will then undergo to alterations under the influence of shearing stresses provoked by the application of the external axial force. Magnitude

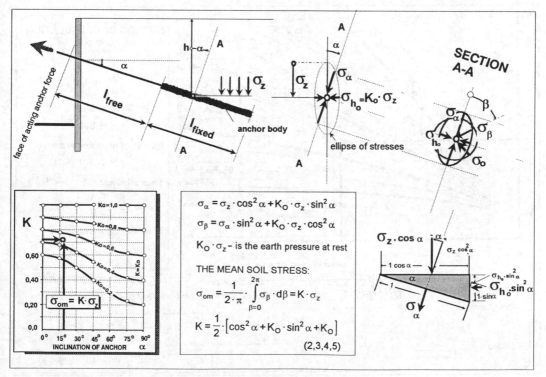

Fig.2. Initial soil stress condition in the soil before injection
Abb. 2. Ausgangs-Spannungszustand im Boden vor der Injektion

79

of arousen shearing stresses can be calculated when we reproduce the spherical distribution of radial compressive stresses around the cylinder.

We are looking for the least breaking off resistance of the cylinder what can be reckoned as the minimum of a product of the potential shearing surface area and the probable shearing strength. Unknown is the radius of this body at the least value of the product.

CROSS-SECTION OF ANCHOR

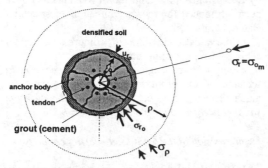

Fig. 3. Cross section of the anchor after grouting
Abb. 3. Anker-Querschnitt nach der Injektion

3.2 *Analysis of the initial stress state (Figure 2.).*

Determination of the spatial distribution of soil stresses around the anchor requires to know the initial soil stress state prior *to making the anchor*, requiring in turn, to examine the radial soil stresses which act in a plane normal to the fictitious anchor axis. In conformity with the equilibrium of forces acting on an elementary soil prism in a plane normal to the anchor axis the mean soil stress (σ_{om}) can be determined.

3.3 *Determination of soil stresses due to grouting*

Injection for the fixed length of the anchor is generally made under high pressure and by using cement mortar.
The quantity of the injected mortar and the applied pressure are in close relation with each other. During injection, the moisture in the grouting material will increase the pore-water pressure in the soil, while the solid part in it will constribute to the increment of effective stresses.
As a result of injection, the major part of the solid grains in the grouting material will fill the voids in

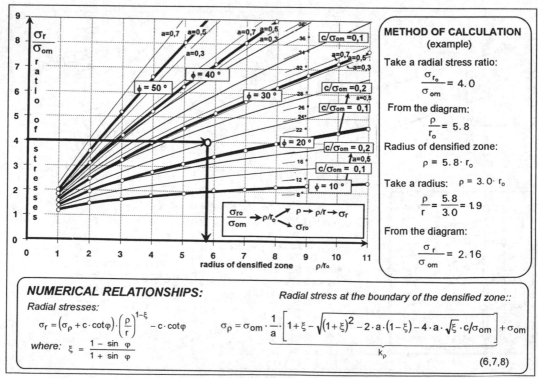

Fig. 4. Diagrams for the calculation of extensions and stress distribution in the densified Zone
Abb. 4. Diagramme zur Ermittlung der Dehnungen und Spannungen im verdichteten Bereich

the soil (whereby the diameter of the anchor body increases) and the minor part of it (2-10%) takes part in stretching the anchor efficiently and permanently into the hole.

This stretching effect iniciates the increment and concentration of compressive stresses in the vicinity of the anchor body and the densification of the soil around the hole.

By making use of the theory of the expanded cylinder (Mecsi 1991-1995) also the spherical distribution of stresses and stains can be determined.

The cross section of the fixed length of an anchorage is shown in Fig.3.

The charts in Figure 4. give information about the extension of the densified zone by the use of various soil parameters. Cohesion is presented in the ratio of the earth pressure at rest. From the diagram we can calculate the stress distribution in relation to the given pressure level for any arbitraily taken radius inside the densified zone.

Figure 5. presents the relationship between stresses and volume changes of the anchor body (strains). From the diagram we can estimate the relationship between radial soil stress and volumetric strain.

3.4 Specific pull-out resistance

The anchor pull-out surface will be approximated by a cylinder shell.

Resistance at the anchor shoulder is slight, hence negligible (with respect to the small acting surface).

Shear strength acting on the shell will be obtained from Coulomb's relationship:

$$\tau_{ult} = \sigma_r \cdot \tan\varphi + c \tag{13}$$

The specific pull-out resistance of an anchor can be calculated by the actual surface area of the pulled anchor and the expected value of the shear strength. Figure 6. presents some details of the calculation.

Experience shows that — while the surface area increases in proportion with the radius of the cylinder to the contrary of the expected value of the shearing strength which decreases — change in the *value of specific pull-out resistance is not significant in most cases in a narrow surrounding around the anchor*. The failure surface is quite close to the hardened soil block.

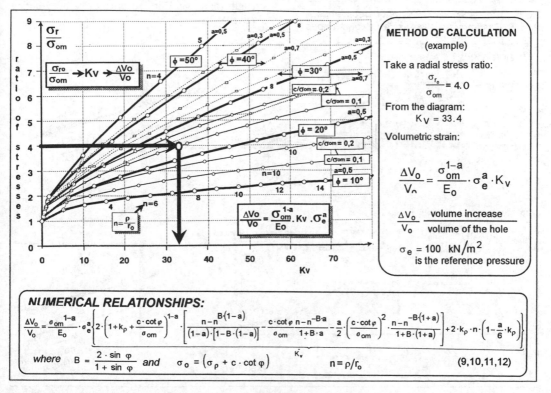

Fig. 5. Diagrams for the calculations of the volume changes of the anchor body
Abb. 5. Diagramme zur Ermittlung der Volumsänderungen des Anker-Körpers

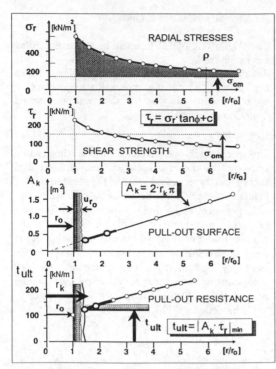

Fig.6. Determination of the pull-out resistance.
Abb. 6. Bestimmung des Auszieh-Widerstandes

3.5 The shear strength activation process

To activate the shear strength, some displacement is needed.

Assumptions:

- Steel strain along of the fixed anchor length is critical.
- Resistance at the anchor shoulder is slight, hence negligible.
- Along the free anchor length, the is anchorage force constant.
- Relative displacements between the pulled anchor body and the soil are determinant in respect of shear stress activation. At higher anchorage forces it is important to give consideration to the possibility of a steel wire being teared-out from the anchor body.

In compliance with the activation process, the fixed anchor length may be divided into two parts:

a.) Activated *shear strength length* (l_o)
b.) Shear *stress length* ($l_{fixed}-l_o$)

In the shear strength zone -where the relative displacement of the anchor body exceeds the ultimate value (Δ_{ult}) - the shear strength value is assumed to be constant, but mind that the shear strength may vary in accordance with grouting circumstances.

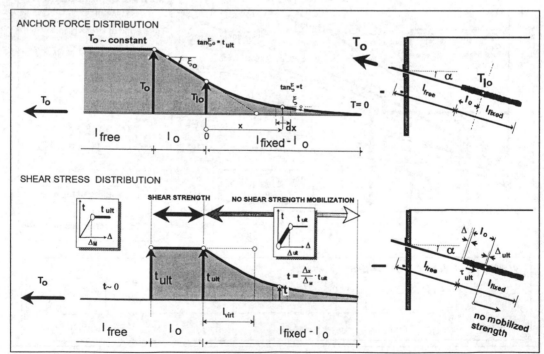

Fig. 7. Analysis of the shear strength activation process
Abb. 7. Berechnung des Vorganges der Aktivierung der Scherfestigkeit

82

Along the shear stress length, force distibution can be calculated from the elastic strain of anchor tendons according to Figure 7.
The anchor elastic elongation will be obtained from Hook's law:

$$E_{steel} \cdot A_{steel} \cdot \frac{d\Delta}{dx} = -T \tag{14}$$

$E_{steel} \cdot A_{steel}$ –is the anchor body rigidity, approximated in terms of tendon rigidity.
The tangent to force distribution in the shear stress length can be expressed as:

$$\tan \xi = -\frac{dT}{dx} = -\frac{\Delta}{\Delta_{ult}} \cdot t_{ult} \tag{15}$$

From the relations (14) and (15), we find the differential equation:

$$\frac{d^2\Delta}{dx^2} = \frac{t_{ult}}{E_{steel} \cdot A_{steel}} \cdot \frac{\Delta}{\Delta_{ult}} \tag{16}$$

From the differential equation can be calculated the force distribution in the shear stress length:

$$T_x = t_{ult} \cdot \frac{1}{k} \cdot \frac{sh[k \cdot (l_{fixed} - l_o - x)]}{ch[k \cdot (l_{fixed} - l_o)]} \tag{17}$$

k - is the anchor "rigidity index"

$$k = \sqrt{\frac{t_{ult}}{E_{steel} \cdot A_{steel} \cdot \Delta_{ult}}} \tag{18}$$

3.6 Analysis of the anchor elongation

Analysis of the anchor elongation fits the anchor force distribution.

a.) Elongation of the free anchor length   (l_{free})
(Constant anchor force)

$$\Delta_{free} = \frac{T_o \cdot l_{free}}{E_{steel} \cdot A_{steel}} \tag{19}$$

b.) Elongation of the shear strength zone   (l_o)
(Distribution of anchor force is linear)

$$\Delta_{l_o} = \frac{T_o \cdot l_o}{E_{steel} \cdot A_{steel}} - \frac{t_{ult}}{2} \cdot \frac{l_o^2}{E_{steel} \cdot A_{steel}} \tag{20}$$

c.) Elongation of the shear stress length   ($l_{fixed} - l_o$)
(displacement which activate the shear strength)

$$\Delta_{(fixed - l_o)} = \Delta_{ult} \tag{21}$$

Knowing the summary of the elongations of the lengths with various stress distribution characteristics, we will be able to determine the force-elongation diagram for the anchorage .
Figure 8. presents the typical anchor force-elongation diagram.

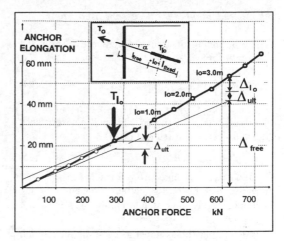

Fig. 8. Anchor force-elongation diagram
Abb. 8. Diagram Ankerkraft-Verlängerung

3.7 Interpreted results from a loading test

In the course of a loading test of an anchor of known geometry and rigidity measured will be the value-pairs of the force (T_o) and elongation (Δ_{total}).
By plotting these values *we will be able to determine the specific pull-out resistance (t_{ult}) and the actual length of the activated shear strength length (l_o).*
Using the Figure 7. any actual anchor force is given by the formula:

$$T_o = t_{ult} \cdot \left[l_o + \frac{1}{k} \cdot th[k \cdot (l_{fixed} - l_o)] \right] \tag{22}$$

and the pertinent total elongation of the anchor can be determined (See 3.6): (23)

$$\Delta_{total} = \frac{T_o \cdot l_{free}}{E_{steel} \cdot A_{steel}} + \frac{T_o \cdot l_o}{E_{steel} \cdot A_{steel}} - \frac{t_{ult}}{2} \cdot \frac{l_o^2}{E_{steel} \cdot A_{steel}} + \Delta_{ult}$$

The two unknown terms (t_{ult} and l_o) in formulas (22) and (23) can be determined.
Theoretically a single point in the load test diagram would suffice to determine the t_{ult} value, but because several measured points are available from the load test, it is possible to lay an equilizer curve onto the points, which curve represents the realiability of our estimate.
Due to the rearrangement of the tendons uncertainities may arise in the measured elongations at low pulling forces. It is suggested to apply a calculated (Formula 22.) elongation value for the lowest pulling force and use this value for correcting the measured elongation.

83

3.8 Determination of the necessary fixed length

At the beginning of shear strength activization, the anchorage force is:

$$T_{lo} = t_{ult} \cdot \underbrace{\left(\frac{1}{k} \cdot th\left[k \cdot \left(l_{fixed} - l_o \right) \right] \right)}_{l_{virt}} \qquad (24)$$

that is a product of the specific pull-out resistance (t_{ult}) and a certain virtual legth (l_{virt}).

$$l_{virt} = \frac{1}{k} \cdot th\left[k \cdot \left(l_{fixed} - l_o \right) \right] \qquad (25)$$

Figure 9. was determined from formula (25).
It shows that beyond a certain elongation of the anchorage body the useful length (l_{virt}) does not change.
This length may be assumed to be at the intersection of curve with the line from the followed equation.

$$\frac{1}{k} \cdot th\left[k \cdot \left(l_{fixed} - l_o \right)_{min} \right] = \left(l_{fixed} - l_o \right)_{min} / 2 \qquad (27)$$

From the equation (27) the minimum length for fixing the anchor body $\left(l_{fixed} - l_o \right)_{min}$ can be determined.

The necessary fixed length $(l_{fixed})_n$ for a given anchor force is:

$$(l_{fixed})_n = \frac{T_o}{t_{ult}} + \frac{\left(l_{fixed} - l_o \right)_{min}}{2} \qquad (28)$$

The safety factor is:

$$\eta = \frac{T_{o_{max}}}{T_o} = \frac{l_{fixed} \cdot t_{ult}}{l_o + \frac{1}{k} \cdot th\left[k \cdot \left(l_{fixed} - l_o \right) \right]} \qquad (29)$$

4. CONCLUSIONS
(Main factors in fluencing pull-out resistance)

- Calculation as well as experience shows that (beyond a rapid increment at the beginning) the volume of injected cement mortar would not increase significantly the ultimate anchor force.

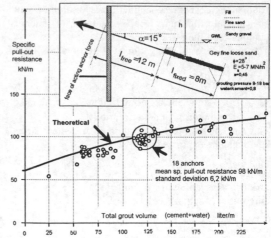

Fig. 10. Relationships between the grout volume and the pull-out resistence
Abb. 10. Beziehungen zwischen dem Injektionsvolumen und dem Auszieh-Widerstand

Fig. 10. shows the specific pull-out resistance values for number 93 anchors at the Határ út Metro station in Budapest which were determined by the presented method.
The anchors were imbedded in the same type of soil with various quantities of cement mortar.
The envelope curve shows the correlation between the injected mortar volume and the ultimate pull-out resistance.

- Soil deformability has high importance in anchor load capacity.
- Depth of the anchorage below the ground level (beyond a certain geostatically justified extent) has no significant influence on the load bearing capacity.

5. ACKNOWLEDGMENT

I would like to thank Mr. Huba Héjj for his professional advices in compounding this article.

6. REFERENCES

Mecsi, J. (1982) Determination of the load bearing capacity of grouted soil anchors from the stress-strain condition around the anchor. *Dr.Tech. Thesis Budapest*

Mecsi, J. (1986) Determination of the Clamping Length and Expectable Test Loading Diagram of injected Anchors *Müszaki Tervezés XXXVI. 9-10. Budapest, Hungary. pp. 51-61.(In Hungarian)*

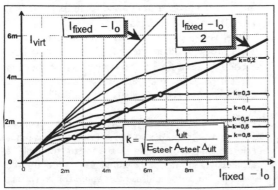

Fig. 9. Analysis of the effective length of the anchor
Abb. 9. Berechnung der wirksamen Ankerlänge

Mecsi J. (1988) Checking the Load-bearing Capacity of grouted anchors *3rd Subway Conference on Underground and City Budapest. 23-25 March. 1988 pp 405-417*

Mecsi J. (1989). Method for the analysis of the tension-strain diagram of grouted soil anchors. *Conference on Foundation '89 Bno*

Mecsi J. (1991 and 94). Stresses, displacements, volume changes around the expansion cylinder in the soil. *10th European Conf. on Soil Mech. and Found Engng Florence May 1991, Vol.,pp. 242-247* Panel intervention *Vol IV. Pages 1268-1272.*

Mecsi J.(1993). Stresses, strains and volume change around an expanded cylinder in the soil. *Dr. Thesis (Ph.D.) Hungarian Academy of Sciences*

Mecsi J. (1994). Stress-strain condition around an expanded cylinder in the soil *Periodica Polytechnica 38/1 1994. TU Budapest Pp. 67-86*

Mecsi J. (1995). Some inspiring ideas for the application of the cylindrical cavity theory *The Pressuremeter and its New Avenues, Ballivy (ed.) 1995. Fourth International Symposium of Pressuremeters (1995 may 16-19. Sherbrook, Quebec, Canada), Balkema, Pages 461-471.*

Mecsi J. (1995). Some aspects of evaluation of pressuremetric data *11th European Conf. on Soil Mech. and Found Engng Copenhagen 27. May - 2. Jun1995, Vol. I. Balkema. pp.*

Petrasovits G. (1981) Factor Affecting Interaction between Soil and Grouted Anchor, *Scan. Int. Conf. SMFE, Stockholm 1981*

AE identification and BEM prediction in pull-out process of anchor bolt

AE-Identifikation und BEM-Voraussage im Ausziehvorgang einer Ankerschraube

M. Ohtsu & M. Shigeishi
Kumamoto University, Japan

A. H. Chahrour
Taisei Corporation, Tokyo, Japan

ABSTRACT: Pull-out process of anchor bolt from concrete is kinematically investigated by acoustic emission (AE) and the boundary element method (BEM). A moment tensor analysis of AE sources could provide information on crack location, crack type, and crack orientation. By applying the analysis to a pull-out test, mixed-mode crack propagation is clarified experimentally. At the crack front, tensile cracks are generated first, and then shear cracks follow. To predict crack trajectories analytically, the two-domain BEM analysis is carried out, on the basis of the maximum tensile stress criterion in the linear elastic fracture mechanics. Crack trajectories analyzed reasonably agree with those of the tests and the effect of lateral constraint is clarified. At the crack tip, tensile motion normally governs except for lateral crack propagation due to shear motion.

ZUSAMMENFASSUNG: Ein Ausziehvorgang einer Ankerschraube aus einem Beton ist durch akustische Emissionsmessung (AE) und die Randelementmethode (BEM) kinematisch untersucht. Eine Momenttensor-Analyse der AE-Entstehungsquellen soll Informationen über Stellung, Type, und Orientierung der Risse geben können. Durch Anwendung dieser Analyse auf die Ausziehversuche ist die Risse-Fortpflanzung in gemischter Form experimentell aufgeklärt. An der Risse-Front werden zuerst Zugrisse erzeugt, und dann folgen Scherungsrisse an der von den Zugrissen gebildeten Oberfläche entlang.

Um die Risse-Trajektorie vorauszusagen, wird eine Zweibereich-BEM-Analyse im Modellstudium ausgeführt. Es wird das maximale Zugspannungskriterium im linear-elastischen Bruchmechanismus angewendet. Die analysierte Rissetrajektorie stimmt mit den Versuchsergebnissen gut überein, und der Effekt der seitlichen Zwangskraft wird aufgeklärt. Während der Rissefortpflanung herrschen Zugbewegungen normalerweise an der Rissespitze. Nur im Falle, wo die Risse sich waagrecht fortpflanzen, sind Scherungsbewegungen vorhanden.

1 INTRODUCTION

To elucidate failure process of concrete, acoustic emission (AE) is applied to identifying the fracture process zone (Mindess 1991) and the crack front (Chen et al. 1992). A technique locating AE sources is also employed for studying fracture of concrete (Berthaud et al. 1991; Nomura et al. 1991). From the arrival time differences at multi-channel observation points, locations of AE sources (crack locations) are determined. Because the accuracy of the locations is limited, the identification of the fracture process zone and the crack mechanisms is still marginal.

Quantitative waveform analysis has been investigated for source characterization of AE (Ohtsu 1982). Thus, a source inversion procedure based on the moment tensor representation is developed, including the location procedure (Ohtsu 1987). It has been sophisticatedly applied to AE waveforms detected at *in situ* hydrofracturing tests, where AE

sources were classified into tensile cracks and shear cracks with the directions of crack motion (Ohtsu 1991).

During the past decade, fracture mechanics has been applied frequently to the analysis of cracking in concrete. In the discrete crack modelling, linear elastic fracture mechanics (LEFM) has been successfully utilized in the analysis of mixed-mode crack propagation. Although the finite element method (FEM) has a long and well-documented history to analyze crack growth, an intrinsic drawback is the need for continuous remeshing to follow crack nucleation. It is expedient to use the boundary element method (BEM) for the analysis, because only the boundary needs to be discretized. It holds a privileged applicability to the discrete crack model. A two-domain BEM technique was studied and satisfactorily applied to simulate mixed-mode crack propagation in a scaled-down model of concrete gravity dam (Chahrour and Ohtsu 1994).

In the present paper, fracture mechanisms of pull-out process is clarified, applying the AE moment tensor inversion to a pull-out test and predicting crack trajectories by the BEM analysis.

2 PULL-OUT TEST

Crack nucleation under the pull-out process of anchor bolt is experimentally investigated. Crack kinematics in concrete is analyzed by AE moment tensor analysis in a pull-out test.

2.1 Experiment

A pull-out test was performed by using a concrete block of dimension 1 m x 1 m x 0.3 m. Compressive strength of concrete was 54.3 MPa at 28-day moisture cure. The velocity of P wave was 4150 m/s. Poisson's ratio was 0.2. An anchor bolt was made of a washer fastened by a nut and a prestressing steel bar. The dimension of the washer was 44 mm diameter and 3.2 thickness. The prestressing steel bar of 17 mm diameter and the nut of 27 mm thickness were employed. An anchor head was embedded at 5 cm depth. To decrease the friction between the prestressing bar and surrounding concrete, the steel bar was wrapped up with polyethylene sheet. An experimental set-up is shown in Fig. 1. The anchor bolt was pulled out by employing a center-hole jack.

During the pull-out process, AE waveforms were recorded, by using six AE sensors attached to the concrete block. Sensors are of resonance type with 150 kHz resonance. Total amplification is 60 dB gain and the frequency range is 30 kHz to 80 kHz. AE waveforms detected at sensors were digitized at 1 MHz sampling frequency and stored.

2.2 SiGMA analysis

The procedure for the moment tensor analysis is already developed as a SiGMA [simplified Green's functions for moment tensor analysis] (Ohtsu, 1991). In the procedure, each waveform recorded is displayed on the CRT screen as shown in Fig. 2. Two parameters of the arrival time (P1) and the amplitude of the first motion (P2) are read. Then, crack location in the three-dimensional coordinates system could be determined by solving equations,

$$R_i - R_{i+1} = v^P \cdot (P1_i - P1_{i+1}), \tag{1}$$

where v^P is the velocity of P wave and distance R_i is defined as,

$$R_i = \sqrt{(x - a_i)^2 + (y - b_i)^2 + (z - c_i)^2}.$$

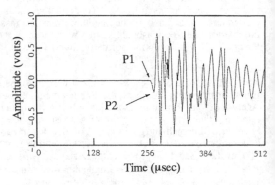

Fig. 2 AE waveform recorded
Aufgenommene AE-Wellenform

It represents the distance from AE source (crack) to the i-th sensor location. The coordinates (x, y, z) denote the unknown location of the crack and (a_i, b_i, c_i) is the coordinates of the i-th sensor. In usual case, eq. 1 is converted into linear equations to be solved. Because a set of three linear algebraic equations is necessary to determine three unknowns: x, y, z, AE system of five-sensor channels is required to solve eq. 1.

Moment tensor components are determined from the simplified relation,

$$P2_i = C \cdot Ref_i \cdot r_p r_q m_{pq}/R_i, \tag{2}$$

where C is the common coefficient and Ref_i is the reflection coefficient of elastic (AE) wave incident to the i-th sensor from AE source. r_p is the direction vector of R_i. Since the moment tensor is symmetric and consists of six independent components, six observation points (six-sensor channels) are at least necessary for the analysis.

For the classification of AE source into a tensile

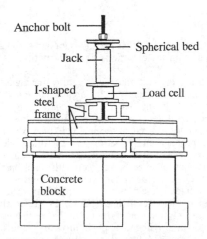

Fig. 1 Experimental set-up of pull-out test
Versuchsaufbau für Ausziehversuch

crack and a shear crack, the eigenvalue analysis of the moment tensor is carried out. By the decomposition of the eigenvalues into the dominant source (Ohtsu 1991), crack type is determined. Then, orientation of crack is determined from the eigenvectors.

3 NUMERICAL SIMULATION OF CRACK PROPAGATION

Theoretical crack trajectories could be computed by the two-domain BEM analysis. Crack propagation in concrete under the pull-out process of hooked anchor is predicted in the two-dimensional problem, which was offered by the RILEM Technical Committee.

3. 1 BEM analysis

The boundary integral equation on traction $t_j(y)$ and displacement $u_j(y)$, both which are defined on boundary S, is given in the following formulation,

$$Cu_i(x) = \int s[U_{ij}(x,y)t_j(y) - T_{ij}(x,y)u_j(y)]dS. \quad (3)$$

Here C is the configuration coefficient and easily calculated by eliminating rigid body motion in eq. 3. In the case of smooth boundary, $C = 1/2$. U_{ij} and T_{ij} are fundamental singular solutions in the two-dimensional BEM (Chahrour and Ohtsu, 1992).
To discretize the integral equation, the boundary S is divided into elements. Tractions are assumed as constant in the element, and displacement are linearly interpolated. Thus, the boundary integral equation is solved numerically.

To analyze crack propagation in arbitrary direction, the two-domain BEM is adopted. As shown in Fig. 3, the whole domain consists of two domains stitched at the common boundary starting from the crack tip. When a crack propagates, the node at the crack tip is separated into two nodes, creating new crack boundary in the direction θ. Based on the criterion in LEFM, the direction of the maximum tangential stress σ_θ is determined from the relation,

$$K_I \sin\theta + K_{II}(3\cos\theta - 1) = 0, \quad (4)$$

where K_I and K_{II} are stress intensity factors of mode I and mode II, respectively. These are determined from displacements on the elements located at the crack tip.

Stress σ_θ when the new crack boundary is created is determined from the relation,

$$\cos\theta/2 \, [K_I \cos^2\theta/2 - 3/2 \, K_{II} \sin\theta] > K_{IC}, \quad (5)$$

where K_{IC} is the critical stress intensity factor of concrete. Then, a new stitching boundary is defined from the new crack tip. The procedure mentioned above is implemented and thus crack propagation is automatically traced in arbitrary directions.

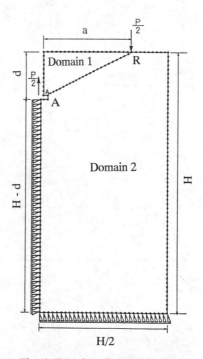

Fig. 4 Two-domain BEM model
Zweibereich-BEM-Modell

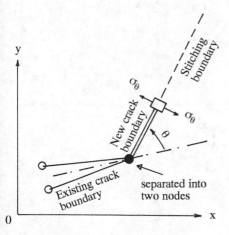

Fig. 3 Crack tip and stitched boundary
Rißspitze und gekoppelte Ränder

3. 2 Analytical models

A configuration of the model analyzed is shown in Fig. 4. Two domains are connected along the stitching boundary. A crack is assumed to start propagating from the upper corner edge of the bolt (point A). The stitching boundary is terminated at the reaction support (point R), and originally linked with the point A by a straight line. Two cases are analyzed. These model sizes and mechanical properties of concrete are given in Tables 1 and 2.

The case 1 is the large model without lateral constraint, while the case 2 is the small model with lateral constraint. In the both cases, the ratios a/d of the distance, a, between the anchor and the reacting support to the depth, d, of the anchor head are varied as 0.5, 1.0, and 2.0. The critical stress intensity factor K_{IC} is determined from the relation:

$K_{IC} = \sqrt{EG_f}$ for the plane stress. These are 1.73 MNm$^{-1.5}$ for the case 1 and 1.71 MNm$^{-1.5}$ for the case 2, respectively.

Table 1. Size of models.

Case	d (mm)	a (mm)	b (mm)	H (mm)
1	150	d/2, d, 2d*	100	900
2**	60	d/2, d, 2d*	80	350

* Three models on the ratios: a/d
** The case with lateral constraint

Table 2. Mechanical properties of concrete

Case	Compressive strength (MPa)	Young's modulus (GPa)	Fracture energy (N/m)
1	30.0	30.0	100
2	34.3	29.4	100

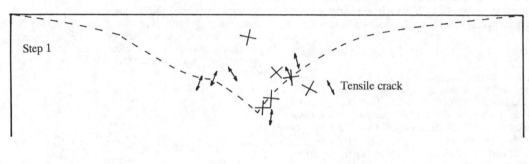

Step 1

Tensile crack

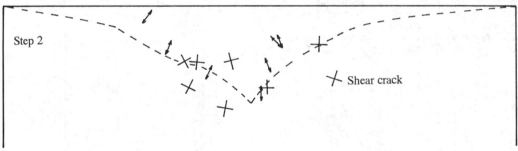

Step 2

Shear crack

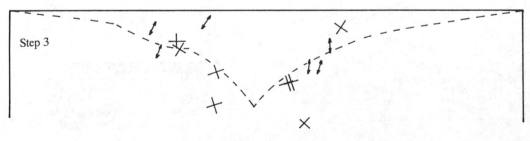

Step 3

Fig. 5 AE sources with crack type and orientation
AE-Quellen mit Rißtyp und -orientierung

These correspond to models in response to the Round Robin Analysis and Tests of Anchor Bolt by RILEM Technical Committee for Fracture Mechanics. Only the right half portion is considered. For the case 1, the boundary element 1.5 cm long is adopted on the external surface, and the crack increment on the stitching boundary is 1 cm. Taking into account the size of the model, the boundary mesh 1 cm long and the crack element 0.5 cm long are employed for the case 2. An initial notch with double nodes is introduced along the straight stitching boundary.

4 RESULTS AND DISCUSSION

Crack kinematics in the pull-out process of anchor bolt is clarified, by applying AE measurement experimentally and by predicting crack propagation in the two-domain BEM analytically. The results are discussed.

4. 1 Crack kinematics by AE

From AE activity, the nucleation of circumferential cracking near the corner edge of anchor head was observed, prior to reaching the peak load. AE sources detected at this stage are classified into three steps and are plotted in Fig. 5. Elevation views are given. The final failure surface is indicated by a broken curve. AE sources are marked at their locations with crack type and crack orientation. A tensile crack is denoted by arrow symbol, and the direction of crack opening is coincident with that of the arrow. A shear crack is indicated by cross symbol. Because either crack normal or crack vector is determined by the SiGMA, two possible vertical directions are given by the cross.

Up to the peak load, AE sources are located closely along the final failure surface. In all steps, both tensile cracks and shear cracks are simultaneously observed in AE cluster. Tensile cracks are dominantly observed at the boundary of the cluster. The directions of crack opening are remarkably vertical to the final failure surface. Many of shear cracks are intensely found inside the cluster. The directions of shear motion are almost parallel to the final failure surface. These results suggest the fracture mechanism that tensile cracks are first generated along the final failure surface and then shear cracks follow. It implies that the governing mechanism at the crack tip is tensile failure in the pull-out process.

4. 2 Crack traces by the analysis

Simulated crack trajectories are shown in Fig. 6 for the models of a = 2d. In all the case except for the case 1 of a = 2d, the crack propagates, directing the reaction support from the edge of anchor head. A typical case is shown in Fig. 6 (b). An inflexion

point is observed, although this is not the case of other models. It implies that cracks attempted to propagate horizontally once, but they directed the reaction support again. As seen in Fig. 6 (a), in contrast, without lateral constraint cracks propagate all the way toward the side of the model, not directing the reaction support. In the case of large ratio a/d, the inflexion point, where cracks tend to propagate horizontally, is observed. Less lateral constraint expedites the propagation in the horizontal direction.

The nucleation of the inflexion point, is studied from the contribution of stress intensity factors. A relation between the normalized stress intensity factors and crack extension is shown in Fig. 7 along with a load versus anchor head displacement curve. This is the result of the case 2; a = 2d shown in Fig. 6 (b). The load-displacement curve simulated shows some abrupt increase or decrease of the loads around the peak load, but it successfully reproduces essential feature of descending branch after the peak load. In the bottom, the variation of the normalized stress intensity factors $K_I{}^* = K_I/K_{IC}$ and $K_{II}{}^* = K_{II}/K_{IC}$ at the crack tip are given. All the way of crack propagation in the pull-out process, mode I factor is almost unity, which means that tensile failure is dominant mechanisms. There exists some stages at which relatively high values of $K_{II}{}^*$ are obtained. These are in good agreement with the stage of the inflexion point. This suggests the change of fracture mechanisms in the pull-out

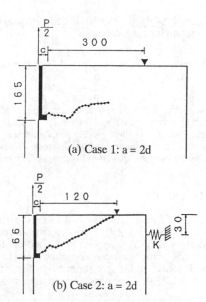

Fig. 6 Crack trajectories by BEM: (a) the case without lateral constraint, and (b) that with lateral constraint

Rißfortpflanzung aus BEM-Analyse (a) ohne und (b) mit seitlicher Stützung

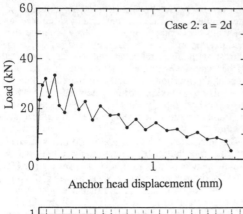

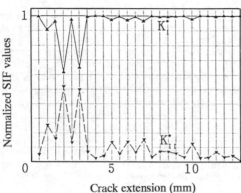

Fig. 7 Load-displacement relation and normalized SIF values under crack extension

Last-Verschiebungs-Diagramm und normalisierte SIF-Werte bei der Riß-Fortpflanzung

process at the inflexion point where pure mode I changes to mixed mode or mode II. After the inflexion point, tensile cracks propagate in the direction of the maximum tangential stress. Without lateral constraint, the direction becomes horizontal.

5 CONCLUSION

The pull-out process of anchor bolt from concrete is studied. Experimentally, AE source inversion procedure by SiGMA is applied to the pull-out test. Mixed-mode crack propagation in concrete is analyzed, based on the maximum tangential stress criterion of LEFM. The results are summarized, as follow:

(1) AE sources are located as close as the final failure surface. Both tensile cracks and shear cracks are simultaneously observed in AE cluster. Tensile cracks are dominantly observed at the boundary of the cluster. Many of shear cracks are intensely found inside the cluster.

(2) These suggest the fracture mechanism that tensile cracks are first generated along the final failure surface and then shear cracks follow.

(3) In the analysis of crack trajectories, the crack propagates, directing the reaction support from the edge of anchor head. In the case of large ratio a/d, it is found that the inflexion point, where cracks tend to propagate horizontally, is generated. Less lateral constraint expedites the propagation in the horizontal direction, not directing the reaction support.

(4) At the inflexion point, pure mode I changes to mixed mode or mode II. After the inflexion point, tensile cracks propagate in the direction of the maximum tangential stress. Without lateral constraint, the direction becomes horizontal.

REFERENCES

Berthaud, Y. et al. 1991. Experimental measurements of localization for tensile tests on concrete. *Fracture Processes in Concrete, Rock and Ceramiccs* : 41-50. London : E & FN Spon.

Chen,, L. et al. 1992. Determination of fracture parameters of mortar and concrete beams by using acoustic emission. *Materials Evaluation.* 50 (7): 888-894.

Mindess, S. 1991. Fracture process zone detection. *Fracture Mechanics Test Methods for Concrete.*: 231-261.London: Chapman and Hall.

Nomura, N. et al. 1991. Properties of fracture process zone and tension softening behavior of concrete. *Fracture Processes in Concrete, Rock, and Ceramics*: 51-60. London: E & FN Spon.

Ohtsu, M. 1982. Source mechanism and waveform analysis of acoustic emission in concrete. *J. Acoustic Emission.* 2(1): 103-112.

Ohtsu, M. 1987. Determination of crack orientation by acoustic emission. *Materials Evaluation.* 45(9):1070-1075.

Ohtsu, M. 1991. Simplified moment tensor analysis and unified decomposition of acoustic emission source. *J. Goephy. Res.* 96(B4): 6211-6221.

Chahrour, A. H. & Ohtsu, M. 1992. Multi-domain BEM implementation for mixed-mode cracking in concrete. *Fracture and Damage of Concrete and Rock*: 196-205. London: E & FN Spon.

Chahrour, A. H. & Ohtsu, M. 1994. Crack growth prediction in scaled down model of concrete gravity dam. *Theoretical and Applied Fracture Mechanics.* 21: 29-40.

Analytical model for the behaviour of bolted rock joints and practical applications

Analytisches Modell für das Verhalten ankerverstärkter Felstrennflächen und praktische Anwendungen

F. Pellet
Bonnard & Gardel, Consulting Engineers Ltd, Lausanne, Switzerland

P. Egger
Swiss Federal Institute of Technology, Lausanne, Switzerland

ABSTRACT : This study deals with a new approach for the stability analysis of rock engineering structures reinforced by passive rockbolts. This approach is based on an analytical model developed to compute the bolt contribution to the shear strength of rock joints. The main characteristics of this model is the account of the interaction of the axial force and the shear force mobilized in the bolt as well as the large plastic displacements of the bolt occurring during the shearing process. The complete curve of the bolt contribution as a function of the displacement along the joint can be computed and the maximum bolt contribution is obtained by dissociating the bolt cohesion and the confining effects. The effects of the most important parameters such as bolt inclination, mechanical properties of bolt material, rock strength and joint friction angle are clearly established. Finally, it is shown how to use this analytical model for the design of rock engineering structures. For a slope defined by the height, the dip angle of the discontinuities and the mechanical properties of both the discontinuities and the intact rock, it is shown how to judiciously select the parameters of the reinforcement system in order to obtain the required safety factor. By the same way, for a tunnel at shallow depth, it is possible to compute the reinforcement system depending on the allowed ground level displacements as well as the mechanical characteristics of the lining support.

ZUSAMMENFASSUNG : Der vorliegende Beitrag stellt eine neue Annäherung für die Stabilitätsberechnung von Bauwerken in mit passiven Ankern verstärktem Fels. Anhand eines analytischen Modells wird der Beitrag eines Ankers zur Scherfestigkeit einer Felstrennfläche in Abhängigkeit der Verschiebung auf der Trennfläche berechnet. Das Modell berücksichtigt die Welchselwirkung zwischen der mobilisierten Normalkraft und Scherkraft im Anker sowie dessen grosse plastische Verformung. Zur Berechnung des maximalen Ankerbeitrags werden effekte Bewehrungskohäsion und Zusammenhalt getrennt behandelt. Auschliessend wird anhand von Beispielen gezeigt, wie die Methode bei Berechnungen von Bauwerken angewandt wird. Für eine Felsböschung, die durch ihre Höhe, die Neigung der Schichten und durch die felsmechanischen Kennziffern definiert ist, können die Kenngrössen für die Ankerverstärkung ermittelt werden, um einem bestimmten Sicherheitfaktor zu erhalten. Ebenso ist es für einen Tunnel mit geringer überdeckung möglich, die Ankerverstärkung in Abhängigkeit der zugelassenen Setzungen an der Terrainoberfläche und der Tragfähigkeit des Ausbau zu bestimmen.

1 INTRODUCTION

The reinforcement of rock masses by passive rock bolts is nowadays widely used to ensure the stability and to restrain the deformation of rock engineering structures. The mechanical action of the bolt is, however, still difficult to determine, especially for bolt installed in fractured rock masses. As a consequence, there is no unanimously approved method for the dimensioning of the reinforcement system.

The present study proposed a new analytical model for the mechanical behaviour of bolted rock joints subjected to shearing. It allows to compute both the strength and the deformability properties of a bolted rock joint, when the failure mechanism involves sliding displacements along the rock joints. Only untensioned fully grouted rock bolts are investigated.

2 OUTLINE OF PAST STUDIES

A lot of experimental programs were performed in order to describe in a qualitative manner the behaviour of bolted rock joint. Dight (1983), Schubert (1984), Spang (1988), Egger and Zabuski (1991), Ferrero (1993), Pellet (1994) carried out experimental studies on different types of rock material reinforced by various elements.

Several analytical expressions have been developed to predict the behaviour of a bolted rock joint. The simplest of them considers only the axial force acting in the bolt (Bjurström 1974). By projection of this force on the joint plane, it is then possible to compute the reinforcement effect of the bolt. This approach does not take into account the shear force mobilised in the bolt nor the deformation of the bolt (i.e., bolt rotation) across the joint.

More sophisticated expressions were developed to take into account the deformation of the bolt during the loading process. Most of them were based on the first order theory of the beams on elastic supports formulated by the small displacements theory. As a consequence, these formulations are valid only for the early stages of the loading process. Dight (1983) proposed an expression to predict the maximum force mobilised in the bolt as well as the associated displacement on the joint. The failure of the bolt is determined by the combination of axial and shear forces, and the displacement is computed taking into account the yields of the grout. Based on Dight's work, Holmberg (1991) proposed a method which gives a good prediction of the maximum bolt contribution when the bolt is inclined to the joint. Furthermore, Spang and Egger (1990) proposed empirical expressions to compute the maximum bolt contribution and the associated joint displacement.

On the same time, some authors (Aydan 1989, Swoboda and Marence 1992, Egger and Pellet 1992) proposed numerical approaches.

3 ANALYTICAL FORMULATION FOR THE BEHAVIOUR OF A BOLTED ROCK JOINT

The analytical description of the behaviour of a bolted rock joint must take into account the multiple interaction phenomena occurring between the bolt the grout and the surrounding medium. The objective is to establish the relation between the force, R_o, acting in the bolt at the joint level and the corresponding displacement, U_o, on the joint (Figure 1). Ultimately, the prediction of both the maximum bolt contribution to the joint shear strength as well as the maximum displacement are determined. A complete description of this model can be found in Pellet (1994).

3.1 Mechanical system and basic hypotheses

As shown in Figure 1, when a bolted rock joint is subjected to a shear displacement, the curvature of the deformed shape of the bolt across the joint is zero. On a mechanical point of view, that means that the bending moment is zero at this point. Therefore, only axial and shear forces act in the bolt.

For purpose of equilibrium analysis, the bolt can be considered as a semi-infinite beam loaded at one end by both an axial force N_o, and a shear force Q_o, which are dependent one to the other. As the loading increases, the surrounding medium (i.e., rock or grout) supplies a reaction on a bolt length which increases until the bolt yields. At this point, one of the most important assumptions concerned the behaviour of the surrounding medium which is considered as a rigid perfectly plastic material.

3.2 Yield limit and failure criterion of the bolt

Considering that the hypotheses of the beam theory are valid, the relation defining the yield limit of the bolt (i.e., upper fibber of the bolt section reach yield stress of the bolt material) is established :

$$Q_{oe} = \frac{1}{2} \sqrt{p_u \, D_b \left(\frac{\pi \, D_b^2 \, \sigma_{el}}{4} - N_{oe} \right)} \quad (1)$$

where,

Q_{oe} : shear force acting at point O at the yield stress of the bolt material

N_{oe} : axial force acting at point O at the yield stress of the bolt material

D_b : bolt diameter

σ_{el} : yield stress of the bolt material

p_u : maximum bearing pressure per unit of length

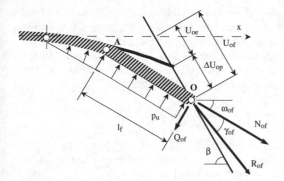

Figure 1 : Displacements and forces acting in the bolt in elastic and plastic conditions - *Verschiebungen und Kräfte im Anker im elastischen und im plastischen Zustand*

For the failure of the bolt material, the Tresca criterion is used. The maximum force acting in the bolt at point O is expressed as a combination of the shear force and of the axial force. This interaction formula, first used by Dight (1983), is expressed here as a function of the shear force :

$$Q_{of} = \frac{\pi \; D_b^2}{8} \; \sigma_{ec} \; \sqrt{1 - 16 \left(\frac{N_{of}}{\pi \; D_b^2 \; \sigma_{ec}} \right)^2} \quad (2)$$

where,

Q_{of} : shear force acting at point O at the failure of the bolt material

N_{of} : axial force acting at point O at the failure of the bolt material

σ_{ec} : failure stress of the bolt material

These two relations are represented in Figure 2. The yield limit (eq. 1) is parabolic, whereas the failure criterion is elliptical (eq. 2).

3.3 Behaviour of the bolt in the elastic stage

When a beam is simultaneously loaded by a shear force and an axial force which are dependent one to each other, the differential equation governing the load-displacement process has no simple solution. Therefore, it is proposed to use a variational approach, based on an energetic concept (Pellet 1994).

Calculating the internal strain energy and the work of the external forces, the expression of the total complementary energy is obtained. Then by minimizing the total complementary energy with respect to the displacements, u_0 and v_0, the relations between forces are expressed by :

$$Q_o = \sqrt[3]{\frac{3 \; N_o \; p_u^2 \; \pi^3 \; tg\beta \; D_b^2}{1024 \; b}} \quad (3)$$

where,

β : angle between the bolt and the joint

b : constant term $= 0.27$

This relation is drawn in Figure 2. The determination of the forces at the yield limit leads to the following third order equation which is the combination of equations 1 and 3 :

$$Q_{oe}^3 + Q_{oe}^2 \left(\frac{3 \; p_u \; \pi^3 \; D_b \; tg\beta}{256 \; b} \right)$$
$$- \left(\frac{3 \; p_u^2 \; \pi^4 \; D_b^4 \; tg\beta \; \sigma_{el}}{4096 \; b} \right) = 0 \quad (4)$$

The solution of this equation provides the shear force and the axial force in the bolt. Knowing the forces acting in the bolt, it is then possible to compute the displacement and the rotation of the bolt.

$$\omega_{oe} = - \frac{2048 \; Q_{oe}^3 \; b}{E \; p_u^2 \; \pi^3 \; D_b^4} \quad (5)$$

$$U_{oe} = \frac{8192 \; Q_{oe}^4 \; b}{E \; \pi^4 \; D_b^4 \; p_u^3 \; \sin\beta} \quad (6)$$

3.4 Behaviour of the bolt in the plastic stage

Beyond the elastic limit, it is assumed that only the axial force in the bolt increases. As a consequence, the shear force remains constant. The determination of the forces acting in the bolt at failure is simply achieved by solving the following equation which is drawn in Figure 2 :

$$N_{of} = \frac{\pi \; D_b^2}{4} \; \sigma_{ec} \; \sqrt{1 - 64 \left(\frac{Q_{oe}}{\pi \; D_b^2 \; \sigma_{ec}} \right)^2} \quad (7)$$

The computation of the displacement and the rotation at the end of the bolt is achieved by the use

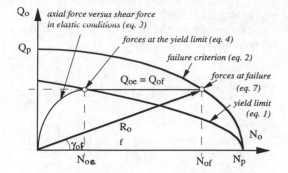

Figure 2 : Shear force versus axial force in the bolt for elastic and plastic conditions - *Scherkraft und Normalkraft im Anker im elastischen und im plastischen Zustand*

of the large displacement formulation. It is shown by simple geometrical consideration that the following expression gives a good approximation of the increment of plastic rotation :

$$\Delta\omega_{op} = \arccos\left[\frac{l_e}{l_f}\sin^2\beta \pm \sqrt{\cos^2\beta\left(1-\left(\frac{l_e}{l_f}\right)^2\sin^2\beta\right)}\right]\ (8)$$

where,

l_e : distance between bolt extremity (point O) and the location of the maximum bending moment (point A) when the yield limit is reached

l_f : length of the part O-A at failure

The solution of equation 8 gives the increment of plastic rotation and then allows for the calculation of the plastic displacement :

$$\Delta U_{op} = \frac{Q_{oe}\ \sin\ \Delta\omega_{op}}{p_u\ \sin\left(\beta\ -\ \Delta\omega_{op}\right)} \tag{9}$$

The total rotation, ω_{of}, and the total displacement, U_{of}, are the sum of the elastic and the plastic components. Additionally, it is assumed that between the elastic limit and the failure point, the axial stiffness of the bolt decreases as an inverse function of the axial force.

3.5 Strength and associated displacement of the bolted joint

At each step of the loading process, both the axial and shear forces acting at the end of the bolt as well as the rotation and the displacement are known. Thus by combining forces and using the Mohr-Coulomb criterion for joint strength, it is possible to compute the bolt contribution, T_b, to the total reinforced rock joint strength :

$$T_b = R_{ot} + R_{on}\ \text{tg}\ \phi_j \tag{10}$$

The additional cohesion, Δc_b, related to the tangential force and the confining effect, $\Delta\sigma_{nb}$, are separately computed as follows :

$$\Delta c_b = \frac{R_{ot}}{A_j} \tag{11}$$

$$\Delta\sigma_{nb} = \frac{R_{on}}{A_j} \tag{12}$$

where, A_j : surface of the joint
R_{on} : force normal to the joint
R_{ot} : force tangential to the joint

3.6 Experimental validations

The reliability of this analytical model was demonstrated by comparison with the results of a large number of tests as well as to the performances of other analytical predictions (see Pellet 1994 and Pellet et al. 1995).

4 APPLICATIONS FOR ROCK ENGINEERING

The analytical model presented above may be applied to the design of rock slopes or underground structures reinforced by fully grouted steel bolts.

4.1 Stability of a reinforced rock slope

Considering an excavated slope with a vertical wall of 15 meters height (Figure 3). The layers are 1.5 meter thick and the dip angle of the layers is 60 degrees. The geometrical and the mechanical characteristics assumed in the calculations are as follows :

Rock: - volumetric weight γ_r = 27 kN/m3
     - compressive strength σ_c = 50 MPa

- joint friction angle $\quad\quad \phi_j = 30°$

Bolts: - failure limit $\quad\quad\quad \sigma_{ec} = 600$ MPa
       - elasticity modulus $\quad\quad$ E=210000 MPa
       - strain at failure $\quad\quad\quad \epsilon_f = 20\%$

Based on the equilibrium equation with respect to sliding, the additional resistance, T, which must be provided by the reinforcement to ensure the slope stability (i.e. safety factor equal to 1), can be simply computed :

$$T = W \left(\sin\theta - \cos\theta \ tg\phi_j \right)$$

W is the weight of the unstable volume of rock.

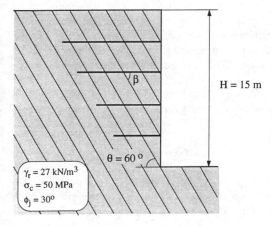

$\gamma_r = 27$ kN/m^3
$\sigma_c = 50$ MPa
$\phi_j = 30°$

Figure 3 : Geometry of slope - *Böschunggeometrie*

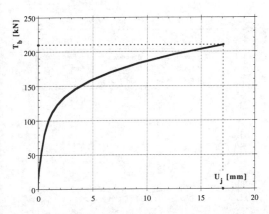

Figure 4 : Contribution of a bolt of 20 mm in diameter, placed at $\beta = 60$ degrees, as a function of joint displacement - *Beitrag eines Anker von 20 mm Durchmesser, in einem Winkel $\beta = 60$, in Abhängigkeit der Verschiebung auf der Trennfläche*

For each of the discontinuities, the following values are obtained:

- joint 1:   H = 15 m   T = 1012 kN/m
- joint 2:   H = 12 m   T = 648 kN/m
- joint 3:   H = 9 m   T = 365 kN/m
- joint 4:   H = 6 m   T = 162 kN/m
- joint 5:   H = 3 m   T = 41 kN/m

For example, for a chosen bolt of 20 mm diameter, placed horizontally ($\beta = 60°$), the bolt contribution versus the joint displacement, is drawn in Figure 4. The detailed calculation of the forces mobilised in the bolt and the associated displacements along the joint is presented in Pellet (1994). The bolt contribution, T_b, and the associated joint displacements, U_j, are equal to :

- at yield limit $T_{be} = 80$ kN $U_{je} = 0.6$ mm

- at failure $T_{bf} = 210$ kN $U_{jf} = 17$ mm

As the bolt contribution to the shear strength of the joint is known, the bolting pattern can be determined. The number of bolts, n_b, installed in one profile as well as their spacing, e, are calculated for a square pattern :

$$n_b = \sqrt{\frac{T\ H}{T_b}} \quad\quad e = \frac{n_b\ T_b}{T} = \frac{H}{n_b}$$

For the present case, one obtains :

- at yield limit : $n_b = 14$ bolts and e = 1.10 m

- at failure : $n_b = 9$ bolts and e = 1.87 m

In case of bolts working at the elastic limit (i.e. e=1.1 m and $n_b = 14$) the safety factor of the bolts, F_b, is equal to :

$$F_b = \frac{T_{bf}}{T_{be}} = \frac{210}{80} = 2.64$$

The global safety factor of the slope against sliding, F, is found to be equal to 2.09.

As the force which loads each individual bolt is known, the displacement of each joint can be determined by utilizing the curve of the bolt contribution (Figure 4). The total vertical displacement between the crest of the slope and the stable part of the mass is obtained by adding the displacements occurring along each individual discontinuity. For the case of the bolts working until failure, we obtain :

$$s = \sum_{n=1}^{n_j} \left(U_j \sin\theta \right) = \mathbf{24.5\ mm}$$

The last point consists in the verification of the bonded length required to prevent pulling-out of the bolts. It must be achieved separately.

4.2   Stability of a tunnel at shallow depth

Let us consider a tunnel of 10 meters of diameter at a depth of 20 meters (Figure 5). The tunnel is excavated in weak rock and a bolting system is required to ensure the roof stability. It is assumed that the volume of unstable rock is delimited by two vertical planes.

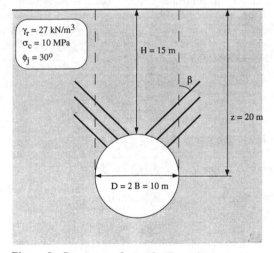

Figure 5 : Geometry of tunnel - *Tunnelgeometrie*

Considering that the residual cohesion is zero and according to the Terzaghi's theory, the supporting pressure may be computed by the following formula.

$$P_v = \frac{\gamma\, B}{ko\, tg\phi}\left[1 - e^{-ko\, tg\phi\, \frac{H}{B}} \right]$$

The computation of the bolt contribution is achieved in the same way that those presented in the

preceding section. For example, for bolts of 30 mm in diameter placed at 45 degrees, we obtain the following values for the bolt contribution and the associated displacement :

yield limit : T_{be} = 296 kN, U_{je} = 2 mm, (10 bolts)

failure : T_{bf} = 487 kN, U_{jf} = 30 mm, (6 bolts)

If the bolting system is associated with a lining support, it is possible to determine the vertical pressure acting on the lining support depending on the displacement. Figure 6 shows the evolution of the vertical pressure as a function of the vertical displacement for elastic and plastic conditions.

The initial supporting pressure decreases as the displacement increases and the contribution of bolts is mobilised. The equilibrium point indicated in the figure 6 corresponds to the yield limit of bolts for the case of 6 bolts working until failure.

As it was already mentioned, the bonded length of bolt must be check separately

4.3   Simplified approach

For purposes of preliminary computations the following figures may be used to determine the bolt contribution as well as the joint displacement.

Figure 7 shows the maximal bolt contribution versus

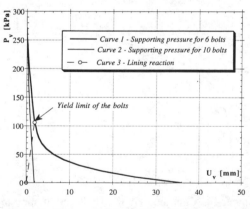

Figure 6 : Pressure acting on the lining support as a function of vertical displacements - *Auf den Ausbau einwirkender Druck in Abhängigkeit der vertikalen Verschiebungen*

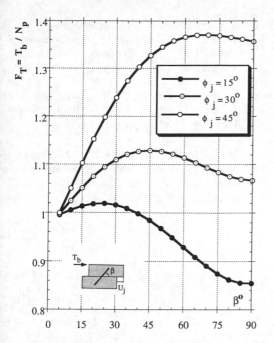

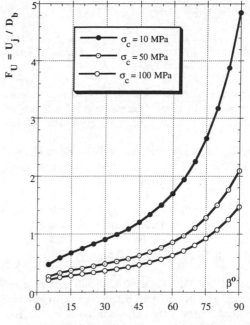

Figure 7. Maximum bolt contribution versus bolt orientation for different values of joint friction angles - *Maximaler Beitrag einer Anker in Abhängigkeit von ihrer Orientierung für verschiedene Reibungswinkel der Trennfläche*

Figure 8. Maximum joint displacement versus bolt orientation for different values of rock strength - *Maximaler Verschiebung auf der Trennfläche in Abhängigkeit von der Orientierung der Anker für verschiedene Werte der einachsigen Druckfestigkeit des Gesteins*

the bolt orientation for different values of the joint friction angle. Bolt contribution is normalised with respect to the tensile strength of the bolt, N_p. It is possible, for a given bolt orientation and a selected type of bolt, to define the bolt contribution and then the bolting pattern.

Figure 8 shows the maximal joint displacement versus the bolt orientation for different values of the strength of the host rock. Joint displacement is normalised with respect to the bolt diameter, D_b. With this Figure, it is possible to determine the joint displacement and the maximum settlement of the crest of the slope.

5   CONCLUSION

The analytical model developed for this work allows the prediction of the complete load-displacement curve for a bolted rock joint subjected to shearing. The maximum bolt contribution to the shear strength

of the bolted rock joint, as well as the maximum displacement on the joint can be obtained.

The additional cohesion as well as the confining effect can be computed separately and the roles of the most important geometrical and mechanical parameters are highlighted.

By comparison of the analytical predictions with the tests results, it is shown that the proposed model improves the prediction of the behaviour of a bolted rock joint, especially when the bolt is nearly normal to the joint.

Based on this analytical development, it is possible to perform the design of the reinforcement system for practical application of rock structures such as rock slopes or underground cavities.

REFERENCES

Aydan Ö., The stabilisation of rock engineering structures by rockbolts, Ph.D. Thesis, Nagoya University, Japan, 1989

Bjurström S., Shear strength of hard rock jointed reinforced by grouted untensioned bolts, Proc. 3rd ISRM Cong., Denver, USA, pp 1194-1199, 1974

Dight P.M., Improvements to the stability of rock walls in open pit mines, Ph.D Thesis, Monash University, Australia, 1983

Egger P., Pellet F., Strength and deformation properties of reinforced jointed media under true triaxial conditions, Proc. 7th ISRM Cong., Aachen, Germany, pp 215-220, 1991

Egger P., Pellet F., Numerical and experimental investigations of the behaviour of reinforced jointed media, Proc. Int. Conf. on Fractured and Jointed Rock Masses, Lake Tahoe, USA, 1992

Egger P., Zabuski L., Behaviour of rough bolted joints in direct shear tests, Proc. 7th ISRM Cong., Aachen, Germany, pp 1285-1288, 1991

Ferrero A.M., Resistenza al taglio di discontinuita' rinforzate, Doctorate Thesis, Technical University of Torino, Italy, 1993

Haas C.J., Analysis of rock bolting to prevent shear movement in fractured ground, Mining Engineering, Vol 33, no 6, pp 698-704, 1981

Holmberg M., The mechanical behaviour of untensionned grouted rock bolts, Ph. D. Thesis, Royal Institute of Technology, Stockholm, Sweden, 1991

Pellet F., Strength and deformability of jointed rock masses reinforced by rockbolts, English Translation of the Doctorate Thesis 1169, Swiss Federal Institute of Technology, Lausanne, Switzerland, 1994

Pellet F., Egger P., Ferrero A.M., Contribution of fully bonded rock bolts to the shear strength of joints : Analytical and experimental evaluation. Proc. 2nd Int. Conf. on Mechanics of Jointed and Faulted Rock, MFJR-2, Wien, Austria, pp 857-862, 1995

Schubert P., Das Tragvermögen des Mörtelversetzten Ankers unter Aufgezwungener Kluftverschiebung, Doctorate Thesis, Montanuniversität, Loeben, Austria, 1984

Spang K., Egger P., Action of fully-grouted bolts in jointed rock and factors of influence, Rock Mech. and Rock Eng., Vol 23, pp 201-229, 1990

Swoboda G., Mareñce M., Numerical modelling of rock bolts in intersection with fault system, Proc. Numerical models in Geomechanics, NUMOG IV, Swansea, U.K., pp 729-738, 1992

Anchors in Theory and Practice, Widmann (ed.) © 1995 Balkema, Rotterdam. ISBN 90 5410 577 1

Shear strength evaluation of reinforced rock discontinuities

Über die Bestimmung des Scherwiderstandes von mit Ankern verfestigte Felsklüften

C. Deangeli, A. M. Ferrero & S. Pelizza
Politecnico di Torino, Dipartimento di Georisorse e Territorio – TUSC Tunnelling and Underground Space Centre, Italy

Abstract: The evaluation of the shear strength of a rock discontinuity reinforced by means of passive bolts (dowels) is needed in order to determine the stability conditions of a reinforced rock mass.
Prestressed steel elements can be easily taken into account in the classical limit equilibrium analysis, whilst the fully bonded dowels present more difficulties. The dowel contribution is, in fact, strictly connected to the rock mass deformation as the different stiffness of the steel and rock material induces stress state within the dowel. The stress state is more complicated when the steel element crosses a discontinuity as both the deformability and strength features of the rock and of the steel influences the dowel deformation. Several experimental and theoretical works have been developed in the field but none of them seem to give a fully satisfactory answer. For this reason, a specific laboratory device, able to measure the shear resistance of a rock discontinuity, reinforced by means of a dowel, has been set up. This device has enabled the performance of several laboratory tests, on full scale specimens, to evaluate the influence of some important parameters: steel type, bar diameter, reinforcement type (bars and pipes) and rock type. Test results are reported in the paper together with a critical comparison with results from technical literature and from an analytical model.

Zusammenfassung: Die Bestimmung der Stabilitätsbedingungen des Gebirges setzt die Schätzung der Scherfestigkeit von druch Ankern verstärkten Kluften voraus. Gespannte, aktive Verstärkungselemente können ohne weiteres durch die Anwendung des Grenzzustandverfahrens im Gleichgewicht berücksichtigt werden. Voll zementierte Anker erschweren eine vernünftige Rechenmodellierung. Der Beitrag der passiven Anker hängt von der Verformung des Gebirges ab, da die Beanspruchung der einbetonierten Stäbe auf den Steifigkeitsunterschied zwischen Fels und Stahl zurückzuführen sein sollte. Die gesamte Spannungsverteilung wird deswegen von einer dazwischenliegenden Felskluft und vom durchgehenden Stahlanker, mit Bezug auf die Verformbarkeit und auf die Festigkeit von beiden, Stahl und Fels, stark beeinflußt. Zahlreiche Versuche und Berechnungen sind auf diesem Gebiet aber mit keiner entsprechenden, einwandfreien Lösung durchgeführt worden. Aus diesem Grund wurde die Laboreinrichtung hergestellt, um den Scherwiderstand von mit Ankern verfestigten Felsklüften messen zu können. Dieses Gerät hat die Verwirklichung zahlreicher Laborversuche mit im Maßstab 1:1 angewendeten Ankern ermöglicht, um die Beeinflussung der Stahleigenschaften, der Stabdurchmesser oder der Rohrquerschnitte und schließlich des Gesteines selbst bestimmen zu können. Die Versuchsergebnisse werden im Bericht beigefügt um sie mit Literaturdaten zu vergleichen und die entsprechenden, analytischen Modelle sind ebenfalls kurz beschrieben.

1 INTRODUCTION

Rock bolts can be classified on the basis of their different anchoring techniques: bolts connected to the rock by a punctual anchor and bolts connected to the rock all along their length. In the first case the bar can be anchored by grouting (mortar or synthetic resin) or by means of mechanical device. The second type of anchor can be realised by mortar or synthetic resin or by friction (Swellex, Split Set).

Rock bolts can be installed untensioned (passive reinforcements) or tensioned (active reinforcements).

An active bolt introduces a force in the rock mass that is defined both in magnitude and orientation; therefore it will improve the rock discontinuity strength by increasing the normal stress acting on the plane. Effective active reinforcement needs a good knowledge of the discontinuity orientation and good quality rock. Poor rock mass is heavily deformed under the bolt tension by causing a loss of stress within the bar. The effect of active bolts on the rock stability can be taken into account by using limit equilibrium analysis. In fact the contribution of active reinforcement can be generally represented by introducing, in the safety factor relationship, the pretension force as an external force of known inclination and modulus.

The passive rock bolt is installed with the purpose of aiding the rock mass to self sustain and its contribution to rock stabilisation depends on rock deformation. The steel element is completely connected to the rock mass by forming a composite material (rock-steel).

Research works carried out in this field have shown how the passive bolt increases the rock mass resistance: the rock and steel system resistance is due to the steel deformation induced by the rock mass strain. The overall strength of a reinforced joint results from both contributions provided by the discontinuity and by the reinforcement element.

The strength of the element depends on two following effects: the increment of the axial force due to the bar deformation; and the dowel effect. The dowel effect represents the resistance due to the shear forces acting in the bar. The increase of the axial force acting in the bar, due to the relative displacement of the two joint walls, can be evaluated by taking into account two different phenomena:

1. the increase of the component of axial force acting perpendicular to the joint which increases friction forces on the joint;

2. the increment of the component of axial force acting parallel to the joint which reduces the active forces.

Experimental works carried out by different authors provide useful suggestions to evaluate the global strength of joints reinforced by passive bolts. A numerical procedure, which takes into account the strength and the deformation features of a rock discontinuity and steel element has not yet been proposed in the technical literature.

The difficulties in studying passive elements from a theoretical point of view have brought the necessity to perform specific experimental work. A laboratory device has been realised by the authors and several tests have been carried out on instrumented reinforced joints.

2 LABORATORY TESTING

Shear tests have been carried out in order to quantify some important phenomena affecting the passive bolting-rock discontinuity mechanical behaviour.

These tests have been performed on specimens of different materials, reinforced by different types of fully connected bolts.

The passive element has been placed perpendicular to the artificial joint plane. In all cases the shear plane was smooth and flat.

Specimens were made by concrete, to simulate a weak rock, and by a hard rock (gneiss). Table 1 shows specimen material features.

Samples dimensions were 30*30*80 cm in order to use full scale bolts. The large size of the specimens have required the design and the setting up of a new laboratory device.

The test device has been designed to allow shear displacements up to 15 cm therefore, allowing the failure of the bolt to be obtained.

Bolts installed inside holes of different diameters (19 mm, 25 mm and 40 mm) consisted of steel bars

Table 1 Specimen material features

Specimen material	Uni. compr. strength σ_c (MPa)	Young modulus (MPa)	Basic friction angle ϕ (°)
Rock (gneiss)	119	40,000	37
Concrete	41	33,000	37

Table 2. Tested bar and tube steel features

Bolt type	Yielding load T_Y (kN)	Failure load T_o (kN)	Ultimate strain ε_o (%)	Yielding stress σ_o (MPa)	Failure stress σ_o (MPa)	Diameter (mm)
Bar steel 1	19.0	28.0	35	380	560	8
Bar steel 2	24.0	34.0	24	480	680	8
Bar steel 3	24.0	36.5	26	480	720	8
Bar steel 3	54.2	81.4	26	480	720	12
Tube 1	66.2	102.5	18	228	353	28(i.d.)/34(o.d.)
Tube 2	36.0	56.2	18	383	596	7(i.d.)/13(o.d.)

and steel pipes of different types and diameters. Table 2 shows bar and tube steel features.

Tests have also been carried out on pipe elements in order to study the behaviour of friction bolts such as Swellex and Split Set.

Bolt grouting has been obtained by means of quick setting mortar.

Shear tests have been performed by using two different procedures:

1. Tests on samples reinforced by non prestressed bolts;

2. Tests on samples reinforced by prestressed bolts, in order to simulate real behaviour of bolts sheared by natural joint (dilatancy effect).

Variation of steel type (ductile or brittle) had the aim of evaluating how the steel behaviour at failure (expressed in terms of tensile strength and extensile strain at collapse) affects maximum shear strength and corresponding shear displacements of a reinforced joint.

Variation of the bolt diameter allows to define the relationship between shear resistance-shear displacement and element size.

Different failure stress states in the bar, depending on different stiffness of surrounding material, have been analysed by varying the sample material.

Tests using variation of the hole diameter (maintaining constant bar diameter) have been performed to study the influence of the mortar stiffness. The mortar placed between the rock and the bar has strength properties lower or equal to the block strength (rock and concrete) and heavily influences the overall behaviour.

Tests using variation in bolt type (bars and pipes) have the aim to study the different behaviour between these different reinforcing systems.

2.1 Test laboratory device

Tests have been carried out by means of the device reported in figure 1. This equipment is constituted by a shear box divided in two sides: one free to move and one fixed. The box is placed between the plates of a conventional press and allows the application of a transverse load to the sample and therefore to shear the bolt.

The reinforced joint is sheared by the device by applying incremental loading, with constant loading steps.

The two steel sides of the shear box are separated by a frictionless contact and lateral displacements between the two sides of the samples are not allowed.

The following quantities have been measured during the tests:

- the shear force, by means of a pressure transducer inserted on the hydraulic circuit of the press;
- the shear displacement, by means of an electric strain transducer;
- the bar deformation, by means of a set of electric strain gauges, placed on different bar sections;
- the tensile stress in the bar by means of a dynamometric cell coaxial with prestressed bolts.

2.2 Shear force-shear displacement results

Shear force (T) at predetermined load and relative shear displacement (D), have been measured. In order to compare results for different types of reinforcement, parameter M has been introduced corresponding to the ratio between applied shear force and tensile strength of the steel reinforcement, i.e. $M=F/T_o$. Table 3 shows results of measured parameters, where: T_o (kN) is the tensile strength of the steel; ε (%) is the steel elongation; F_p (kN) is the peak shear strength of the bolted joint; F_u (kN) is the failure shear strength of the bolted joint; $M_p=F_p/T_o$ is the ratio between peak shear strength of the bolted joint and tensile strength of the element; $M_u=F_u/T_o$ is the ratio between failure shear strength of the bolted joint and tensile strength of the element; D_p (cm) is the peak shear displacement; D_u (cm) is the failure shear displacement; P (kN) is the pretension of the bar; abbreviations indicate the block material: R (rock), C (concrete); steel type: 1,2 and 3, for bar of diameter 8 mm, 12 mm; pretension P; reinforcement type B (bar), T (tube); reinforcement diameter: 8 mm, 12 mm, 13 mm and 34 mm.

Figure 2 shows the diagrams corresponding to the test performed on the three different kinds of steels.

The mechanical behaviour of sheared reinforced joints varies according to the different features of the tested material:

- for given testing conditions, shear resistance is proportional to the square of the bar diameter, whilst the shear displacement is proportional to the bar diameter;
- the maximum resistance value was obtained for the most ductile of the tested steel (the reinforced joint showed a maximum strength equal to 130 % of the tensile strength of the bar alone, whereas with the most brittle of the steels the maximum resistance was equal to 115% of the bar tensile strength), and this steel allows large shear movements between the joint surfaces;
- ductile steel reinforcements reach the ultimate tensile resistance of the material at failure; less

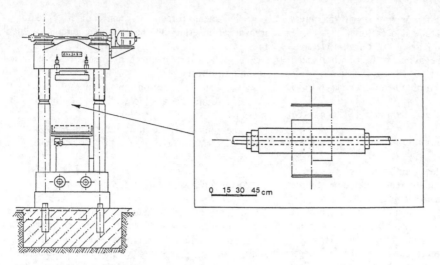

Figure 1   Test laboratory device   Bild 1   Laborapparat

Table 3. Shear tests results.

test n.	T_o (kN)	ε_o %	F_p (kN)	F_u (kN)	M_p	D_p (cm)	D_u (cm)	P (kN)	type
2.1	28.0	35.0	37.0	37.0	1.32	3.54	3.54	0.0	CB1/8
2.2	28.0	35.0	35.0	35.0	1.25	3.34	3.34	0.0	CB1/8
2.3	28.0	35.0	36.0	30.0	1.29	2.51	3.10	0.0	RB1/8
2.4	28.0	35.0	34.0	33.0	1.22	2.38	2.64	0.0	RB1/8
2.5	28.0	35.0	36.5	36.5	1.30	2.55	2.35	3.18	CB1P/8
2.6	56.2	18.0	60.0	50.0	1.07	2.68	2.96	0.0	CT/13
2.7	56.2	18.0	62.0	50.0	1.10	3.15	3.68	0.0	CT/13
2.8	28.0	35.0	33.0	33.0	1.19	1.95	1.90	0.0	RB1/8
2.9	28.0	35.0	37.0	37.0	1.32	2.29	2.60	10.60	CB1P/8
2.10	34.0	24.0	39.0	39.0	1.15	1.60	1.60	0.0	CB2/8
2.11	28.0	35.0	35.0	32.0	1.25	1.48	1.93	11.60	CB1P/8
2.12	28.0	35.0	35.5	34.0	1.27	2.30	2.45	10.50	CB1P/8
2.13	28.0	35.0	37.5	37.5	1.34	3.60	3.60	0.0	CB1/8
2.14	56.2	18.0	54.5	54.5	0.97	3.00	3.19	0.0	CT/13
2.15	36.5	26.0	45.5	40.0	1.25	2.14	2.49	0.0	CB3/8
2.16	36.5	26.0	44.5	40.0	1.22	2.09	2.42	0.0	CB3/8
2.17	34.0	24.0	40.0	40.0	1.18	1.73	1.73	0.0	CB2/8
2.18	36.5	26.0	45.0	45.0	1.23	1.77	2.27	0.0	CB3/8
2.19	80.0	26.0	96.0	96.0	1.20	2.75	2.75	0.0	RB3/12
2.20	80.0	26.0	99.0	99.0	1.24	2.83	2.83	0.0	RB3/12
2.21	80.0	26.0	110.0	110.0	1.37	2.86	2.86	0.0	CB3/12
2.22	80.0	26.0	100.0	100.0	1.25	2.84	2.84	0.0	CB3/12
2.23	102.5	18.0	123.0	50.0	1.20	4.53	6.15	0.0	CT/34
2.24	102.5	18.0	122.0	50.0	1.19	4.60	6.30	0.0	CT/34
2.25	102.5	18.0	120.0	50.0	1.17	4.55	5.08	0.0	RT/34
2.26	102.5	18.0	119.0	50.0	1.16	3.90	5.00	0.0	RT/34

104

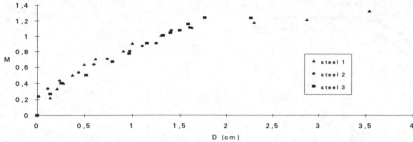

Figure 2   Laboratory shear test result diagrams for different kinds of steels. Shear displacements (D) vs the
ratio between shear force and the bar tensile strength (M) (after Ferrero, 1995)
Bild 2   Laborscherversuchsergebnisse für verschiedene Stahlsorten. Scherverschiebungen (D) vs.Scherkraft
durch Zugfestigkeitsverhältnis in den Stäben (M) (nach Ferrero, 1995).

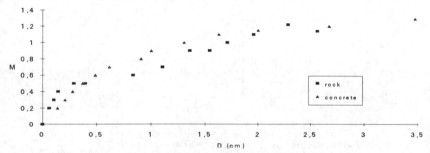

Figure 3   Laboratory shear test result diagrams for different specimen materials. Shear displacements (D)
vs. the ratio between shear force and the bar tensile strengh (M) (after Ferrero, 1995)
Bild 3   Laborscherversuchsergebnisse für verschiedene Probenstoffe. Scherverschiebungen (D) vs.
Scherkraft durch Zugfestigkeitsverhältnis in den Stäben (M) (nach Ferrero, 1995).

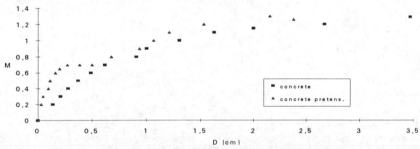

Figure 4   Laboratory shear test result diagrams for prestressed bars and dowels. Shear displacements (D)
vs. the ratio between shear force and the bar tensile strength (M) (after Ferrero, 1995).
Bild 4   Laborscherversuchsergebnisse für mit Zug verspannte Stäbe und Dübel vs Scherkraft durch
Zugfestigkeitsverhältnis in den Stäben (M) (nach Ferrero, 1995)

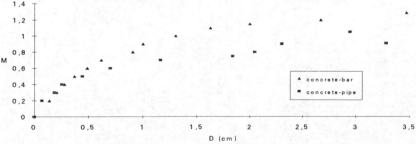

Figure 5   Laboratory shear test result diagrams for bars and pipes. Shear displacements (D) vs the ratio
between shear force and the bar tensile strength (M) (after Ferrero, 1995).
Bild 5   Laborscherversuchsergebnisse für Stäbe und Röhren vs. Scherkraft durch Zugfestigkeitsverhältnis in
den Stäben (M) (nach Ferrero, 1995).

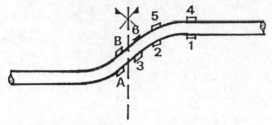

Figure 6   Strain gauges location.
Bild 6   Einrichtung der Meßstreifen

z (cm)

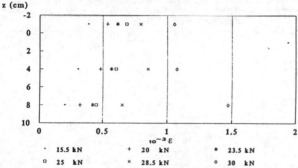

· 15.5 kN   + 20   kN   * 23.5 kN
□ 25   kN   × 28.5 kN   ◊ 30   kN

Figure 7   Distance from shear plane vs. axial strain at different shear forces (after Deangeli, 1992)
Bild 7   Abstand zwischen Scherebene und Achsenverschiebungen für verschiedene Scherkräfte (nach Deangeli, 1992)

F (kN)

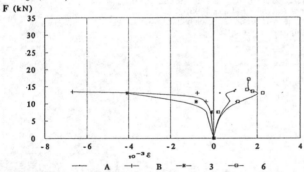

——— A   —+— B   —*— 3   —□— 6

Figure 8 Strain behaviour at a distance of -1 cm, +1 cm from shear plane vs. shear force (after Deangeli, 1992)
Bild 8   Verschiebungsverteilung im Abstand von -1 cm, +1 cm zwischen Scherebene und Scherkraft (nach Deangeli, 1992)

F (kN)

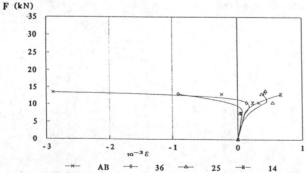

—×— AB   —◊— 36   —△— 25   —*— 14

Figure 9   Axial strain behaviour at a distance of -1 cm, +1 cm, 4 cm, 8 cm from shear plane vs. shear force (after Deangeli, 1992)
Bild 9   Achsenverschiebungsverteilung im Abstand von -1 cm, +1 cm, 4 cm, 8 cm zwischen Scherebene und Scherkraft (nach Deangeli, 1992)

ductile steel fails due to exceeding of the ultimate strain of the material;

• stronger and stiffer rock material leads to a higher shear stress in the bars and consequently a lower global resistance of the reinforced joint, as the steel shear resistance is much lower than the steel tensile resistance (figure 3);

• pretension does not influence the ultimate resistance of the system, but it determines a stiffer system and it heavily influences the magnitude of the shear displacements (peak displacements equal to 4.4 bar diameters for non prestressed bars and 2.7 bar diameters for prestressed bars) (figure 4);

• the standard ratio between the bar diameters and the hole diameters is 0.5, with a lower ratio (i. e. larger mortar thickness) leading to - in the case of hard rock - a decrease of shear stress in the bar;

• pipes show lower shear resistance (figure 5), when compared with the other tested reinforcement types, in all the examined conditions. Pipes also exhibit a remarkable softening behaviour. All the tested pipes were of a stiff type.

Pipes allow a useful comparison with friction dowels (such as Swellex). The friction dowels are installed into the rock by means of pressurised water that makes the steel swell against the rock. Friction dowels are made of a kind of steel which is similar to the steel of the tested pipes and with the same cross section. The test results show how a more ductile steel could give better results even for this type of reinforcement.

2.3 *Stress - strain results*

Six bars have been provided with electric strain gauges and have been tested in order to measure the bar deformation and to understand better the shear behaviour of bolts.

Strain gauges have been placed along the element axis as showed in figure 6. The used strain gauges can supply deformation values up to 3000 $\mu\varepsilon$. The steel employed for tests yields at 2200 $\mu\varepsilon$, therefore strain measurements were possible till the bar yields.

Recorded strain data have been interpreted by analysing three different kinds of diagrams.

Diagram [z-ε_a] (figure 7) reports the distance from the shear plane versus the axial deformation at increasing loads. Deformation decreases with the distance from shear plane. Sections close to the shear plane yield at low loading levels, while further sections reach plasticity at higher loads.

Diagram [F-ε] (figure 8) shows an antimetric behaviour of bar deformation. Nevertheless deformation provided by strain gauge 3 (close to the

shear plane) is available for only low loads. This phenomenon is probably due to mortar crushing around the bar at the discontinuity intersection. Strain gauges A-6 always provide positive deformation values, i.e. elongations; while strain gauges B-3 provide negative deformation values for low loads and then positive deformation values. This fact can mean the formation of two plastic hinges. For high loading levels, deformation values show tensile stresses in all the sections of the bar.

Diagram [F-ε_a] reports shear force versus axial deformation results. Figure 9 shows that strain gauges at a distanceof more than 4 cm from shear plane, always exhibit always a positive deformation (tensile stress). Close to the shear plane, the bar is subjected to bending (flexural stress) at low applied load, and to extensile strain (tensile stress) at higher load.

3 CONCLUDING REMARKS

A theoretical study of the dowel failure mechanics has been analysed on the basis of the laboratory results in order to set up a computational scheme to be applied for the design of the rock mass by means of a dowel.

Both the dowel effect (Q) and the tensile load in the bar (T) increase the global joint resistance which can be expressed as:

$$F = T_r \cos\alpha - Q \operatorname{sen}\alpha - (T_r \operatorname{sen}\alpha - Q\cos\alpha)\tan\varphi$$

where ϕ is joint friction angle, T_r is the traction induced in the bar, Q is the shear force induced in the bar, α is the bar inclination with the discontinuity and F global reinforced joint resistance.

Both the dowel and the rock work in the elastic field for small shear displacements; dowel failure can occur, at higher loads, following two different kinds of mechanisms depending on the prevalent kind of stress (prevalent shear and tension or prevalent tension and bending moment). These different kinds of failure correspond to different rock and steel deformability and strength features.

Two different computation methods have been set up (Ferrero '95) in order to determine the ultimate reinforced joint resistance for the two failure mechanisms:

1. the first method foresees a failure due to the combination of the axial and shear force acting at the bar-joint intersection;

2. the second method foresees a failure due to the axial force (due to the exceeding the ultimate tensile load or the ultimate steel strain) after the formation of two plastic hinges symmetric with respect to the shear plane.

The first failure mechanism is typical of hard rock (compressive strength over 50 MPa), while the second mechanism is observed in weaker rocks.

Finally, for very weak rock failure occurs due to a pull-out mechanism.

Research has shown that the steel strain behaviour at failure plays a very important role in providing the discontinuity reinforcement. This fact seems very important in weak rock where larger shear displacements at failure, compared with displacement in hard rock, occur. Only a very ductile steel can allow such deformation without failing by reaching the ultimate steel strain. For this fact empirical relationships proposed in literature give a good understanding of the phenomenon for hard rock whilst, in weak rock are obtained less realistic solutions as they do not take into account the steel stress-strain behaviour.

Pipes show a lower resistance compared to bars probably due to the steel type and not their shape.

The work will develop with further tests on different types of bolts and of particular interest to the authors is the comparison of pipes of different steel types in both hard and weak rock.

REFERENCES

Azuar, J.J. et al. 1979. Le renforcement des massifs rocheux par armatures passives. Proc. 4th Cong. ISRM, Montreux., 1, pp. 23-29.

Deangeli, C. 1992. Studio sulla resistenza al taglio di rocce rinforzate. Tesi di Laurea. Politecnico di Torino.

Dight, P.M. 1982. A case study of the behaviour of a rock slope reinforced with fully grouted rock bolts . Int. Symp. Rock Bolting, Abisko, pp. 523-538.

Di Prisco, M. 1989. Sul Comportamento a taglio delle barre d'armatura nel calcestruzzo armato. Tesi di Dottorato di ricerca in ingegneria delle strutture. Politecnico di Milano.

Dulascka, H. 1972. Dowel action of reinforcing crossing cracks in concrete. ACI Journal, 69 / 12, pp. 754-757.

Egger, P. & Fernandez H. 1983. Nouvelle presse triaxiale-Etude de modèles discontinus boulonnés. 5th Cong. ISRM, Melbourne, A171-A175.

Egger, P.& Spang K. 1987. Stability investigations for ground improvements by rock bolts at a large dam. 6th Cong. ISRM, Montreal, 1, pp. 349 -354.

Egger, P. & Pellet F. 1990. Behaviour of reinforced jointed models under multiaxial loading. Int. Cong. on Rock joints, Loen, Norway, pp. 191-195.

Farmer, I.W. 1975. Stress distribution along a resin grouted rock anchor. Int. J. Rock Mech. Min. Sci., 12, pp. 347-352.

Ferrero, A.M. 1995. The shear strength of reinforced rock joints. Int. J. Rock Mech. Min. Sci., vol. 32, n. 6.

Haas, C.J. 1981. Analysis of rock bolting to prevent shear movement in fractured ground. Journal of Mining Engineering, 33/6, pp. 698-704.

Holmberg, M. 1991. The mechanical behaviour of untensioned grouted Rock bolts. Ph.D. Thesis, Royal Institute of Technology, Stockholm.

Holmberg, M. & Stille H. 1992. The mechanical behaviour of a single grouted bolt. Int. Symp. on Rock support, Sudbury, Ontario, Canada, pp. 473-481.

Ludvig, B. 1983. Shear tests on rock bolts. Int. Symp. Rock bolting, Abisko, pp 193-203.

Pells, P.J.N. 1974. The behaviour of fully bounded rock bolts. 3rd Int. Conf. ISMR, Denver., IIB, pp. 1212-1217.

Schubert, P. 1984. Das tragvermögen des mörtelversetzten Ankers unter aufgezwungener Kluftverschiebung. Ph.D. Thesis, Montan-Universität, Leoben.

Spang, K. 1988. Beitrag zur rechnerischen berücksichtigung vollvermörtelter anker bei der sicherung von felsbauwerken in geschichtetem oder geklüftetem gebirge. These No 740. Ecole Polytechnique Federale de Lausanne.

Spang, K. & Egger P. 1990. Action of the fully grouted bolts in jointed rock and factors of influence. Rock Mech. and Rock Eng., 23, pp. 201-229.

Stillborg, B. 1984. Experimental investigation of steel cables for rock reinforcement in hard rock. Ph.D. Thesis, Lulea University, Sweden.

Das Arbeitsvermögen geankerten Gebirges bei großen Konvergenzen

The strain energy capacity of bolted rock at large tunnel deformations

R. Poisel, A. H. Zettler & M. Egger
Technische Universität Wien, Austria

KURZFASSUNG: Obwohl Felsanker im Tunnelbau weltweit eingesetzt werden, sind Wesen und Wirkung der Anker nach wie vor ungeklärt und der in der Praxis beobachtete Vorrang der Ankerung gegenüber Spritzbeton und Bögen kann in numerischen Modellen von Gebirgshohlraumbauwerken noch immer nicht nachvollzogen werden. Mittels kontinuumsmechanischer, diskontinuumsmechanischer und bruchmechanischer Rechenmodelle wird gezeigt, daß Anker in erster Linie Scherverschiebungen an bzw. das Aufgehen von Klüften und Neubrüchen vermindern. Die spezielle, strenge Kinematik des Gebirges um einen Gebirgshohlraum wird dadurch behindert, die Festigkeit des Gebirges erhöht und die Entfestigung im Post-Failure Bereich vermindert. Das Arbeitsvermögen des Verbundsystems Gebirge - Anker wird damit entscheidend erhöht.

ABSTRACT: Rock bolts have been successfully used in tunnelling for more than half a century by now in order to improve rock conditions. They have helped to tackle extremely bad rock conditions. The mode of action of rock bolts has, however, remained fairly mysterious up to now and the increased effectivity of bolting in comparison to shotcrete and steel arches as observed in tunnelling has not yet been satisfactorily simulated by numerical models of rock excavations. This paper tries to show the effectivity of rock bolts with the help of both the stress-strain curves of the bolted rock surrounding an excavation as well as its strain energy capacity (which is highly dependent on its strength and its post-failure behaviour). The stress-strain relations were calculated by the use of continuum, discontinuum and fracture mechanic models. Since it is not possible to describe the behaviour of jointed or broken rock by these different approaches using the same material parameters (e.g. joint friction angle, joint cohesion), the stress-strain curves and the strain energy capacity can not be compared directly. Thus the increase of strength and strain energy capacity calculated by each particular model was evaluated separately to verify the influence of rock bolts.

Continuum mechanic calculations have shown that the influence of the support resistance exerted by rock bolts on the strain energy capacity of the bolted rock mass is rather small. Investigations of jointed rock with the help of the distinct element method (UDEC) have revealed that bolting does not merely affect the support resistance but also reduces shear displacements and opening of joints. Thus the kinematics of rock blocks which determine the behaviour of a jointed rock mass around a rock excavation are obstructed. Strength is increased and loosening of the rock at large displacements is reduced. The strain energy capacity of the rock is thus increased dramatically. As the rock structure becomes less important in deep tunnels, new fractures assume a dominant role. Depending on the nature of the rock, extension or shear fractures are formed when rock strength is exceeded. These new fractures determine the special kinematics of broken rock around a tunnel. Fracture mechanic models enable investigations of fracture propagation. In such models bolts crossing a fracture reduce the stress intensity factor of the rock at the end of the fracture and reduce the danger of fracture propagation. Thus rock bolts obstruct the kinematics of the rock mass around a rock excavation, increase the strength and reduce loosening at large deformations of rock masses by limiting fracture propagation. Consequently the strain energy capacity of the rock mass is increased significantly.

1 EINLEITUNG

Felsanker werden im Bergbau bereits seit mehr als 100 Jahren vor allem zum Anheften einzelner Blöcke bzw. zur Verhinderung lokaler Ablösungen verwendet. Die Neue Österreichische Tunnelbauweise setzt Anker seit fast einem halben Jahrhundert in Form von Systemankerungen zur Gebirgsvergütung ein und hat damit schwierigste Gebirgsverhältnisse bewältigt. Wesen und Wirkung der Anker sind aber nach wie vor ungeklärt und der in der Praxis beobachtete Vorrang der Ankerung gegenüber Spritzbeton und Bögen kann in numerischen Modellen von Gebirgshohlraumbauwerken noch immer nicht nachvollzogen werden.

Die vorliegende Arbeit versucht, die Wirkung der Ankerung mittels des Zusammenhanges in-situ Spannung - Konvergenz („Arbeitslinie") und des Arbeitsvermögens des den Gebirgshohlraum umgebenden Gebirges darzustellen. So wie die Kennliniendarstellung soll dieses Diagramm vor allem das Verständnis für das Zusammenwirken von Gebirge und Ausbau fördern und nicht ein Bemessungsdiagramm darstellen.

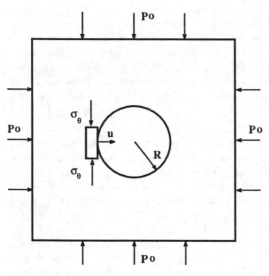

Abbildung 1: Prinzipskizze der numerischen Modelle

p_0 In-situ Spannung (isotrop)
u Verschiebung des Hohlraumrandes (Konvergenz)
σ_0 Beanspruchung des Gebirges in tangentialer Richtung
R Hohlraumradius

Figure 1:   Sketch of numerical models

p_0 In-situ stress (isotropic)
u Displacement of the tunnel wall
σ_0 Tangential stress
R Tunnel radius

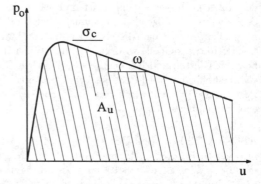

Abbildung 2: Zusammenhang in-situ Spannung - Konvergenz („Arbeitslinie" des den Hohlraum umgebenden Gebirges)

p_0 In-situ Spannungen (isotrop)
σ_c Maximale Festigkeit
ω Winkel der Festigkeitsabnahme im Post-Failure Bereich
A_u $A_u = AR$
A Verzerrungsenergiedichte (Arbeitsvermögen)
R Hohlraumradius

Figure 2:   In-situ stress vs displacements of the tunnel wall

p_0 In-situ stress (isotropic)
σ_c Peak stress
ω Angle of stress reduction after peak leavel (post failure behaviour)
A_u $A_u = AR$
A Strain energy density
R Tunnel radius

Betrachtet man die in-situ Spannung p_0 als Maß für die Beanspruchung des Gebirges in tangentialer Richtung (σ_θ) und die Verschiebung des Hohlraumrandes als Maß für die tangentiale Verzerrung ($\epsilon_\theta = u/R$) des Gebirges, kann der Zusammenhang in-situ Spannung - Konvergenz als Arbeitslinie des den Hohlraum umgebenden, durch die Anker bewehrten Gebirges interpretiert werden. Die unter der „Arbeitslinie" liegende Fläche stellt daher ein Maß für die im Gebirge gespeicherte Verzerrungsenergiedichte (Arbeitsvermögen) dar (Abb. 2). Alle Größen wurden an zweidimensionalen Modellen im ebenen Verzerrungszustand unter isotropem in-situ Spannungszustand bestimmt (Abb. 1). In allen Modellen werden Anker als elastisch-ideal plastische Elemente, die über ihre ganze Länge mit dem Fels schubweich verbunden sind, modelliert. Die Kennwerte dieser Elemente werden aus den jeweiligen Stahl- und Mörtelkennwerten ermittelt. Wie bruchmechanische Betrachtungen (Poisel, Steger, Zettler, 1995) zeigen, ist die Modellierung des Mörtelverbundes zwischen Gebirge und Anker über die ganze Ankerlänge besonders wichtig. Anker behindern in erster Linie Scherverschiebungen an und das Aufgehen von Klüften bzw. Neubrüchen nach Überschreitung der Mörtelfestigkeit im Kluft- bzw. Bruchbereich. Die wirksame Ankerlänge ist in diesen Bereichen klein, die Ankersteifigkeit und die Ankerwirkung daher groß. Wird der Verbund Gebirge - Anker über die ganze Ankerlänge im Modell nicht berücksichtigt und der Anker nur als einfache Feder zwischen Ankerkopf und Ankerende abgebildet, ist die wirksame Ankerlänge zu groß, die Ankerfestigkeit und damit die Ankerwirkung klein. Nach Überschreitung der Festigkeit des Verbundsystems Gebirge - Anker- und Felskennwerten (d.h. im Post-Failure Bereich) wurde der Zusammenhang in-situ Spannung - Konvergenz unter der Bedingung einer gleichbleibenden Verschiebungsgeschwindigkeit du/dt ermittelt. Diese Verschiebungsgeschwindigkeit mußte in den jeweiligen Modellen so klein gewählt werden, daß die „unbalanced force" (dies ist die maximale, noch nicht ausgeglichene Knotenkraft des Finite Differenzen Netzes auf Grund des im Programm implementierten Lösungsalgorithmus) im untersuchten System einen Maximalwert nicht überschritt, um die Stabilität der Lösung zu gewährleisten. Diese Vorgangsweise wird auch in der Materialprüfung gewählt (z.B. bei Bestimmung des Post-Failure Verhaltens von Fels mittels dehnungsgesteuerter Druckversuche).

2 KONTINUUMSMECHANIK

Die Berechnungen mittels eines kontinuumsmechanischen Ansatzes wurden mit dem Finite Differenzen Programm FLAC unter Zugrundelegung eines homogenen, isotropen Materials durchgeführt (Abb. 2).

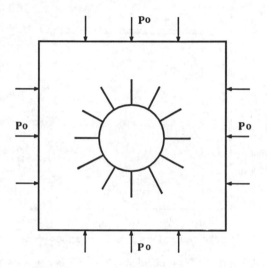

Abbildung 3: Untersuchtes kontinuumsmechanisches System mit Ankern

Figure 3: Continuum mechanics model with rock bolts

Wie Abb.3 zeigt, kann die in der Praxis beobachtete, entscheidende Wirkung der Anker mit kontinuumsmechanischen Rechenmodellen nicht nachvollzogen werden. In diesen Modellen wirken die Anker nur über die Aufbringung eines Ausbauwiderstandes, der nur zu einer geringfügigen Anhebung der Festigkeit und damit auch nur zu einer geringfügigen Steigerung des Arbeitsvermögens führt. Auch die Art des Materialverhaltens (elastisch-ideal plastisch, strain softening behaviour etc.) sowie die Berücksichtigung der Anker mittels einer Erhöhung der Kohäsion des Gebirges führt nicht zu einer entscheidenden Erhöhung des Arbeitsvermögens des geankerten Gebirges.

Wie verschiedene Autoren (z.B. Hoek et. al., 1995) zeigen, ist der Einfluß der Anker mit kontinuumsmechanischen Ansätzen nicht nachvollziehbar. Entstehende Entfestigungs- bzw. Bruchzonen werden in der Rechnung durch Anker nicht verhindert oder deren Lage aufgrund der Dominanz des Spannungsfeldes nicht beeinflußt.

gewählt, die jener in einer Studie von Vogele, Fairhurst und Cundall (1978) ähnlich ist (siehe auch Poisel, Engelke, 1994). Die Autoren untersuchten das Verhalten dieses Kluftkörperverbandes bei niedrigen in-situ Spannungen. Dabei konnte vor allem das Aufgehen von Klüften in der Firste beobachtet werden.

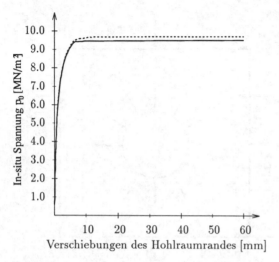

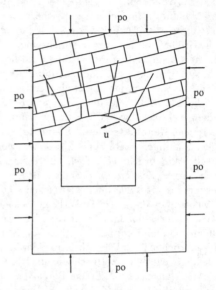

Abbildung 4: Zusammenhang in-situ Spannung - Konvergenz für das untersuchte kontinuumsmechanische System

Figure 4:   In-situ stress vs tunnel wall displacements for the continuum mechanics model

Abbildung 5: Untersuchtes diskontinuumsmechanisches System (mit 4 Ankern)

Figure 5:   Discontinuum mechanics system (with 4 rock bolts)

Die von verschiedenen Autoren (z.B. Pande, 1993) vorgeschlagene Einführung komplexerer, äquivalenter Materialmodelle zur kontinuumsmechanischen Beschreibung von Diskontinuen (z.B. das sogenannte multilaminate model) könnte die rechnerische Modellierung der Ankerwirkung noch verbessern, scheitert derzeit aber noch an den nicht verfügbaren Materialparametern.

3 DISKONTINUUMSMECHANIK

Für die numerischen Untersuchungen des diskontinuumsmechanischen Modelles wurde das Diskrete Elemente Programm UDEC verwendet. Für diese Untersuchungen wurde ein Kluftkörperverband mit einer Trennflächengeometrie (Abb.4)

Das Verhalten dieses Kluftkörperverbandes unter großen in-situ Spannungen wird im Gegensatz dazu in erster Linie durch einen keilförmigen Kluftkörper im rechten Kämpferbereich bestimmt. Der untere Querschnittsbereich wurde daher nicht weiter in Kluftkörper zerlegt; diese Vereinfachung des Systemes hat für die hier untersuchte Fragestellung keine Bedeutung.

Die Berechnungen haben gezeigt, daß es einen exakt bestimmten Reibungswinkel gibt, bei dem es zum Gleiten des keilförmigen Kluftkörpers und damit zum „Bruch" des Gebirges kommt. Wird dieses Gleiten geringfügig behindert (z.B. durch Anker), kommt es erst bei wesentlich höheren in-situ Spannungen zum Erreichen des Tragvermögens. Wie die Abbildung 5 zeigt, führt eine Erhöhung der Anzahl der Anker über diese Behinderung der Kinematik sowohl zu einer Erhöhung der Festigkeit des Verbandes als auch zu einer Verminderung der Festigkeitsabnahme im Post-Failure Bereich (Egger, 1979).

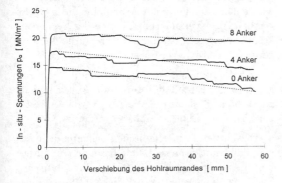

Abbildung 6: Zusammenhang in-situ Spannung - Konvergenz für das untersuchte diskontinuumsmechanische System

Figure 6:   In-situ stress vs tunnel wall displacement for the discontinuum mechanics system

Dies bedeutet eine deutliche Erhöhung des Arbeitsvermögens des Gebirges bei Einbau von Ankern. Druckversuche an Modellkluftverbänden (Pellet, 1994; und Pellet, Egger und Ferrero, 1995) haben nicht eine so deutliche Steigerung des Arbeitsvermögens erbracht. Ein Grund dafür könnte sein, daß die kinematischen Zwänge im Druckversuch nicht so ausgeprägt sind wie jene in den Bereichen um einen Tunnel. Behinderungen dieser Zwänge führen daher im Falle des Tunnels zu einer deutlicheren Erhöhung des Arbeitsvermögens.

4 BRUCHMECHANIK

Im Gegensatz zum kontinuumsmechanischen Modell, das verschmierte Klüfte betrachtet, und zum diskontinuumsmechanischen Modell, das vollständig durchgerissene Klüfte untersucht, beschäftigt sich die Bruchmechanik mit nicht vollständig

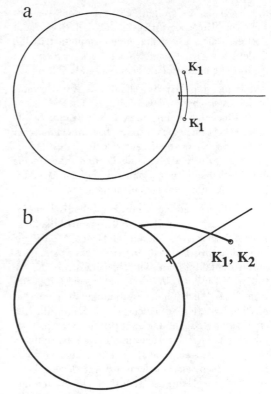

Abbildung 7: Untersuchtes bruchmechanisches System

a   Anker quert einen Spaltbruch; Reduzierung des Spannungsintensitätsfaktors K_1

b   Anker quert einen Scherbruch; Reduzierung der Spannungsintensitätsfaktoren:
K_2 zufolge Scherverschiebungen und
K_1 zufolge Dilatation

Figure 7:   Fracture mechanics system

a   Rock bolt crosses a spalling fracture; Reduction of stress intensity factor K_1

b   Rock bolt crosses a shear fracture; Reduction of stress intensity factors:
K_2 due to shear displacements and
K_1 due to dilatation

durchgerissenen Klüften, also mit der Ausbreitung von Rissen bzw. Brüchen (Poisel, Steger und Zettler, 1995). Bei tiefliegenden Tunneln tritt die Bedeutung des Trennflächengefüges zurück und Neubrüche nehmen eine beherrschende Rolle ein.

Abhängig vom Bruchmechanismus des Gesteins bilden sich bei Überschreitung der Festigkeit auch im Gebirge um einen Felshohlraum Spalt- bzw. Scherbrüche aus. Mittels bruchmechanischer Methoden kann die Beanspruchung des Gebirges am Bruchende untersucht und die Gefahr der Bruchausbreitung (Spannungsintensitätsfaktoren K_1, K_2 etc.) angegeben werden.

Berücksichtigt man im bruchmechanischen Modell, wie in Abb. 6 dargestellt, einen Anker, ergibt sich bereits in diesem einfachen Modell für einen Spaltbruch eine Reduzierung des Spannungsintensitätsfaktors an der Rißspitze von $K_1 = 2,69 MN/m^{3/2}$ auf $K_1 = 1,96 MN/m^{3/2}$ d.h. um rund 25 %. Auch bei der Anwendung bruchmechanischer Kriterien auf einen Scherbruch ergibt sich eine starke Reduzierung der Spannungsintensitätsfaktoren zufolge der Ankerung. Berücksichtigt man in einem solchen Modell auch die Dilatanz von Scherbrüchen, also die Volumsvergrößerung beim Brechen des Gebirges, reduziert ein Anker den Spannungsintensitätsfaktor K_1 von $0,88 MN/m^{3/2}$ auf $0,54 MN/m^{3/2}$ und K_2 von $1,40 MN/m^{3/2}$ auf $0,85 MN/m^{3/2}$, d.h. beide Spannungsintensitätsfaktoren um rund 40%. Betrachtet man den Spannungsintensitätsfaktor als Maß für die Standsicherheit, da er die Gefahr des Weiterreißens angibt, hat die Ankerung in einem solchen Modell einen bedeutenden Einfluß auf das Arbeisvermögen des Gebirges.

5 ZUSAMMENFASSUNG

Die vorliegende Arbeit versucht die in der Praxis beobachtete große Wirkung von Ankern im Felshohlraumbau rechnerisch nachzuvollziehen. Mittels kontinuumsmechanischer, diskontinuumsmechanischer und bruchmechanischer Ansätze wurde das Arbeitsvermögen von geankerten Gebirgskörpern in der Leibung eines Gebirgshohlraumes untersucht. Ein direkter Vergleich dieser Betrachtungsweise ist auf Grund der Verschiedenartigkeit der Ansätze nicht möglich. Vor allem die bei

einem Kluftkörperverband beobachteten Materialkennwerte wie Kluftreibungswinkel, Kluftkohäsion etc. sind auf ein Kontinuum sehr schwierig zu übertragen. Setzt man in ein Kontinuumsmodell die für das ungestörte Gebirge maßgeblichen Werte ein, wird das Kluftverhalten nicht berücksichtigt. Setzt man andererseits die in den Klüften beobachteten Materialkennwerte ein, ergibt sich meist ein für das Gebirge zu weiches Gesamtverhalten. Es wurde daher die Form der Arbeitslinien des Gebirges bzw. ihre Veränderung zufolge der Systemankerung untersucht. Im besonderen wurde die Steigerung des Arbeitsvermögens betrachtet.

Kontinuumsmechanische Rechenmodelle zeigen, daß der Einfluß des durch die Ankerung aufgebrachten Ausbauwiderstandes auf das Arbeitsvermögen des Gebirges äußerst gering ist.

Diskontinuumsmechanische Rechenmodelle zeigen, daß Kluftkörperverbände besonders unter hohen in-situ Spannungen starken kinematischen Zwängen unterliegen. Wird die Kluftkörperkinematik auch nur geringfügig behindert, führt diese zu einer Verminderung der Entfestigung im Post-Failure Bersich. Es konnte rechnerisch gezeigt werden, daß Anker eine wesentliche Erhöhung des Arbeitsvermögens bewirken können.

Bruchmechanische Untersuchungen von Neubrüchen, haben gezeigt, daß Anker die Beanspruchung des Gebirges am Ende von Spalt- und Scherbrüchen und damit die Ausbreitung solcher Brüche stark vermindern. Es scheint daher, daß die in der Praxis beobachtete große Ankerwirkung nur bei Berücksichtigung der strengen Kinematik in der Leibung eines Gebirgshohlraumes rechnerisch nachvollziehbar ist. Dazu sind vor allem diskontinuumsmechanische und bruchmechanische Verfahren sehr gut geeignet.

Die Behinderung dieser Kinematik durch Anker führt zu einer Anhebung der Festigkeit und zu einer geringeren Festigkeitsabnahme im Post-Failure Bereich. Diese Wirkung ist dem Effekt einer Stahlfaserzugabe zu Beton sehr ähnlich (Lauffer H., 1994). Eine Erhöhung des Stahlfasergehaltes bewirkt eine Erhöhung der Festigkeit sowie eine verminderte Festigkeitsabnahme im Post-Failure Bereich von Beton. So wie das Arbeitsvermögen von Beton durch die Zugabe von Stahlfasern wird auch das Arbeitsvermögen von Fels durch die Bewehrung mittels Ankern erhöht.

LITERATURVERZEICHNIS

Egger P., 1979. *Diskussionsbeitrag zu „La Bullo-natura in Sotterraneo".* Associazione Mineraria Subalpina 16/3 Turin (in italienisch)

Hoek E., Kaiser P.K. und Badwen W.F., 1995. *Support in Underground Excavations in Hard Rock.* Balkema, Seiten 125-126

Lauffer H., 1994. *Die Entwicklung der NÖT im Spannungsfeld zwischen Theorie und Praxis.* Felsbau 5 /1994, Seite 307-311

Pande G.N., 1993. *Constitutive Models for Intact Rock, Rock Joints and Jointed Rock Masses.* Comprehensive Rock Engineering ed. J.A.Hudson Vol. 1, Seite 427-441, Pergamon Press

Pellet F., 1994. *Strength and Deformability of Jointed Rock Masses Reinforced by Rock Bolts.* Dissertation ETH Lausanne

Pellet F., Egger P. und Ferrero A.M., 1995. *Contribution of fully bondet rock bolts to the shear strength of joints; Analytical and experimental evaluation.* In: Mechanics of Jointed and Faulted Rock, ed. Rossmanith H. P., Balkema, Seiten 873-878

Poisel R., Steger W. und Zettler A.H., 1995. *Neue Ansätze für Standsicherheitsuntersuchungen von Tunneln.* Felsbau 3/1995.

Poisel R., Engelke H., 1994. *Zu den Konzepten der NÖT.* Felsbau 5/1994, Seiten 330-332

Vögele M., Fairhurst Ch. und Cundall P.A., 1978. *Analysis of tunnel support loads using a large displacement, distinct block model.* Proc. 1st Int. Syposium "Rockstore", Stockholm, 1977, Vol.2,pp 247-252

Passive resistance of strip anchors

Passiver Widerstand von Plattenankern

P. Regenass & A. H. Soubra
École Nationale Supérieure des Arts et Industries de Strasbourg, France

ABSTRACT: This paper describes an upper bound method in limit analysis for calculating the ultimate load of plate anchors. The present analysis considers the general case of a frictional and cohesive soil with an eventual surcharge loading on the ground surface. Three translational failure mechanisms are considered for the calculation schemes. The numerical results obtained from the different mechanisms are presented in the form of non dimensional coefficients. They show that the increase of the embedment depth significantly increases the ultimate load. The same phenomena is also valid for the anchor inclination. Finally, the effect of the anchor inclination on the critical slip surface is presented. It shows that the slip surface tends to a planar surface when the anchor inclination decreases.

ZUSAMMENFASSUNG: Diese Studie stellt ein Berechnungsverfahren vor, das es erlauben soll, die Belastungsgrenze von Plattenankern rechnerisch zu bestimmen. Es handelt sich dabei um einen kohärenten, schwimmenden Boden, bei dem eine zusätzliche Belastung an der Bodenoberfläche berücksichtigt werden kann. Drei Übertragungsmechanismen werden dabei in Betracht gezogen. Die an Hand dieser Mechanismen erzielten Ergebnisse werden in Form adimensionaler Koeffiziente ausgedrückt. Wir zeigen dabei, daß die Belastungsgrenze wesentlich mit der Verankerungstiefe ansteigt. Ähnlich verhält es sich mit der Neigung der Verankerung. Der Einfluß der Neigung der Verankerung wird auf der äußersten Gleitlinie dargestellt. Dabei hat sich herausgestellt, daß sich diese Gleitlinie bei einer geringen Neigung der Verankerung einer Geraden nähert.

1 INTRODUCTION

The problem of the passive resistance of anchors has been widely studied in literature: Smith (1962), Smith & Stalcup (1966), Ovesen & Stroman (1972), Neely et al (1973), Das & Seely (1975), Rowe & Davis (1982), Dickin & Leung (1983) and Murray & Geddes (1989). In this paper, we focus our study on the determination of the ultimate load of strip anchors.

In fact, this problem belongs to the stability problems in geotechnical engineering and it has been modelled by either the limit equilibrium methods, the slip line methods or the limit analysis methods.

While the limit equilibrium methods are simple, the slip line methods are more complicated since they require the establishment of a stress field in the plastically deformed region. Notice however that these methods do not allow to know if the solution obtained is an upper or a lower bound one with respect to the exact solution for an associated flow rule material.

The limit analysis method is based on the limit theorems of Drucker et al (1952) and it is employed to obtain upper and lower bounds of the collapse load using the upper and lower bound methods.

It is to be noted that the lower bound method in limit analysis has not been widely used in geotechnical engineering since it is so complex for obtaining the statically admissible stress field in the soil mass. However, due to the facility of establishment of kinematically admissible mechanisms, the upper bound method has been used by Chen (1975) who presented the solutions for many stability problems in geotechnical engineering by using different failure mechanisms.

In this paper, we present an upper bound

method in limit analysis to calculate the ultimate load of strip anchors. Three failure mechanisms are considered for the calculation schemes : These mechanisms are of the translational type as it will be shown later in the following sections. They are a generalisation of three failure mechanisms considered by Chen (1975) in the passive earth pressure problem and reviewed by Murray & Geddes (1989) for the calculation of the ultimate failure load of strip anchors. Notice that Murray & Geddes (1989) have considered the case of a cohesionless soil.

The present analysis will consider the general case of a frictional and cohesive (c, φ) soil. The ultimate failure load due to soil weight, soil cohesion and surcharge loading on the ground surface is given in the form of non dimensional anchor force coefficients M_γ, M_c and M_q.

2 HYPOTHESES

For the problem of computation of the ultimate failure loads of strip anchors, the following assumptions will be adopted here:

- As shown in figure (1), the anchor plate is characterised by its breadth h, its embedment depth H and its inclination ψ.

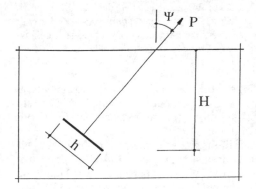

Figure 1 : Strip Anchor, Notations
Abbildung 1 : Plattenanker, Bezeichnungen

- The soil is assumed to be an associated flow rule Coulomb material obeying Hill's maximal work principle. It is characterised by its angle of internal friction φ and its cohesion c.
- The angle of friction δ at the soil-structure interface is assumed to be constant. This hypothesis is in conformity with the

kinematics assumed in this paper as it will be shown later in this paper.

- An eventual uniform surcharge loading can act at the soil surface which is assumed to be horizontal.
- The assumption of a sliding by friction is adopted at the soil-structure interface. Hence, the velocity at this interface is tangent to the anchor plate.

3 THEORETICAL ANALYSIS

According to the upper bound theorem in limit analysis, for a kinematically admissible velocity field, an upper bound of the exact collapse load can be obtained by equating the power dissipated internally to the power expended by the external loads.

A kinematically admissible velocity field is one that satisfies the flow rule, the velocity boundary conditions and compatibility. During plastic flow, power is assumed to be dissipated by plastic yielding of the soil mass, as well as by sliding along velocity discontinuities where jumps in the normal and tangential velocities may occur.

Note that the velocity field at collapse is often modelled by a mechanism of rigid blocks that move with constant velocities. Since no general plastic deformation of the soil mass is permitted to occur, the power is dissipated solely at the interfaces between adjacent blocks, which constitute velocity discontinuities. This kind of velocity field will be used herein. Finally, note that in the case of the ultimate load of strip anchors, the upper-bound theorem gives an unsafe estimate of the failure load.

In this paper, three failure translational mechanisms are considered for the calculation of the ultimate failure load. These mechanisms will be refereed to as the M_1, M_2 and M_3 mechanisms. They are described in the following sections.

3.1 Mechanism M_1

As shown in figure (2), this mechanism is composed of a single rigid bloc moving with velocity V_1. The anchor plate moves with the velocity V_0 and V_{01} represents the relative velocity at the soil-structure interface.

This mechanism is kinematically admissible since the velocity along the velocity

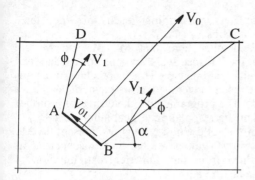

Figure 2 : Mechanism M_1
Abbildung 2 : Mechanismus M_1

discontinuities AD and BC makes an angle ϕ with these surfaces : It is characterised by a single parameter α.

3.2 Mechanism M_2

As shown in figure (3), this mechanism is composed of two rigid blocs ABC and ACDE moving respectively with velocities V_1 and V_2. V_{12} represents the relative velocity at the discontinuity surface AC while V_{01} is as defined above.

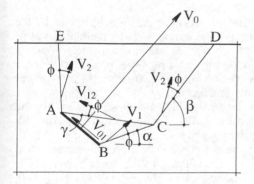

Figure 3 : Mechanism M_2
Abbildung 3 : Mechanismus M_2

This mechanism is kinematically admissible since it verifies all the kinematical constraints mentioned above. This collapse mechanism depends on three angular parameters α, β and γ. These angular parameters describe completely the failure surface.

3.3 Mechanism M_3

As shown in figure (4), this mechanism is composed of a radial shear zone AEF sandwiched between two rigid blocs ABE and AFCD.

The radial shear zone is limited by a log spiral slip surface EF. The log spiral slip surface is tangent to lines BE and FC respectively at points E and F.

The rigid blocs ABE and AFCD move respectively with the velocities V_1 and V_2.

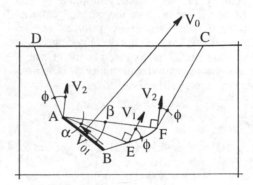

Figure 4 : Mechanism M_3
Abbildung 4 : Mechanismus M_3

It is to be noted here that due to the fact that the log spiral is tangent to lines BE and FC, there is no velocity discontinuities along lines AE and AF. It was shown by Chen (1975) that the velocity distribution along the log spiral slip surface is given by:

$$V(\theta) = V_1 \cdot \exp((\theta - \alpha)\tan\phi) \qquad (1)$$

This mechanism is kinematically admissible and it is completely defined by two angular parameters α and β. It will be named the log sandwich mechanism as made by Chen (1975) for the earth pressure problem.

3.4 Work equation

For each of the three failure mechanisms, one can write the work equation by equating the rate of external work done by the external forces to the rate of internal energy dissipation along the plastically deformed surfaces.

Finally, one can obtain the critical failure load after extremization of the 'potential' failure load as it will be shown in the following sections.

The incremental external work due to an external force is the external force multiplied by the corresponding incremental displacement or velocity.

The external forces contributing in the incremental external work consist of the load anchor, the weight of the soil mass and the surcharge q at the ground surface.

The incremental external work due to self weight in a region is the vertical component of the velocity in that region multiplied by the weight of the region. The incremental external work for the different external forces can be easily obtained. They are not presented herein.

The incremental energy dissipation per length unit along a velocity discontinuity or a narrow transition zone can be expressed as follows:

$$\Delta D_L = c.\Delta V.\cos\phi \qquad (2)$$

where ΔV is the incremental displacement or velocity which makes an angle ϕ with the velocity discontinuity according to the associated flow rule of perfect plasticity, and c is the cohesion parameter. The incremental energy dissipation along the different velocity discontinuities can be easily calculated. They are not presented herein.

Finally, it is to be noted that along the soil-structure interface where we have adopted the assumption of sliding by friction, the energy dissipation is given as follows:

$$\Delta D_L = P.\tan\delta.V_{01} \qquad (3)$$

By equating the total external work to the total internal energy dissipation, we have:

$$P = \frac{1}{2}\gamma.h.H.M_\gamma + c.h.M_c + q.h.M_q \qquad (4)$$

where M_γ, M_c and M_q are the non dimensional anchor force coefficients.

4 NUMERICAL RESULTS

The failure load of anchors can now be obtained by minimisation of P with respect to the angular

parameters describing each of the three mechanisms.

Remember here that M_1, M_2 and M_3 are described respectively by one, three and two angular parameters. Three computer programs have been developed with equation (4) as a basis. The programs give the critical slip surface and the corresponding critical failure load P.

In the following sections, we present the M_γ, M_c and M_q values for the three mechanisms as obtained from the numerical extremisation of these coefficients.

4.1 Influence of the embedment depth on the anchor force coefficients

Figure (5) shows the variation of the anchor force coefficients M_γ, M_c and M_q as function of ϕ for two values of the embedment ratio H/h (H/h=1 ; 5) when $\Psi=90°$ and $\delta/\phi=2/3$. The solutions presented in these figures concern the results obtained from the three mechanisms M_1, M_2 and M_3.

From these figures, one can easily see that the M_1 mechanism highly overestimates the anchor force coefficients especially for the high embedment ratio H/h and for the high ϕ-values. For example, the M_1 mechanism overestimates the M_c value by about 270% compared to the M_2 and M_3 mechanism when $\Psi=90°$, $H/h=5$, $\phi=40°$ and $\delta/\phi=2/3$.

The comparison of the solutions given by the M_2 and M_3 mechanisms shows that the results of the M_γ coefficient are approximately identical for both mechanisms. The percent difference does not exceed 5% for $\Psi=90°$, $H/h=5$, $\phi=40°$ and $\delta/\phi=2/3$.

However, for the M_c and M_q coefficients, the numerical results (figure 5) show that the M_2 mechanism gives better solutions than the M_3 mechanism since the corresponding upper bound solution is smaller. This remark is only valuable for high values of the embedment ratio H/h. Notice however, that both mechanisms give approximately the same results for small values of the embedment ratio H/h.

4.2 Influence of the anchor inclination on the anchor force coefficients

To show the influence of the anchor inclination on the anchor force coefficients, we first present the results of the anchor plate for $\Psi=45°$.

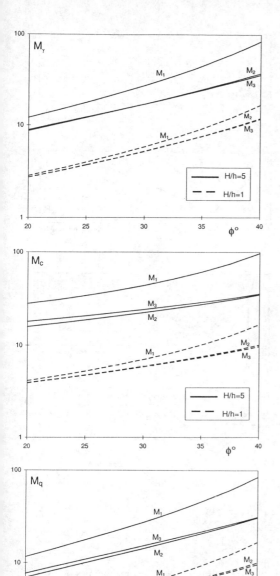

Figure 5 : Anchor force coefficients M_γ, M_c and M_q (Ψ =90°, δ/ϕ=2/3)

Abbildung 5 : Belastungskoeffiziente der Verankerung M_γ, M_c and M_q (Ψ =90°, δ/ϕ=2/3)

Secondly, we present the comparison with the results given above for Ψ =90°.

Figure (6) shows the variation of the anchor force coefficients M_γ, M_c and M_q as function of ϕ for Ψ =45°, H/h=5 and δ/ϕ = 2/3. As mentioned in the preceding section, the solutions presented in these figures concern the results obtained from the three mechanisms M_1, M_2 and M_3.

The comparison of the solutions of the M_1 and M_3 mechanisms shows that the results of the M_γ factor are approximately identical for both mechanisms. The difference between the results of both mechanisms slightly increases for the M_c and M_q factors. As in the preceding section, the M_1 mechanism continues to overestimate the anchor force coefficients.

Finally, notice that the numerical results of all coefficients M_γ, M_c and M_q show that the M_2 mechanism gives better solutions than the M_3 mechanism since the corresponding upper bound solution is smaller.

The comparison of the force anchor coefficients for different anchor inclinations show that these coefficients decrease with the anchor inclination decrease. For example, the percent reduction of the M_γ value as given by the M_3 mechanism is about 58% when the anchor inclination varies from 90° to 45° for H/h=5, ϕ=40°, δ/ϕ=2/3.

4.3. Critical slip surfaces

Figure (7) show the critical slip surfaces of the M_1, M_2 and M_3 mechanisms as obtained from the numerical extremisation of the anchor force coefficient M_γ with respect to the angular parameters. These surfaces concern the case of an anchor with the following characteristics: Ψ = 45° & 90°, H/h=2, ϕ=20° & 40° and δ/ϕ=2/3.

As it is well known in the passive earth pressure problem, the M_1 mechanism highly overestimates the failure load due to the fact that the slip surface is far from a planar surface especially for the high values of Ψ, ϕ and δ.

The M_2 and M_3 mechanisms allow the slip surface to develop below the horizontal direction passing through the bottom of the plate anchor. These mechanisms can estimate the anchor force coefficient with good accuracy especially for high values of ϕ, δ and Ψ.

Finally, one can easily see that the slip surfaces tend to planar surfaces when the anchor

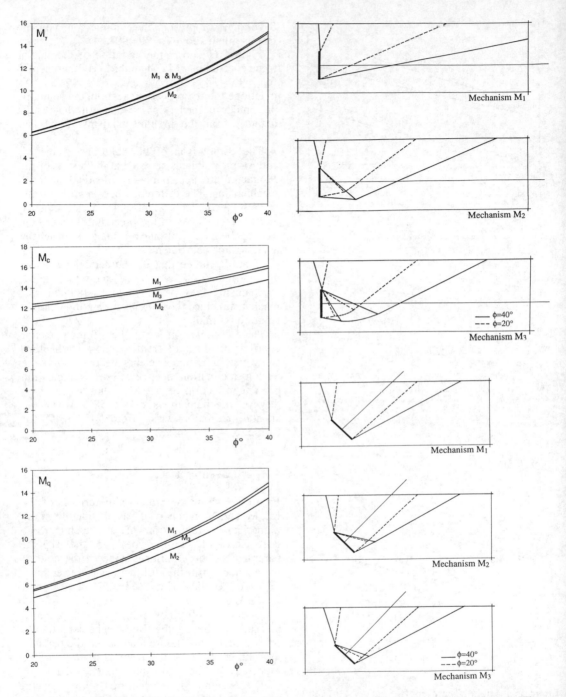

Figure 6 : Anchor force coefficient M_γ, M_c and M_q (Ψ =45°, H/h=5, δ/ϕ=2/3)

Abbildung 6 : Belastungskoeffiziente der Verankerung M_γ, M_c and M_q (Ψ =45°, H/h=5, δ/ϕ=2/3)

Figure 7 : Critical slip surfaces (H/h=2, Ψ =45° & 90°, ϕ=20° & 40°, δ/ϕ=2/3)

Abbildung 7 : Kritische gleitflachen (H/h=2, Ψ =45° & 90°, ϕ=20° & 40°, δ/ϕ=2/3)

inclination decreases. The radial shear zone of the log sandwich mechanism M_3 disappear. For the two-bloc mechanism M_2, the velocity discontinuity between the two blocs vanishes. Hence, the mechanisms M_2 and M_3 tend to the case of the unique bloc mechanism M_1 for small values of the anchor inclination.

5 CONCLUSION

The upper bound method in limit analysis is used to calculate the anchor force coefficients of a plate anchor. Three failure mechanisms are considered in this analysis. The unique rigid bloc highly overestimates the load anchor for high values of ϕ, δ and Ψ. However, the two-blocs and the log-sandwich mechanisms give results which are in reasonable agreement.

REFERENCES

Chen, W.F. 1975. Limit Analysis and Soil Plasticity. Amsterdam, *Elsevier*, 637p.

Das, B.M. & Seely, G.R. 1975. Pullout Resistance of Vertical Anchors. *J. Geotech. Engng. Div. Am. Soc. Civ. Engrg.*, 101, GT1, 87-91.

Dickin, E.A., & Leung, C.F. 1983. Centrifugal Model Tests on Vertical Anchor Plates, *J. of Geotech. Engng., ASCE*, Vol 109, N° 12, 1503-1525.

Drucker, D.C., Greenberg, H.J. & Prager, W. 1952. Extended limit design theorems for continuous media, *Q. Appl. Math.*, 9, 381-389.

Murray, E.J., & Geddes, J.D. 1989. Resistance of passive inclined anchors in cohesionless medium. *Géotechnique*, Vol 39, N° 3, 417-431.

Neely, W.J., Stewart, J.G., & Graham, J. 1973. Failure Loads of Vertical Anchor Plates in Sand. *J. Soil Mech. Fdns. Div. Am. Soc. Civ. Engng.*, 99 SM9, 669-685.

Ovesen N.K. & Stroman H., 1972. Design methods for vertical anchor plates in sand. *Proceedings of Specialty Conference on Performance of Earth at Earth Supported Structures*, New-York, ASCE, 1481-1500.

Rowe, R.K., & Davis, E.H. 1982. The Behaviour of Anchor Plates in sand. *Géotechnique*, Vol 32, N° 1, 25-41.

Smith, J.E., 1962. Deadman anchorages in sand. *Technical report R199. US Naval Civil Engineering Laboratory*, Port Hueneme.

Smith, J.E., 1966. Deadman anchorages in various soil mediums. *Technical report R434. US Naval Civil Engineering Laboratory*, Port Hueneme.

Anchors in Theory and Practice, Widmann (ed.)© 1995 Balkema, Rotterdam. ISBN 90 5410 577 1

Rock reinforcement by integrated system of bolting and grouting

Felssicherungen mit einem integrierten System für Ankerungen und Injektionen

Richard Snupárek
Institute of Geonics, Czech Academy of Sciences, Ostrava, Czech Republic

Libor Paloncy
Ankra, Petrvald, Czech Republic

ABSTRACT: Two technologies of rock reinforcement are mostly widespread - bolting and grouting. These two methods are realised separately in spite of the similar basic operations - drilling of boreholes with small diameter. The new integrated system consists of hydraulically anchored tube bolts which involve both functions - bolting and grouting.

ZUSAMMENFASSUNG: Im Felsbau sind der Einbau von Ankern und Injektionen die am meisten angewandten Sicherungsmethoden. Beide Methoden werden in der Regel getrennt angewandt, obwohl beide die Herstellung von Bohrungen erfordern.
Das neue integrierte System besteht aus Rohrankern, die hydraulisch fixiert werden und auch zur Injektion des umgebenden Gesteins dienen.
Der Beitrag befaßt sich sowohl mit den gebirgsmechanischen Voraussetzungen für den optimalen Einsatz des Systems als auch mit dem Einsatz in der Praxis und den bisherigen Erfahrungen.

Collaboration of the support of underground openings with surrounding rock mass and employing rock properties for stability control create base of recent methods of driving and supporting in mining and civil engineering.

The integral part of the methods are technologies of reinforcement of rock mass around openings. The most frequent technologies are bolting and grouting. Hitherto the two technologies have been executed separately in spite of the fact, that the basic operation for both technologies is drilling of holes with small diameters (mostly up to 40 mm).

The practice of separate anchoring and grouting (when both technologies are used) is disadvantageous from both theoretical and practical point of view.

The measurements of the distribution of longitudinal strain in grouted bolt (Sun Xuey, 1983) led to the establishment of the mechanical model of the fully grouted rockbolt (fig. 1).

Adequate results were obtained by Janas (1994) from a mathematic model of rock displacement along the fully grouted rockbold depended on time or on the distance of moved face (fig. 2). The results are not only important for design of bolt parameters, but also for evaluating of rock mass around the opening.

Rock along the bolt can be divided into two parts: part where the bolt is anchored (the character of relative displacement is elongation) and part where the bolt with pad compresses rock beds (relative displacement is compression). The two parts are divided by the neutral point, where the relative displacement of rock to the bolt is zero. According to in situ measurements as well as the mathematic model, the neutral point occurs approximately in one third of the common length of the bolt (from the outer end).

If we consider combined methods of rock reinforcement by bolting and grouting, we find, that the grouting is effective mostly in the part of rock mass, where bolts are anchored (above the neutral point). The later described combined anchoring and grouting system makes it possible to fulfill this requirement.

Recently the reinforcing system has been developed and practically proved. It combines the two basic technologies - bolting and grouting. The system is called Boltex and its elements are given in figure 3.

The basic element is hydraulically expansible tube bolt made of the enfolded steel tube of special shape

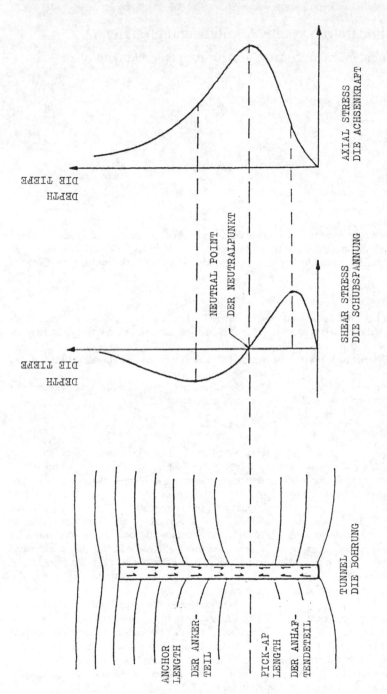

FIG. 1 STRESS DISTRIBUTION MODEL FOR GROUTED BOLTS (AFTER LIT. 1)

EIN MECHANISCHES MODEL DES KLEBENANKERS (NACH 1)

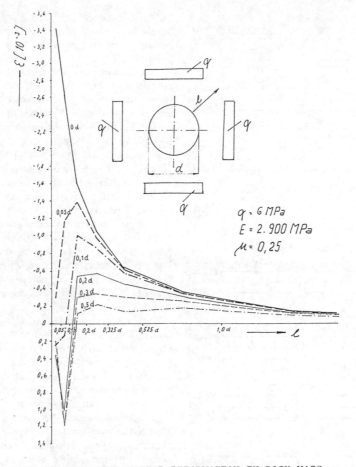

FIG. 2 RELATIVE RADIAL DEFORMATION IN ROCK MASS
ALONG THE GROUTED BOLT (AFTER LIT.2)

DIE RELATIVE RADIALE DEFORMATIONS IM MASSIV
LANGST DES KLEBENANKERS (NACH 2)

(principle of Swellex). When a liquid medium under high pressure is conveyed into the closed space of the pipe bolt, the bolt expands and presses the wall of the borehole. The relation between the length of anchoring and results of pull-out tests for two types of bolts anchored in limestone are given on figure 4.

The inner end of the bolt is provided with the injection membrane. It is designed in such way so that - after the following increasing of the medium pressure - the membrane cracks and makes it possible to grout rock mass through the bolt. The value of grouting pressure can differ from the pressure of bolting.

The part of the grouting system are packers, preventing the grout flowing out through the mouth of borehole. There are two types of packers: either rubber sleeve on the pipe bolt, which seals (after expansion of the bolt) the space between the bolt and the borehole wall, or special packers, which are noted with appreciable capacity of expansion and are able to pack the borehole also in damaged parts (caverns, fissures).

The principle of packing is given on figure 5. The shape of pipe bolt retains - after expansion - shallow grooves, which serve to transport the grouting material from the inner end of the bolt along the borehole. To prevent grout flowing out of the hole mouth it is necessary to seal this grooves in proper point. It is provided by raw rubber, which fills the groves and the sleeve from scorched rubber put on the pipe bolt. The length of the packing sleeve is mostly 30 to 50 cm, i.e. much shorter than the length of bolt. According to the position of the sleeve on the bolt it is possible to change the point of the packing

127

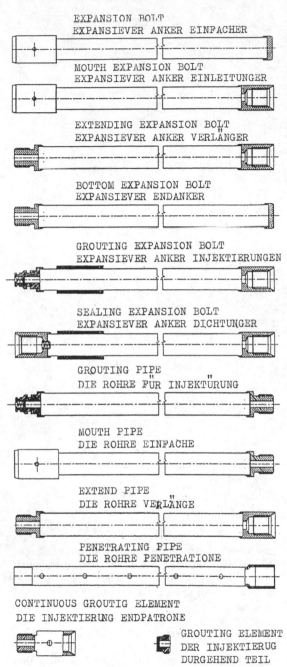

EXPANSION BOLT
EXPANSIEVER ANKER EINFACHER

MOUTH EXPANSION BOLT
EXPANSIEVER ANKER EINLEITUNGER

EXTENDING EXPANSION BOLT
EXPANSIEVER ANKER VERLÄNGER

BOTTOM EXPANSION BOLT
EXPANSIEVER ENDANKER

GROUTING EXPANSION BOLT
EXPANSIEVER ANKER INJEKTIERUNGEN

SEALING EXPANSION BOLT
EXPANSIEVER ANKER DICHTUNGER

GROUTING PIPE
DIE ROHRE FÜR INJEKTÜRUNG

MOUTH PIPE
DIE ROHRE EINFACHE

EXTEND PIPE
DIE ROHRE VERLÄNGE

PENETRATING PIPE
DIE ROHRE PENETRATIONE

CONTINUOUS GROUTIG ELEMENT
DIE INJEKTIERUNG ENDPATRONE

GROUTING ELEMENT
DER INJEKTIERUG
DURGEHEND TEIL

FIG. 3
ANCHORING AND GROUTING SYSTEM BOLTEX
ANKERN UND INJEKTIERUNGSYSTEM BOLTEX

and to influence the active grouting length of borehole. According to the former analysis it is advantageous to locate the packer into the neutral point of bolt.

Accessories of bolting and grouting system contain also set of pads, which are situated on the outer part of bolt. Its construction ensures certain automatic prestression between the pad and the rock.

The bolting and grouting system Boltex makes it possible to use different kinds of grouting materials, e.g. cement or two-part polyurethane resins. Grouting can be realised basically in two ways:
- as a low pressure injection, when grouting pressure is smaller than pressure necessary for the expansion of bolt. Then the bolt is expanded and the injection membrane is broken by water from bolting pump and in second step proper grouting is realised with low pressure from special injection pump;
- as a high pressure injection, when the expansion of bolt, breaking of the membrane and proper grouting is realised directly with grouting material from only one pump.

Hitherto experience from using of the combined bolting and grouting system in coal mines and underground structures are encouraging. It was proved, that bolts reinforce rock mass ring on the perimeter of excavation and prevent from the creation and increasing of fissures, so that the amount of waste grouting material flowing into the excavation decreases. So bolts can partly or fully replace the first - packing - step in two-step injection procedure. The combined technology makes it possible to use higher injection pressure. Due to immediately anchored bolts the total amount of grouting material also decreases in 20 - 30 %. This type of bolts can be anchored in weak rocks too.

Practice confirms, that combined bolting and grouting system Boltex brings not only economical effect due to cancellation of certain operations, but it is advantageous from the geomechanical point of view too. The important advantages of the combined bolting and grouting technology can be summarized:
- very quick and easy installation,
- low requirements on the accuracy of the borehole diameter,
- immediate bearing capacity of bolts,
- prestressing between pad and rock,
- large range of injection pressure,
- unlimited length of bolt column,
- arbitrary position of packers along the length of bolt,
- brick-box character of the Boltex makes it possible to adapt the system to the real conditions of using.

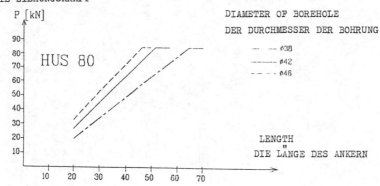

FORCE
DIE ZIEHUNGSKRAFT

P [kN]

DIAMETER OF BOREHOLE
DER DURCHMESSER DER BOHRUNG

HUS 80

LENGTH
DIE LÄNGE DES ANKERN

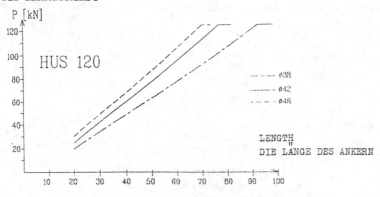

FORCE
DIE ZIEHUNGSKRAFT

P [kN]

HUS 120

LENGTH
DIE LÄNGE DES ANKERN

FIG. 4 PULL-OUT TEST IN LIMESTONE
DIE PRÜFUNGEN DER FESTIGKEIT ANKERN IM KALKSTEIN

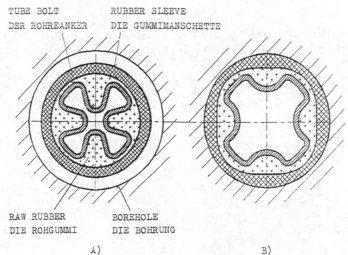

TUBE BOLT   RUBBER SLEEVE
DER ROHREANKER   DIE GUMMIMANSCHETTE

RAW RUBBER   BOREHOLE
DIE ROHGUMMI   DIE BOHRUNG

A)   B)

FIG. 5 SCHEME OF PACKING
SCHEMA DER DICHTUNG

129

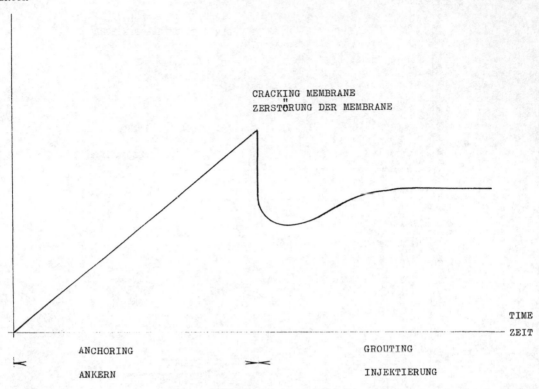

FIG. 6 PROCESS OF ANCHORING
VORGANG DES ANKERNS

REFERENCES

Sun, X. 1983. Grouted rock bolt used in underground engineering in soft surrounding rock or in highly stressed regions. Proc. International Symposium on Rock Bolting Abisko, Balkema.

Janas, P. 1994. Mathematic modelling of different bolts in rocks. Report Research Mining Institute Ostrava.

Supárek R., Paloncy, L. 1994. Bolting and grouting system Boltex. Proc. International Conf. Underground Structures Prague.

Anker in Theorie und Praxis, Widmann (Herausgeber) © 1995 Balkema, Rotterdam. ISBN 90 5410 577 1

Darstellung der Ankertechniken mit physikalischen Modellverfahren

Presentation of bolt techniques by physical modelling

Michael Würtele
Deutsche Montan Technologie, IGH, Essen, Germany

ZUSAMMENFASSUNG: Handelte es sich bei Einsatzorten von Ankern im deutschen Steinkohlenbergbau bisher oft um Abbaustrecken, so soll die Ankertechnik nun vermehrt in Gesteinsstrecken in großen Teufen genutzt werden. Dort sind Anker bezüglich ihres Versagens auf eine kombinierte Zug- Scherbelastung so auszulegen, daß sie auch in Partien mit großen Gebirgsverformungen lange Standzeiten garantieren. Maßstäbliche physikalische Modellversuche, in denen das umgebende Gebirge zu modellieren ist, geben eine wertvolle Hilfe, Ankeranwendungen und Ankerverhalten im Vorfeld zu testen und wirkungsvoller zu gestalten. Die Versuche werden bei DMT auf dem Strekkenmodellprüfstand durchgeführt, bei dem scheibenförmige Modelle zum Einsatz kommen.

ABSTRACT:Bolts should not only be longer used for gate roads, but also for other roadwaysystems in big depths in the German coal-fields, where the range of bolt applications should be increased. In these depths bolts have to be designed for combinations of huge tensile- and shearingarrogations and they should also be able to guarantee long standing times in rockparts of big stratadisplacements. Especially for this problems, physical modelling can be used to present strata boltings systems and their effects on strata movement. So nearly 300 two dimensional physical modeltests - at the scales of 1:10 up to 1:20 and the size of 2m*2m*0,4m - were carried out over many years by DMT and have brought about the possibility of reproducing support behaviour and fracture mechanism of layerd rock mass.

1. EINLEITUNG

Um kosten- und zeitaufwendige in-situ-Versuche, soweit diese überhaupt möglich sind, auf das notwendigste zu reduzieren, werden neue Ankertechniken sowie deren Auswirkungen auf die geomechanischen Abläufe im Gebirge im Vorfeld mit physikalischen Modellverfahren untersucht.

Für physikalische Modelluntersuchungen ste-

hen bei der DMT scheibenförmige und räumliche Modelle der Größen 0,4m*2m*2m bzw. 3m*2,5m*4m zur Verfügung. Um bei physikalischen Modelluntersuchungen quantitativ und qualitativ richtige Ergebnisse zu erlangen, müssen alle relevanten physikalischen Parameter des Ankerausbaus,aber auch die des umgebenden Gebirges maßstabsgerecht dargestellt werden.

2. Darstellung des Gebirges in der physikalischen Modelltechnik

Das als anisotrop zu bezeichnende Steinkohlengebirge des Ruhrgebietes besteht meist aus einer geschichteten Abfolge von Sandsteinen, Sandschiefern, Schiefertonen sowie Kohlen unterschiedlichster physikalischer Eigenschaften. Aufgrund dieser Schichtung und der mehr oder weniger unregelmäßig ausgebildeten Kluftsysteme stellt sich das Ruhrkarbon als ein Diskontinuum dar, dessen Belastbarkeit durch die Berührungs- und Kluftflächeneigenschaften der Kluftkörper und in geringerem Maße durch physikalische Eigenschaften der einzelnen Schichten gegeben ist. Daraus folgt, daß die sogenannte Verbandsfestigkeit die eigentliche Gebirgsfestigkeit darstellt.

Darüberhinaus lassen sich die streckennahen Gebirgsschichten in angeschnittene und nicht angeschnittene Schichten einteilen, wobei sich erstgenannte unter Beanspruchung anfangs nur bankparallel, die anderen lediglich bankrecht ausdehnen können.

Aus den genannten Gründen sind unter Tage bestimmte Bruchbilder vorherrschend. So liegt das Ziel der Modellierung darin, eben diese Bruchsituationen als Funktion bestimmter Beanspruchungen nachzubilden. Maßstabsgetreue Simulationen bankrechter Druckfestigkeit und bankparalleler Biegezugfestigkeit, verbunden mit einer feinen Bankung der Modellschichten reichen aus, um mit Modellwerkstoffen naturgetreue Bruchbilder zu erzeugen.

Seit 1958 wurden bei der DMT mehr als 300 körperliche Modelle untersucht, die i.d.R. aus Portland- Tonerdeschmelz-, einem kalziumaluminathaltigen Feinzement und Elektrofiltersche sowie Wasser bestanden.

Horizontale Trennflächen werden im Modell zum einen durch Gleitlösen, die ein Gleiten der Schichtpakete aufeinander erlauben, simuliert, zum anderen durch Lösen, die die Schichtpakete in feine Bankungen < 5mm unterteilen. Die Gleitlösen und Lösen zeichnen sich durch unterschiedliche Reibungsbeiwerte aus, die für Gleitlösen bei μ = 0,2 bis 0,3 sowie für Lösen bei μ =0,6 bis 0,8 liegen.

Verwendet werden die Modelle ab dem Zeitpunkt, ab dem die erste und die zuletzt gegossene Modellschicht annähernd gleiche Druck- und Biegezugfestigkeiten ausweisen, d.h. daß die Aushärtung weitestgehend beendet ist.

3. Darstellung der Anker in der physikalischen Modelltechnik

Heute gibt es eine große Anzahl von verschiedenen Ankern, die sich durch ihr jeweils charakteristisches Zug- und Scherverhalten unterscheiden. Bis zu einer kritischen Dehnung von 10% werden Anker als starr angesehen, darüberhinaus als nachgiebig. Das gewünschte Verhalten ist bei Modellankern durch die Verwendung eines Materials, jedoch auch durch die Kombination verschiedenster Materialien zu erreichen. Die Auslegung von Ankern sowie eine praxisnahe Auswertung der Ankerkennlinien kann nur dann erfolgreich sein, wenn die Wirkungsweise eines Ankers bzw. einer Systemankerung bekannt ist.

Nach DIN 21521 sind Gebirgsanker "Bauteile", die in Bohrlöchern im eingebauten Zustand durch Aufnahme von Zugkräften oder von Zug- und Scherkräften Gebirgsteile miteinander oder Konstruktionselemente mit dem Gebirge verbinden." Mit Ankern wird also eine Verbesserung der Gebirgseigentragfähigkeit erreicht, die unter folgenden Aspekten kurz dargelegt werden.

Aufhängen von einzelnen Kluftkörpern durch Einzelankerung.

Aufhängen einer flächenhaften Sicherung (Verzug).

Aufhängen von größeren Gebirgsteilen durch Systemankerung.

Stabilisierung von Gebirgszonen durch Einbringen zusätzlicher Normalkräfte zur Erhöhung der Reibung an natürlichen Trennflächen (Vernagelung).

Erhöhung der inneren Reibung einer Gebirgszone durch Vorspannen der Anker oder Injektion von Verfestigungsmitteln (Balkenbildung).

Erhöhung der inneren Reibung und der Festigkeit durch Ankervorspannung bei gewölbten Hohlraumgeometrien (Gewölbebildung).

In die körperlichen Modellen wurden bisher folgende Modellanker eingebaut:

Starre Anker:

M-24 bzw. M-27 Anker mit und ohne Profilierung sowie Seilanker

Nachgiebige Anker:

Kombianker sowie M-27 Anker mit Freispielstrecke

Bei der modelltechnischen Ankernachbildung kommt es weniger auf das exakte Nachbauen jedes einzelnen Ankerbauteils an, als vielmehr auf das Nachbilden der Ankerkennlinien, die entweder aus in-situ Messungen oder aber aus, bei der DMT im Maßstab 1:1 durchgeführten Ankerzug und -scherversuchen, bekannt sind.

Bild 1 zeigt die in reinen Zugversuchen an Modellankern - Maßstab 1:15 - ermittelten Zugkräfte und die dabei aufgetretenen Dehnungen. Hier ist in erster Linie Anker 4 - der Kombianker - zu nennen, der bei geringer Dehnung bis zum Bruch des Hüllrohres hohe Zugkräfte und anschließend bis zum Versagen bei ca. 30% Dehnung Zugkräfte aufnehmen kann.

Seit kurzer Zeit werden nun Modellanker im Kleinmodellprüfstand kombinierten Scher- und Zugversuchen unterzogen, deren Ergebnisse für Kombianker Bild 2 zu entnehmen sind.

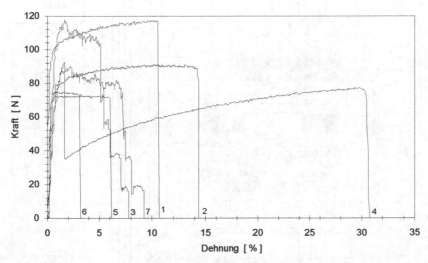

Bild 1: Kennlinien der Zugversuche im Maßstab 1:15: 1.M-27 mit Freispiel 2.Bündelanker
    3.Seilanker 4.Kombigleitanker 5.M-24 6.M-24, profiliert 7.M-27, profiliert
Fig. 1: Tensile test for bolts in a reduced scale of 1:15
    1. M-27 with a part of free movement 2. Flexbolt 3. Ropebolt 4. Kombibolt
    5. M-24 6. M-24,with shaping 7. M-27, with shaping

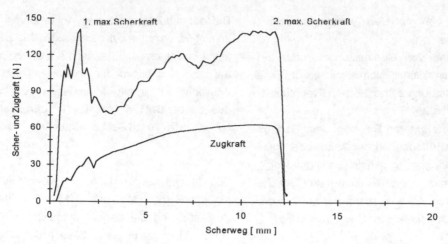

Bild 2: Kennlinien des Kombigleitankers bei einem Scherversuch im Maßstab 1:15 unter einem Scherwinkel von 90°

Fig. 2: Shearing- and tensileforces for a flexible Kombibolt

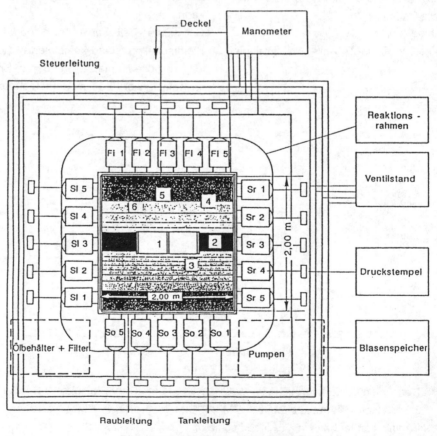

Bild 3: Der Streckenmodellprüfstand

Fig. 3: The roadway testing stands

Wie Bild 3 zeigt, liegt ein 2m*2m*0,4m großer Modellkörper mit lotrechter Streckenachse auf einer waagerechten Grundplatte flach in dem aus vertikalen und horizontalen Rahmen bestehenden Streckenmodellprüfstand. Jede der Stirnseiten des Modellblocks wird mit jeweils fünf Stempeln über Kalotten- und Ausgleichsplatten mit Druck beaufschlagt. Die acht Eckzylinder, vier Mittelzylinder sowie die acht restlichen Zylinder verfügen als Zylindergruppen über jeweils eine eigene Ringleitung.

Um ein Aufbrechen der Modelloberfläche nach oben zu verhindern, wird auf diese über eine mit vier Stempeln besetzte Deckelplatte Druck aufgebracht. Jeder der 24 im Prüfstand verwendeten Zylinder kann bei einem Zylinderinnendruck von 320bar eine maximale Stempelkraft von 1MN aufbringen. Bei 320 bar führt dies im Modellinneren zu einer Spannung σ_{Mi} von 6,84MPa, auf der Modelloberfläche σ_{Mo} zu 1MPa. Beim Fahren des Versuches liegt das Verhältnis von σ_{Mo} zu σ_{Mi} bis zu einer Spannung σ_{Mi} von 2,14MPa bei ca. 0,45. Damit sind Rißbildungen, die waagrecht zur Modellebene verlaufen und das o.g. Aufbrechen der Modelloberfläche hervorrufen, zu verhindern. So läßt sich eine maximale Vergleichsteufe - der Maßstab 1:15 vorausgesetzt - von 4102m simulieren. Diese Vergleichsteufe gibt dabei die Teufe an, in der der jeweilige Druck als Überlagerungsdruck ansteht. Die Umrechnung basiert auf der Annahme, in 1000m Teufe eine Spannung von 25MPa anzutreffen, eine mittlere Dichte des Ruhrkarbons von 2,5t/m³ vorausgesetzt.

4. Modellankersysteme im direkten Vergleich

Im folgenden Kapitel werden zwei Bogenstrecken miteinander verglichen, die unterschiedliche Ankerarten sowie Ankerdichten aufweisen. Die Modellblöcke stammen aus der gleichen Herstellungsserie, weisen zudem die gleiche Geologie auf und dienen daher bestens zu Vergleichszwecken.

Während der Ausbau in Modell 283 aus einem 4-teiligen Bogenausbau und 66 starren M-24 Ankern besteht, weist Modell 286 eine neue Art der Verzugsverstärkung auf. Letzterer Ausbau wird zudem durch starre Anker und Kombianker verstärkt, so daß der Ausbaustützdruck alleine durch die Anker bei 550kN/m² liegt und somit doppelt so hoch ist wie in Modell 283.

Das folgende Bild 4 zeigt das Konvergenzverhalten beider Modelle. Leicht läßt sich an den Verläufen der erhöhte Ausbaustützdruck erkennen. So liegt das Ausbauversagen in Modell 283 bei einer Vergleichsteufe von 1550m und einer vertikalen Konvergenz von 11% vor, während der Ausbau in Modell 286 erst bei 2100m sowie 20% Vertikalkonvergenz versagt. Interessant ist aber auch ein Vergleich bei ähnlicher Vergleichsteufe, bei der für Modell 286 geringere Gebirgsverformungen feststellbar sind.

Die ersten Brüche machen sich in beiden Modellen in Form des Sohlenaufbruchs bemerkbar. Bei Modell 283 lag das Ausbauversagen in der zweiten Teststufe vor; bereits in der nachfolgenden Teststufe weiteten sich Brüche bis in das 0,6-fache des Streckendurchmessers ins Liegende aus. Auch wurden höhere Konvergenzen erreicht, freilich ohne Vergrößerung der aufgebrachten Gebirgsdruckes. Während des vierten Versuchsstufe kam es in den Stößen zu ersten Keilbrüchen und zum Abscheren der Anker, die die Gleitlösen durchörtern.

Das Ankerverhalten von Kombiankern sowie starren Ankern sollte anhand des Modells 286 näher untersucht werden, in dem 130 Kombianker und 70 starre Anker integriert waren. Nach dem Beenden aller Versuche erfolgte die Freilegung einzelner Ankerebenen, um Auf-

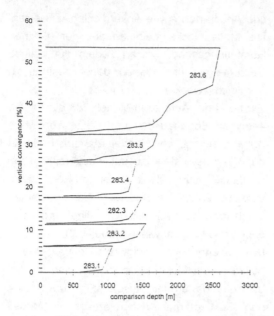 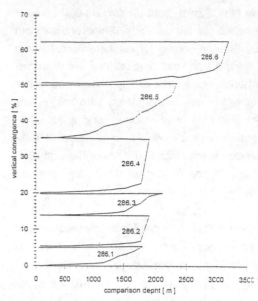

Bild 4:Vertikalkonvergenzen der Streckenmodelle 283 (links) und 286 (rechts)

Fig. 4:Vertical convergence of the roadways in model 283 (on the left) and in model 286 (on the right)

Bild 5:1.Freigelegte Ankerebene in Modell 286

Fig. 5:Exposing of the first bolt plane

esterharzen einwandfrei und schaumfrei verklebt und wiesen Bruchformen auf wie sie aus untertägigen Untersuchungen bekannt sind. Während die starren Ankern in den drei untersten Ankerreihen wenig verformt, aber dennoch gerissen waren, liegen die Kombianker teilweise in sehr stark verformtem, jedoch intaktem Zustand vor.

Ein Vergleich dieser beiden Systeme zeigt deulich die Möglichkeiten der physikalischen Modelltechnik auf. Zum einen ist es möglich das umgebende Gebirgenachzubilden, zum anderen beinahe beliebige Ausbaukomponenten - in diesem Fall waren es Anker - maßstabsgerecht zu modellieren. Auf diese Weise lassen sich über das Maß der gebirgsmechanischen Beeinflussungen durch Ausbausysteme in qualitativer sowie quantitativer Hinsicht Aussagen treffen..

schluß über das Ankerverhalten und den Ankerzustand zu erlangen. Wie Bild 5 zeigt, waren in der ersten Ankerebene alle Anker mit Poly-

136

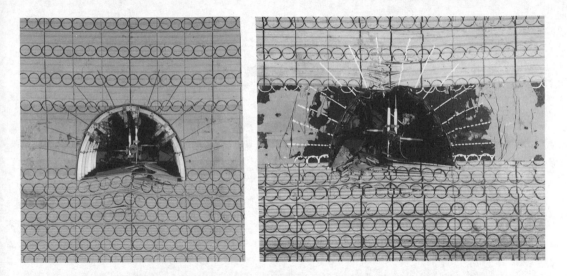

Bild 6: Zeitpunkt des Ausbauversagens in Modell 283 (links) und 286 (rechts)
Fig. 6: Point of time of support failure in model 283 (on the left) and 286 (on the right)

Anchors in Theory and Practice, Widmann (ed.) © 1995 Balkema, Rotterdam. ISBN 90 5410 577 1

Static service behaviour of rock bolts subjected to blast loadings in tunnelling

Das statische Betriebsverhalten von Tunnel-Sprengungen ausgesetzten Felsankern

H.Xu
Université de Sherbrooke, Que., Canada

A.A.Rodger & D.C.Holland
University of Aberdeen, UK

G.S.Littlejohn
University of Bradford, UK

ABSTRACT: This Paper describes the static service behaviour of rock bolts subjected to close proximity blasting during the construction of Pen y Clip Tunnel in North Wales. Twenty-four 6 m-long two-speed resin bonded rock bolts were installed at distances of 1.1 to 5.7 m from the blast face, and were prestressed to loads varying from 3 to 108 kN. The residual static service behaviour of the rock bolts during tunnelling was continuously monitored for up to 47 days. The effects of blasts, prestress load, and bolt distance from the blast face on the residual static service behaviour of the rock bolts are discussed in the Paper. Based on the test results, a distance of 4 m from the blast face and a correlated peak particle velocity of 225mm/s were established for the safe installation of permanent resin bonded rock bolts at Pen y Clip Tunnel.

ZUSAMMENFASSUNG: Ein an den Universitäten Aberdeen und Bradford durchgeführtes Forschungsprogramm soll ein grundlegendes Verständnis des dynamischen und statischen Verhaltens von Sprengungen ausgesetzten Tunnel-Felsankern bereitstellen. Die Arbeit besteht aus Finite-Elemente-Modellierungen, Labor-Experimenten und Tests vorort. Die praktischen Tests wurden bei Bauarbeiten (Bohr-Spreng-Methode) zu zwei neuen Tunneln in Nord Wales durchgeführt. Diese Arbeit beschreibt das statische Kurz- und Langzeit-Betriebsverhalten der Felsanker in einem der Tunnel bei Pen y Clip Headland, nahe Conwy.

Vierundzwanzig 6m lange Harzverbundanker wurden in einer Entfernung von 1,1 bis 5,7 m von der Sprengstelle in Mikrodiorit installiert und mit 3 bis 108 kN vorgespannt. die Einflüsse von Anker-Vorspannung und Entfernung zur Sprengstelle auf das Betriebsverhalten bis 47 Tage nach der Sprengung werden diskutiert. Es wird gezeigt, daß das Anker-Verhalten nach Sprengungen in vier Zeitabschnitte eingeteilt werden kann:

1) Unmittelbare Kraftänderung
2) Kraftverlust
3) Kraftzunahme und
4) Zufällige Kraftänderungen

Es wurde gefunden, daß die Entfernung zur Sprengstelle einen signifikanten Einfluß auf das Kurz- und Langzeit-Verhalten der Felsanker hat. Höhere Kraftänderungen wurden von Ankern erhalten, die Nahe der Sprengung installiert waren. Die Höhe der Vorspannung zeigte lediglich einen Einfluß auf das Kurzzeit-Verhalten (<24 Stunden nach der Sprengung), wobei eine höhere Kraft eine niedrigere Kraftänderung bewirkte. Auf den Untersuchungen basierend wurde für den Pen y Clip Tunnel eine Entfernung zur Sprengstelle von vier Metern für die sichere Installierung von permanenten Harzverbund-felsankern etabliert. Die damit verbundene Spitzen-Partikelgeschwindigkeit beträgt 225mm/s.

1 INTRODUCTION

Rock bolting has been widely used in civil and mining engineering as an economic, effective and simple means of rock support and reinforcement. In many hard rock tunnels constructed using the drill and blast method, the need for safe and early support often necessitates that the rock bolts be installed adjacent to the advancing tunnel face in close proximity to the blast source. This gives rise to the possibility of bolt damage resulting from the blasting operations and the necessity of defining a safe distance from the blast face beyond which bolts will not diminish their capacity for permanent support.

Due to the dearth of published information on research into this complex problem, current design criteria are believed to be conservative in estimating the safe distance for the installation of permanent resin bonded rock bolts subjected to blast loading. In the absence of theoretical or predictive methods for estimating the dynamic response of rock bolts to blast loading, current design practice uses a safe distance derived from precedent practice, or a limiting parameter of rock response to blasting, normally peak particle velocity (PPV). For example a PPV threshold of 200 mm/s and an associated safe distance of 5 m were established for the preliminary design of the permanent rock bolts in rhyolite at Penmaenbach Tunnel in North Wales (Littlejohn et al. 1989). If rock bolts need to be placed closer to the tunnel face than the specified safe distance, the bolts are deemed to be temporary and may be replaced after blasting. This leads to costly duplication of the bolts which may be unnecessary.

To develop an improved design approach, a research programme has been conducted by the Universities of Bradford and Aberdeen in two phases over the period of 1986-1992. The first phase was conducted in three parts from 1986 to 1989 and included a full scale field investigation at Penmaenbach Tunnel in North Wales (Littlejohn et al. 1989; Rodger et al. 1988; Holland 1994; Mothersille 1989), laboratory model tests (Mothersille 1989), and finite element studies (Chin 1994), of the dynamic response of resin bonded rock bolt systems. The second phase of the research programme, part of which is represented in this paper, was conducted over the period of 1989-1992 and involved a full scale field investigation at Pen y Clip Tunnel in North Wales (Rodger et al. 1993; Xu 1993; Holland 1994), laboratory model tests (Xu 1993), and finite element studies (Chin 1994), of the dynamic and static response of a resin bonded rock bolt system. The field programme of the latter phase was devised to: (1) examine the validity of the results obtained at Penmaenbach Tunnel for a different rock mass classification; (2) study the influence of dynamic loading, prestress loading, and bolt distance to the blast face, on the dynamic and static behaviour of rock bolts; and (3) investigate in detail the dynamic load distribution along the bolt length due to blasting.

This paper describes the residual static service behaviour of rock bolts subjected to close proximity blasting at Pen y Clip Tunnel, North Wales. Twenty-four 6 m-long two-speed resin bonded rock bolts were installed at distances of 1.1 to 5.7 m from the blast face, and were prestressed to loads between 3 and 108 kN. The residual static service behaviour of the rock bolts during tunnelling was continuously monitored for up to 47 days. The effects of blasts, prestress load, and bolt distance from the blast face on the residual static service behaviour of rock bolts are discussed.

2 ROCK MASS CHARACTERISTICS AND TUNNEL SUPPORT

The Pen y Clip Tunnel was one of the final stages of the upgraded A55 coast road through the Pen y Clip Headland, North Wales. The tunnel is 930 m long, 10 m wide, and 8 m high. The Headland is a steeply rising igneous intrusion and measures approximately 3000 m by 4000 m in an east-west direction. The headland is composed mainly of dark grey, fresh or surface stained, fine grained microdiorite. All the rock excavation described in this Paper was performed through this material using the drill and blast method. A burn cut blasting pattern was used with charges placed typically in 138 drilled holes and detonated in 23 arrays. A typical face advance for each blast was 4m.

The geotechnical properties for both intact rock and rock mass are listed in Table 1. The intact rock material is generally described as very strong to extremely strong with joint spacing of typically from 0.02 to 0.6 m. The rock mass was classed as fair and poor according to the tunnelling quality index (Q) (Barton et al. 1974) and the Rock Mass Rating (RMR) scheme (Bieniawski 1989), and localized weakness zones comprising spaced open fractures with clay infillings were interspersed throughout the tunnel site. Full details of the site geology are described elsewhere (Xu 1993). Three classes of the rock were established according to the Index Q (Barton et al. 1974) and the RMR (Bieniawski 1989).

Table 1. Geotechnical properties of microdiorite
 Geotechnische Eigenschaften von Mikrodiorit

Parameter	Range	Mean
Intact rock		
Bulk density (Mg/m3)	$2.46 \sim 2.86$	2.77
Compressive strength (MPa)	$73.7 \sim 406.5$	254.3
Point load index (MPa)	$7.73 \sim 19.20$	13.21
Static elastic modulus (GPa)	$55.7 \sim 93.5$	75.9
Dynamic elastic modulus (GPa)	$80.5 \sim 92.5$	85.3
Poisson's ratio	$0.19 \sim 0.34$	0.23
Sonic velocity (m/s)	$5655 \sim 6220$	5906
Rock mass		
Rock mass rating (RMR)	$36 \sim 56$	46.4
NGI tunnelling quality index (Q)	$0.70 \sim 2.22$	1.35

The most competent rock, requiring least support was designated as Class II, whilst Classes III and IV were defined as less competent rock masses. Rock bolting and shotcrete were the principal forms of tunnel support. Rock classes II and III were supported with a standard form of patterned rock bolting and sprayed concrete of 100 mm thickness at sidewalls and were combined with welded steel mesh reinforcement at the tunnel roof. Rock class IV and the zones near the portals, where rock cover is shallow or the quality is very poor, were supported using steel ribs. Bolt length, fixed anchor length, and bolt spacing were related to rock class with the location of rock bolts on the tunnel perimeter being as shown in Table 2. Two-speed resin bonded rock bolts were selected due to their resilience to blasting.

Table 2. Standard forms of rock support at Pen y Clip
Standardformen von Felsbefestigungen im Pen y Clip- Tunnel

Position in tunnel	Bolt length (m)	Fixed anchor length (m)		Bolt spacing (m)	
		Class II	Class III	Class II	Class III
South sidewall	3.5	1.5	2.0	2.0	1.5
Haunch	3.5	1.5	2.0	1.5	1.0
Crown	7.0	2.0	3.0	1.5	1.0
Haunch	3.5	1.5	2.0	1.5	1.0
North sidewall	3.5	1.5	2.0	2.0	1.5

3 EXPERIMENTAL PROGRAMME

The experimental two-speed resin bonded rock bolts consisted of 6 m-long standard bolts, and fully instrumented bolts fitted with 5 load cells along the bolt length. The bolt tendon consisted of a 25 mm diameter galvanised deformed rebar. The test arrangement of a fully instrumented rock bolt is shown in Figure 1. The field tests commenced at a tunnel chainage of 1281m, where rock class II was encountered with Q values of 0.47~2.22 and RMR values of 36~56. The location of experimental rock bolts associated with the tunnel face chainage is shown in Figure 2. The field programme was divided into two phases. In Phase I, only one standard bolt was installed in test 1 in order to check the monitoring systems and verify the settings of tape recorders and amplifiers. After test 1, two standard rock bolts were installed at approximately 2 m centres to coincide with a typical face advance of 4 m per blast. In total one fully instumented, and ten standard, rock bolts were installed, and up to six rock bolts were monitored dynamically at the same time. In Phase II, seven fully instrumented and six standard rock bolts were installed at approximately 1.3 m centres. The simultaneous dynamic monitoring was limited to two fully instrumented and three standard rock bolts for each test.

To examine the influence of prestress load (Tw), the bolts were post-tensioned to a specified load from 3 to 108 kN, after the fast setting resin had set (approximately 2 minutes). The bolt tension was then locked-in by the slow setting resin (approximately 11 minutes), except for a 0.7 m length below the bolt bearing plate (Figure 1). This length was decoupled with grease impregnated tape to permit some load adjustment in the event of surface spalling of the rock after blasting. All experimental bolts were installed at 1.1 to 5.7 m from the blast face near the springing line and instrumented with load cells and accelerometers on the bolt heads (Figure 1).

4 INSTRUMENTATION SYSTEMS

Figure 3 shows the instrumentation systems employed which comprised: (1) the load cell system

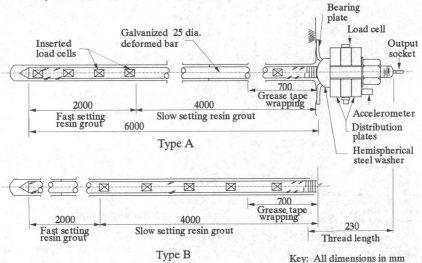

Type A

Type B

Figure 1 Fully instrumented rock bolt types A and B

Key: All dimensions in mm

Abbildung 1 Vollinstrumentierte Felsanker der **Typen A und B**

141

for monitoring the dynamic and static load fluctuation on the bolt head during and after each blast; (2) the accelerometer system for monitoring vibration movement of the bolts and surrounding rock; and (3) the specially instrumented rock bolt system for monitoring instantaneous dynamic and static load distributions along the bolt length. Details of the instrumentation systems of items 2 and 3 involved and the protection of instrumentation and cables have been described elsewhere (Xu 1993; Rodger et al. 1993). Only the load cell system is briefly described herein.

The load cell system comprised annular Glotzl hydraulic load cells and a load amplifier. The load cells have a capacity of 0-250 kN with a 20% overload capability, an accuracy of 1%, and a frequency response of up to 2 kHz. Inside the load cells, electronic pressure transducers were fitted to allow remote reading.

Test number	Face chainage (m)		Chainage of experimental bolt (m)	Layout
1	1285.2	1	1283.60	● 1
2	1289.6	2	1286.40	● ● ●
		3	1287.35	3 2 1
3	1294.0	4	1288.54	● ● ● ● ●
		5	1290.66	5 4 3 2 1
4	1298.3	6	1294.28	● ● ● ● ● ● ○
		7	1295.78	7 6 5 4 3 2 1
5	1302.2	8	1298.40	● ● ● ● ● ● ○
		9	1299.98	9 8 7 6 5 4 3
6	1306.5	10	1303.30	B ● ● ● ● ● ○
		11	1304.72	11 10 9 8 7 6 5

(a) Phase I

				Layout
7	1311.0	12	1306.80	A ● B ● ● ○ ○
		13	1309.60	13 12 11 10 9 8 7
8	1315.5	14	1310.74	A B ● a ● ○ ○
		15	1312.08	16 15 14 13 12 11 10
		16	1312.96	
9	1319.7	17	1315.73	A B ● a b ○ ○
		18	1316.77	19 18 17 16 15 14 13
		19	1318.02	
10	1323.8	20	1320.58	A B ● a b ○ ○
		21	1321.47	22 21 20 19 18 17 16
		22	1322.73	
11	1327.9	23	1325.01	● ● A B ● ○ ○
		24	1326.13	24 23 22 21 20 19 18

(b) Phase II

● Standard rock bolt with load cell and accelerometer
○ Standard rock bolt only with load cell
A or B Bolt types A or B dynamically monitored
a or b Bolt types A or B only load cell & accelerometer monitored

Figure 2   Position of Experimental bolts at Pen y Clip Tunnel

Abbildung 2 Lage der Versuchsanker im Pen y Clip Tunnel

Each load cell was located between a hemispherical hardened steel washer in order to obtain axial readings and secured in place by a lock nut (Figure 1).

Prior to the tests, all load cells were calibrated over their full load range in the laboratory using the actual field instrumentation system. In total 500 m long cables were used from the monitoring station to the last test position. To prevent excessive signal loss in this length of cable, an amplifier housed in a steel cabinet was used to energise and amplify the outputs of up to ten rock bolts. The amplifier, situated about 50 to 90 m from the test bolts, was able to drive the composite cables connecting the load cells to the monitoring station some distance away, typically 400 m, with a flat response from DC to 5 kHz, and had a low drift with respect to time and temperature.

The static load variations of rock bolts were monitored before and after blasting using an automatic computer data collection system (Xu 1993; Rodger et al. 1993). The system was devised to monitor the critical load variation with time at pre-set time intervals, which were: (1) before each blast; (2) every 2 minutes for the first 10 minutes after blasting; (3) every 10 minutes between 10 and 60 minutes; (4) every 30 minutes between 1 and 4 hours; and (5) every 1 hour between 4 and 24 hours. After the remote monitoring system had been disconnected, a digital transducer meter was used to read the load cells directly.

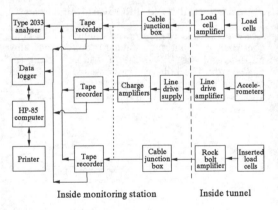

Figure 3   Schematic of instrumentation systems at Pen y Clip Tunnel

Abbildung 3 Schema der Instrumentierungs-systeme im Pen y Clip Tunnel

5   RESULTS AND DISCUSSION

5.1   *Immediate residual static behaviour of rock bolts*

The response of the rock bolts was influenced by the blasts, bolt distance to the blast face, the prestress load, and the rock mass condition. The typical residual static service behaviour within 24 hours of blasting is shown in Figure 4.

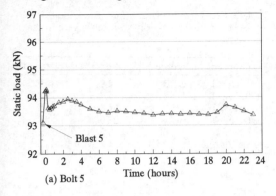

(a) Bolt 5

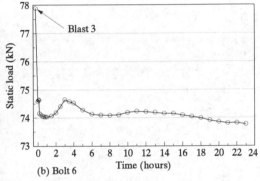

(b) Bolt 6

Figure 4   Typical static residual load variation of
rock bolts immediately after blasting
Abbildung 4 Typische Veränderung der statischen
Ankerrestkraft unmittelbar nach einer
Sprengung

Generally, the load variations of rock bolts can be divided into four periods:

1.   Period of immediate load variation. Immediately after blasting, 41 out of 61 rock bolts had increased loads, while 20 out of 61 rock bolts had decreased loads. When a blast occurred, elastic stress waves and borehole gases spread out into the rock mass, creating new micro-cracks and enlarging the apertures of existing fractures. As a result, the rock mass was expanded. The rock bolts were thus extended, and their loads were increased. However, the inner rock mass structure was also affected by blasting, and the loads of the rock bolts might suddenly reduce due to the closure of fractures. These load variations were usually observed over the first 10 minutes after blasting.

2.   Period of load loss. After the initial expansion of the rock mass, the microcracks and fractures tended to close, therefore the loads (57 out of 61 bolts) decreased. This behaviour was observed between 10 and 60 minutes after blasting.

3.   Period of load gain. In this period, the variations of load were dominated by tunnel convergence. Blasts created a new space inside the tunnel, around which the stress would redistribute, and the main amount of convergence would rapidly occur in a short period of time (usually less than 12 hours as recorded). As a result, the loads in the rock bolts would increase with increasing convergence. This behaviour usually took place between 1 and 5 hours after blasting.

4.   Period of random load variation. In this period the loads tended to be stable. The load variations, if any, were controlled by many factors, such as the convergence of rock mass, and nearby construction activity. This behaviour was observed between 5 and 24 hours after blasting.

The rock bolt response corresponding to periods 1 to 3 above may be defined as the *immediate* residual static behaviour, period 4 the *short term* behaviour, and that beyond 24 hours, the *long term* residual static behaviour.

The influence of the prestress load and the bolt distance to blast face on the immediate residual behaviour of rock bolts is shown in Figures 5 and 6, where the rock bolt and blast numbers are marked out for load variations greater than ±10% of prestress load. Although prestress loading has been found to have a significant effect on the dynamic response of rock bolts (Rodger et al. 1993) it was found to have no discernible influence on the immediate residual static behaviour. Generally, the load variation of rock bolts was between -27.2% and +6.7% of prestress load as shown in Figure 5. It is believed that the rock mass discontinuities had a significant effect on the immediate residual behaviour of the rock bolts. The immediate load increases were mainly caused by the movement of the rock mass such as expansion; and the sharp load decreases are related to the readjustment of rock mass, e.g. fracture closure.

As shown in Figure 6, the bolt position to the tunnel face has a significant effect on the immediate residual behaviour of the bolts. Within 3.8 m from the tunnel face, the rock bolts were affected by the blasts, with load variations between -27.2% and +6.7% of prestress load. Between 3.8 and 7 m, the load variations were between -12.2% and +4.8% of prestress load. Between 7 and 11 m, the blast did not affect the rock bolts significantly, and the load varied only between -0.1% and +3.7% of prestress load. Beyond 11 m, the blast could be considered not to have any effect on the static load behaviour of resin bonded rock bolts.

With the exception of bolts 18 and 19, all bolts positioned greater than 4 m from the tunnel face experienced an immediately static load variation of less than ±10% of prestress load which is acceptable in practice according to BS8081 (1989). It was observed that the load losses of bolts 18 and 19 were mainly caused by rock mass movement other than was induced by blast 10. Also, bolts 18 and 19 sustained only marginal greater load fluctuations of -12.2% and -11.2% of prestress load, respectively and continued to perform reliably. As a consequence, a safe distance of 4 m was proposed for design

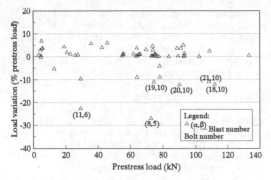

Figure 5   Influence of prestress load on the
immediate residual load of rock bolts

Abbildung 5 Einfluß der Vorspannkraft auf die
unmittelbare Restkraft der Felsanker

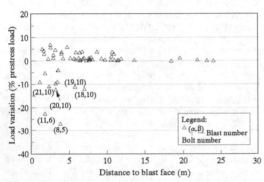

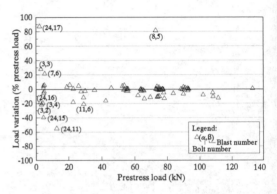

blast numbers are marked out for the load variations
greater than ±20% of prestress load.

Table 3. Statistical load variation of rock bolts
Statische Ladungsvariation von Felsankern während
des Betriebes

Load variation	Immediate %	Short-term %	Long-term %
>-20%T	3.3	4.8	16.7
-(20~10)%T	6.5	15.7	29.2
-(10~0)%T	23.0	51.8	25.0
+(0~10)%T	67.0	22.0	8.3
+(10~20)%T	0.0	0.0	8.3
>+20%T	0.0	4.8	12.5

Figure 6   Influence of distance to blast face on
immediate residual load of rock bolts

Abbildung 6 Einluß des Abstandes vom Sprengort
auf die unmittelbare Restkraft der
Felsanker

Figure 7   Influence of prestress load on the short
term residual load of rock bolts

Abbildung 7 Einfluß der Vorspannkraft auf die
Kurzzeit-Restkraft der Felsanker

purposes at Pen y Clip Tunnel. Even within 4 m from
the tunnel face, the load variations of 90.5% of rock
bolts were still within the limit of ±10% of prestress
load. A check lift on all bolts within 4 m would
enable static residual load fluctuations greater than
±10% to be corrected, if required. For a 4 m safe
distance, the correlated mean PPV of 225 mm/s
would be appropriate (Xu 1993; Rodger et al 1993).
The statistical residual load variation of rock bolts
immediately after blasting is shown in Table 3.

5.2   *Short-term Residual Static Load Variation*

The short-term (<24 hrs.) residual static behaviour of
the rock bolts was controlled by blast, rock mass
structure, prestress load, and bolt distance from the
tunnel face. The influence of prestress load and bolt
distance from the blast face on the short-term
residual static load variation of the rock bolts are
shown in Figures 7 and 8, where the rock bolt and

Generally, higher prestress loads resulted in
lower load variations, except bolt 8. Although bolt 8
had a high initial prestress load of 72.8 kN, it
suffered a load increase of 58.5 kN (+80.4% of initial
prestress load). For the bolts with less than 15 kN
prestress load, the load variations within 24 hours of
blasting were between -55.4% and +86.7% of
prestress load; for bolts with prestress load between
15-30 kN, the load variations were between -21.6%
and +5.4% of prestress load; for bolts with greater
than 30 kN prestress load, the load variations were
only between -16.9% and +4.4% of the prestress
load. Bolts 3, 7, and 24, with less than 13 kN
prestress load, could be considered as virtually
passive rock bolts. Small load changes could result in
very high percentage load variation as shown in
Figure 7.

Generally, the static load variations of rock
bolts tended to decrease with increasing distance
from the tunnel face. When the bolts were within 3 m
of the tunnel face, the load variations within 24 hours
of blasting were between -22.1% and +2.7% of
prestress load; when the bolts were within 3 ~ 7.6 m
from the tunnel face, the load variations were
between -18.6% and +5.4% of prestress load; when

the bolts were situated beyond 7.6 m from the tunnel face, the load variations were between -3.7% and 5.4% of prestress load, except bolts 3, 7, and 24 (Figure 8) due to their low prestress load. For example, bolt 24 with a prestress load of 2.2 kN sustained a load variation as high as +86.7% of prestress load although it was situated 27 m away from the tunnel face. The statistical data on short-term residual load variation is shown in Table 3.

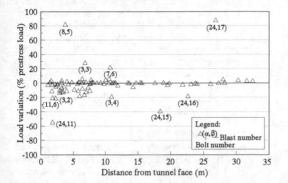

Figure 8   Influence of distance to blast face on the short term residual load of rock bolts

Abbildung 8 Einfluß des Abstandes vom Sprengort auf die Kurzzeit-Restkraft der Felsanker

5.3   Long-term Residual Static Load Variation

The static loads of rock bolts were monitored for 3 to 47 days. A typical long-term load variation (bolt 22) with time is shown in Figure 9, where the load fluctuations during each blast can be seen clearly. It is believed that due to wide fractures in the rock mass, blasting might have caused the separation and relative sliding movements in the rock mass, and thus

sudden changes in bolt loads such as during blasts 10 and 11. Except in the case of blast 10, which resulted in a load decrease of -9.2% of prestress load, blasts 11 to 17 all caused load increases due to the expansion of rock mass immediately after blasting. Some blasts had an irreversible effect on the rock bolts such as rock bolt 8.

The influence of prestress load and bolt distance to the tunnel face on the long-term residual behaviour of rock bolts are shown in Figures 10 and 11. Six rock bolts gained loads, while 18 rock bolts lost loads. The load variations of rock bolts were between -63.8% and +59.8% of initial prestress load at the end of the test programme.

Figure 10 indicates that, in similar manner to the immediate residual static behaviour of rock bolts, the prestress load has no effect on the residual static load variations of rock bolts. Even the bolts with high prestress load (>75 kN) still suffered high load variations, varying from -47.6% to +42.7% of initial prestress load up to 47 days.

In Figure 11, the rock bolts near the tunnel face, suffered higher load variations than those further away. Within 3 m, the load variations are from -63.8% to +59.8% of initial prestress load; in contrast to those situated between 3 and 4 m from the blast face, having load variations from -48.7% to +42.7% of initial prestress load. For the rock bolts 4 m away from the tunnel face, the loads were relatively stable, with load variations between +2.1% and -6.8% of initial prestress loads. This might be associated with the 4 m tunnel advance, or the 4 m safe distance, since the rock mass beyond this distance from the tunnel face had already sustained two close proximity blasts before bolt installation with consequential stress redistribution. The statistical data on long-term static load variations of rock bolts during service are summarised in Table 3.

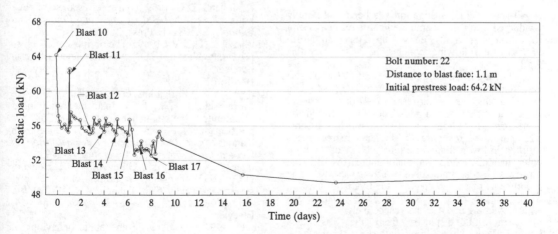

Figure 9   Typical residual static service behaviour

Abbildung 9 Typisches Betriebsverhalten der statischen Restkraft eines Felsankers

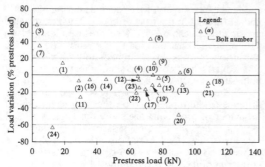

Figure 10   Influence of prestress load on the long
term residual load of rock bolts

Abbildung 10 Einfluß der Vorspannkraft auf die
Langzeit- Restkraft der Felsanker

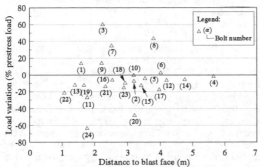

Figure 11   Influence of distance to blast face on the
long term residual load of rock bolts

Abbildung 11 Einfluß des Abstandes von der
Sprungstelle auf die Langzeit-Restkraft
der Felsanker

6   CONCLUSIONS

A full scale field investigation has been successfully
conducted into the dynamic and static responses of a
rock bolt system at Pen y Clip Tunnel in North
Wales. The following conclusions arise from this
work:

The amplitude of prestress load (3 to 108 kN)
was found not to influence the immediate (3~4 hours
after blasting) and long-term (3~47 days) residual
static load behaviour of rock bolts, but affected the
short-term (4~24 hours) load variation. In the short
term period of response the higher prestress loads
resulted in lower load variations. 90.5% of the rock
bolts sustained load variations within ±10% of initial
prestress loads immediately after blasting. The
proportions of rock bolts with load variations within
±10% of prestress loads were 74.7% and 33.3%
within 24 hours and 47 days respectively.

The bolt distances from the blast face
influenced not only the immediate residual behaviour,
but also the short-term and long-term load variations
of rock bolts following blasting. The closer to the
blast face the bolts, the higher the load variations.

The results indicated that a safe distance of 4 m and a
correlated mean PPV of 225 mm/s could be used for
the permanent rock bolt installation at Pen y Clip
Tunnel. Furthermore, rock bolts within 4 m could
still be used since 90.5% of the rock bolts sustained a
load variation within ±10% of prestress load. A
check lift on all bolts within 4 m would enable
residual static load fluctuations greater than ±10% to
be corrected, if required.

Blasting created new micro-cracks, enlarged
the apertures of existing fractures, and caused the
separation of closure of the rock mass. Generally,
load increases were considered to be due to the
expansion or separation of the rock mass, while load
decreases were mainly due to fracture closure. The
load increases were much lower than the load
reduction. Over the range of 1.1 to 5.7 m from the
blast face, all bolts performed well and were capable
of holding load after blasting.

ACKNOWLEDGEMENTS
The authors acknowledge the assistance received
from the U.K. Science and Engineering Research
Council (EPSRC), the Welsh Office, Travers Morgan
& Partners, Howard Humphreys & Partners, John
Laing Construction Ltd., Cementation Mining Ltd.

REFERENCES
Barton,   N.,   Lien,   R.Q.   &   Lunde   J.
1974.Eng.classification of rock mass for the
design of tunnel support. *Rock Mech. and Rock
Eng. J.*, 6(4): 189-236.
Bieniawski, Z.T. 1989. *Engineering rock mass
classification*. John Wiley and Sons, New York.
Chin, T L. 1995. A finite element investigation into
the response of rock bolts to close proximity
blasting. M.Phil. Thesis, Univ. of Aberdeen.
Holland, D.C. 1994. The behaviour of resin bonded
rock bolts subjected to close proximity blasting.
Ph.D. Thesis, Univ. of Aberdeen, UK.
Littlejohn, G.S., Rodger, A.A., Mothersille, D.K.V.
& Holland, D.C. 1989. Dynamic response of rock
bolts. *Proc. 2nd Int. Conf. on Found. and
Tunnels*, Eng. Tech. Press, 2: 57-64.
Mothersille, D.K.V. 1989. The influence of close
proximity blasting on the performance of resin
bonded bolts. Ph.D. Thesis, University of
Bradford.
Rodger, A.A., Littlejohn, G.S., Holland, D.C. &
Mothersille, D.K.V. 1988. Instrumentation used
to monitor the influence of blasting on the
performance of rock bolts at Penmaenbach
Tunnel. *Proc. Int. Conf. on Instrumentation in
Geotechnical Eng.*, TTL, London, 267-279.
Rodger, A.A., Littlejohn, G.S., Xu, H. & Holland,
D.C. 1993. Dynamic response of rock bolt
systems at Pen y Clip Tunnel in N. Wales, *Int.
Tunnelling Assoc., Dev. in Geotech. Eng.*, 74,
Elsevier.
Xu, H. 1993. The dynamic and static behaviour of
resin bonded rock bolts in tunnelling. Ph.D.
Thesis, University of Bradford .

Anker in Theorie und Praxis, Widmann (Herausgeber) © 1995 Balkema, Rotterdam. ISBN 90 5410 577 1

Wirkungsmechanismen von Ankern in Tunneln

Effects of rock bolts in tunneling

U.Zischinsky
DMT-Gesellschaft für Forschung und Prüfung mbH, Essen, Germany

ZUSAMMENFASSUNG: Gebirgsanker wirken vor allem durch ihren Einfluß auf die Bruchstrukturen des Gebirgsmantels. Für die realistische Dimensionierung einer Ankerung müssen daher zunächst die im Einzelfall maßgebenden Bruchmechanismen geklärt werden. Dazu muß der Planer über eine Sammlung von Bruchmechanismen verfügen, deren Zutreffen er für sein Projekt prüft. Der Bericht bringt eine solche Sammlung von Bruchmechanismen für nachbrüchigen bis druckhaften Fels.

ABSTRACT: Common methods of dimensioning tunnel linings consider rock bolts to apply supporting forces on the tunnel surface or shear forces within the rock mass or they consider rock bolts as reinforcement against dilatation. All these effects are within continuum mechanics.

On the other hand, for more than 20 years it is discussed that by these means the influence of bolts on rock mass behaviour is not described realistic. And this complies with the experience of DMT gained on sites and by numerous physical models:

Rock bolts mainly act by their influence on the pattern of fractures and by this on the residual strength of the rock mass surrounding a tunnel. And this is beyond continuum mechanics.

Searching for the real effect of rock bolts in a given application we are to find which mechanisms of failure rule the situation. With that knowledge only we are able to design rock bolts adequately. For this we need a collection of failure mechanisms to refer to in each individual case. Such a collection is based on experience and therefore incomplete on principle.

The paper is presenting a first attempt of such a collection ranging from fractures caused by dead loads to the failure of hard rock under heavy pressure

In den üblichen Konzepten zur Bemessung von Tunnelauskleidungen wirken Gebirgsanker v.a. durch die von ihnen erzeugten Stützkräfte, durch die Erhöhung des Scherwiderstandes im Gebirge oder als Bewehrung gegen Querdehnung im Rahmen kontinuumsmechanischer Modellvorstellungen. Seit über 20 Jahren wird aber beklagt, daß zumindest für Fels der tatsächliche Einfluß der Anker auf das Gebirgsverhalten auf diese Weise nicht ausreichend beschrieben wird. Nach den Beobachtungen der DMT unter Tage und nach zahlreichen Versuchen mit körperlichen Modellen ist das auch zu erwarten: Gebirgsanker wirken vor allem durch ihren Einfluß auf die Bruchstrukturen und damit auf die Rest-festigkeit des Gebirgsmantels „jenseits" einer kontinuumsmechanischen Betrachtung.

Will man für einen Anwendungsfall die tatsächliche Wirkung von Gebirgsankern ermitteln, so muß man daher erkennen, welche Bruch-mechanismen im konkreten Fall maßgebend sind und die Anker daraufhin auslegen. Für ein solches Vorgehen ist es notwendig, über einen Katalog von Bruchmechanismen zu verfügen, deren Zutreffen im Einzelfall abgefragt wird. Ein solcher Katalog beruht auf Erfahrung und ist notwendigerweise immer unvollständig.

BRUCHMECHANISMEN IN NACHBRÜCHIGEM UND GEBRÄCHEM FELS

In der Literatur bekannt ist die Verwendung von Gebirgsankern gegen das Niederbrechen von Kluftkörpern oder von Platten, die an einem kompetenten Hangenden aufgehängt werden (Bild 1). Das Aufhängen von Schichtbänken ist z.B. die maßgebende Wirkung der Anker in hessischen Salzbergwerken mit flacher Lagerung. Bei einer Ankerlänge von nur 1,2 m, einer Bruchlast von 116 bis 149 kN und Ankerplatten von 8 x 8 cm genügt dort 1 Spreizanker pro 6 m², um die 14 bis 18 m breiten Abbauräume sicher zu beherrschen.

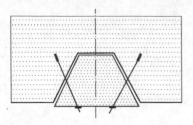

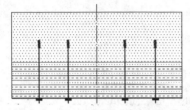

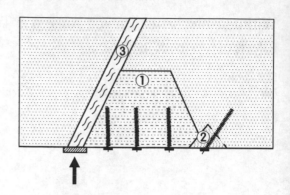

Bild 1: Aufhängen von Kluftkörpern und
Schichtplatten an einem kompetenten Hangenden
Suspending of blocks and of layers at competent rock

Bild 3: Der Hauptkluftkörper (1) bricht nieder, wenn
der Schlüsselstein (2) wegbricht, oder die Störung (3)
ausläuft, oder wenn sich seine Schichten einzeln
ablösen.
The main block (1) will cave in, if the key block (2)
fails, if the material of fault zone (3) runs out or if its
beds peel off individually

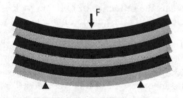

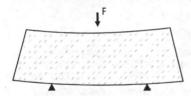

Bild 2: Geringe Durchbiegung einer kompetenten
Bank, starke Durchbiegung eines Stapels dünner
Einzelschichten
Slight bending of a competent layer, strong bending
of a pile of thin beds

Ein weiterer Bruchmechanismus, der häufig zitiert
wird, ist das Durchbiegen/Durchbrechen
geringmächtiger Einzelschichten. Gelingt es, mit
Hilfe von Gebirgsankern einen Stapel von
n Schichten zu einem kompetenten Balken zu
verdübeln, so verringert sich dadurch die
Durchbiegung auf $1/n^2$, wenn das Modell des Trägers
auf zwei Stützen zutrifft (Bild 2).
Nach einer alten Erfahrung ist es wichtig, das

Herunterfallen des „ersten Steines" zu verhindern,
weil sonst der Sicherungsaufwand wesentlich größer
wird. Hinter dieser Beobachtung steht eine ganze
Gruppe von Bruchmechanismen (Bild 3):
Zum einen gibt es den Fall, daß der Hauptkluftkörper aus geometrischen Gründen erst nachbrechen
kann, wenn vorher ein „Schlüsselstein" versagt.
Wenn dieser Schlüsselstein entsprechend verzahnt
oder nicht vollständig von der Umgebung abgetrennt
ist, so genügt ein geringer Aufwand an Stützmitteln,
um das gesamte System zu sichern. In anderen Fällen
werden die gefährlichen Kluftkörper von einer
Störung begrenzt. Nur wenn man zuläßt, daß diese
Störung ausläuft oder ausgespült wird, kann sich der
Hauptkluftkörper bewegen.
In beiden Fällen ist allerdings zu überlegen, welches
Sicherungsmittel zweckmäßiger ist: Sollen „erste
Steine" geankert werden, so müssen sie dem Planer,
je nach Konzept der Ankerung auch der Vortriebs-
mannschaft, nach Geometrie und Lage bekannt sein.
Bei Verwendung von Spritzbeton ist das nicht
notwendig. Der „kennt" selber jeden „ersten Stein".
Wenn der Schlüsselstein klein, die Störungszone
mobil, die gesamte Situation unübersichtlich ist, so
wird daher vielfach Spritzbeton geeigneter sein als
eine Ankerung.
Ein dritter Bruchmechanismus dieser Gruppe ist
das sukzessive Auflockern des Hauptkluftkörpers,
z.B. durch Abblättern einzelner Schichten, die sich
auf einen weichen Ausbau legen. Im Beispiel Bild 4
führte das zu einem schlagartigen Bruch des
Türstockausbaus, obwohl der Bruchkörper auf der
rechten Seite gar nicht durch eine Kluft begrenzt war.
Relativ kurze Vollverbundanker hätten diesen Unfall
mit Sicherheit verhindert.

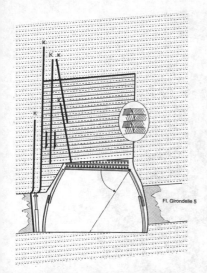

Bild 4: Verbruch in einem Stollen mit weichem Türstockausbau. Der Bruchkörper war nur links von Störungsklüften begrenzt. Er konnte sich trotzdem lösen, weil er aus Schichten mit geringer Haftung bestand und weil der Türstockausbau das Auflockern dieser Einzelschichten nicht verhindern konnte wie das bei Vollverbundankern der Fall gewesen wäre

Roof fall in a tunnel supported by weak frames. The rock mass was limited by joints only on the left side. Nevertheless it collapsed because it consisted of beds with low adhesion to each other. The frame was not able to prevent the indiviual loosening and fracture of these beds as fully bonded rock bolts could have done

In allen bisher besprochenen Bruchmechanismen wird die Wirkung frühzeitig gesetzter Anker noch verstärkt, wenn die an der Bewegung beteiligten Trennflächen verzahnt sind; und dies ist der Normalfall. Mit Hilfe vorgespannter oder wegen des Vollverbundes steil auflastender Gebirgsanker kann dann die Spitzenscherfestigkeit der Trennflächen ausgenutzt werden; dies wird noch besser erreicht, wenn die Gebirgsanker die Auflockerung verhindern, damit den Kluftkörper starr erhalten und dadurch eine Dilatation an der Scherfläche verhindern.

Besonders die letztgenannte Wirkung der Gebirgsanker setzt allerdings voraus, daß das Stabilisieren der Kluftkörper im wesentlichen ein Eigengewichtsproblem ist; daß die Kluftkörper nicht vom Gebirgsdruck ausgequetscht werden. Die bisher besprochenen Wirkungsmechanismen von Gebirgsankern sind daher Mechanismen, die in nachbrüchigem bis gebrächem Gebirge maßgebend sind. In druckhaftem Gebirge bilden sie nur mehr einen Teilgesichtspunkt für die Bemessung einer Ankerung.

BRUCHMECHANISMEN IN DRUCKHAFTEM FELS

Wird geschichteter Fels druckhaft, so muß man zwei Grundfälle unterscheiden (Zischinsky 1984). Im Fall 1 werden die Schichten senkrecht durchörtert, z.B. wenn ein Schacht in der flachen Lagerung geteuft wird. In diesem Falle entstehen dieselben Bruchstrukturen wie in massigem Fels, nämlich Scherbrüche nach Bild 5, sofern die beiden senkrecht zur Schachtachse stehenden Komponenten des Gebirgsdruckes gleich groß sind und ihr Betrag höher ist als die halbe Festigkeit des Gebirges. Zusätzlich zu den Scherbrüchen, je nach der Sprödigkeit des Gesteins auch an ihrer Stelle, entstehen unter Tage oberflächenparallele Spaltbrüche. Die Relativbewegungen an diesen beiden Systemen von Rissen sind klein im Verhältnis zu den Abmessungen des Schachtes, sie sind über die gesamte Bruchzone gleichmäßig verteilt, ein Ausdehnen der Bruchzone ist nur durch das Aufreißen neuer Risse möglich. In diesem Fall kann die Verformung mit guter Näherung als die einer plastischen Zone in einer elastischen Scheibe, also mit Begriffen der Kontinuumsmechanik beschrieben werden.

Für die Praxis bedeutet dies, daß der Schacht mit einem vorläufigen Ausbau aus Spritzbeton oder Spreizankern gesichert werden kann, deren Länge kleiner ist als ein 1/4 des Schachtdurchmessers und daß in Teufen von mindestens 1600 m noch ein endgültiger Ausbau aus Ortbeton möglich ist. Alternativ können auch Vollverbundanker der o.g. Länge als endgültiger Ausbau verwendet werden, sofern sie bereits auf der Schachtsohle eingebaut werden. Ein Bemessen dieser Anker nach der Theorie der elasto-plastischen Scheibe z.B. nach Kastner 1962 ist angemessen.

Werden die Schichten parallel zum Streichen durchörtert (Fall 2), so muß mit einem Bewegungsbild nach Bild 6 gerechnet werden. Die Schichtflächen des abgebildeten Modells haben einen Reibungsbeiwert von etwa 0,9, an den mit Kreisen markierten „Gleitlösen" beträgt die Reibung etwa 0,3. In der Firste war dieser Stollen mit nachgiebigen Ankern ausgebaut, die Sohle war nicht ausgebaut. Die Schichten unterhalb der Sohle wurden aufgefaltet (vgl. Bild 7). Diese Bewegungen sind vor allem das Ergebnis einer horizontalen Verschiebung an einem oder mehreren der vorgegebenen „Gleitlösen". Die Schichten, die von der Falte horizontal entspannt wurden, bilden eine Einheit mit denen, die unmittelbar vom Stollen angeschnitten sind. Sie werden gemeinsam durch die Vertikalspannungen zerquetscht und in den Hohlraum geschoben. Dadurch werden die Horizontalverschiebungen und damit auch die Faltenhöhe verstärkt. Dieser Ausquetschmechanismus ist nicht in der Firste aufgetreten; in diesem Bereich sind anstelle einer Falte Scherbrüche ausgebildet. Ein solcher

Bild 5: Bruchstrukturen um einen druckhaften
Schacht in der flachen Lagerung
Pattern of fractures around a shaft in level formations
under heavy pressure

Bild 6: Bruchstrukturen um einen druckhaften
Stollen in der flachen Lagerung
Pattern of fractures around a tunnel in level
formations under heavy pressure

Bild 7: Falte in der Sohle eines Pilotstollens (vgl.
Bild 6)
Fold in the floor of a pilot tunnel (c.f. Fig. 6)

Bild 8: Stufen in der Wand eines Stollens
(vgl. Bild 6)
Steps in the side walls of a roadway under heavy
pressure (c.f. Fig. 6)

Scherbruch bewirkt eine wesentlich höhere
bankparallele Restfestigkeit als eine Falte
(Buschmann 1970). Infolgedessen liegt das den
Ausquetschmechanismus begrenzende Gleitlösen
unterhalb der Firste und tritt direkt in den Stollen aus.
An dieser Austrittsstelle ist in der Stollenwand eine
Stufe zu sehen, wie sie auch aus Untertage-
beobachtungen bekannt ist (Bild 8).

Gestein ist ein sprödes Material. Damit es
zerquetscht und d.h. damit seine Mächtigkeit
verringert werden kann, müssen keilförmige
Scherbrüche entwickelt und die entstandenen Keile
ineinander geschoben werden. Unter sonst gleichen
Bedingungen ist die Tiefe dieser Keilbruchzone
proportional zu ihrer Mächtigkeit (Schmücker und
Zischinsky 1995). Sie ist ausbautechnisch schwierig
zu beeinflussen. Wenn Anker etwas bewirken sollen,
so müssen sie mindestens die zweite, besser noch die
dritte Keilfläche durchstoßen. Die ersten beiden
Flächen sind nämlich kaum zu verhindern. Das
bedeutet aber, daß die Anker eine Länge haben
müssen, die mindestens der 1,5 fachen Mächtigkeit
der ausgequetschten Schicht entspricht. Und es
bedeutet, daß die Anker durch die Verschiebungen an
den vorderen Keilflächen nicht zerstört werden
dürfen; daß sie also gegenüber Scherung nachgiebig
sein müssen.

Die wichtigste Maßnahme gegen ein
tiefgreifendes Zerkeilen des Gebigsmantels ist daher,
die Mächtigkeit der zerkeilten Schichten klein zu
halten. Wie Bild 6 zeigt, ist das möglich, wenn die

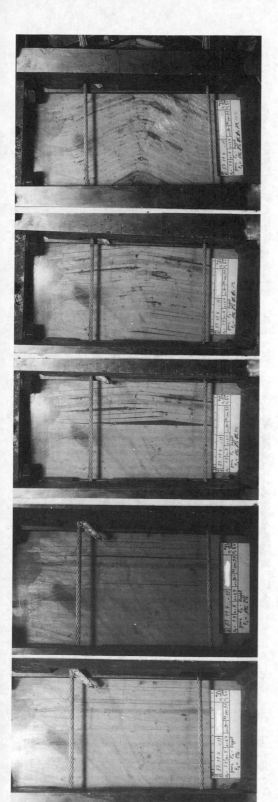

Falte verhindert und statt dessen ein Scherbruch erzwungen wird. Ältere Modellversuche haben gezeigt, daß dazu ein Mindestwert des Ausbaustützdruckes erforderlich ist, der vom Gebirgsaufbau abhängt und der im Augenblick des ersten Bruches bereits wirksam sein muß (Reuber 1979). Gebirgsanker eignen sich zum Realisieren eines solchen frühzeitig wirkenden Ausbaustützdruckes besonders gut, weil sie vorgespannt oder über die gesamte Länge mit dem Gebirge verbunden werden können. Der zweite Weg ist der unter Tage normalerweise verwendete. In diesem Falle werden dem Gebirgsanker bei einem Öffnen von Klüften oder bei Verschiebungen an Klüften lokal Längenänderungen aufgeprägt, die wegen des Vollverbundes von kurzen Dehnlängen aufgenommen werden müssen. Dies führt zu einem steilen Anstieg der Ankerkräfte genau an der Stelle, an der eine Verformung verhindert werden soll. Schon bei kleinsten Deformationen wird auf diese Weise die Kraft des Ankers an der Streckgrenze aktiviert. Dies ist die Erklärung dafür, daß in Versuchen gezeigt werden konnte, daß bei Verwendung von Gebirgsankern dann das gleiche Ergebnis wie bei einem frühtragenden Unterstützungsausbau erzielt wird, wenn das Produkt aus Ankerdichte und Kraft des Ankers an der Streckgrenze gleich dem Ausbaustützdruck des Unterstützungsausbaus ist.

Diese Berechnung des Ausbauwiderstandes wird auch in der Literatur üblicherweise verwendet. Sie hat allerdings zur Voraussetzung, daß die Anker ausreichend lang sind. Welche Ankerlänge ausreichend lang ist, wird seit Jahren in der Literatur diskutiert. Fast alle Autoren gehen dabei von kontinuumsmechanischen Vorstellungen aus, die in geschichtetem Fels in keiner Weise zutreffen.

Diese Überlegungen werden durch die Fotos von Bild 9 anschaulich gemacht. Sie zeigen die Entwicklung einer Falte in einem Teilmodell, das z.B. der Sohle des Pilotstollens in Bild 7 entsprechen könnte. In 8a sieht man das erste Aufbeulen der obersten Schichten. In Foto b ist eine offene Spalte entstanden. Im nächsten Stadium (Foto c) ist diese Spalte wieder geschlossen und dafür haben sich neue Spalten entwickelt. In Foto d ist beinahe das ganze Paket ausgeknickt und zuletzt (Foto e) ist eine ausgeprägte Falte entstanden; ohne offene Spalten im Maßstab der Falte, aber offensichtlich geschwächt durch eine große Anzahl von „Mikrorissen".

Bild 9: Teilmodell der nicht angeschnittenen Schichten eines Stollens mit 5 Stufen der Entwicklung einer Falte. Abfolge von unten (a) nach oben (e)

Partial model of strata tangential to a tunnel showing five stages of the development of a fold (from bottom (a) to top (e))

Ein Vollverbundanker kann diesem Aufwölben an jeder einzelnen Schichtfläche entgegenwirken und zwar selbst dann, wenn der Abstand zwischen den zwei Enden des Ankers konstant bleibt; und selbst wenn er an einer Trennfläche reißt, kann er dem Öffnen der nächsten Trennfläche noch entgegenwirken. Werden Falten mit Vollverbundankern ausgebaut, so reicht aus diesem Grunde eine Ankerlänge von 1/3 bis 2/3 der Knicklänge aus. Werden dagegen Freispielanker verwendet, so werden diese nur geringfügig auflasten und daher nur einen geringen Ausbaustützdruck entwickeln, wenn sie nicht länger als 1 x Knicklänge der Falte ist. Vor allem aber können die einzelnen Schichten innerhalb der Falte ausknicken, ohne daß der Anker nennenswert weiter belastet und d.h., ohne daß dieses Ausknicken behindert wird. Entsprechende Überlegungen gelten verstärkt für die Wirkung eines Unterstützungsausbaus.

Das Bewegungsbild des Gebirgsmantels von Stollen in der flachen Lagerung hat also 2 Haupt-elemente: Die Falten- oder Scherbrüche in den nicht angeschnittenen Schichten und die Kombination von Gleitlösen und Keilbrüchen in den vom Hohlraum angeschnittenen Schichten (Jacobi 1981). Es ist wesentlich, daß die Gebirgsbewegung an vorgegebenen Schwachflächen stattfindet. Aus diesem Grunde ist hier der Konvergenzzuwachs mit steigendem Gebirgsdruck wesentlich höher als in Schächten. Ein Ausbau aus wenigen und kurzen Gebirgsankern und Spritzbeton, wie er im Schacht ausreicht, genügt nur bei geringen Gebirgsbeanspruchungen.

Es gibt die alte Erfahrung, daß man rechteckige Querschnitte nur bei niedrigen Gebirgsdrücken anwenden kann und bei hohen Gebirgsdrücken auf Bogen- und letztendlich auf Kreisquerschnitte übergehen muß.

Im geschichteten Gebirge werden die Konsolen eines Bogenquerschnittes von Stapeln vorkragender Einzelschichten gebildet. Werden diese Kragträger von stollenparallelen Klüften zerlegt, so werden die Konsolen niederbrechen (Bild 10). Dadurch entstehen erneut rechteckige Querschnitte mit großen Knicklängen der tangierenden Schichten. Dieser Bewegungsablauf entsteht aber auch, wenn keine stollenparallelen Klüfte vorhanden sind, wie das z.B. in den Bildern 6, 12 und 13 gezeigt wird. Im Modell des Bildes 11 war der Kreisquerschnitt des Stollens nicht ausgebaut. Die ersten Risse, die von den Gebirgsspannungen erzeugt wurden, haben die Gewölbekonsolen isoliert, diejenigen in der Firste wurden rotiert bzw. fielen herunter. Die Knicklänge der Firstschichten beträgt nunmehr mindestens das 3-fache der ursprünglichen und als Folge entstand der Ansatz zu einer Falte.

Auch in diesem Falle ist es von besonderer Bedeutung, ohne Ausbauverspätung und mit

Bild 10: An Schichtflächen und Klüften nachgebrochene Gewölbekonsolen eines Stollens
Downfall of the console in the arch of a tunnel isolated by bedding planes and joints

Bild 11: Gebrochene Gewölbekonsolen ermöglichen die Entwicklung einer Falte
Fractured consoles of a circular tunnel allow the development of a fold

ausreichender Steifigkeit auszubauen. Mit relativ geringem Aufwand kann dann die Vergrößerung der Knicklänge und damit das Entstehen einer Falte verhindert werden.

Zur Demonstration dieser Bruchmechanismen und der beschriebenen Aussagen zu den Gebirgsankern wurden 3 Modellversuche im Maßstab 1 : 15 durchgeführt. In allen 3 Modellen wurde ein kreisrunder Stollen mit einem Durchmesser von D = 6m (alle Angaben in Untertage-Maßen)

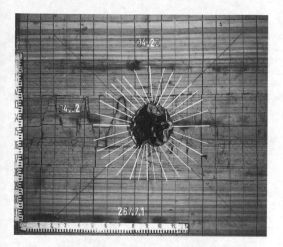

Bild 12: Modell eines geankerten Stollens in geschichtetem Fels. Der Ausbauwiderstand beträgt 560 kN/m², die Ankerlänge beträgt 2/3 des Stollendurchmessers
Model of a bolted tunnel in stratified rock. Support resistance 560 kN/m². The length of the bolts equals 3/3 of the tunnel diameter

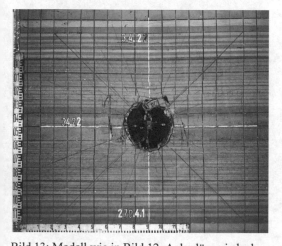

Bild 13: Modell wie in Bild 12, Ankerlänge jedoch 1.5 Durchmesser
Model as shown in fig. 12 however with bolt lengths of 1.5 diameters

hergestellt. Der Gebirgsaufbau war in den 3 Versuchen ident, die mittlere Gesteinsfestigkeit des Gebirgsmantels betrug β_D = 56 MPa. Der Ausbauwiderstand betrug jeweils 560 kN/m². Er war damit ausreichend hoch, um das Entwickeln einer Falte in Firste oder Sohle zu verhindern und Scherbrüche zu erzwingen. Variiert wurde die

Ankerlänge von 1/3 D, über 2/3 D (Bild 12) bis 1,5 D (Bild 13).

Nach Aussage der Modellversuche genügt bei dem gewählten Ausbauwiderstand eine Ankerlänge von 1/3 D bis höchstens 2/3 D, um dem Gebirge in Firste und Sohle günstige Bruchformen aufzuprägen. Die Keilbildung in den Stößen kann mit diesen Ankerlängen jedoch nur bis zu einer Teufe von rd. 1300 m bzw. 1800 m verhindert werden. Ab Teufen von 1800 bzw. 2000 m erreichen die Keilbrüche bereits den Modellrand (Tiefe der Keilbruchzonen ≥ 2 D). Mit einer Ankerlänge von 1,5 D wurden die Stöße dagegen selbst in einer Teufe von 2700 m und mehrfacher Belastung nur bis zu einer Tiefe von 1 D zerkeilt.

Insofern haben die Versuche bestätigt, was oben zum Bruchmechanismus der Falten und Keile gesagt wurde. Es zeigte sich jedoch, daß der ausbautechnische Zustand des Stollens und auch die Entwicklung der gemessenen Konvergenzen in keiner Weise den erkennbaren Unterschieden in der Bruchmechanik des tieferen Gebirgsmantels entsprach. Zwei zusätzliche Bruchmechanismen haben sich als dafür maßgebend herausgestellt:

Zum einen gelingt es mit Gebirgsankern nur sehr schwer, die Bewegung an Gleitlösen zu verhindern, die in geringem Abstand (≤ 0,15 D) über der Firste bzw. unter der Sohle verlaufen. Dies aber führt zu einem Ausquetschen von „Säcken", die sowohl in sicherheitlicher Hinsicht als auch im Hinblick auf die Querschnittsverminderung des Stollens den positiven Effekt der verhinderten Großfalte wieder aufheben. Vergleichbare Ergebnisse sind unter Tage aus Abbaustrecken bekannt.

Zum zweiten wurde deutlich, welche Bedeutung der Mechanismus der abbrechenden Konsolen hat. Diese Konsolen der angeschnittenen Schichten werden regelmäßig an Gleitlösen und an Trennbrüchen oder an den ersten Keilbruchflächen von ihrer Umgebung abgelöst. Die oft dreieckigen Bruchkörper verhalten sich wie isolierte Starrkörper und führen unabhängig von der Verformung des tieferen Gebirgsmantels Verschiebungen und Rotationen aus. Auch diese Bewegungen können weder in sicherheitlicher Hinsicht noch im Hinblick auf den Querschnittsverlust hingenommen werden.

Das Beherrschen der „Säcke" und der „Konsolen" gewinnt damit ausbautechnisch ein eigenes Gewicht, das dem Gesichtspunkt Beherrschen der Falten und der Bruchkeile gleichwertig ist. Für diese Aufgaben sind zusätzliche Gebirgsanker mit Längen zwischen 1/3 D und 2/3 D notwendig.

In großen Teufen ist ein Ausbauwiderstand von mehr als 1 MPa erforderlich, selbst wenn man Deformationen von mehreren Prozent des Stollendurchmessers in Kauf nimmt. Die Gebirgsanker müssen dann einen Scherbetrag von mindestens 3 Stangendurchmessern ohne Versagen aufnehmen.

Der Ankerausbau kommt hier an seine Grenzen. Gegebenenfalls muß er mit einem nicht gebirgsverbundenen Unterstützungsausbau kombiniert werden, der in der Lage ist, durch Scherung entstandene Stufen zu überbrücken.

LITERATUR

Buschmann, N., 1970. Bruchverformung und Ausbauwiderstand in rechteckigen Flözstrecken nach Modellversuchen. *Glückauf Forschungsh.* 31: 133-144
Jacobi, O., 1981. *Praxis der Gebirgsbeherrschung.* 2. Auflage, Essen: Glückauf
Kastner, H., 1962. *Statik des Tunnel- und Stollenbaus.* Berlin: Springer
Reuber, U., 1979. *Modellversuche zum Verformungswiderstand und zum Nachbruchverhalten von Schichtpaketen.* Diplomarbeit Univ. Bochum
Schmücker, H. & Zischinsky, U., 1995. Die Bruchmechanik im Stoß von Strecken. *Glückauf Forschungsh.* 56 2/3: 89-93
Zischinsky, U., 1984. Bruchformen und Standfestigkeit von Stollen im geschichteten Gebirge. *Felsbau* 2/3: 125-136

2 History cases

Ausführungsbeispiele

Anker in Theorie und Praxis, Widmann (Herausgeber) © 1995 Balkema, Rotterdam. ISBN 90 5410 577 1

Bau von schachtnahen Großräumen des Bergwerkes Göttelborn/Reden

Large scale excavation workings for Göttelborn/Reden collieries

Michael Feld
Saarbergwerke AG, Bergwerk Göttelborn/Reden

ZUSAMMENFASSUNG: Im Zuge der Realisierung des Verbundbergwerkes Göttelborn/Reden wurden am neuen Schacht Göttelborn 4 in rd. 1 100 m Tiefe zwei Großräume mit bis zu 220 m² Ausbruchquerschnitt auf-gefahren und in Anker-Spritzbeton-Technik ausgebaut. Neben den betrieblichen Randbedingungen und der ge-planten Nutzung wird die Entscheidungsfindung für diese an der Saar neue Technik beschrieben. Der Ausbruch wurde abschnittsweise aus der Schachtröhre heraus erweitert und in Teilabschnitten endgültig ausgebaut. Durch meßtechnische Begleitung wurde die Qualität der Ausführung überwacht und bestätigt. Erhebliche Zeit- und Kosteneinsparungen unterstreichen den Erfolg der Maßnahme.

ABSTRACT: Within the framework of compounding of Göttelborn/Reden collieries, two largescale workings of up to 220 m² excavation cross section were headed in round about 1 100 m of depth on the new Göttelborn No. 4 shaft, and supported by bolting and shotcreting. The decision-making for this support technology and the framework of general operation conditions for the new pit bottom as well as its intended use are described. The excavation was headed sectionwise from the shaft, each section was given its final support. By comprehensive measurements carried out synchronuously with excavation support work the support quality was monitored and confirmed. The success of this support technology was underlined by substantial savings in time and costs.

1 VERBUNDKONZEPT/BERGWERKSPLANUNG

Das Verbundbergwerk Göttelborn/Reden stellt neben dem Verbundbergwerk Warndt / Lui-senthal und dem Bergwerk Ensdorf den dritten wesentlichen Baustein des sogenannten "Drei-Standorte-Konzeptes" der Saarberg-werke AG dar.

Die Schaffung dieses Verbundbergwerkes im Ostraum des Saarreviers ist verbunden mit dem betrieblich - organisatorischen Zusam-menlegen der ehemals selbständigen Berg-werke Göttelborn, Reden und Camphausen an einem zentralen Förderstandort in Göttelborn.

Als wesentliche Realisierungsmaßnahmen sind zu betrachten :
- untertägige Arbeiten zum fördertechnischen Anbinden der Baufelder Reden an Göttel-born,
- Erweiterungs- und Erneuerungsinvestitionen über Tage im Bereich der Aufbereitung sowie
- Neubau des Schachtes Göttelborn 4 zu ei-nem leistungsfähigen Dienstleistungs- und Förderschacht.

Der Schacht übernimmt im Endzustand so-wohl die gesamte Produktenförderung des Bergwerkes als auch die Aufgaben eines zen-tralen Frischwetter-, Material- und Seil-fahrtsschachtes.

Im Folgenden sind die wesentlichen Kenn-daten zusammengestellt:
- lichter Durchmesser: 8,3 m
- Teufe: 1.160 m
- Förderkapazität: max. 24.000 t$_{Roh}$ /d
- Transportkapazität: max. 470 Einheiten/d
- Seilfahrt: max. 1.200 Personen/d, ein- und ausfahrend
- Langteiltransport bis 10 m, Schwerlasten bis 34 t Einzellasten
- 2 Förderanlagen mit 1.000 t$_{Roh}$ / h bzw. 700 t$_{Roh}$ / h, Hauptseilfahrtanlagen
- Hilfsfahranlage, mittlere Seilfahrtanlage
- verschiedene Rohrleitungen bis Druckstufe PN 160 und DN 500
- Energie- und Fernmeldekabel.

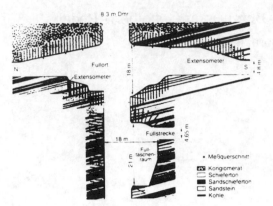

Bild 1: Schnitt durch den Bereich des Füllortes und des Fülltaschenraumes

Fig. 1: Section through the pit bottom zone and the skip-filling bunkers

2 GROßRÄUME 6. SOHLE

2.1 Geometrie

Die genannten Kenndaten wirken sich unmittelbar auf die Geometrie der Großräume, d.h. "Füllort und Fülltaschenraum" im Bereich der 6. Sohle aus (Teufe: rd. 1.100 m).

Der größte Querschnitt im Füllort als (Haupt-) Material- und Seilfahrtsanschlag beträgt ca. 220 m², bei einer Breite von max. 14 m und einer Höhe von max. 18 m.

Der unmittelbar unterhalb angeordnete Fülltaschenraum zur Aufnahme der beiden Füllanlagen besitzt einen Querschnitt von rd. 200 m², bei einer Höhe von max. 21 m und einer Breite von max. 12 m (Bild 1).

Die sich durch die geplante Nutzung ergebenden Abmessungen, die langfristige Bedeutung (> 20 a) und die verhältnismäßig große Teufe stellten den Ausgangspunkt unserer Überlegungen dar :

Wie sind derartige Grubenräume sicher, schnell (ein extrem enger Zeitrahmen ist kennzeichnend für die Gesamtmaßnahme!) und schließlich kostenoptimal überhaupt herzustellen?

Oder :

Welches Ausbaukonzept gewährleistet die Einhaltung der beschriebenen Randdaten?

2.2 Ausbaukonzept, Entscheidungsfindung

Das Thema dieses Berichtes nimmt das Entscheidungsergebnis unserer Untersuchungen vorweg: die Ankerspritzbetontechnik läßt gegenüber einem "konventionellen" Unterstützungsausbau am ehesten die von uns gesteckten Ziele erreichen.

Welche Gründe führten nun im Einzelnen zu dieser Entscheidung?

Größe, Teufe und Lebensdauer der Grubenräume erfordern standsichere Ausbausysteme. Dafür war das Langzeitverhalten der Großräume durch geeignete Prognoseverfahren rechnerisch zu erfassen und vorherzusagen. Darüber hinaus waren mögliche Ausbauverspätung sowie Art und Umfang der Sicherungsmaßnahmen oder des vorläufigen Ausbaus gegenüberzustellen und zu bewerten.

Diese Kriterien waren bei eingehender Betrachtung mit Unterstützungsausbau nicht befriedigend zu erfüllen.

Berücksichtigt man zudem, daß Ausbaukonzepte, die auf den Grundsätzen der NÖT beruhen, mehrfach auch im (Steinkohle-) Bergbau erfolgreich eingesetzt wurden und damit Stand der Technik sind, ist eine Entscheidung in Richtung Ankerspritzbetontechnik als zukunftsweisend zu betrachten.

Nachdem schließlich eine Untersuchung des Institutes für Gebirgsbeherrschung und Hohlraumverfüllung (IGH) der DMT, Essen, die grundsätzliche Ankerbarkeit des anstehenden Gebirges ergeben hatte, stand dem Projekt nichts mehr im Wege. D.h. fast nichts mehr:

Ein Kriterium darf nämlich nicht unerwähnt bleiben: "Erfahrung und Qualifikation der Teufmannschaft"!

Der Schacht 4 einschließlich der hier beschriebenen Großräume wurde durch eine Arbeitsgemeinschaft der Firmen Deilmann-Haniel, Thyssen-Schachtbau und Saarberg-Interplan geteuft, wobei sich die Teufmannschaft unter deutscher Leitung aus polnischen Arbeitskräften der Firmen PbsZ und Kopex rekrutierte.

Bei der Erstellung der Bauwerke waren Größe, Teufe, Lebensdauer und ein recht enger Zeit- und Kostenrahmen verbunden mit der Tatsache, daß die Ankerspritzbetontechnik erstmalig an der Saar zum Einsatz kommen sollte, verbunden mit betriebsinternen Vorbehalten bis hin zum Erreichen der Zustimmung der Genehmigungsbehörden.

Während Anker- und Spritzbetonarbeiten als gängige bergmännische Techniken weniger in der Ausführung Probleme aufwerfen, ist der Ankerspritzbetonausbau, vor allem in Grubenräumen der beschriebenen Größenordnung, hinsichtlich seines Anspruches an Bauleitung und Aufsichten für Arbeitsvorbereitung und Überwachung als sehr anspruchsvoll zu bewerten.

Im vorliegenden Fall hatte die örtliche ARGE - Bauleitung bereits einen großen Erfahrungsschatz aufzuweisen und war von Beginn an zusammen mit dem Bergwerk und den Gutachtern der DMT in die Projektentwicklung eingebunden, so daß auch Bedenken bezüglich möglicher Risiken und Ausführungsprobleme ausgeräumt werden konnten.

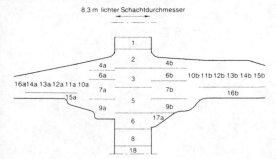

8.3 m lichter Schachtdurchmesser

Bild 2: Ausbruchplan
Fig. 2: Excavation plan

3. BAU - / PROJEKTABWICKLUNG

Nachdem die Risikoabschätzung zugunsten des ASB - Projektes gefallen war, wurde das IGH der DMT mit der Erstellung eines Detailgutachtens beauftragt.

Das Gutachten sollte im wesentlichen :
- die Darstellung des Dimensionierungsverfahrens,
- Aussagen bezüglich Standsicherheit und Langzeitstabilität der Grubenräume,
- die notwendigen Anforderungen an Qualität und Einbau der Stützmittel und Ausbaukomponenten,
- Empfehlungen für den Bauablauf sowie
- eine Beschreibung der während und nach Abschluß der Auffahrung notwendigen meßtechnischen Überwachung
beinhalten.

Auf die inhaltlichen und systematischen Grundlagen, die gewählten rechnerischen Ansätze etc., die Eingang in das Gutachten gefunden haben, geht Herr Dipl.-Ing. Kammer in seinem Vortrag ein.

Sowohl die gutachterliche Arbeit als auch die spätere Umsetzung war gekennzeichnet durch einen ständigen Informationsaustausch aller Beteiligten. So ist z.B. die Hohlraumgeometrie entstanden aus Eckwerten, die durch die geplante Maschinentechnik minimal erforderlich waren, in Verbindung mit den Erfordernissen einer ausbautechnisch machbaren Formgebung.

Desweiteren basieren Art und Umfang der verschiedenen Ausbauphasen auf den vom "Teufer" gewählten Randbedingungen, wie Ausbruchplanung, Betriebsmitteleinsatz etc.

Diese Vorgehensweise, die nicht zuletzt auch terminlich notwendig erschien, führte dazu, daß von Beginn an alle Belange im Gutachten ihren Niederschlag fanden.

Der weitere Bauablauf stellt sich wie folgt dar:

3.1 Herstellen der Ausbrüche

Gemäß der entwickelten Hohlraumgeometrie wurde ein Ausbruchplan erstellt (Bild 2).

Nach Erreichen der Füllortfirste im Schachtregelprofil (ca. 1.048 m Teufe) wurde der Schachtdurchmesser um rd. 0,7m vergrößert. Dies war notwendig, um Tragkranz und Umsetzschalung, die unterhalb der Großräume wieder zum Einbringen des Schachtbetons benötigt wurden, durch Füllort und Fülltaschenraum verfahren zu können.

Dieser bereits in ASB ausgebaute Kreisquerschnitt (lichter Durchmesser: ca. 9 m) wurde in der Folge abschnittsweise zur Teufe und bis 6 bzw. 8 m beiderseits in Füllortachse erweitert. In dieser ersten Phase entstand damit die komplette Schachtglocke, d.h. der Bereich der größten Abmessungen, wobei die Vorgabe der Ausbauplaner, den Sohlenschluß so früh wie möglich zu erreichen, Berücksichtigung fand.

Nachfolgend wurden die hierbei entstandenen Ortsbrustflächen, die wegen ihrer Höhe von rund 18 m ebenfalls mittels Anker und Spritzbetonschale gesichert wurden, von unten nach oben hereingesprengt. Mit dem dabei anfallenden Haufwerk sollten die für die weitere Auffahrung notwendigen Arbeitsniveaus (wieder-) hergestellt werden.

In einer zweiten Bauphase wurden die beiden Füllortstümpfe ca. 33 m nach Norden und 43 m nach Süden aufgefahren. Abschließend erfolgte der Übergang auf Unterstützungsausbau mit einem Querschnitt von 27 m².

Die beschriebene, zur zweiten Auffahrungsphase notwendige Haufwerksanböschung hat leider nicht zum Erfolg geführt. Entgegen der ursprünglichen Einschätzung konnte betriebsbedingt nicht soviel Haufwerk zwischengebunkert und angeböscht werden, um die Füllortfirste wieder zu erreichen. Deshalb mußte zwischenzeitlich von hohen Gerüsten aus gearbeitet werden.

Nach Fertigstellung des Füllortes wurde der Schacht ca. 10 m tiefer geteuft, um mit der Herstellung des Fülltaschenraumes zu beginnen. Dieser hat im oberen Bereich die Gestalt einer Viertelkugel. Hier hinein mündet die südöstlich angeordnete "Füllstrecke". Der Firstbereich jenes im Grundriß ovalen Grubenraumes wurde nach unten in mehreren Scheiben weiter aufgefahren, wobei jeweils der gesamte Querschnitt, einschließlich des Schachtquerschnittes gelöst wurde.

An dieser Stelle sei angemerkt, daß die gesamte Maßnahme von dem Gutachterteam der DMT bauüberwachend begleitet wurde, so daß fallweise aufgetretene Nachbesserungen und Plankorrekturen zeitnah berücksichtigt werden konnten.

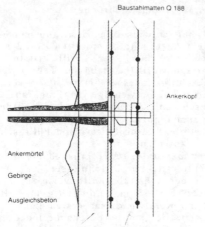

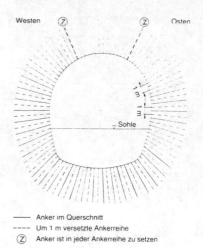

1. Spritzbeton-Schale 10 cm   2. Spritzbeton-Schale 10 + 5 cm

——— Anker im Querschnitt
- - - - Um 1 m versetzte Ankerreihe
(Z) Anker ist in jeder Ankerreihe zu setzen

Bild 3: Aufbau der Anker - Spritzbeton - Schale
Fig. 3: Structure of the bolt and shotcrete shell

Bild 4: Anordnung der Anker im Querschnitt
Fig. 4: Bolt pattern, cross section

3.2 Ausbaukomponenten

Im Zuge der Ausbruchserstellung wurde je nach Freilegen der endgültigen Konturen sofort die Ausbauschale eingebracht (Bild 3).

Die ausbautechnische Sicherheit im Vorortbereich gewährleistete eine erste, (mindestens) 10 cm dicke Spritzbetonschale. Diese "Konsolidierungsschicht" diente gleichzeitig dazu, die endgültige Formgebung der Hohlräume einzustellen.

Nach Erreichen einer Festigkeit von ca. 5 MPa, nach ca. 5 h, konnte das Ort wieder betreten werden. Danach wurden die Traganker eingebohrt und vollvermörtelt sowie eine erste Mattenlage (Baustahl, Q 188) eingebaut. Die sich anschließende Spritzbetontragschicht besteht wiederum aus zwei 10 bzw. 5 cm dicken Schalen. Dazwischen erfolgte der Einbau der zweiten Baustahllage.

Grundlage für die Ankerung waren entsprechende Ankerschemata (Bild 4).

Zur Optimierung des Arbeitsablaufes und Überbrückung der während dem Abbinden des Betons anfallenden Zeiträume wurde im jeweils gegenüberliegenden Großraumbereich weiter gearbeitet.

Die folgende Auflistung gibt einen Überblick über die verwendeten Ausbaukomponenten:
- Spritzbeton:
  - Festigkeitsklasse: frühtragend
  - Festigkeitsentwicklung:
    02 MPa nach   1 h
    05 MPa nach   5 h
    12 MPa nach 12 h
  - Endfestigkeit: mindestens entsprechend B 25 - Qualität
- Anker:
  - M 33 bzw. M 27
  - überwiegend 5 m lang
  - Typ RAM 700, gerippt, Kugelbundmuttern
  - Bergbauzulassung erforderlich
- Ankermörtel:
  - Konsistenz verarbeitbar
  - ohne weitere Prüfung
  - Bergbauzulassung erforderlich
- Baustahlgewebe: Q 188

Die Überwachung des Spritzbetones erfolgte sowohl im Labor (Eigenüberwachung des Herstellers) als auch stichprobenartig vor Ort. Dabei kamen verschiedene Prüfverfahren zum Einsatz, die jedoch zumindest bei der Kontrolle der geforderten Frühfestigkeiten mehr oder minder überfordert waren. Letztlich mußten deshalb die Herstellerangaben verbunden mit einer entsprechenden Dokumentation die Eignung des verwendeten Baustoffes belegen.

Der Einsatz von Injektionsmitteln war nicht erforderlich.

Zur Sicherung der Langzeitstabilität wurden lediglich die Spritzbeton - Tragschicht sowie die Mörtelanker berücksichtigt. Demzufolge tragen die übrigen Ausbaukomponenten verbunden mit Betonüberstärken und der letztlich nachgewiesenermaßen Endfestigkeit > 35 MPa zusätzlich zur Standsicherheit der Bauwerke bei, was sich bislang auch voll bestätigt hat.

160

4. MEßTECHNISCHE ÜBERWACHUNG

Wesentlicher Bestandteil des beschriebenen Ausbaukonzeptes ist eine intensive meßtechnische Begleitung:
- Einmal dienen Konvergenzmessungen, kurz- und langfristig, dazu, die sich einstellenden Gebirgsbewegungen zu erfassen, zu bewerten und damit die Betriebssicherheit zu garantieren.
- Weiterhin ist die Einhaltung der Ausbauschemata und Sollkonturen zu überwachen.
- Schließlich sind die zur Feststellung und Verrechnung der Auffahrleistungen notwendigen Eckdaten aufzunehmen.
- Dazu war ein aufwendiges Meßprogramm erforderlich, daß jedoch gleichzeitig so anzulegen war, daß der Baufortschritt durch Vermessungstätigkeiten nicht über Gebühr beeinträchtigt wurde.

In der Abfolge von Ausbruch - Konsolidierung - Ausbau gab es letztlich nur einen Zeitpunkt, an dem alle die Sicherheit und Ausführungsgenauigkeit bestimmenden Parameter aufgenommen werden konnten : nach Fertigstellung der ersten Spritzbetonschale und vor bzw. während dem Einbringen der Anker.

Durch die Markscheiderei Göttelborn wurden aufgenommen :
- die Profilgenauigkeit (Toleranz: +20 cm / -10 cm)
- die Dicke der Konsolidierung durch stichprobenartige Bohrlochmessung
- die Dicke der Tragschicht, entsprechend der Ankerüberstände
- die Einhaltung der Ankerschemata und Ankerdichte sowie
- evtl. aufgetretene Konvergenzen an vorher fertiggestellten Bauabschnitten.

_Zu dem zuletzt genannten wurden in Absprache mit dem Gutachter insgesamt 7 Meßquerschnitte im Füllort und 3 im Fülltaschenraum eingerichtet, zeitlich versetzt im Zuge der Auffahrung, und regelmäßig kontrolliert.

Solange die zugehörigen Meßanker zugänglich waren konnte eine maximale Stoßwanderung von lediglich rd. 25 mm an dem als erstes eingerichteten Meßhorizont im Bereich des größten Querschnittes festgestellt werden.

Nach Fertigstellung der Großräume und nachdem keine Kontrollen vor Ort mehr möglich waren, wurde die Konvergenzüberwachung durch Extensiometer übernommen, die über elektronische Meßwertaufnehmer zudem die Übertragung an die Grubenwarte ermöglichte. Nach fast 2,5 Jahren konnte diese Langzeitüberwachung abgelegt werden, ohne daß über die geringe Auffahrkonvergenz hinaus weitere Bewegungen meßtechnisch nachgewiesen werden konnten.

Die beschriebene meßtechnische Überwa-

chung hat die Prognosen der Ausbauplanung eindrucksvoll bestätigt.

Der Vollständigkeit halber sei erwähnt, daß im Ausbauschema keine Änderungen vorgenommen wurden, auch dort nicht wo die Dimensionierung und/oder die Meßergebnisse dies ermöglicht hätten. Diese letztlich auch in den Grundsätzen der NÖT begründeten Spielräume wurden aus Standardisierungsgründen nicht genutzt.

5. KOSTEN, BAUZEIT, LEISTUNGSDATEN

Die im folgenden vorgestellten Daten zu Bauzeit, Kosten und Leistungen sollen den Erfolg der Maßnahme unterstreichen:
- eingebrachte Spritzbetonmenge (ohne Rückprall): ca. 1.350 m^3
- mittlere Spritzbetondicke: rd. 32 cm
- ausgebaute Oberfläche: rd. 4.500 m^2 (einschl. der Ortsbrustflächen)
- eingebrachte Anker: rd. 4.900 Stück.
- Ausbruch: rd. 15.000 fm^3
- Leistung: rd. 1,55 fm^3/MS.
  (Bild 5)

Die gesamte Maßnahme konnte nach rund 8 Monaten Bauzeit abgeschlossen werden

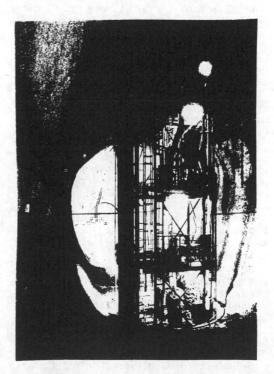

Bild 5: Blick in das fertiggestellte Füllort
Fig. 5: View on the ready pit bottom

161

(April bis Nov. '92). Eine Gegenüberstellung mit den Plandaten des ursprünglich vorgesehenen Unterstützungsausbaus einschließlich aller dazu notwendigen Arbeiten für vorläufigen Ausbau, Hinterfüllung etc. ergeben eine Bauzeitersparnis von rd. 32 % sowie eine Kostenreduzierung von rd. 17 %.

6. ERFAHRUNGEN, STAND DER ARBEITEN

Welche Erfahrungen konnten nach Abschluß der eigentlichen Maßnahme mit dem für die meisten Beteiligten neuen Ausbausystem gesammelt werden?

Beispielhaft sei ein in der Praxis wesentlicher Punkt betrachtet: das Fehlen von geeigneten Aufhängemöglichkeiten. Unterstützungsausbau kann relativ leicht zum Anhängen, Umschlag und Transport von Lasten genutzt werden, wobei die Flexibilität in Wahl und Ausführung von Aufhängepunkten sehr hoch ist. Bei den Geometrien und Bedingungen des ASB jedoch war es frühzeitig notwendig, die z.B. zur Montage der maschinellen Einrichtungen erforderlichen Anschlagpunkte festzulegen und mittels Klebeanker und Ketten vorzubereiten.

Außerdem führte die z.T. komplexe, zumindest jedoch räumliche Vorstellung voraussetzende Geometrie bei der Planung der Einrichtungen immer wieder zu Verständnisschwierigkeiten. So war z.B. bei dem Stahlbau der Füllanlage kaum eine Standardisierung möglich, so daß jeder Horizont entsprechend der Ist - Situation eigens betrachtet und bewertet werden mußte.

Gleichzeitig ermöglichte die ASB-Bauweise quasi an beliebiger Stelle Auflager, Fundamente u.ä. herzustellen, ohne die Ausbauwirkung zu beeinträchtigen.

Zusammenfassend läßt sich vielleicht folgendes sagen: Anker-Spritzbeton setzt insbesondere bei einem Stahlbaukonstrukteur sicherlich ein weit intensiveres Verständnis für Hohlraumgeometrien und Ausbaubedingungen voraus, als dies bei Unterstützungsausbau oder herkömmlichem Betonbau der Fall ist. Jedoch beinhaltet diese Technik aus meiner Sicht eine weitaus größere Flexibilität und Kreativität.

Bislang konnten alle Verständnisprobleme erfolgreich bewältigt werden. Die Füllanlage im Fülltaschenraum, Schachtstuhl und Beschickung im Füllort stehen kurz vor ihrer Inbetriebnahme.

Das bislang Gesagte läßt sich in folgendem Satz zusammenfassen:

Die Herstellung der Großräume auf der 6. Sohle Schacht Göttelborn 4 in der Technik des Anker-Spritzbeton-Ausbaus wurde insgesamt ein großer Erfolg und ist verbunden mit einem erheblichen Know-how-Zugewinn für den operativen Bereich der Saarbergwerke AG.

Anchors in Theory and Practice, Widmann (ed.) © 1995 Balkema, Rotterdam. ISBN 90 5410 577 1

Recent applications of soil nails and cables in the UK

Neue Anwendungsmöglichkeiten für Bodennägel und -kabel in Großbritannien

J. S. Harper, B. G. D. Smart & J. M. Sommerville
Department of Petroleum Engineering, Heriot-Watt University, Scotland, UK

M. L. Davies & I. M. Spencer
BRC Mining and Land Reinforcement, Stafford, UK

Abstract: Since 1975 soil nailing has been a recognised method of stabilisation of embankments in the USA and mainland Europe, however, in Britain, engineers have been reluctant to adopt this technique until relatively recently. In this paper four different soil nail type anchors are discussed. Each of the case histories illustrates a different type of soil nail and represents examples of current soil nailing practices in use in the UK. The case histories represent some of the most recent applications of soil nailing as a safe, cost effective method for the stabilisation of slopes. The examples give an insight into how soil nailing is a maturing practice in the UK, and that techniques used for many years in the rest of Europe are being adopted and adapted by British engineers with a high degree of confidence.

Zusammenfassung: Es werden vier verschiedene Verankerungen in der Art von Bodennägeln diskutiert. Jedes der Fallbeispiele demonstriert einen anderen Bodennagel-Typ und bringt Beispiele für die derzeitige Bodennagel-Praxis in Großbritannien. Diese Fälle sind die neuesten Anwendungsbeispiele für Bodennägel als kostengünstige Methode für die Stabilisierung von Hängen. Die Beispiele erlauben eine Einsicht in die reifende Praxis des Einsatzes von Bodennägeln in Großbritannien und zeigen die Tatsache auf, daß Techniken, die im restlichen Europa seit vielen Jahren praktiziert werden, heute mit wachsendem Selbstvertrauen von britischen Ingenieuren übernommen werden.

Introduction

Soil nailing is the term used to describe the reinforcement of the ground with small diameter bars (typically 20mm to 30mm) installed in such a manner that they control ground movement by reacting primarily to axial extensive strains, inducing tensile stress in the nail and reinforcing the soil. Cablenail is the term given to a series of small, interlocking bars typically 6-8mm in diameter. The cable strands, usually 8 or 9 off, are arranged in an open weave configuration. The nails and cables may also react to shear stress developed in a radial direction across the nail, but due to the currently incomplete understanding (certainly as presented in the public domain) of the manner in which the nails reinforce in shear, and the likelihood that the nails are much less efficient in shear than tension, soil reinforcement by shear tends to be ignored in design. Accordingly the nails are installed most frequently so as to cut across the likely shear surface at as acute angle as possible. The nail or cable may be bonded to the soil along all or part of its length using a combination of mechanical interference and chemical or cement adhesives. In addition, the nail is coupled at its free end to a soil confining member or structure. This helps to distribute a soil confining action over the face of the excavation. The action of soil nails installed in this manner is demonstrated in Figure 1 Some unstable slopes may be stabilised by installing GRP nails to act primarily in shear, as shown in Figure 2.

Types of Soil Nail and Cables

Steel Rebar Soil Nails

This type of soil nail consists of a single length of steel Type II rebar, nominally 25mm (T25) in diameter with a 200mm x M24 thread at the proximal end. The steel is manufactured in lengths up to 10m which can be cut to the required size easily on site. The nails are centralised in a 112-150mm borehole with ABS plastic lantern spacers.

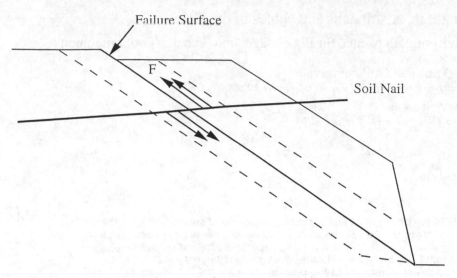

Figure 1　Illustration of soil nails installed to act primarily in tension
Bild 1　Darstellung von Bodennägeln, die primär bei Spannung eingesetzt werden

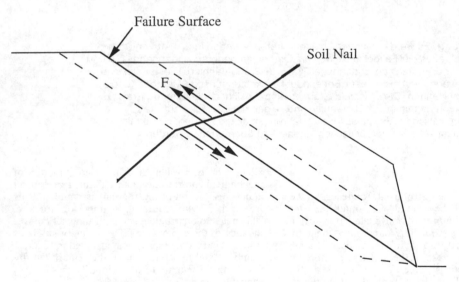

Figure 2　Illustration of soil nails installed to act primarily in shear
Bild 2　Darstellung von Bodennägeln, die primär bei Abscherungen eingesetzt werden

A tremie pipe, consisting of a 13mm plastic tube, is inserted in the borehole with the nail for grouting, and is gradually withdrawn as grouting proceeds. The nails are grouted through their tremie pipes with 0.45 w/c Ordinary Portland Cement (OPC) to give full column bonding between the nail, the grout and the soil. The surface restraint usually consists of a Geotextile Fabric and Reinforcing Mesh positioned over the proximal ends of the nails and secured by a steel plate, conical seat and M24 nut. The nails are torqued off with a nominal holding load, typically 10-20 kN.

For added corrosion protection (double protection system), nails can be pre-constructed in a 61mm diameter (nominally) plastic tube. The rebar nail is inserted in a plastic tube with centralising spacers attached and a tremie pipe. A plastic end piece is attached to the distal end of the plastic tube and sealed. The tube and nail is then raised vertically and grouted using OPC, with the tremie pipe being withdrawn as grouting proceeds. The nail is then vibrated at a high frequency until all the air is removed from the column. The remaining void is topped up and the grout left to cure for three days. The nail, fully bonded within the tube, is sealed at the proximal end with a rubber compound.

The nail is installed in a pre-drilled borehole and centralised using lantern spacers. It is then grouted using the same technique as the standard soil nails.

Ischebeck Titan ™ Soil Nails

This anchor consists of a single tube of steel (typically 4m long) with a coarse rebar type thread. The nail has a nominal OD ranging from 30-103mm and an ID ranging from 16-78mm respectively. The continuous thread permits extension of the anchor with couplers. The anchors are produced from fine grain constructional steel which can be cut easily on site if required. Ischebeck nails require no borehole to be pre-drilled. The anchor is drilled into the soil using a weak OPC flush and grouted off with a standard 0.45 w/c OPC Grout. The proximal end of the nail is completed with a load distribution plate, a wedge disc and a spherical collar nut. Again the nails are torqued off with a nominal holding load.

Glass Reinforced Plastic Injection Tubes

Glass Reinforced Plastics (GRP) have been used for some time in civil engineering, although in small quantities. Their high strength to weight ratio and excellent resistance to electrochemical corrosion make them attractive materials for soil stabilisation.

GRP soil nails are formed from fine glass fibre filaments, usually in the range of 3-5 microns in diameter, arranged in a matrix and impregnated with a resin. The glass fibres are mainly composed of silicon, and the compound lime alumino-borosilicate is the glass fibre most often used for its high strength and deformation characteristics. Other silicates, which have superior strength and stiffness characteristics, may be used, but generally are not economically viable.

The GRP soil nails are pultruded to form rods with a central borehole. It is through this central bore that the grout is pumped, thereby dispensing with the need for a tremie pipe and increasing the confidence of a full-column grout bond.

GRP bolts are manufactured world-wide by a host of manufacturers including Pultrall Inc., (Canada), Cousin Fréres (France) and Weidmann (Switzerland). A typical illustration of GRP soil nails is given by Weidmann below:

Weidmann GRP Nail Technical Specification
• Approx. 75% Glass / 25% Resin
• Structured to provide a high nail/grout bond
• Threads, Nuts and Bearing Plates formed from plastic materials
• Manufactured to ISO 9001/ EN 29000
• Lengths up to 10m easily accommodated
• Lengths over 10m accommodated by coupling

The Technical Data for Weidmann GRP nails is given in Table 1.

Table 1 Technical Data

Tensile Strength	1200 N/mm^2
Young's Modulus	50000 N/mm^2
Injection bolt OD	22 mm
Injection bolt ID	10 mm
Weight of bolt	0.75 kg/m
Breaking Load of Injection bolt	310 kN
Bending Radius	2.5 m

Advantages of GRP

GRP rockbolts and soil nails operate in the same manner as traditional steel bolts / nails but offer additional advantages:

Corrosion Resistance - eliminating expensive surface coating.

Weight Reduction - one quarter the weight of steel bolts - they are easily handled, transported and fixed into position, thereby dispensing with expensive lifting equipment and has obvious advantages in inaccessible locations.

Flexibility - having a bending radius of approximately 2.5m, allows setting in inaccessible locations.

Adaptability - where necessary, bolts can be severed so as not to impede the removal of consolidated material, permitting the enlargement of shafts, tunnels and cuttings.

Drawbacks of using GRP

The main disadvantage of GRP reinforcing bars is their price. They are between 5 to 15 times more expensive than ordinary rebar type steel bars. However, they are more cost effective if consideration is given to the costly maintenance often required for steel reinforcements.

The areas where soil nailing with GRP is particularly attractive financially are:
• In northern areas where due to the climate, large quantities of de-icing salts are used.
• In tropical areas where the soil acidity is particularly high.
• In coastal areas where steel corrosion by sea water is a common consideration.
• In other areas where the soil may be contaminated by domestic or industrial waste.

The other disadvantage with GRP

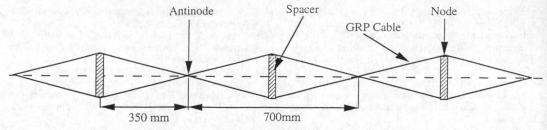

Figure 3   Node/Antinode configuration of GRP cable nails
Bild 3   Knoten-/Anti-Knoten-Konfiguration von GRP-Kabelnägeln

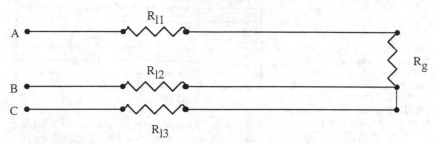

Figure 4   Three wire bridge configuration of strain gauges
Bild 4   Brückenkonfiguration mit drei Drähten auf Meßstreifen

reinforcing bars is that relatively little is known about the long term life expectancy of GRP when compared to steel rebar reinforcement. However, experience over the last 20 years of using GRP, has led to a greater understanding regarding the long term life expectancy.

GRP Cablenails

Cablenails are formed from several small diameter rods (soil nails). The composition is exactly the same as GRP nails, the only differences being the diameter, typically 6mm, and the cables not having an inner bore. Most often the rods are arranged in a configuration such that they have nodes and antinodes as shown in Figure 3. One cablenail may have between 6 to 12 rods
Cablenails are inserted into pre-drilled boreholes and are tremie pipe grouted with OPC or in some cases resin. They have both a high load capacity and, due to the node/antimode arrangement, a high pull out resistance. GRP cablenails are lightweight, (62g/rod/metre) extremely flexible, with a bending radius of about 1m, therefore are ideal for restricted environments where lengths of up to 22m can be accommodated.

Instrumented Soil Nails

Soil nails instrumented with strain gauges positioned at various locations in a reinforced slope

can be used to determine the strains subjected on a installed nail. They are useful tools to determine how the reinforced slope is reacting as conditions change, prediction of possible failure and determination of where the failure may be occurring.

A typical monitor nail construction methodology is given below. This is the method used in the construction of the monitor nails used in Case History 1.
The stain gauges are wired in a 3 wire bridge configuration, as shown in Figure 4, to account for the resistance (R_{l1}, R_{l2}, R_{l3}) of the leadout cables, when determining the strain gauge resistance (R_g) using a Wheatstone Bridge system.
The two strain gauge cluster arrangements used for the monitor nails constructed in Case History 1 are shown in Figure 5.

1. 120 Ohm strain gauges were selected with

• integral leads and matching terminals

• thermal characteristics and strain range compatible with the material on to which they were mounted

• a gauge length of 30mm

2. An adhesive was selected which was compatible with the nail material, the gauges and terminals.

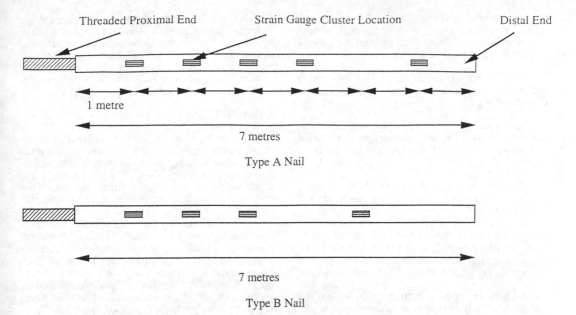

Threaded Proximal End   Strain Gauge Cluster Location   Distal End

1 metre

7 metres

Type A Nail

7 metres

Type B Nail

Figure 5   Illustration of strain gauge locations on instrumented soil nails
Bild 5   Darstellung der Lokalisation von Dehnungsmeßstreifen auf instrumentierten
           Näglen.

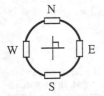

Figure 6   Crossection of nail showing a four strain gauge cluster
Bild 6   Querschnittdarstellung eines Nagels mit vier Meßstreifen

3. The flats were ground on to the nails at specified locations . The flats were oriented in diametrically opposite pairs in a N-S E-W configuration as shown in Figure 6. The width of the flats were 4mm greater than the width of the strain gauges. The length of the flats were 60mm

4. The flats were thoroughly cleaned using solvents compatible with the nail material and the adhesive.

5. The gauges and terminals were mounted on the clean flats by picking them up on low-tack tape applying adhesive to the gauge and clamping it to the flats. The clamping force, applied evenly through thin rubber pads, was great enough to create an acceptably thin glue line.

6. On removal of the low-tack tape, the terminals were cleaned of adhesive and the gauge leads soldered to the terminals. A heat sink was placed on the leads between the terminal and the gauge.

7. Cores of a screened cable were connected to the terminals of each pair of strain gauges in a mode which measures either axial strain or bending of the nail, as required. (Note that if bending is assumed to occur in only the vertical plane, the pair of gauges mounted on the top and bottom of the nail can be connected to give both bending and axial strain). Opposite pairs of gauges were connected via loops to make a three wire bridge configuration as shown in Figure 4

8. The cable was secured to the nail in the immediate vicinity of the strain gauges with three plastic cable ties. The cable ties were placed around the intact cable.

9. A water-proof coating of non-corrosive silicon rubber was applied over a 150mm length of nail covering at least 50mm either side of the gauges and the stripped end of the cable.

10. Two layers of self-amalgamating tape were applied, lapping at least 50mm either side of the coated length.

11. One layer of heat shrink was applied, lapping at least 50mm over the length of the tape. The heat shrink was then sealed with non-corrosive silicon rubber.

12. The cable was secured at regular intervals along the length of the nail with cable ties. Slack was left between the ties.

13. At the outside end of the nail, all the cables from that nail were fed through protective sheathing to a terminal strip located in ABS plastic waterproof enclosures.

14. An indexing mark was filed on the end of the nail, indicating the orientation of the strain gauges, and the nail encapsulated in the double protection system as discussed above.

15. Checks were made on the electrical continuity and resistance of the strain gauges after steps 6-14.

Recent Applications

Case History 1 - A31 Ashley Heath Grade Separated Junction.

Introduction

This project involved the installation of some 2900 no. single protected steel rebar type soil nails to stabilise the road improvement works on the A31 near Southampton. The client was Dorset County Council, the Contractors were Wimpey Engineering and Construction for the Department of Transport, with the Consultant being Dorset Engineering Consultancy Services Ltd. The soil nailing programme was carried out by BRC Mining and Land Reinforcement.

Design Criteria.

The calculations for the design referred to the Draft British Standard BS 8006. "Code of Practice for Strengthened/Reinforced Soils and Other Fills" July 1991, and the British Standards Institution publication BS 8081, "British Standard Code of Practice for Ground Anchorages" 1989.

Nail Design Calculations

The project involved the stabilisation of the vertical embankment below the A31 on/off ramps. The

design was produced by the Highways Agency for Dorset County Council. The soil nailing scheme was designed using the Limit Equilibrium Method. Calculations were made to determine the length of nail required to stabilise the slope. It was calculated that 7m long nails installed at an angle of 20° would need to be installed.

Soil Nailing Design Configuration

The soil nails were installed through a 100mm thick gunite sprayed vertical wall excavated through free flowing sand and gravel. The nails, 7m in length were installed in a grid pattern with 1.5m horizontal and 1.0m vertical spacing. The wall height varied up to a maximum vertical height of 8m.

The 25mm diameter rebar nails were hot dipped galvanised, pre-formed in a 61mm diameter corrugated PVC tube, and were full column grouted in a 140mm diameter borehole. The boreholes were drilled using a rotary percussive drill rod with air as the flushing medium inclined at an angle of 20° to the horizontal. The nails were faced off with $225mm^2$ x 12.5mm thick steel face plate, tapered washer and nut.

As part of the design requirements instrumented soil nails had to be installed at two locations in the soil nailed walls. The nails were constructed as discussed above. A total of 10 no. 7m nails were constructed, 6 off Type A with four gauges at 5 locations and 4 off Type B with four gauges at four different locations as illustrated in Figure 5.

The Face Retaining System

The face of the embankment was reinforced with a mesh and sprayed with concrete to an approximate depth of 100mm. The construction sequence was as follows:

1. Excavate to depth of strip and trim face
2. Fix mesh reinforcement and concrete spray face
3. Drill borehole, install and grout nail
4. Repeat steps 1-3, until ultimate depth is attained.

The grout was allowed to reach a minimum material strength of 10N/mm2 prior to being torqued to a nominal holding load. The face was then shuttered and completed with reinforced concrete to a thickness of 450mm.

Pull-out Testing

Prior to initiation of the soil nailing programme, pull-out tests were performed on a trial panel. A total of 12 nails were tested.

The first 9 were standard pull-out tests performed on 7m long T25 rebar preformed in the 61mm diameter PVC corrugated pipe. These nails were loaded to destruction, with a P_{ult} of 20-22T before the bar yielded.

The remaining 3 were standard pull-out tests performed on 3m long T42 rebar preformed in the 61mm diameter PVC corrugated pipe. These nails were loaded to destruction, with a P_{ult} ranging from between 19 to 30T, with failure occurring in the bond between the grout and the soil mass.

Case History 2 - British Rail Coventry

Introduction

This project involved the stabilisation of a failed retaining wall and the improvement of the factor of safety against failure of the remaining retaining wall adjacent to the Rugby - Birmingham railway line near Coventry.

It was proposed that the failed retaining wall and the remaining retaining wall be supported using Ischebeck Titan™ soil nails.

Manstock Geotechnical Consultancy Services Ltd. were the Consulting Engineers, Charles Gregory Engineering the Contractors, with Manstock Geotechnical Consultancy Services Ltd. providing the soil nailing design with Heriot-Watt University providing the third party soil nailing design, for BRC Mining and Land Reinforcement who implemented the soil nailing programme.

Design Criteria.

The design was divided into two sections. Section 1 was the stabilisation of the failed retaining wall, Section 2 being the design for the upgrading of the factor of safety of the remaining retaining wall.

The calculations for Section 1 of the design referred to the Design Manual for Roads and Bridges, Volume 4, Section 1, Part 4, HA 68/94, "Design Methods for the Reinforcement of Highway Slopes by Reinforced Soil and Soil Nailing Techniques", February 1994.

The calculations in Section 2 were based on the equations developed by R.T. Murray in the Transport Research Laboratory Working Paper "A Review of Soil Nailing" using Limit Equilibrium Analysis on a two part wedge failure mechanism, these equations are also cited in BS 8006, and also are based on the equations published in BS 8081 : 1989, "British Code of Practice for Ground Anchorages".

The soil strength design parameters were based on the Geotechnical Report on Ground Investigation, "Coventry Station Retaining Wall", No. 114088, May 1994, compiled by Exploration Associates Limited.

Nail Design Calculations

Section 1 was designed using the Limit Equilibrium Method, calculations were made to determine the length of nail required to stabilise the slope.

This design was based on the worst case

where the slope was taken as 70° and was some 4m high. No surcharge was added to the crest of the slope, which was assumed to be horizontal. The design consisted of four rows of 5m long nails per metre run of slope, all the rows having a spacing of 2m horizontally and 1m vertically, the uppermost nail having been installed some 0.75 m below the crest of the slope. All nails were installed at 10° to the horizontal.

The retaining wall height was some 4.5m and it was calculated that six rows of 6m long nails per metre run of slope, would need to be installed for stability. In this design the retaining wall angle was taken as 90°(vertical). No surcharge was added to the top of the slope. All the rows had a spacing of 2m horizontally and 0.75 m vertically, the uppermost nail being installed some 0.75m below the crest of the slope. All nails were installed at 10° to the horizontal.

Soil Nailing Design Configuration

The Ischebeck soil nails used had nominal OD of 30mm and an nominal ID of 16mm. As Ischebeck nails require no borehole to be pre-drilled, the anchor was drilled directly into the soil using a 48mm bit and air flush. The soil type was generally rubble infill with backfill of clay/marl.

The nails were completed at the proximal end with 300 x 300 x10 mm bearing plates which were mortared to the face and torqued with a nominal holding load.

The Face Retaining System

The face was completed with sprayed concrete to a depth of 100mm. A 20B16 mesh reinforcing fabric was placed centrally in the concrete.

Pull-out Testing

Prior to the initiation of the soil nailing project, 3 suitability pull-out tests were performed on nails identical to those used in the soil nailing scheme. All the nails failed with a P_{ult} of between 20 and 24T.

Acceptability pull-out tests were performed on site to determine the in situ pull-out strength between the nail and the ground. A minimum pull-out strength of 1T/m was required for the design calculations to be valid, although the design parameters had to be verified by continued on-site testing and monitoring before the slope design could be guaranteed and insured. Pull-out strengths in excess of 1T/m were achieved.

Case History 3 - A 42 Lay-by

Introduction

The A42 Birmingham to Nottingham route, required two emergency lay-bys to be constructed near Castle Donnington. It was proposed that this be implemented by steepening the existing roadside embankments and supporting them with soil nails and a geotextile geogrid. Scott Wilson Kirkpatrick were the Consulting Engineers, Charles Gregory Engineering the Contractors, with BRC Mining and Land Reinforcement in conjunction with Heriot-Watt University providing the soil nailing design and BRC Mining and Land Reinforcement performing the soil nailing.

The project required the installation of 310 GRP injection nails to support the excavated face to allow lay-by construction. There were two areas which required soil nailing. In both cases the existing cutting required to be steepened in order to widen the road by 3.1m.

Design Criteria.

All contract works were undertaken following the ICE Conditions of contract, 6th Edition, January 1991. The soil nailing design had to conform to published standards, including Department of Transport Advice Note HA 43/91.

The design had to remain stable throughout the installation and have a design life in excess of 60 years. Furthermore, the effect which frost action may have on the nails had to be considered.

Additionally, the following factors of safety were required:
Soil nail characteristic strength = 2.0
Grout / bar interface shear strength = 3.0
Grout / soil interface shear strength = 3.0
Long term stability of the cut face = 1.5

It was decided that the slopes could be stabilised by using Glass Reinforced Plastic (GRP) nails. The full column injection soil nail relies on soil displacement to transfer tensile load into the tendon. Using this type of nail, all of the above factors of safety could easily be achieved with the additional benefits of being more economic and impervious to frost action.

GRP Nail Design Calculations

Using the Limit Equilibrium Method, calculations were made to determine the length of nail required to stabilise the slopes.

The embankment height at Lay-by No 14 was some 6m and it was calculated that it required three 6m long nails to be installed at 7° to the horizontal for stability.

The embankment height at Lay-by No 11 was some 4m and it was calculated that two, 6m long nails and one 4m long nail, all at 7° to the horizontal, would need to be installed for stability.

Soil Nailing Design Configuration

The structure consists of a series of soil nails, manufactured from 22mm diameter GRP, with a nail head thread at the proximal end. The nails were centralised using ABS plastic lantern spacers fixed to the nails at discrete locations. The boreholes

were drilled using a rotary percussive drill rod with air as the flushing medium at an angle of 7° to the horizontal. The nails were installed manually in the 42mm OD open boreholes using continuous air flush to prevent blocking of the central bore. It was calculated that three rows of nails were required at a horizontal spacing of 1m and a vertical spacing of 0.5m. To ensure non-interference between the three rows, the central row of nails was staggered by 0.5m horizontally, resulting in a five spot pattern as shown in Figure 7.

When positioned, cementitious grout (Ordinary Portland Cement) was pumped through the central bore of the injection tube until the grout filled the annulus to the surface to give full column bonding between the nail / grout and the soil. Thereafter the boreholes were plated up and the grout allowed to cure. Once the grout had cured the geogrid was connected, a bearing plate was attached to the nail head and torqued up.

The Geogrid Retaining System

The geotextile used was Enkamat S20. The geogrid was positioned over the whole of the slope, with 200mm wide GRP straps running horizontally along the rows, and held in position by the nail heads (See Figure 7).

Pull-out Testing

Initially a provisional design was proposed dependent on the grout/soil bond strength being achieved. The bond strength was confirmed by on site pull-out tests on the nails at various locations along the slope. The equipment consisted of a 30T hydraulic ram attached to a distribution plate and support against which the nail head was stressed. The displacement of the nail head was monitored using a dial gauge. As the nail was stressed, the load and displacement were recorded.

The results of the pull-out tests gave pull-out strengths in excess of 15kN/m bond length. Once the required pullout values had been achieved on site, the final design was confirmed, although the design parameters had to be verified by continued on-site testing and monitoring before the slope design could be guaranteed and insured.

Case History 4 - Danbank

Introduction

This project involved the stabilisation of a steeply inclined slope on the A626 Stockport Road at Danbank. The slope comprised mainly of glacial clays and sands, however pulverised fuel ash had been used as infill below the A626. The inadequate roadside drainage had created a shallow slip circle, upon which the slope had mobilised, creating tension cracks on the road. The site itself was very inaccessible due to the heavy plant overgrowth and the steepness of the valley side, which was approximately 60° to the horizontal in places.

The contractors were Shepherd Hill Construction Ltd., and the client was Stockport Metropolitan Borough Council. The Engineers were EDGE Consultants UK Ltd.

The soil nailing programme was implemented by BRC Mining and Land Reinforcement. The contract period to install 450 no. cablenails varying in length between 6 and 9m was 10 weeks.

Design Criteria.

The design was developed by Manstock Geotechnical Services Ltd., and referred to the Design Manual for Roads and Bridges, Volume 4, Section 1, Part 4, HA 68/94, "Design Methods for the Reinforcement of Highway Slopes by Reinforced Soil and Soil Nailing Techniques", February 1994.

Cablenail Design Calculations

The soil nailed section ranged from 8m to 3m vertically and it was calculated that nails ranging from 9m to 6m respectively would be required to stabilise the slope. The nailing angle was 20°. In order to achieve the required length cablenails were used.

The cablenails were installed in a diamond pattern with a 2m horizontal spacing and a 1m vertical spacing.

Soil Nailing Design Configuration

The cablenails used comprised of 9 strands of GRP arranged in a node/antinode configuration with a removable tremie pipe. Each individual strand was 6mm in diameter, with the maximum diameter of the node/antinode configuration of approximately 50mm.

The nails were inserted into 100mm diameter boreholes and when centralised Sulphate Resisting Grout (SRG) was pumped through the tremie pipe (whilst the pipe was withdrawn) until the grout filled the annulus to the surface to give full column bonding between the nail / grout and the soil. Thereafter the boreholes were plated up and the grout allowed to cure. Once the grout had cured the geogrid was connected, a bearing plate was attached to the nail head and torqued up to a load of 1T.

The Geogrid Retaining System

The geogrid comprised of a GM4 geotextile secured with 500 x 500 x 10mm galvanised steel plates. The plates were mortared into position using a 3:1 sand : cement mix.

Pull-out Testing

Two different pull-out tests were required at

Danbank. The first suite were suitability tests which identical cablenails were stage loaded for a period of 3hours, and were deemed to have failed if over 1mm displacement occurred on any given holding value. All cablenails passed the suitability tests.

The second suite of tests were acceptance tests. The cablenails were installed and had to endure loading up to 120% of their normal working load for a period of 15 minutes. All cablenails passed the acceptance tests.

Conclusions

Soil nailing, although still in its infancy in the UK compared with the rest of Europe, has developed into a cost effective means of stabilising soil.

British engineers are using soil nailing with increasing confidence, and this is supported by the four different soil nailing programmes given above. The four methods show that the basic principle of soil nailing has matured to be applicable in a wide range of different circumstances, with varying ground conditions, often in tight locations and with little environmental impact.

Soil nails have the added benefit of being installed quickly. Typically, contract periods are in the order of weeks. This has positive implications over other stabilisation methods where transport routes may be affected. Additionally if urgent remedial works need to be effected, soil nails can be rapidly installed as either temporary or permanent works.

Acknowledgements

The principal author is indebted to the staff of BRC Mining and Land Reinforcement, Stafford, England, for their time, support and co-operation.

References

1.   Bruce DA, Jewell RA (1986) Soil nailing: Application and practice - Part 1. *Ground Engineering* Nov, 1986.

2.   Bruce DA, Jewell RA (1987) Soil nailing: Application and practice - Part 2. *Ground Engineering* Nov, 1987.

3.   Draft British Standard 8006 (1991) Code of Practice for Strengthened/Reinforced Soils and Other Fills. British Standards Institution.

4.   Mitchell JK, and Villet CB (eds) (1987) *Reinforcement of earth slopes and embankments*   NCHRP Report 290, Transportation Research Board, Washington.

5.   Murray RT (1990). A review of soil nailing

Transport Research Laboratory Working Paper WP/SAU/2. Can be obtained from: Structural Analysis Unit, Structures Group, Transport and Road Research Laboratory, Crowthorne Berkshire

6. "Geotechnical Considerations and Techniques for Widening Highway Earthworks" Department of Transport Highways, Safety And Traffic Departmental Advice Note HA 43/91 (1991). Department of Transport.

7. "British Standard Code Of Practice For Strengthened/Reinforced Soils And Other Fills" Draft BS 8006: 1991. British Standards Institution.

8. H. Weidmann AG, 8640 Rapperswil, Switzerland. Technical Data Sheets KMG/CRM E2.91.

9. " Soil Nailing and Rock Bolting Information Catalogue" BRC Mining and Land Reinforcement, Lichfield Road, Stafford, England. April 1994.

Dimensionierung des Anker-Spritzbeton-Ausbaus für die schachtnahen Großräume des Bergwerks Göttelborn/Reden

Dimensioning of bolt/shotcrete support for large-scale pit bottom excavations of Göttelborn/Reden colliery

Willi Kammer
DMT-Gesellschaft für Forschung und Prüfung mbH, Institut für Gebirgsbeherrschung und Hohlraumverfüllung (Institute Strata Control and Cavity Filling), Essen, Germany

ZUSAMMENFASSUNG: Auf dem Verbundbergwerk Göttelborn/Reden wurden im neuen Förderschacht 4 in 1.100 m Tiefe zwei Großräume mit bis zu 220 m² Ausbruchquerschnitt aufgefahren. Diese mit großen Tunnelbauwerken bzw. Kavernen vergleichbaren Räume wurden mit Ankern und Spritzbeton ausgebaut. Der Ausbau wurde als wartungs- und korrosionsfreier Ausbau für die Lebensdauer des Bauwerkes ausgelegt. Die Standsicherheit des Ausbaus wurde nachgewiesen. Die Großräume wurden unter meßtechnischer Überwachung der Gebirgsbewegungen in nur acht Monaten hergestellt. Die Herstellungskosten waren geringer als bei Anwendung von Unterstützungsausbau.

ABSTRACT: Two large-scale workings of up to 220 m² excavation cross section were realized in the new No. 4 shaft of Göttelborn/Reden compound colliery in 1,100 m depth. These excavations, comparable to large tunnel/cavern workings were supported by bolting and shotcreting. This support configuration was, on the one hand, cheaper especially in view of the excavations's size, however, required exact planning and carrying out. DMT's institute for strata control and cavity filling was entrusted with the support's planning and dimensioning. According to the customer's specifications the bolt-and-shot-crete support was designed as maintenance and corrosion-free system for the intended useful life of the workings. The proof of stability for the support was furnished. The support consisted of fully grouted M 27 and M 33 bolts of 4 - 5 m of length, two shot-crete layers of 10 and 15 cm thickness respectively, and of two layers of structural-steel mats of Q 188 quality. According to the excavations plan, the workings were excavated - with simultaneous monitoring by measurements of strata movement - in eight months only. This corresponds to approx. 2/3 of the time schedule which had been necessary for conventional standing support. The excavation costs were lower than they had been in case of conventional support.

1. GEOLOGISCHE SITUATION

Die geologischen Voruntersuchungen ergaben, daß die zu errichtenden Großräume in rund 1.100 m Teufe in geschichtetem Gebirge liegen, bestehend aus einer Wechsellagerung von Konglomerat, Sandstein, Schieferton und dünnen Kohleflözen (Bild 1).

Die Gesteine zeigten große Festigkeitsunterschiede mit folgenden einaxialen Druckfestigkeiten:

Konglomerat	100 MPa
Sandstein	80 bis 100 MPa
Schieferton	40 bis 60 MPa
Kohle	15 bis 25 MPa

Das Einfallen der Gebirgsschichten betrug rd. 10 Gon.

Die Lage der Großräume im Gebirge konnte noch in der Planungsphase in geringem Maße vertikal verschoben werden. Deshalb wurde zunächst eine Optimierung im Hinblick auf die festen Gebirgsschichten in der Firste sowie auf den Durchtritt von weicheren Gebirgsschichten an geometrisch günstigen Stellen der Großräume durchgeführt. Aus dieser Optimierung und aus

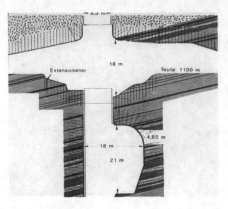

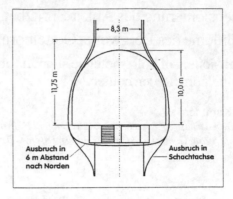

Bild 1: Schnitt durch den Bereich des Füllortes und
des Fülltaschenraumes

Fig. 1: Section of the pit bottom and the filling
bunkers

Bild 2: Schnitt durch den Schacht und das Füllort

Fig. 2: Section of shaft and pit bottom

dem Aufbau der Gebirgsschichten sowie
aus der Festigkeit der Gesteine ergab sich
die grundsätzliche Anwendbarkeit von
Anker-Spritzbeton-Ausbau. Wasserzuflüsse
oder spätere Abbaueinwirkungen, die den
Gebirgsdruck beeinflussen, waren nicht zu
erwarten.

2. OPTIMIERUNG DER HOHL-RAUMFORM

Der erste Entwurf der Grubenräume war
entsprechend der geplanten Nutzung
erstellt worden. Viele Hohlraumbegrenzun-
gen, wie auch die Räume unterhalb des
Sohlenniveaus, waren in Rechteckform
ausgeführt (Bild 2). Eine solche Ausbruch-
form kann nicht für formstabilen Ausbau genutzt werden. Weiterhin ist die Herstellung solcher Räume mit vorspringenden
Ecken in geschichtetem Gebirge mit
Sprengarbeit problematisch, weil im Bereich der Ecken häufig Mehrausbruch
entsteht.

Für einen möglichst stabilen Ausbau sind
Rundungen erforderlich. Entsprechend
dieser Forderung wurde die Hohlraumform,
in Zusammenarbeit mit dem Auftraggeber,
optimiert. In diese Optimierung wurde auch
die Planung der Einbauten, wie die der
Beschickungsanlage für die Gefäßförde-
rung, einbezogen. Geringe Änderungen der
Anlage führten unter anderem bei gleichem
Ausbruchvolumen zu einer deutlichen Verbesserung der Formstabilität.

Einerseits konnte der Ausbruchquerschnitt durch Verlegen des Rohrkanals
außerhalb des Füllortes und durch Anpassung der Beschickungsanlage für die Ge-
fäßförderung an die gerundete Ausbruch-
form verringert werden. Andererseits
wurde damit der an einigen Stellen erforderliche Mehrausbruch ausgeglichen, so
daß das gesamte Ausbruchvolumen auch
nach der Optimierung der Hohlraumform
etwa gleich groß blieb.

3. AUSBAUDIMENSIONIERUNG
3.1 Grundlagen

Die Planung und Bemessung des Anker-
Spritzbeton-Ausbaus wurde nach dem
Verfahren der DMT (1, 2 und 3) durch-
geführt. Dabei wird davon ausgegangen,
daß der Ausbau den zwei folgenden
Anforderungen genügen muß: Er muß den
gestützten Querschnitt offenhalten, also die
infolge des Gebirgsdruckes entstehenden
Gebirgsbewegungen in einer zulässigen
Größenordnung halten, sowie die
Belegschaft während der Herstellung und
während des Betriebes mit ausreichender
Sicherheit gegen Steinfall schützen.

Die Sicherheit gegen das Niederbrechen
von isolierten Totlasten in Form von Kluft-
und Rißkörpern wird im Steinkohlenbergbau mit einem rechnerischen Stand-
sicherheitsnachweis geprüft (1). Der erforderliche Stützdruck zum Offenhalten des
Querschnittes wird nach dem hierfür
entwickelten Rechenverfahren (2) ermittelt,
das die Einflüsse von Gebirgsdruck,

174

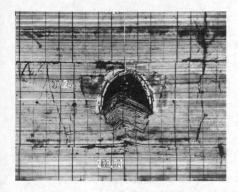

Bild 3: 45% Konvergenz in einer Strecke mit Stahlbogenausbau

Fig. 3: 45% convergence in an arch-supported roadway

Gebirgsaufbau und zulässiger Konvergenz berücksichtigt. Nach beiden Verfahren werden die Stützmittel dimensioniert; die jeweils strengere Anforderung an den Ausbau wird unter Tage verwirklicht (3).

Die Grundlagen für die Ausbaubemessung wurden aus umfangreichen Untertagemessungen im deutschen Steinkohlenbergbau sowie aus ebenen und räumlichen Gebirgsmodellen im Labor abgeleitet. Die Untertagemessungen stammen aus Strecken mit sohlenoffenem Stahlbogenausbau. Dieser Ausbau (Bild 3) hat einen Ausbauwiderstand von 0 (offene Sohle) bis 0,1 MPa. In geringer Teufe und in festem

Nebengestein weisen die so aufgefahrenen Strecken keine Konvergenz auf. Sie sind standfest. Das Bild 4 zeigt die Grenzbedingungen für das Auftreten von Konvergenz. Diese Grenzbedingungen kennzeichnen wegen des sohlenoffenen Ausbaus auch den Übergang von standfestem zu druckhaftem Gebirge. Danach wird z. B. das Gebirge in einer im Schieferton stehenden Strecke in Teufen von mehr als 900 m druckhaft.

Das geplante Füllort wurde ebenfalls unter Bedingungen aufgefahren, die Konvergenz in druckhaftem Gebirge erwarten ließ. Die Berechnung des erforderlichen Ausbaustützdruckes zur Vermeidung der Konvergenz wurde mit Hilfe der Gleichung vorausberechnet, die aus den Untertagemessungen abgeleitet wurde:

$$ AS = \frac{a \cdot P}{b + K} - c \cdot \sqrt{\beta_D} $$

AS = Ausbaustützdruck
P = Gebirgsdruck
K = Konvergenz
β_D = Gesteinsfestigkeit
a, b, c = Empirisch ermittelte Koeffizienten

Mit dieser Gleichung wird in eindrucksvoller Weise die Fenner/Pacher-Kurve (4), eine der wesentlichen Grundlagen der NÖT, grundsätzlich bestätigt.

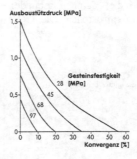

Bild 5: Erforderlicher Ausbaustützdruck für eine Teufe von 1.600 m in Abhängigkeit von Gesteinsfestigkeit und Konvergenz

Fig. 5: Required support pressure for a depth of 1,600 m as a function of rock strength and convergence

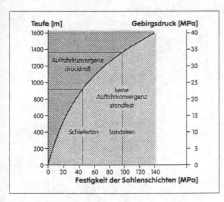

Bild 4: Bedingungen für das Auftreten von Konvergenz bei der Streckenauffahrung

Fig. 4: Condition for convergence occuring during roadheading

Im Bild 5 ist der Zusammenhang, entsprechend der o.a. Gleichung, zwischen den Größen *Gebirgsdruck, Gebirgsfestigkeit, Konvergenz* und *Ausbaustützdruck* grafisch dargestellt. Die Kurven gelten für Strecken mit sohlenoffenem Stahlbogenausbau bei verschiedenen Gesteinsfestigkeiten und einem Gebirgsdruck von 40 MPa, was einer Teufe von 1.600 m entspricht. Danach sind für die Verminderung der Konvergenz - die elastische Entspannung des Gebirges und die Konvergenz bis zum Einbringen des Ausbaus müssen zugelassen werden - Ausbaustützdrücke von 0,4 bis 1,5 MPa erforderlich. Mit den sohlenoffenen Stahlbögen können bei dem Stützdruck von rd. 0,1 MPa Konvergenzen von 50 % der Streckenausgangshöhe und mehr zugelassen werden.

Die dargestellten Kurven sind Teilabschnitte der Fenner/Pacher-Kurve. Dies soll am Beispiel der Kurve für die Gesteinsfestigkeit von 28 MPa erläutert werden (Bild 6). Im Bild ist sowohl die für das Ruhrkarbon quantifizierte Fenner/Pacher-Kurve als auch die Konvergenzentwicklung für Strecken mit sohlenoffenem Stahlbogenausbau dargestellt. Die Konvergenz, die bei der relativ hohen Gebirgsbeanspruchung in 1.600 m Teufe auftritt, stellt sich nach einer Standzeit von rd. 150 Tagen ein. Sie beträgt rd. 56 % der Streckenausgangshöhe. Diese Konvergenz wird Auffahrkonvergenz genannt. Die Standzeit von 150 Tagen ist die Zeit, die das Gebirge benötigt, um sich auf den Hohlraum und die damit verbundene Druckumlagerung einzustellen. Das Gebir-

ge wird dabei bis zu einem Streckendurchmesser und mehr von der Bruchverformung erfaßt.

Nach Erreichen der Auffahrkonvergenz geht das Gebirge in eine Kriechverformung über. Die sogenannte Kriechkonvergenz wächst linear mit der Zeit.

Die in Bild 6 dargestellte Fenner/Pacher-Kurve gilt, wie o. a., für sohlenoffenen Stahlbogenausbau mit einem Ausbauwiderstand von 0 bis 0,1 MPa. Mit diesem Ausbau werden im deutschen Steinkohlenbergbau Konvergenzen bis 100% der Streckenausgangshöhe, bei ständigem Senken der aus der Sohle nachwandernden Gebirgsschichten, beherrscht.

Durch Einbringen eines geschlossenen Ausbaus mit einem Ausbauwiderstand von z. B. 1 MPa kann die Konvergenz auf ein geringes Maß beschränkt werden (Bild 6). Mit einem begrenzt nachgiebigen Ausbau mit einem Stützdruck von 0,5 MPa kann die Auffahrkonvergenz einer mit Stahlbögen ausgebauten Strecke halbiert werden.

3.2 Erforderlicher Ausbaustützdruck

Die Dimensionierung des erforderlichen Ausbaustützdruckes für das Füllort des Bergwerks Göttelborn/Reden wird im folgenden ebenfalls an der einer Fenner/Pacher-Kurve dargestellt (Bild 7).

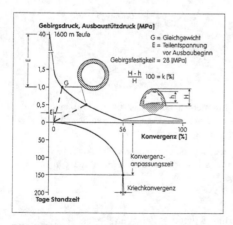

Bild 6: Gebirgs- und Ausbaukennlinien: Teufe von 1.600 m, Gesteinsfestigkeit 28 MPa

Fig. 6: Strata and support characteristics: depth = 1,600m, rock strength = 28 MPa

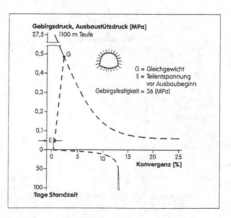

Bild 7: Gebirgs- und Ausbaukennlinien für das Füllort

Fig.: 7 Strata and support characteristics for the pit bottom

Die für Stahlausbau zu erwartende Konvergenz betrug rd. 14% der Streckenausgangshöhe. Sie war wegen der relativ geringen Gebirgsbeanspruchung bereits etwa 50 Tage nach der Auffahrung zu erwarten. Durch die Anwendung von Anker-Spritzbeton-Ausbau mit dem vorausberechneten Stützdruck von 470 kN/m^2 wurde die Konvergenz auf rd. 2 bis 3 % begrenzt.

3.3 Standsicherheitsnachweis

Der zweite Schritt der Ausbaudimensionierung ist, wie o. a., das Sichern des Gebirges gegen niederbrechende Gebirgskörper während der Herstellung der Grubenräume. Hierbei wurde von Lastannahmen ausgegangen (1), die den von der Geometrie des Bauwerkes her größtmöglichen Bruchkörper berücksichtigt, weil z. Z. der Ausbauplanung keine Aufschlüsse über die Klüftung des Gebirges vorlagen. Diese Bruchkörper können durch natürliche Fugen, wie Klüfte, Störungen und Schichtflächen sowie durch Risse begrenzt sein und müssen vom Ausbau getragen werden. Für den Vorortbereich wird angenommen, daß sich eine Platte entsprechend der Streckenbreite und einer Höhe von 0,3 m aus der Firste lösen kann.

Unter Berücksichtigung dieser Vorgaben wurden alle möglichen Kluft- und Rißkörper, auch für den Übergangsbereich des Schachtes zum Füllort (Bild 8), konstruiert und deren Gewicht errechnet. Die sich daraus ergebenden Lasten müssen vom Ausbau in jeder Phase der Auffahrung sicher getragen werden.

Die endgültige Dimensionierung des Ankerausbaus erfolgte mit einem EDV-Programm, das auch Unterkluftkörper berücksichtigt. Um die optimale Anordnung der Anker zu finden, war das Durchrechnen vieler Varianten erforderlich.

3.4 Ausbaumittel und Ankerschema

Die Planungen ergaben unter Berücksichtigung des erforderlichen Ausbaustützdruckes zur Beherrschung der Konvergenz folgenden Ausbau:

4 bis 5 m lange vollvermörtelte Anker M 27 und M 33.
Zwei Lagen Spritzbeton (10 und 15 cm dick).
Zwei Lagen Baustahlmatten Q 188.

Die so aufgebaute Anker-Spritzbeton-Schale ist im Bild 9 dargestellt. Die Sicherheit im Vorortbereich wurde durch die erste 10 cm dicke Spritzbeton-Schale erreicht. Wenn diese eine Festigkeit von 5,2 MPa hatte, konnte das Ort wieder betreten werden. Alternativ hierzu konnten auch kurze, mit Kunstharz vermörtelte Anker verwendet werden. Durch die Anwendung dieser Anker entfielen Wartezeiten, die jedoch wegen der Entzerrung der Arbeitsvorgänge nur selten anfielen.

Die Anordnung der Anker ist im Längs-

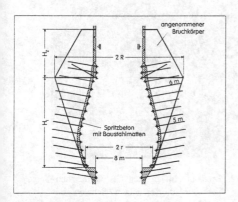

Bild 8: Lastannahme für Anker-Spritzbeton-Ausbau einer Schachtglocke

Fig. 8: Load assumption for bolt-and-shot-crete support of the shaft curb

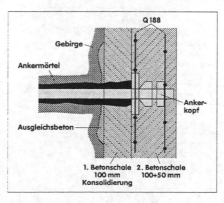

Bild 9: Aufbau der Spritzbetonschale

Fig. 9: Configuration of shot-crete shell

177

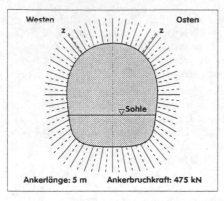

Bild 10: Ankerschema für den Streckenquerschnitt von 187 m^2

Fig. 10: Bolting pattern for a 187 m^2 roadway cross section

schnitt des Füllortes und des Füll-taschenraumes beispielhaft zu erkennen (vgl. Bild 1). In den vertikalen Bereichen des Fülltaschenraumes wurden die Anker so angeordnet, daß sie sowohl einzelne Schichtpakete durchstoßen als auch durch ihre nach oben weisende Richtung, den Vertikalkräften des Wandkluftkörpers, möglichst effektiv entgegenwirken. Das Bild 10 zeigt beispielhaft die Anordnung der Anker in einem Querschnitt 6,3 m aus der Schachtachse nach Norden. Die Ankerdichte beträgt 1,1 Anker je Quadrat-meter.

Aus der Vorgabe, den Sohlenschluß so früh wie möglich zu erreichen, ergab sich

nach der ersten Bauphase zu jeder Seite des Schachtes eine rd. 18 m hohe Ortsbrust, die ebenfalls mit Anker-Spritzbeton-Ausbau gesichert wurde. Das hierfür erforderliche Ankerschema zeigt das Bild 11.

Die Ortsbrust wurde gewölbeartig mit einem Stichmaß von 1 m hergestellt und mit 3 m langen Ankern mit einer Bruchkraft von 320 kN (M 27) gesichert. Der Spritz-beton hatte eine Dicke von 10 cm.

4. ERFAHRUNGEN BEI DER BAU-AUSFÜHRUNG

Nach Beendigung der Auffahrung des Füllortes wurde der Schacht tiefer geteuft und rd. 10 m unterhalb der Füllortsohle mit der Herstellung des Fülltaschenraumes be-gonnen (vgl. Bild 1).

Die Auffahrung im oberen Bereich dieses Hohlraumes schien zunächst wegen der kugeligen Hohlraumform aufwendig, erwies sich jedoch wegen der guten Beherr-schung der weichen Gebirgsschichten im Bereich der Flözdurchtritte als sehr vorteilhaft.

Die weitere Auffahrung des Fülltaschen-raumes erfolgte im gesamten Querschnitt, einschließlich des Schachtquerschnittes. Das Bild 12 zeigt die Teufensohle etwa 5 m unterhalb der Sohle der Füllstrecke. Hier sind die einzelnen Schalen des Ausbaus gut zu erkennen.

Bild 12: Teufsohle im Bereich des Fülltaschen-raumes; hier sind die einzelnen Schalen des Anker-Spritzbeton-Ausbaus zu erkennen

Fig. 12: Excavation level, loading bunker excava-tion; the individual shells of the bolt-and-shot-crete support may be identified

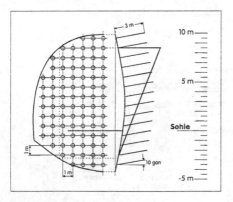

Bild 11: Ausbau der Ortsbrust

Fig. 11: Support of the heading face

Der Ausbruch des Fülltaschenraumes konnte, bis auf zwei Stellen, profilgenau hergestellt werden. Diese Stellen, an denen Mehrausbruch entstand, waren die „vorspringenden Ecken" am Übergang von der Sohle der Füllstrecke zum Fülltaschenraum sowie am Übergang von der Sohle des Fülltaschenraumes zum Schacht (vgl. Bild 1). An diesen Stellen bestätigte sich, daß eckige Ausbruchsformen weniger stabil sind.

Insgesamt verlief der Bau der Großräume problemlos. Es waren nur einige Risse an Übergängen von einem Spritzbetonabschnitt zum anderen entstanden, die nachbehandelt werden mußten. Der Mehrausbruch war relativ gering. An den „vorspringenden Ecken" wurde der Mehrausbruch durch mehrere Lagen Spritzbeton und Baustahlmatten sowie mit Ankern ausgeglichen.

Quellennachweis

1) Götze, Wilhelm:
Grundlagen für den Ankerausbau in bogenförmigen Streckenquerschnitten
in: Glückauf 113 (1977), Nr. 15, Seite 753-755

2) Kammer, Willi:
Der erforderliche Ausbaustützdruck zur Streckenbeherrschung
in: Glückauf-Forschung 49 (1988), Nr. 6, Seite 267-271

3) Zischinsky, Ulf:
Bemessung und Erfahrungen mit Anker-Spritzbeton-Ausbau
in: Ankerausbau im Steinkohlenbergbau. Essen: Verlag Glückauf (1988)

4) Rabcewicz, L., J. Gosler und E. Hackl:
Die Bedeutung der Messung im Hohlraumbau, Teil I
in: Bauingenieur 47 (1972), Seite 225/34

Anker in Theorie und Praxis, Widmann (Herausgeber) © 1995 Balkema, Rotterdam. ISBN 90 5410 577 1

Die Bedeutung der Verbundanker für den Untertagebau

The importance of compound anchors for underground excavations

H. Lauffer
Allgemeine Baugesellschaft- A. Porr AG, Wien, Austria

KURZFASSUNG: Nicht vorgespannte Verbundanker gehören zu den wichtigsten Stützmitteln des Untertagebaues. In der Regel erfüllen sie gleichzeitig mehrere Aufgaben, von der Abdeckung von Querzugspannung bei Querschnittsänderungen bis zur
Bereitstellung eines Ausbauwiderstandes am Hohlraumrand. Eine überragende Bedeutung für die Entstehung eines Untertagebauwerkes haben Verbundanker als Gebirgsvergütung. Sie ermöglichen es dem Gebirge die Spannungen aus der Hohlraumher-
stellung ohne großen Festigkeitsabfall umzulagern. Aus dieser Erkenntnis lassen
sich wertvolle Hinweise für Planung und Ausführung von Untertagebauwerken gewin-
nen.

ABSTRACTS: Non-prestressed compound anchors play an import part in underground
excavation. They are used to secure isolated blocks and largely stress free rock
parts of cross sections which deveate from elyptic or circular forms. Anchor
reinforce the bond between rock and shotcrete. They contribute to the support
resistance at the excavation surface, they increase the strength of the anchored
rock and they help to conserve the strength of rock which is subjected to big
deformations.

It is an outstanding quality of compound anchors that the fullfil many or all of
these tasks at the same time. This fact has a favourable effect on the total
economy of rock anchoring. The most important effect of rock anchoring is the
big increase in failure energy of the anchored rock, so that the stresses generated by the excavation process can be redistributed.

Rock anchors should be placed early in the excavation sequence. Some changes of
the construction routine are necessary to achieve this goal.

1 EINLEITUNG

Eine umfassende ingenieurmäßige und
rechnerische Begründung des Einsatzes
von Ankern im Untertagebau ist trotz
des Fortschrittes der Wissenschaft und
der Rechentechnik auch heute noch
schwierig bis unmöglich. Eine der Ursachen dafür ist, daß die Bestandteile
eines Tunnelbauwerkes, das Gebirge und
die Summe der Stützmaßnahmen, sich in
einer Grenzsituation des Spannungs-
und noch mehr des Verformungsniveaus
befinden können. Im normalen Inge-
nieurbau werden solche Grenzsituatio-
nen nur im Katastrophen- bzw. Versage-
nsfall erreicht. Die besondere Bedeutung der Anker liegt aber gerade in
ihrem günstigen Einfluß auf das Verhalten des Untertagebauwerks bei hohen
Spannungs- und Verformungsgrößen.

Die Probleme mit einer ingenieurmäßigen Erfassung der Gesamtheit der möglichen Ankerwirkung hat den Einsatz
von Ankern in der Vergangenheit verzögert und erschwert ihn, insbesondere
im Ausland, noch heute. Über die ge-
schichtliche Entwicklung des Ankerein-
satzes im Rahmen der NÖT wurde von
Lauffer 1994 berichtet. Einige Zitate
seien hier wiederholt. Seeber berichtet 1979 über die Verwendung von Felsankern beim Bau des Druckstollens des
Kaunertalkraftwerkes (1960 bis 1964)
folgendes: "Die Kombination Spritzbeton mit Felsankern wurde vorwiegend in
den gebrächen Gebirgsklassen eingesetzt, während in druckhaftem Gebirge
vor allem Stahlbogen mit Spritzbeton
vorgesehen waren. Der Wirkung der Felsanker im druckhaften Gebirge schenkte
man trotz der guten Erfahrungen von
Imst noch nicht volles Vertrauen".

Die hervorragende Wirkung einer Systemankerung wurde jedoch von den
Praktikern unter den Tunnelbauern zunehmend erkannt und umgesetzt. So sagt
Detzelhofer 1969 folgendes über einen
druckhaften Schiefergneisabschnitt
einer Bachüberleitung des Kaunertalkraftwerkes (1960-1964):
"Aufgrund dieser Beobachtung wurde in
der Folge daher beim Vortrieb nur mehr
eine dünne Schicht Sicherungs-
spritzbeton aufgebracht. Die sofortige
Nagelung beim Ausbruch mit Perfoankern
wurde weiterhin beibehalten. Sie
stellte den eigentlichen wirksamen
Einbau dar. Ausreichend nachgiebig,
aber dabei zähen Widerstand leistend,
verhinderte er das Bruchfließen und
die Auflösung der Randzone und führte
bald zur Stabilisierung".
Sehr große gedankliche Schwierigkeiten
macht auch heute noch die Tatsache,
daß ein Untertagebauwerk als Folge des
Ausbrechens des Hohlraumes (der Hohlraum ist Luft und kein Bauwerk) mit
mehr oder weniger großer Unterstützung
durch den Tunnelbauer in dem den Hohlraum umgebenden Gebirge entsteht und
eben nicht gebaut wird. Dieser dynamische, zeitabhängige und verformungsab-
hängige Entstehungsvorgang ist nur
beschränkt beeinflußbar und mit üblichen statischen Dimensionierungsaufga-
ben nicht vergleichbar. Bei diesem
Entstehungsvorgang haben Anker als
Bewehrung des Gebirges eine hervorragende Bedeutung.

2   ANKERARTEN

2.1 Anker mit und ohne Verbund

Bei Ankern ohne Verbund
(Freispielanker) führen örtliche Verformungen des Gebirges (Rißbildung,
Abscherungen etc.) zu keiner wesentlichen Änderung der Ankerkraft. Sie setzen diesen Verformungen keinen großen
Widerstand entgegen. Schlaffe Anker
ohne Verbund (Keilanker) werden daher
nur für untergeordnete Aufgaben eingesetzt. Vorgespannte Anker ohne Verbund
werden dort eingesetzt, wo es ausschließlich um die Einleitung einer
Vorspannkraft geht. In der Folge ist
daher immer nur von Verbundankern die
Rede.

2.2 Anker mit und ohne Vorspannung

Die Erfahrung hat gezeigt, daß in fast
allen Anwendungsfällen schlaffe Anker
ausreichen. Sie sind wegen ihrer Ro-
bustheit, ihrer hohen Verformungsre-
serven und ihrer Wirtschaftlichkeit
vorgespannten Verbundankern vorzuziehen. Im Untertagebau werden vorgespannte Anker nur in Ausnahmefällen
eingesetzt. Dementsprechend ist in der
Folge nur von schlaffen Verbundankern
die Rede.

2.3 Schlaffe Verbundanker

Es gibt derzeit im Untertagebau drei
Haupttypen: Den Mörtel-Klebeanker
(z.B. SN), den Injektionsanker und den
Verbundanker mit mechanischem Kraftschluß (z.B. Swellex). Sie unterscheiden sich nur unwesentlich in ihrer
Wirkung. Die Wahl ist vom ange-
troffenem Gebirge und von den Gesamt-
kosten bestimmt.
Die Zeitspanne vom Einbau bis zur vollen Wirksamkeit hängt bei Mörtel- und
Injektionsankern von der Erhärtungs-
zeit des verwendeten Ankermörtels bzw.
des Injektionsmaterials ab.

3   ANKERWIRKUNGEN

Die Wirkung von Ankern kann eine sehr
vielfältige sein. Um die Bedeutung der
Verbundanker für den Untertagebau her-
auszuarbeiten, wird versucht die möglichen Ankerwirkungen getrennt zu beschreiben (Wirkungsmodelle), obwohl in
der Wirklichkeit immer Über-
schneidungen der einzelnen Wirkungen
auftreten. Am Hohlraumrand kommt es
zusätzlich zu einem Zusammenwirken mit
anderen Stützmittel, insbesonder mit
einer Spritzbetonschale.

3.1 Örtliche Hohlraumsicherung

Einzelne durch Klüfte und Schichtflächen abgetrennt Gebirgselemente mit
freier, unbehinderter, Bewegungsmög-
lichkeit zum Hohlraum werden wirtschaftlich durch einzelne Anker im
Gebirgsverband gehalten (Bild 1). Auch
für diese Anwendung sind Verbundanker
einem Freispielanker (Keilanker) vorzuziehen.

3.2 Rückhängen von "spannungslosen"
     Gebirgsteilen

Bei einem an sich standfesten Gebirge

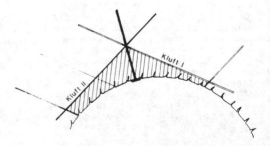

Bild 1: Sicherung einzelner Gebirgs-
        elemente (Blöcke)
        Securing of isolated blocks

kann eine Verankerung von "spannungslosen" Gebirgsbereichen erforderlich
sein.(Bild 2)

Tunnel-Stollen

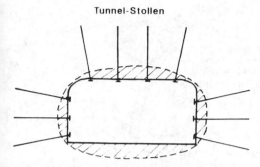

Kaverne

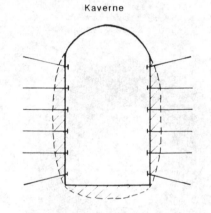

Nische

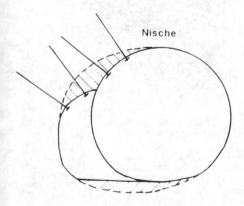

Bild 2: Rückhängung von Gebirgsteilen
        bei Abweichung des Querschnit-
        tes von einer Ellipsen-Kreis-
        form
        Securing rock parts with cross
        section deveating from an
        eliptic or circular form

3.3   Abdeckung von Querzugspannungen

Abrupte Änderungen des Hohlraumquer-
schnittes und damit des Spannungsflus-
ses haben Querzugspannungen zur Folge,
die ohne den Einsatz von Ankern kaum
abzudecken sind. Derartige Querschnittsänderungen treten bei nicht
geschlossenen Spritzbetonschalen auf,
beispielsweise im Bereich der Fußpunk-
te von Kalottenschalen oder auf beiden
Seiten von Verformungsschlitzen. Ähnliches gilt für Rißbildungen in einer
Spritzbetonschale. (Bild 3)
Der Einsatz von Ankern in diesen Bereichen garantiert die Verbundwirkung
zwischen dem Gebirge und Teilen einer
Spritzbetonschale und damit die Tragfähigkeit des Verbundsystems.

3.4   Ausbauwiderstand

Anker leiten über den Ankerkopf radiale Kräfte in die Hohlraumoberfläche
ein. (Bild 4). Sie leisten damit einen
Beitrag zum Ausbauwiderstand und erhöhen dadurch die Tragfähigkeit des
Hohlraumrandes.

3.5   Verbessern der Festigkeitseigen-
       schaften

Durch das systematische Versetzen von
Verbundankern entsteht ein Verbundsystem aus Gebirge und Ankern. (Bild 5)
Dieses Verbundsystem besitzt eine entsprechend höhere Zugfestigkeit
(Kohäsion) und damit auch eine erhöhte
Tragfähigkeit (Mohr'sche Grenzlinie).
In einem inhomogenen Gebirge mit vor-
gebildeten Gleitflächen (Klüfte,
Schichtflächen) werden diese Schwäche-
zonen durch eine entsprechende Ankerung über Zug- und Dübelwirkung ge-
sperrt und die Festigkeit des Ge-
steinsmaterials aktiviert.
(Homogenisierung des Gebirges).

3.6   Erhalten der Tragfähigkeit des
       Gebirges bei großen Verformungen

Bei größeren Überlagerungen ist eine
(wirtschaftliche) Herstellung von Untertagebauten ohne Gebirgsmitwirkung
nicht möglich. Ein hohes relatives
Spannungsniveau mit Überschreitung der
elastischen Tragfähigkeit des Gebirges
bewirkt Spannungsumlagerungen, die mit
bleibenden Verformungen einhergeben.
Dabei können in Extremfällen Stauchun-
gen des Hohlraumrandes von bis zu 20 %
auftreten.

Bei großen Verformungen ist mit Brüchen (Abfall der Kohäsion) und Entfe-
stigungen (Verringerung Reibungswinkel)im Gebirge zu rechnen. Verstellun-
gen des Gesteinskörpers führen zur
Zerstörung des Gebirgsverbandes und
schließlich zur Auflockerung.

183

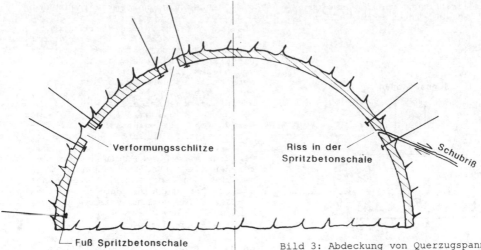

Verformungsschlitze

Riss in der
Spritzbetonschale

Schubriß

Fuß Spritzbetonschale

Bild 3: Abdeckung von Querzugspannungen bei Querschnittsänderungen
(Krafteinleitung)
Shear reinforcement with
changes of cross section

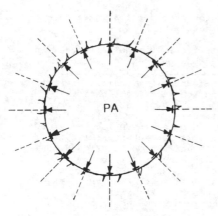

PA

Bild 4:
Erzeugen eines Ausbauwiderstandes
Generation of support resistance

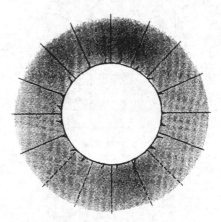

Bild 5: Gebirgsvergütung
Rock improvement

Der Einbau einer Gebirgsvergütung in
Form von Verbundankern bewirkt für den
geankerten Gebirgsbereich eine entscheidende Erhöhung des Arbeitsvermögens. Dies bedeutet, daß das geankerte
Gebirge in die Lage versetzt wird große Verformungen zu ertragen, ohne daß
die Festigkeitseigenschaften entscheidend abfallen. Am Hohlraumrand sind
meist ergänzende Maßnahmen erforderlich (Spritzbetonsicherung, Ausbauwiderstand).

4   DETAILDISKUSSION DER
     ANKERWIRKUNGEN

Es ist eine herausragende Eigenschaft
von Ankern, daß sie bei einer entsprechenden Dichte und Länge alle Wirkungen entsprechen Pkt. 3.1 bis 3.6
gleichzeitig abdecken können. Dies
erklärt auch ihre hervorragende vor
Ort beobachtete Wirkung und erklärt
auch, warum sie trotz vergleichsweiser
hoher Kosten im großen Umfang Verwendung finden. Eine überschlägige Berechnung der Kosten, die zur Erzeugung
eines Ausbauwiderstandes aufzuwenden

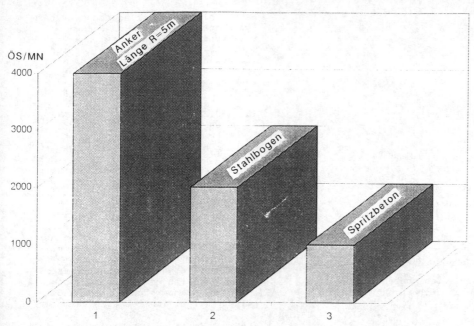

Bild 6: Kosten Ausbauwiderstand für Hohlraum mit Radius R = 5,00 m Cost of support
resistance for tunnel with radius R = 5,0 m

sind, zeigt nämlich, daß Anker im Vergleich zu einer (geschlossenen)
Spritzbetonschale die vierfachen, im
Vergleich zu einem (geschlossenen)
Stahlbogen die doppelten, Kosten verursachen.(Bild 6).
Anker haben sich als unverzichtbare
Stützmittel zur wirtschaftlichen Beherrschung von Vortriebssituationen
erwiesen, wo hoher Gebirgsdruck sich
in großen bleibenden Verformungen äu-
ßert. Unter diesen geomechanischen
Umständen muß es das vorrangige Ziel
jeder Maßnahme sein, zuerst die Tragfähigkeit (Festigkeitseigenschaften)
des Gebirges zu erhalten (Pkt. 3.6).
Erst dann ist es sinnvoll die Erhöhung
der Tragkraft bzw. die Kompensation
des Abfalls der Tragfähigkeit (Pkt.
3.5) oder eine Erhöhung des Ausbauwiderstandes am Hohlraum anzustreben
(Pkt. 3.4). Die Erfahrung (z.B. am
Tauerntunnel, Herbeck 1975) haben gezeigt, daß es nicht möglich ist einen
wegen fehlender Ankerung (zu kurze
Anker) zerstörten Gebirgsbereich durch
später eingebaute Anker zu sanieren.
Anhand von schematischen Gebirgskenn-
linien (Pacher-Fenner Kurven) lassen
sich diese Überlegungen gut zeigen
(Bild 7 und Bild 8).
Ein mit zunehmenden Verformung zunehmender Festigkeitsabfall und even-
tuelle Gewichtslasten aus der Auflockerung, ergeben einen schnell
zunehmenden Bedarf an Ausbauwiderstand
(Bild 7).

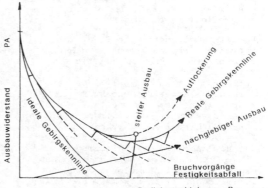

Bild 7: Reale Gebirgskennlinie mit
Festigkeitsabfall
Realistic deformation characteristics of rock with
strength reduction

Dabei ist noch nicht berücksichtigt,
daß Auflockerungsbereiche nicht nur
eine Gewichtlast darstellen, sondern
daß sie zusätzlich die Geometrie des
Untertagebauwerkes ungünstig verändern.

Umgekehrt läßt sich zeigen (Bild 8),
daß sich bei Erhaltung einer idealen
Gebirgskennlinie (kein Festigkeitsab-

185

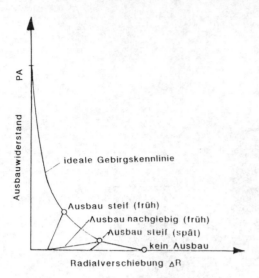

Bild 8: Ideale Gebirgskennlinie
Ideal deformation characteristic of rock

fall) ein Gleichgewicht ohne Ausbauwiderstand einstellt. Ein zusätzlicher
Ausbauwiderstand hätte nur noch die
Begrenzung der Hohlraumverformung zum
Ziel. Es ist dabei unerheblich, ob
dieser Ausbau steif oder nachgiebig
ausgebildet wird.

Es ist immer zuerst die Frage zu be-
antworten, ob und welche Maßnahmen
erforderlich sind, um die Tragfähigkeit des Gebirges zu erhalten. Die
Erfahrung zeigt, daß der Einbau entsprechend langer Verbundanker die wir-
kungsvollste Maßnahme ist. Die Vergütung (Bewehrung) des Gebirges durch
Anker verhindert, daß vorhandene
Trennflächen im Gebirge aktiviert werden und daß sich die Verformungen auf
diese Trennflächen konzentrieren und
daß es zu einer Ausbildung durchgehender Versagensflächen kommt. Die Ankerung verhindert außerdem die Neubil-
dung von Bruchflächen durch Unterdrük-
kung der Rißausbreitung. (Poisel 1993
und 1995). Das Verformungs- bzw. Arbeitsvermögen des Gebirges wird durch
die Ankerung um ein Mehrfaches erhöht.
Diese Erhöhung des Arbeitsvermögens
durch eine Bewehrung ist u.a. durch
Versuche an dem gesteinsähnlichen Baustoff Beton erwiesen. (Lauffer 1994).

Untertagebauwerke mit einer entsprechenden Systemankerung haben in der
Regel ein gutmütiges Versagensverhal-
ten. Dies bedeutet, daß es bei örtlichen Überlastungen zu Lastumlagerungen
kommt, oder daß zumindest Zeit zum
Räumen der Arbeitsstellen oder sogar
Zeit zum Einbau von Verstärkungsmaß-

nahmen bleibt. Untertagebauwerke ohne
Systemankerung zeigen dagegen eher
Versagensmechanismen, die ohne Vorwär-
mung rasch ablaufen. Diese Tatsache
ist bei allen Überlegungen für anker-
lose Stützungen von untertägigen Hohlräumen zu berücksichtigen.

5.   SCHLUSSFOLGERUNGEN FÜR PRAXIS
      UND THEORIE

Die zur Erhaltung der Tragfähigkeit
des Gebirges, einschließlich des Hohlraumrandes, erforderlichen Maßnahmen
sind frühzeitig zu setzen. Eine Systemankerung ist daher grundsätzlich
in vollem Umfang im vordersten Arbeitsbereich einzubringen, wobei auf
ein schnelles Erreichen der Ankerwirkung zu achten ist (Erharten des Ankermörtels und des Injektionsmaterials). Beim zyklischen konventionellen)
Vortrieb sollten alle Anker wirksam
sein, bevor ein neuer Abschlag begonnen wird. Nachankerungen sind ungünstig und daher möglichst zu vermeiden.

Eine dem Ausbruch vorauslaufende Ankerung würde eine weitere Verbesserung
bewirken. Dabei ist immer auch die
Kostenfrage zu betrachten. Eine gewisse Voraussicherung ist durch den Einbau schräg nach vor (Kalotte) oder
schräg nach unten (Strosse) gerichte-
ter Anker möglich.

Falls ein Pilotstollen im Ausbruchquerschnitt vorweg hergestellt wird,
kann aus diesem heraus eine Vorausan-
kerung für den nachfolgenden Großquer-
schnitt eingebaut werden. Dadurch
könnten zugleich die negativen Auswirkungen des Ausbruchs eines Pilotstollens auf den Ausbruch des nachfolgenden Querschnittes vermindert werden.

Bei sehr vielen Vortrieben wird die
Gebirgssicherung in den Sohlebereichen
der Teil- und Endquerschnitte vernach-
lässigt. Die Ursache dafür ist sicher,
daß Verformungen und Gebirgszerstörun-
gen im Sohlbereich kaum auffallen, und
daß die Sicherheit der Mannschaft und
des Hohlraumes in der Regel dadurch
nicht unmittelbar beeinträchtigt wird.
Es wird sehr oft überhaupt keine Sicherung eingebaut (Teilquerschnitte),
oder der Einbau erfolgt sehr lange
Zeit nach dem Ausbruch der Kalotte
(endgültiges Sohlgewölbe). Anker werden im Sohlbereich nur in Sonderfällen
eingebaut. Es ist nun äußerst wahrscheinlich, daß durch diese Vorgangsweise das Gebirge bei großen Verformungen in den Sohlbereichen geschädigt
wird und daß dadurch die Gesamtverfor-
mungen des Untertagebauwerks stark
zunehmen. Es ist davon auszugehen, daß
eine frühzeitige eingebaute Ankerung

im Sohlbereich das Verhalten des Gesamtquerschnittes entscheidend verbessern würde. Dazu müßten entsprechende Bauabläufe entwickelt werden, die es erlauben ohne allzu große Mehrkosten eine Sohlsicherung in kurzer Entfernung hinter dem Kalottenausbruch einzubauen.

An Wissenschaft, Planung und Baudurchführung stellen sich u.a. folgende Fragen:
- Auswirkung von zeitlich und räumlich unterschiedlichem Einbau einer Sohleankerung
- Wirkung von Voraussicherungen bei großen und bei geringen Überlagerungen
- Arbeitsvermögen verschiedener Gebirgsarten mit und ohne Ankerung und bei verschieden hohem Ausbauwiderstand
- Einfluß von Ankerquerschnitt [cm²/m²] und Ankerdichte [Stk/m²] auf das Arbeitsvermögen

Eine Beantwortung dieser Fragen ist nur möglich, wenn die wissenschaftlichen Überlegungen und die Modellversuche durch Großversuche bei aktuellen Vortriebsarbeiten ergänzt und parallel die baubetrieblichen und wirtschaftlichen Konsequenzen untersucht werden.

Quellennachweis:
1. Detzlhofer, H.: Erfahrungen bei der Sicherung von Stollenausbrüchen in gebrächen und druckhaften Gebirgsstrecken. In: Felsmechanik (1969), S. 166
2. Herbeck,H.: Arbeitsgemeinschaft Tauern-Scheiteltunnel, Bauarbeiten 1970 bis 1975. In: PORR-Nachrichten (1975) Nr.63 S. 23-24
3. Lauffer,H. Wien: Die Entwicklung der NÖT im Spannungsfeld zwischen Theorie und Praxis. In: Felsbau 12 (1994) Nr.5
4. Poisel,R.,Steger W. Zettler A.H.: Neue Ansätze für Standsicherheitsuntersuchungen von Tunneln. In: Felsbau 13 (1995) Nr. 3
5. Poisel,R.,Zettler A.H., Egger H.: Das Arbeitsvermögen geankerten Gebirges bei großen Konvergenzen. Proceedings Symposium, Anker in Theorie und Praxis, Salzburg (1995)
6. Seeber, G.: Entwicklung und derzeitiger Stand der Neuen Österreichischen Tunnelbauweise (NATM). In: Österreichische Wasserwirtschaft, 31 (1979), Heft 5/6, Seite 116.

Anchors in Theory and Practice, Widmann (ed.)© 1995 Balkema, Rotterdam. ISBN 90 5410 577 1

Some procedures for installation of anchoring systems and the evaluation of capacity

Einige Verfahren über das Einbringen und die Tragbewertungen der Ankersysteme

Uroš Bajželi, Jakob Likar & Franc Žigman
Institute for Mining Geotechnology and Environment, Ljubljana, Slovenia

ABSTRACT: Anchoring systems may be used as supporting elements in the construction of underground openings either independently or in combination with other procedures for the purpose of ensuring the necessary stability conditions. The effects of the excavation method used during face advancement in an underground area depend, on the one side, on the geotechnical conditions in the nearby rock and, on the other side, on the technological characteristics and quality of procedures used. To facilitate the planning of procedures for the installation of anchoring systems, studies and measurements were conducted and some static and dynamic effects on installed anchoring systems were analyzed. Correlations between the use of advanced theoretical procedures for determining the technical parameters of anchoring systems and the measured values were established, which will enable improvements in the planning of support measues.

ZUSAMMENFASSUNG: Ankersysteme können als Ausbauelemente beim Bau von Untergrundhohlräumen unabhängig oder in Kombination mit anderen Verfahren verwendet werden, um die erforderliche Standfestigkeit zu erreichen. Die Auswirkungen der Aushubverfahren hängen einerseits von den geotechnischen Bedingungen im nahen Felsbereich und andererseits von den technologischen Eigenheiten und der Qualität der Verfahren ab. Um die Planung für den Ankereinbau zu erleichtern, wurden Studien und Messungen ausgeführt und einige statische und dynamische Effekte an eingebauten Ankersystemen analysiert. Beziehungen zwischen dem Gebrauch von fortschrittlichen theoretischen Verfahren zur Bestimmung technischer Parameter von Ankersystemen und den gemessenen Werten wurden aufgezeigt, welche eine Vervollkommnung der Planung von Stützungsmaßnahmen erlauben.

1. INTRODUCTION

The relations between stresses and deformations in the rock - support system during the excavation or construction of underground openings continue to be the subject of study and analysis from different viewpoints.

The objectives of this kind of works are more or less known, while the mathematical description of relations and the reqired reliability of parameters which describe and represent individual technical terms are a more difficult task

and, in particular, need to be supported with relevant measurements.

This is why the upgrading of the description and verification of calculated parameters of effectiveness of anchoring systems, both as independent and combined supporting measures, by means of in situ measurements remains a convenient method for the optimalization and determination of the actual effectiveness of the said procedures. The technological particularities applied in the installation of passive as well as active anchoring systems have been adapted to the changing rock properties and other goetechnical conditions prevailing at individual locations where such works are being performed.

Our discussion shall be focused primarily on the method of installation and analysis of the actual performance of passive anchors and anchoring systems in supporting permanent and temporary underground openings in coal mines in the Republic of Slovenia.

A few examples of the installation of cable bolts for securing excavation areas in tectonically affected sandstone and the results of corresponding measurements will be presented. Static and dynamic loads acting on the cable bolts in the mine were measured seperately.

Some results of bearing capacity measurements of the combined supports performed during deepening of the Hrastnik 2 shaft into tectonically affected rock masses of poor bearing capacity, which are mainly comprised of pseudozilian slate rock, are presented separately. The installation of these anchors was based on the use of a simple technological method and a locally developed machine for the preparation of cement binder with a low water-cement ratio.

The last presented example includes the results of measurements of changes in stress and deformation states of the lignite layer in the vicinity of roadways obtained with the help of CSIRO measuring cells and measuring anchors made of trival. Geomechanical monitoring was conducted in areas with frequent pillar bursts in order to determine the adequacy of stress release procedures.

2. INSTALLATION AND STUDY OF THE BEARING CAPACITY OF LONG CABLE BOLTS UNDER STATIC AND DYNAMIC LOADS

Long cable bolts were used to reinforce and support rock masses in the construction of underground openings in tectonically affected grey and red sandstone. Characteristic of sandstone is that the uniaxial compressive strength, which is about 40 MPa in normal states, increases in intact condition.

The machine equipment used in the installation of single, double or even triple cable bolts ($7 \times \Phi 5$ mm) having a length of up to 25 m was developed at our Institute in collaboration with the BELT company and experts from the Žirovski Vrh uranium mine. The equipment was tested in this mine and used in the production process until the mine was closed, after which it was periodically employed within the scope of maintenance works.

The cable bolt installation procedure is comprised of the following work phases:
- drilling and cleaning of drill holes with a diameter of 38 mm or 52 mm;
- preparation of cement grout with a water-cement ratio of $W/C = 0.3$ and the filling of drill holes;
- impressing of the load-bearing parts of anchors into the holes previously filled with cement grout along their entire length.

The anchor installation technology described above presents a basis for the use of passive elements in supporting underground openings. The system's bearing capacity is activated by the development of deformations in rock strata surrounding the opening, provided, however, that the deformations do not overmatch the maximum strength of the surrounding rocks.

2.1 Distribution of axial forces along single and double cable bolts in dependence of installation depth

The measuring method used to determine the distribution of axial and possibly other forces in anchors has been presented in previous contributions. Only certain results of measurements of the distribution of axial forces along single and double cable bolts shall be presented here.

By increasing the axial force acting on the head of an anchor, only a part of the process activating passive anchors can be simulated, since the time development of deformations is an essentially nonuniform process, not to mention other changes. Another question that arises is how to consider the shear and combined deformation processes, which also contribute to additional loads on passive anchoring systems. Let us limit our discussion to the distribution of axial forces along the length of the cable bolt. An analysis of the results of measurements of stress distribution in the described anchor has shown the nonlinear relations between measured stresses and geometrical parameters. To illustrate the results of measurements, Figure 1 shows the allocation of deformation sensors, while Figures 2 and 3 show the distribution of axial forces along the length of the single and double cable bolts.

Studies of the depth distribution of axial forces are justified because they allow us to determine the required extension of anchors with respect to the estimated shifts in the rock mass surrounding the openings, i.e. to determine the size of the plastic zone.

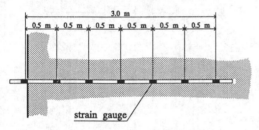

Figure 1 Allocation of measuring points on the anchor
Bild 1 Die Einteilung der Meßpunkte am Anker

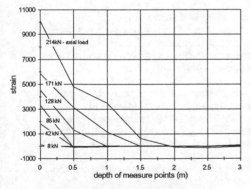

Figure 2 Distribution of axial forces along the length of the single cable bolt
Bild 2 Die Kräfteverteilung entlang dem Einseilanker

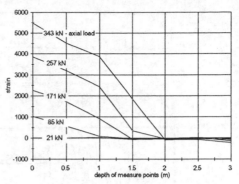

Figure 3 Distribution of axial forces along the length of the double cable bolt
Bild 3 Die Kräfteverteilung entlang den Zweiseilankern

2.2 Dynamic stresses in long cable bolts

Three 10 m long cable bolts furnished with deformation sensors were installed in the ceiling immediately behind the face front advancing in tectonically affected sandstone. The distribution and location of measuring anchors are schematically presented in Figure 4.

Measurements of dynamic effects occuring during blasting were performed using a twelve-channel bridge amplifier, which enables the transfer and recording of amplified signals with the help of a thirteen-trace tape-recorder. The tape recorder was used to simutaneously record stress changes in the anchors. These changes manifested themselves mainly in the form of additional stresses of short duration.

In Figure 4 the time intervals in which additional stresses were observed as a consequence of blasting are plotted on the horizontal axis. The additional loads observed in time intervals of a few milliseconds were found to have a maximum intensity of around 20 % of the allowable stresses in individual anchors.

3. MEASUREMENTS OF THE UTILIZATION LEVEL OF BEARING CAPACITY OF THE ANCHORING SYSTEM IN THE HRASTNIK 2 SHAFT

The construction of the Hrastnik 2 shaft, which is being conducted in accordance

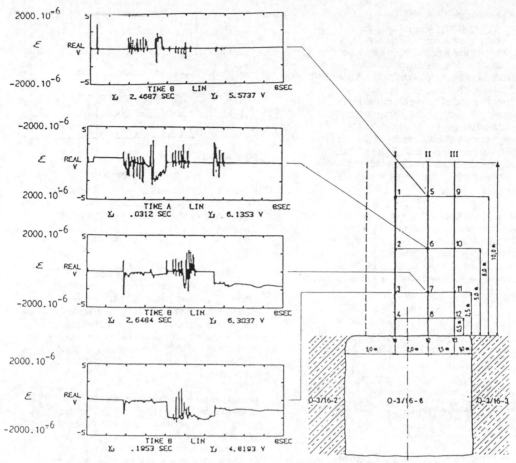

Figure 4   Results of measurements of dynamic load effects on cable bolts during blasting

Bild 4   Die Meßresultate über die Einwirkungen der dynamischen Belastungen an den Seilankern während der Sprengstoffdetonation

with NATM principles, involves the use of combined supporting elements, such as shotcrete, passive anchors and steel supports. Shaft deepening is conducted in rocks of poor bearing capacity which cannot support additional loads resulting from excavation works.

Some geotechnical parameters, particularly convergence, were monitored during the entire construction process, and the occurrences in the vicinity of the shaft were analyzed in detail using measuring anchors. By determining the utilization level of bearing capacity and the distribution of axlal forces in the anchor system, we were able to optimize the lengths of a secondary anchoring

system. Figure 5 shows, in schematic form, the support measures used in shaft construction. Attention should be given to the relative thicknesses of the primary and secondary shaft linings.

During the making of a primary lining, four measuring anchors were installed at a depth of 251 m. Their location is shown in Figure 6 and was conditioned by the tectonic structure of surrounding rock.

Strain gauges were combined into electric bridges by connection through an amplifier and registrator. The positive voltage on the output corresponded to tensile loads and the negative voltage to compressive loads acting in individual

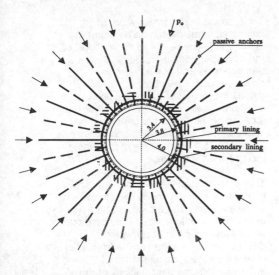

Figure 5   Schematic presentation of support measures used in the deepening of the Hrastnik 2 shaft
Bild 5   Die schematische Darstellung vom Schachtausbau - Schacht Hrastnik 2

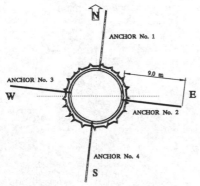

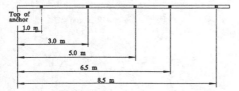

Figure 6   Locations of measuring points on the measuring anchors in the Hrastnik 2 shaft
Bild 6   Die Meßstellen an den Maßankern im Schacht Hrastnik 2

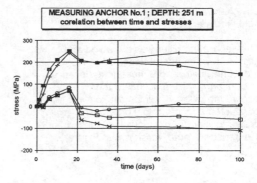

Figure 7   Time development of stresses in the measuring anchor
Bild 7   Die zeitliche Entwicklung des Spannungszuwachses in den Meßankern

measuring cross-sections. The measuring strain gauges were installed in the measuring anchor facing two directions, thus enabling the observation of enventual bending loads.

The results of measurements presented in Figure 7 are a typical example of such monitoring. This graphic presentation of the results obtained shows that the anchoring system is exposed to tensile loads up to a depth of around 5 m, while compressive loads are dominant in the remaining part.

The same conclusion can be made on the basis of an analysis of the results of measurements of the distribution of axial forces along the measuring axis, as shown in Figure 8. In the given case, the so-called neutral point is located at a depth of around 5 m, which means that the selected lengths of anchors in the system are adequate and meet the requirements for ensuring the long-term stability of the structure.

4. DEFORMATION AND STRESS CHANGES AROUND TRANSPORTING MINE ROADWAYS DURING LIGNITE EXTRACTION IN BURST AREAS

Geomechanical measurements in the pillar between panels E and D on level +40 D were performed during exvacation of panel D with the purpose of recording any changes in the stress-deformation field in this area.

The separating pillar was observed and provided with measuring points. 4 CSIRO measuring cells and 9 measuring anchors (2.5 m long) were placed into the separating pillar. Their installation was conducted according to a previously prepared plan from various locations in the transport route, as shown in Figure 9.

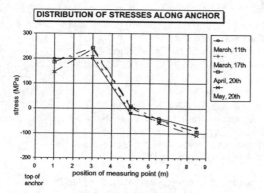

Figure 8   Stress distribution along the measuring anchor
Bild 8   Die Spannungsverteilung entlang dem Meßanker

It was found that a stress wave of increased intensity can be detected at a distance of approx. 64 to 45 m ahead of the front, depending on the local particularities and speed in front advancement. The obtained stress values in the part of the pillar subjected to additional load gradually changed due to local particularities in the structure of lignite blocks and the additional measures carried out for the purpose of unloading the transport route area.

This process is presented in the diagram in Figure 10, which shows the increase of stresses ahead of the roadway face at measuring profile no. 3. It is evident that substantial changes occur within a distance of 65 m ahead of the front and continue to increase as the front advances into the direct vicinity of the measuring point. The diagram shows the nonuniform increase in stress as a consequence of occurrences in the surroundings of the measuring point. However, the general increase in stress with the decreasing distance of the front is clearly noticeable.

During exvacation of the said panel, serveral dynamic occurrences were observed. These were accompanied by releases of energy in the form of small collapses in the surroundings of mine

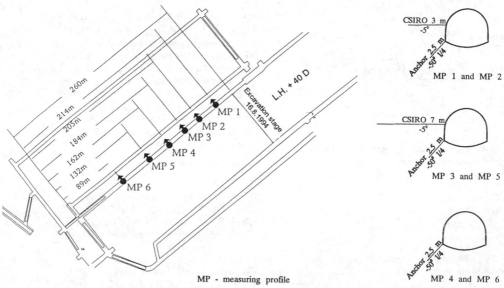

Figure 9   Location of measuring cross-sections during excavation of level k +40
Bild 9   Die Meßprofile während der Abbauarbeiten an der Etage k +40

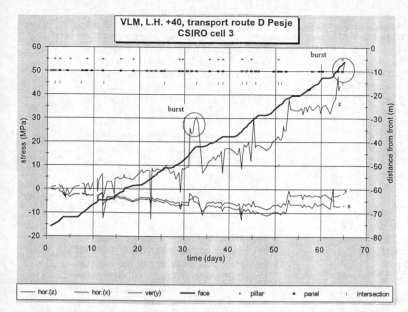

Figure 10 Stress changes in the pillar in dependence of the distance from the face front

Bild 10 Die Spannungsveränderungen im Pfeiler abhängig von der Entfernung der Abbaustelle

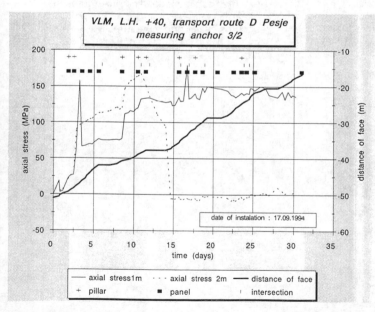

Figure 11 Development of deformations in the measuring anchor in dependance of roadway face advancement and applied unloading measures

Bild 11 Die Deformationsentwicklungen in den Meßankern abhängig von dem Abbaufortschritt und den Entlastungsmaßnahmen

works, as well as on higher levels and at more distant locations. No direct effect on the changes in stress states in the observed areas was detected, except in a few cases, when the stresses in the pillar fell for a short time interval and then rose to their previous or a higher value.

Among the most important measures carried out during excavation with the purpose of reducing stress peaks in rocks in the direct vicinity of the transport route, intersection and the front is stress relief blasting.

In most cases the stress relief blastings were successful, which is also proven by the results of measurements. The results of stress relief blasting are presented in Figure 10, which clearly shows a substantial decrease in stress for a short period of time after blasting. In approximately two days the stresses rose to their previous values, but no collapses or hazardous dynamic phenomena occured in the roadway area, since the direct surroundings of the roadway had evidently been sufficiently unloaded. Figure 11 shows similar results obtained using measuring anchors, which also prove that the stress release depth exceeds 2 m. In practise this means that the repeated application of such measures in specific areas will bring successful results.

5. CONCLUSION

The development technology for the installation of anchoring systems for the reinforcement and additional protection of underground openings was, together with a BERIL 325, which enabled the installation of single, double and triple cable bolts, successfully applied in practice at the Žirovski Vrh Mine in the Republic of Slovenia.

The studies of the distribution of axial forces along the length of cable bolts shown that the effective depth is at least 2 m at allowable and greater loads. In other words, the obtained results have demonstrated the advantages of longer anchors because of their greater effectiveness.

Rope anchors installed in the rock mass in the direct vicinity of blastings and exposed

to dynamic loads develop a satisfactory bearing capacity shortly after their installation. The increases in dynamic loads of short duration which were observed during blasting and correspond to approx. 20 % of the allowable bearing capacity, do not essentially affect the bearing abilities of the anchors.

The determination of the utilization rate of the bearing capacity of the anchoring system installed during sinking of the Hrastnik 2 shaft provided firm grounds for the preparation and implementation of support measures in ensuring the required stability conditions.

The efficiency of the applied stress release procedures was assessed by the geotechnical monitoring of occurrences in pit works close to lignite excavations in areas with frequent rock outbursts. The result of combined measurements, which included the observation of changes in stress states and the level of stress releases in the immediate surroundings of roadways with the help of measuring anchors, also served as a basic for wellfounded decisions related to the extent of applied unloading procedures in hazardous areas. Such an approach ensures easier and more effective planning and conduction of works in hazardous areas.

Prestressed soils using high capacity anchor systems

Vorgespannte Böden durch die Verwendung von Ankern mit großer Tragkraft

M. P. Luong
CNRS-LMS, École Polytechnique, Palaiseau, France

ABSTRACT: Environment, cost, performance and reliability constraints have been the driving force in this investigation for a new concept of prestressed foundation for overhead line towers. The paper presents the concept of prestressed foundations for overhead line towers that are subjected to uplift, compressive loads and/or overturning, induced by severe climatological conditions. Tests on scale models have been performed in the centrifuge, under forced vibrations at fixed frequencies for different force excitations. The comparative analysis conducted in the centrifuge has shown that (1) the gravity foundation worked correctly when subjected to low amplitude of motions and (2) the pile foundation presented a more stable mechanical behaviour. However when subjected to greater motion excitations, the pile foundation would fail at a given loading threshold. Finally (3) the prestressed foundations using prestressing anchoring devices have shown the predicted best mechanical behaviour and this for different loading intensities. This centrifuge simulation offers the great advantage of suggesting a ready choice either of excitation or prestressing force within the stability domain of the prestressed foundation where the mechanical behaviour is quasi elastic. The prestressing force hardens the soil, thus increasing the elastic soil modulus and decreasing the settlements induced within the foundation by the service loadings. The mechanical behaviour of such prestressed foundations for electric pylons already in service appears to be very satisfactory in different geotechnical sites presenting gravelly, sandy or clayey soils.

ZUSAMMENFASSUNG: Umwelteinflüsse, Kosten, Leistungsfähigkeit und Zuverlässigkeit waren bei der Errichtung eines neuen Konzeptes für druckspannungsbeanspruchte Fundamente von Hochspannungsmasten zu berücksichtigen. In zahlreichen Fällen wurden die Fundamente von horizontalen Kräften, beispielsweise infolge Wind, beansprucht, wodurch Zug- und Druckkräfte ausgeübt wurden. Eine Beschreibung des Verhaltens des Fundamentes erfordert die genaue Kenntnis mechanischer und geotechnischer Größen vor Ort, die für eine Installation und den Betrieb erforderlich sind. Ein steiniger und sandiger Untergrund zeigt ein schwer kontrollierbares, rheologisches Verhalten. Der Stabilitätsbereich des Untergrundes wird mit internen Materialparametern beschrieben, die in mehraxialen Laborversuchen ermittelt werden. Im Gegensatz dazu kann bei einem ausgedehnten Stabilitätsbereich, welcher eingeprägte Kräfte enthält, ein stabiles Fundament errichtet werden. Dies wird durch Vorspannungskräfte erreicht. Mittels einer Zentrifuge wurden die experimentellen Daten ermittelt, die für ein deutliches Verständnis des Schwingverhaltens von Hochspannungsmasten in vorgespannten Fundamenten erforderlich sind. Die Versuche wurden bei erzwungenen Schwingkräften bei einer bestimmten Frequenz und unterschiedlichen Lastamplituden durchgeführt. Eine vergleichende Analyse hat gezeigt, daß (1) das schwerkraftausgerichtete Fundament bei einer niedrigen Lastamplitude korrekt arbeitet, (2) das stapelige Fundament ein demgegenüber stabileres mechanisches Verhalten zeigt und (3) die Konstruktion mit ankerförmigen Elementen demgegenüber auch bei einer hohen Beanspruchung am stabilsten ist. Die Ergebnisse gestatten, das Entwurfsverhalten vorgespannter Fundamente zu modifizieren, welches an mehreren Hochspannungsmasten von "Electricité de France" angewandt wurde. Die Vorspannungskräfte härten den Untergrund, erhöhen damit den Elastizitätsmodul und verringern Ablagerungen innerhalb des Fundamentes, die unter Betriebsbeanspruchung auftreten. Das mechanische Verhalten derart vorgespannter Untergründe von Hochspannungsmasten, die bereits in Betrieb sind, erscheint sehr zufriedenstellend an unterschiedlichen Orten steiniger oder sandiger Untergründe.

1 INTRODUCTION

In several cases, horizontal forces, for instance generated by winds, when loading slender, high rise and light structures such as electric pylons or transmission line towers, induce onto their foundations both compressive and tensile loadings (Gagneux 1986).

Traditional building techniques (Plumelle & Lapeyre 1986) employ huge heavy concrete blocks, large buried concrete plates or costly pile groups (Bustamante et al 1986) able to resist punching or pulling loads. When subjected to severe and diverse climatological variations or earthquake loading, the geotechnical behaviour of these slender structures may degenerate leading to failure. In addition, environment, cost, performance and reliability constraints have been the driving force in this investigation for a new concept of foundation for overhead line towers.

The design of common foundations often requires a precise knowledge of mechanical and geotechnical characteristics, appropriate for a chosen location, necessary for their installation and their capacity to support the expected service loads. These conditions determine their proper foundation choice. Several design methods do exist (Berthomieux 1976), but they are often developed in very specific cases neglecting the actual stress-strain response of the soil foundation and the three-dimensional nature of loading conditions.

This paper proposes a new foundation principle, already used for overhead line towers in France. It is based on a prestressing technique, applied on the soil foundation to harden it, thus ensuring for the foundation a quasi elastic response to service loads with higher deformation modulus and much smaller displacements or settlements.

2 SOIL MECHANICS BACKGROUND

The salient feature of soil is its particulate nature. The soil particles are solid and they are relatively free to move with respect to one another. Hence the individual particles deform as the result of contact forces. Observed macroscopic deformations are derived essentially from their structural modifications, that is, rearrangements of the constituent grains. Volume changes depend on particle associations and arrangements, caused by either generalised contraction or expansion of solid skeleton without modification of soil structure, or variations of grain arrangements, grain orientations, particularly sensitive in shear tests or during the first

hydrostatic loading. Distortional strains are governed by friction - which consists of microscopic interlocking due to roughness of contacting particle surfaces -, and interlocking friction or physical restraint to relative particle translation afforded by adjacent particles. The deformation of a soil mass is thus controlled by interactions between individual particles, especially by sliding between particles. Because sliding is a non linear and irreversible deformation, the expected stress-strain behaviour is strongly non linear and irreversible.

Three main points are evident: (a) strain hardening reflecting the difference of stress-strain behaviour between first loading on virgin soil and following cycles of unloading-reloading, (b) hysteresis of soil under cyclic loading due to an irreversible behaviour during each load cycle, and (c) stress path enhancing mechanisms of deformation associated either with great volumetric strain or with predominant distortional strain.

Usually the ground at a site in consideration is not ideal from the soil engineering point of view. If another site cannot be chosen, an approach to the problem of bad soils is to improve the soil or to adapt the design for the conditions at hand.

2.1. Compacted soil
Verdichteter Boden

The most common and important method of soil improvement is densification by (a) compaction with mechanical equipment, (b) preloading by placing a temporary load and/or (c) dewatering with removal of pore water.

2.2. Reinforced soil
Bewehrter Boden

The concept of reinforcing masses with strips of metal or sheets of synthetic fabric is a relatively new development (Vidal 1969). Laboratory and field tests on sandy soil reinforced with randomly distributed, discrete fibres and oriented, continuous fabric layers show that all reinforcement systems increase strength and modify the stress-deformation behaviour of soil in a significant manner. The strength of compacted reinforced soil is greater than for compacted plain soil. The stress-strain curves for reinforced soil specimens show that the reinforced soil failed at a greater strain than the plain soil. They are similar at low strain, but the reinforced soil is able to hold together for more deformation and therefore higher stress at failure.

2.3. Prestressed soil
Vorgespannter Boden

It is known in soil mechanics that when a soil mass is prestressed, its domain of mechanical stability can be extended in a controlled manner. Soil often exhibits a rheological behaviour that is difficult to control, a situation that makes the design of any associated structure a rather delicate affair. A mechanical hardening below the characteristic threshold (Luong 1980) ensures an extended domain of pseudo elastic stability (elasticity or accommodation) in which any external load generates only recoverable and stable strains.

Conventionally the stability domain of soil foundation is bounded by an intrinsic curve deduced from triaxial tests. If the service loads are not included in this domain, the foundation cannot be built. On the contrary, a stable foundation can be realised when the stability domain is large enough to include the imposed loads. Thus the more extended the stability domain, the more various and severe loads can be sustained by the foundation. This is obtained by means of soil prestressing. It can be seen on the stress-strain curve that a loading in the under characteristic domain followed by unloading and reloading readily defines a hardening threshold below which the rheological behaviour of the material is practically recoverable, i.e. stable. In the stress space, the mechanical hardening generates a domain of stability and integrity for the material, that is bounded by the corresponding hardening curves. In this hardened stress domain, the material presents advantageously a quasi elastic behaviour, which may easily be taken into account in design codes.

The requirements in service define, for a structure, a loading domain that must be confined in the hardened domain bounded by the hardening curves. The more stable the behaviour of the whole soil-structure, the more distant the loading domain from the limits of intrinsic stability. This can be obtained by soil prestressing. Graphically, this corresponds to a horizontal displacement of the vertical axis of Mohr diagram toward the tensile region unacceptable for soils. The magnitude of the displacement vector represents the prestressing force, required to withstand the imposed loads.

Overhead line foundations are principally subject to uplift, compressive loads and/or overturning. They are thus essentially stretched, compressed or bent. These requirements define a loading domain to be imposed on the foundation. This domain determines the magnitude of the recommended prestressing force used firstly to harden the foundation material and secondly to consolidate its stability domain.

3 CENTRIFUGE TESTING

To establish the correct design by means of numerical techniques is rather delicate because of the geometrical interaction of neighbouring supports loaded in tension and in compression. This is a tridimensional configuration difficult to deal with. In addition, the self-weight-induced stresses, appropriate to the prototype earth structures, exert in this case very strong influences (Mandel 1962). Experimental modelling and physical simulation are therefore vital to a fuller understanding of the problem.

An experimental study has been conducted in the centrifuge, to determine the basic data required for a clear understanding of the dynamic, vibratory, cyclic and transient responses of pylons resting on their prestressed foundations. This may lead to an economical construction and the satisfactory performance of overhead line towers.

Tests have been carried out on the CESTA's centrifuge located near Bordeaux, France. Its main characteristics are: 100 g (g denotes the natural gravity) = 1,000 $m.s^{-2}$ acceleration, 2,000 kg payload and 10 m radius. The swinging basket of the centrifuge is equipped with an electromagnetic vibrator. The usual design methods, employing models of an elastic continuum, present an application field limited to relatively small displacements. The aim of this study is to examine the diverse physical phenomena that occur in centrifugal model tests under large displacements and then to validate the obtained experimental data with full-scale tests.

Test analysis, based on frequency responses of the scale models, permits a ready interpretation of nonlinearity occurrence. The pylon model was shaken by a horizontal force supplied by the vibrator mounted on the swinging basket. The instrumentation was composed of three tridimensional accelerometers (# 1-2-3), two 1D accelerometers (# 11-12), one displacement sensor (# 13) and one force sensor (# 14). They are affixed as shown on Figure 1.

As the total centrifuged mass exceeded the allowable limit, the radial acceleration level was set at 25, 50, 75 and 90 g. The centrifugation process was performed with steps at 25, 50, 75, and 90 g in order to enable accurate measurement of the fine sand pack settlement under its own weight.

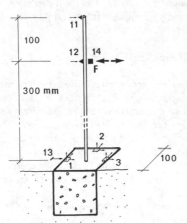

Figure 1. Centrifugal model of gravity foundation for overhead line tower.
Abb. 1. Modell eines Freileitungsmastes mit herkömmlicher Fundierung für Versuche in einer Zentrifuge

Full scale extrapolation requires the following coefficients (Table 1) deduced from scaling conditions:

Acceleration	1/25	1/50	1/75	1/90
Force	625	2,500	5,625	8,100
Moment	15,625	125,000	421,875	729,000
Displacement	25	50	70	90
Frequency	1/25	1/50	1/70	1/90

Table 1. Full-scale extrapolation coefficients.

Three types of foundations for overhead towers were chosen for this comparative study.

3.1 Gravity foundation

The stresses of each pylon leg are distributed to a stepped or pedestal footing designed to satisfy the limit total displacement to an acceptable small amount and eliminate differential settlements between parts of a structure as nearly as possible. As the pylon may be subjected to overturning forces, the footing has been designed on the assumption of linear variation of soil pressure and constructed with sufficient resistance to deformation. Stress in the longitudinal direction of the pylon leg is transferred to its pedestal by extending the longitudinal steel into the support. The stress transfer bar projects into the base a sufficient compression-embedment distance to transfer the stress in the column bar to the base concrete. A cubic mass of concrete of 100 mm in size simulates the foundation weight that essentially ensures the structure stability. In the middle of its upper part is installed a rigid rod of 10 mm in

diameter and 400 mm in length representing a rough pylon scale model. A vibratory harmonic horizontal force is applied in flight on the top of the rod at different intensities and frequencies.

3.2 Pile foundation

Piles have been used to resist uplift and overturning moments developed through friction along the sides of the piles and to distribute loads over a sufficiently large vertical area of relatively weak soil to enable it to support the loads safely. The safe friction values to use for the project has been determined by uplift tests. The compressive loads are supported through bearing at the tip, friction along their sides, adhesion to soil, or a combination of these means. Four groups of four piles simulate the foundation of this pylon scale model that rests on a squared metallic plate of 100 mm in size and 5 mm in thickness. Vibratory loading is applied on the rod model as in the previous case.

3.3 Prestressed foundation

A high capacity anchoring system is used to load the prototype base resting on the soil. The prestressing forces are applied on the model by means of four springs (Figure 1). This loading generates a mechanical hardening on the soil foundation. This results in a supporting soil stiffer in appearance, whose mechanical behaviour is quasi static. The response to forced vibration is recorded by acceleration, displacement and force transducers. A force transducer, placed between the electromagnetic exciter and the pylon model, gave the force loading signal. A displacement transducer and three 3D-accelerometers were installed on the anchored plate.

The analysis of the results has been focused on the appearance of non linear phenomena that announce the beginning of stability loss of the foundation. Tests were carried out under forced vibrations at fixed frequency for different force intensities. The response/excitation ratio is expressed as a complex number so as to contain phase-shift information. Phase shift is important theoretically in the analysis of mechanical systems and practically as a symptom of resonance when resonance frequencies caused by the occurrence of the soil softening behaviour are being identified experimentally.

The comparative analysis conducted in the centrifuge has shown that the gravity foundation worked correctly when subjected to motions of low amplitude, that the pile foundation presented a more stable mechanical behaviour. But when subjected to greater excitory motions, the pile foundation would fail at a given loading threshold. Finally the prestressed foundations using four 160 kN (16 tons) prestressing forces have shown the predicted best mechanical behaviour and this for different intensities of loading excitation.

For sake of ready comparison between the three types of foundations (Figure 2), horizontal, transverse and vertical displacement measurements as recorded in flight have been calculated as for full scale prototype (Table 1) in order to obtain a better subsequent analysis with field tests on actual pylons in service. F_p denotes the prestressing force, F the vibratory amplitude of the applied load, b a characteristic size of the pylon (in the centrifuge b is the width of the metallic plate of the pylon model) and d the measured displacements. The vibratory motion of the pylon is characterised by bending B, rotation R and stamping S.

Performed tests have permitted the assessment of the influence of the prestressing force on the mechanical behaviour of the prestressed foundation. A significant influence of the prestressing force F_p is observed on the amplitude F of the excitation within the stability domain. The experimental results obtained from centrifuge simulation evidence the following principal trends: (a) the domain of mechanical stability of the prestressed soil mass is enlarged with the magnitude of the prestressing force, (b) the rheological behaviour of the prestressed soil within the stability domain is quasi elastic, and (c) with a small amplitude prestressing force used for the tests (160 kN), the mechanical behaviour of the prestressed foundation proves to be better than the pile group foundation and much more reliable than the buried concrete block foundation.

This centrifuge simulation offers the great advantage of suggesting a ready choice either of excitation force or prestressing force within the stability domain of the prestressed foundation where the mechanical behaviour is quasi elastic.

4 WORKING UP PRESTRESSED FOUNDATION

The centrifuge study reinforces the conceptual predictions of this new type of foundation. It forms the basis of a design method of a prestressed foundation that has been efficiently installed on several overhead lines in France.

The anchoring system used (Figure 3) belongs to a family of high performance anchorage devices suitable for any type of soil (Habib et al 1982), which are easy to use with known driving in (Figure 4) or propulsion processes. It comprises (1) an anchorage element adapted for penetrating into the ground under the driving action, and (2) one articulated element pivotably connected to the main body by a flexible connection. It further combines the following characteristics: a cranted shape of the articulated element and the location of the centre of gravity between the fastening point being situated on each side of the driving plane. They are especially designed for low energies when driving into the soil. Anchoring ultimate resistance was defined as that maximal tension load causing admissible displacement on the anchoring line.

In fact, this value is given by the soil uplift resistance where the anchoring device must withstand tensile forces. It depends mainly on the mechanical properties of surrounding soil, embedment, active anchoring area and roughness of the surface (Agatz & Lackner 1977, Sutherland 1988).

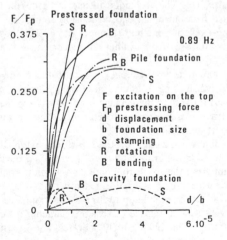

Figure 2. Comparison of the three responses
Abb. 2. Vergleich der Versuchsergebnisse für die drei Arten der Fundierung

Figure 3. High capacity anchoring system
Abb. 3. Ankersystem für Anker mit hoher Zugtragfähigkeit

Figure 4. Driving high holding power anchorings in subsoil
Abb. 4. Rammen eines Ankers mit hoher Zugtragfähigkeit in den Untergrund

Several laboratory and field tests show that at shallow embedment ratios of up to three, uplift resistance exhibits a well-defined peak resistance. For embedment ratios greater than five, the anchoring system resistances exhibit oscillatory behaviour at large displacements. This demonstrates the existence of a critical depth from which the anchoring mechanism is located entirely in the subsoil. The holding capacity of the tested anchoring device embedded in a cohesionless soil is then approximately given by $T_p = \gamma' DSA_p$ where γ' is the effective unit weight of soil, D the embedment, S the effective area of the anchoring system and A_p a dimensionless coefficient dependent on soil mechanical characteristics and on relative embedment.

To work up the prestressed foundation for an electric pylon, a concrete slab is laid on the ground and a prestressing force is applied between the anchoring device and the slab. The prestressing force hardens the soil modulus and decreasing the settlements induced within the foundation by the service loading. This has been demonstrated both on centrifugal fine sand and on various types of subsoils.

5 CONTROL TESTING OF PRESTRESSED SOIL

Overhead line towers are commonly exposed to dynamic loading from several sources, including high winds, earthquake ground motions and others. The design task is made quite challenging by inherent constraints of economics, demand for extreme reliability and considerable uncertainty in defining the dynamic loading which the structure must endure.

The major difficulties encountered in the application of the modern control techniques to structural systems may be listed as follows:

i. Active control requires the ability to generate and apply large controlled forces to the structure.

ii. Modern control theory often leads to feedback control laws, thus requiring on-line measurement or estimation of all the system state variables.

iii. On-line control requires that both measurement and control be performed in real time.

From a practical standpoint, while the application of large control forces to a structure does not raise insurmountable difficulties, the generation of such forces over sustained periods of time, as necessitated by continuous optimal feedback control theory, may cause the concept of active control to become impracticable. To bypass this possible drawback, the approach under consideration attempts to use pulses of relatively short duration to control the structural system. The objectives of this vibratory non destructive testing were:

(a) to characterise the soil-structure interaction of overhead line towers resting on different types of foundations such as traditional concrete footings, piles or prestressed foundations,

(b) to verify the geotechnical performance of these foundations subjected to dynamic loading, and

(c) to investigate and develop simple experimental testing in order to readily obtain a vibratory signature for the control and inspection of the mechanical behaviour of the transmission tower.

The physical nature of the transmission line system restricted the dynamic excitation alternatives for testing. The classical procedures were sine dwell, sine sweep, fast sine sweep or chirp, random, impulse, etc. The technique in use is based on the release from an initial chosen tension similar to the twang-excitation method (Kemper et al 1981) based on a release from initial displacement of the structure.

An experimental frequency response technique by impulsive slacking from the top of the pylon has been applied for testing the behaviour of several overhead line towers of Electricité de France resting on three types of foundations.

A cable was attached on the top of the pylon. It is then fixed in the ground by a small anchoring device, distant from the pylon foundation. A tension system allows to increase the cable tension to a relatively low value for example up to 30 kN. An explosive tie allows a sudden release of tension, inducing a rapid excitation. The pylon freely vibrates after tension release (Figure 5).

Several 3D-accelerometers fixed at different locations on the pylon record the vibratory motions. Data reduction by fast Fourier transforms give frequency response spectra that are of course the vibratory signature of the pylon resting on its prestressed foundation (Figure 6).

Series of tests on many different pylons resting on their foundations have evidenced that the prestressed foundation presents a quasi elastic behaviour. These plots of frequency versus vibrational amplitude in arbitrary units provide vibration signatures of the soil-structure interaction characterising the mechanical performance of the tested transmission tower. Of course, this non destructive test can be applied also to other types of slender structures (Figure 7), because the vibration signature plots pinpoint vibration frequencies and indicate conditions such as poor workmanship, damage in structure or in subsoil. As a periodic maintenance activity, changes in subsequent plots permit early detection and identification of damage.

In many situations, the impulse technique for structural frequency response testing is the simplest and fastest of the various techniques commonly used today. The transient excitation offers good estimates of the required frequency response information. However the nature of the excitation and response signals in the impulse technique requires especial signal processing technique that has been proposed by Luong (1992), using Volterra series analysis.

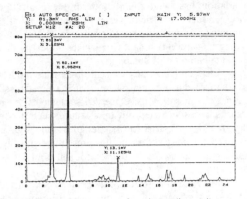

Figure 6. Frequency response of a pylon resting on its prestressed foundations
Abb. 6. Frequenzspektrum für Mast mit vorgespannter Fundierung

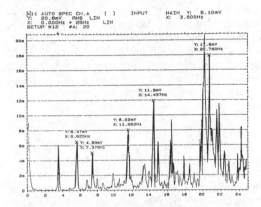

Figure 7. Frequency response of a pylon resting on its concrete footings.
Abb. 7. Frequenzspektrum für Mast mit Betonfundamenten

6 CONCLUDING REMARKS

Centrifuge testing on overhead line tower models has been very useful for evaluating the geotechnical performance of the corresponding prototypes. The main results of these experimental studies are:

6.1 Subject to excitations, the gravity foundation, both in centrifuge pylon model and field tests, presents a great amount of non linearities caused by inelasticity in the soil mass and at the interface soil-concrete. The recorded frequency response spectra are highly amplitude dependent.

6.2 The pile foundation behaves in a better manner. But the pylon motion under vibratory loading both in the centrifuge pylon model and field tests, is rapidly dependent on the excitation amplitude.

Figure 5. Tension releasing test on a transmission line tower
Abb. 5. Schwingungsversuch durch plötzliche Entlastung an einem Freileitungsmasten

6.3 The quasi linear elastic response of prestressed foundation has been found both in the centrifuge pylon model and field tests as shown by the frequency response spectra to excitations where only loading or natural resonant frequencies appear with a very small amount of damping.

6.4 Tests on laboratory or centrifuge scale models have shown the potential value of using prestressed foundations in soft soils. Observed settlements, even in cases of tests carried out until failure, seem to be very small if compared to the strength values. This is readily obtained in different sites presenting gravelly, sandy and clayey soils, thanks to soil foundation hardening.

6.5 Several non destructive tests have been carried out on overhead pylons in use. Experimental determination of the vibratory signature of such a slender structure resting on its prestressed foundations, subsequent to a sudden release of tension at its summit, reveals to be adequate for integrity control and inspection of the evolution of the pylon behaviour as regard to fatigue or damage.

6.6 The concept of prestressed soil reveals to be very promising for many applications in civil engineering: soil improvement, site effects, protection against vibrations and earthquakes.

REFERENCES

Agatz, A. and E.Lackner 1977. *Erfahrungen mit Grundbauwerken*, Springer-Verlag, 385-405.

Berthomieux, G. 1976. *Modèles tridimensionnels de fondations de pylônes sollicités horizonta-lement*, Thèse 3e cycle, Université de Grenoble.

Bustamante, M., A.Frossard, D.Gouvenot et D.Sage 1986. Nouvelles foundations de supports aériens, *Journées d'Etudes sur les Fondations des Supports Aériens*, ESE, Gif sur Yvette 27-11-1986.

Gagneux, M. 1986. Evolution dans le calcul et choix des fondations des supports aériens HT, THT à l'Electricité de France, *Journées d'Etudes sur les Fondations des Supports Aériens*, ESE, Gif sur Yvette, 27-11-1986.

Goyder, H.G.D. 1984. Foolproof for frequency response measurements, *Proc. 2nd Int. Conf. Recent Advances Structural Dynamics*, South-ampton, April 1984.

Habib, P., P.LeTirant and M.P.Luong 1982. Anchors and models tests of high capacity anchorings. Geotechnics in a marine environment, *Bull. Tech. Bureau Veritas*, Jan. 1982, 5-15.

Kemper, L.Jr., R.C.Stroud and S.Smith 1981. Transmission line dynamic/static structural testing, *J. Struct. Div. ASCE*, 107(ST10), October 1981.

Luong, M.P. 1980. Stress-strain aspects of cohesion-less soils under cyclic and transient loading, *Proc. Int. Symp. on Soil under Cyclic and Transient Loading*, Balkema, Rotterdam, 315-324.

Luong, M.P. 1986. Simulation des forces de masse, *Ann. ITBTP, Essais et Mesures*, 442, Février 1986.

Luong, M.P. 1992. Centrifuge testing. *In Earthquake Engineering, AFPS*, OEPA, VI(6), 781-794.

Luong, M.P. 1992. Safety evaluation of pylon using implulse testing, *Int. Workshop on Safety Evaluation Based on Identification Approaches Related to Time Variant and Non Linear Structures*, Lambrecht, Germany, 6-9 September, Ed. H.G. Natke, G.R.Tomlinson and J.T.P.Yao, Vieweg Int. Scientific Book Series, 222-237.

Mandel, J. 1964. Essais sur modèles réduits en mécanique des terrains - Etude des conditions de similitude, *Rev. Industrie Minérale*, 44(9).

Plumelle, C. et J.L.Lapeyre 1986. Les fondations des supports aériens, *Journées d'Etudes sur les Fondations des Supports Aériens*, ESE, Gif sur Yvette, 27-11-1986.

Ramsey, K.A. 1975. Effective measurements for structural dynamics testing: Part I. *Sound and Vibration*, 9(11).

Sutherland, H.B. 1988. Uplift resistance of soils, *Géotechnique*, 38(4), 493-516.

Vidal, H. 1969. The principle of reinforced earth, *Highway Research Record*, 282, NCR-HRB, Washington D.C., 1-16.

Problematics of underground works drifting in condition of high side stress

Die Problematik unterirdischen Streckenvortriebes unter Bedingungen großen Seitendruckes

Vladimir Petros
Technical University of Mining and Metallurgy, Ostrava, Czech Republic

ABSTRACT: The way of unsettling of underground works made in solid rocks under high horizontal stresses is described in this paper. As it has been found out, the unsettling takes place in a flat triangle shape at the top of an underground work. This is explained by the stress distribution in a rock sample by the compressive laboratory tests. The unsettling can occur as high-energetic "off-flaking" that can endanger workers. The possible way of prevention of those events is also suggested.

ZUSAMMENFASSUNG: Zur Zeit wurden oft Überlegungen angestellt, Untertagebauten in Teufen von ungefähr 1000 m in gutem Fels auszuführen. Solche Hohlräume können zum Beispiel als Untertagedeponien dienen. Die Ausführung solcher Arbeiten ist in Granit mit einer Festigkeit von 150 MPa ohne Probleme möglich, soferne die Spannungen im Gebirge nur vom Eigengewicht der Hangenden kommen. In massivem Fels können auch hohe Horizontalspannungen wirken, die zu Zerstörungen führen und die Arbeiter vor Ort bedrohen. Der Ausbruch hat eine Dreiecksform der meistens nicht sofort, aber im weiteren Fortschreiten entsteht. Dieser Ausbruch ist ähnlich dem, der im Labor bei einem einachsigen Druckversuch beobachtet werden kann. Durch die Reibung, die zwischen Probe und Lastplatten des Belastungsmechanismus auftritt, brechen die Seiten der Probe ungefähr dreiecksförmig aus. Den Anfang des Bruches können wir durch den Verlauf der Volumendeformation feststellen. Im Gebirge beginnt der Bruch bereits dann, wenn die Horizontalspannung etwa den halben Wert des einachsigen Druckversuches erreicht. Gegen diese Brucherscheinungen können wir uns mittels Felsauflockerungssprengungen oberhalb und unterhalb des geplanten Hohlraumes schützen. Die Felsauflockerungssprengungen bewirken eine Abschirmung vor großen Horizontalspannungen.

1. INTRODUCTION

In recent time, the creation of underground works in a solid rock mass in depths of approximately 1.000 m has often been considered. Such underground works should serve as waste storage sites, for instance. The making of those works, e.g. in granite, is supposed to be without any problems due to the strength of that rock. (The uniaxial compressive strength of such rocks usually exceeds 150 MPa). The drifting of underground works would be easy providing that only the stress caused by the weight of overburden acted in the rock mass. In the solid rock mass, however, high horizontal stresses can also act. If there is no enhanced horizontal stress, the underground cavern is stable and there is no need to support it. On the contrary, the enhanced horizontal stress can cause unsettling of rocks, mainly at the top of an underground work. This can occur as "off-flaking", which can heavily endanger safety of workers. The explanation of that unsettling is to be found in the rules of stress distribution by compressive loading. The energy of off-flaked rock fragments can be evaluated in laboratory with the help of power-balancing of compressively loaded rocks. Since the potential energy transforms to the kinetic energy by the off-flaking, we can call this phenomenon a rock burst event.

2.THE CAUSES OF ROCK BURST EVENTS

The general view of the causes of rock burst events concerns the way of unsettling of underground work periphery. The unsettling takes place mainly at the top of an underground work, mostly by the formation of a triangle shaped vault. This clearly indicates the direction of maximum acting stress. This maximum stress acts probably in the horizontal direction, which fact is supported by present experience resulting from a variety of measurements performed in various places on Earth, by which the average horizontal stress meassured were 2.5 times bigger than the vertical stresses. Moreover, the horizontal stresses varied strongly. The geostatic stress itself depends on depth and deformation properties of rocks and cannot cause rock burst events. The vertical geostatic stress can reach approximatelly 26 MPa in the depth of 1.000 m. The horizontal stress induced by the vertical stress can be evaluated from equation

$$\sigma_h = \sigma_v \frac{\mu}{1-\mu} \qquad (1)$$

where μ .... Poisson´s ratio

The avarege value of Poisson´s ra-

tio is 0.19 for granite. Then, if vertical stress is 26 MPa, horizontal stress is 6,1 MPa according to (1). This if we suppose the value of rock strength to be about 150 MPa, would ensure the stability of underground cavern. However, in the case of triangle shaped unsettling of the top of an underground work, the real horizontal stress must be of about one order higher than the value calculated according to (1).

The real values of acting stresses can be obtained e.g. by the hydrofracturing method. As it has been found out, the horizontal stress is about twice as high as the vertical stress in unstable parts of an underground work-fig.1.

In general, a rock mass element is loaded with the vertical stress that induces horizontal stress . An additional stress caused by geological processes by the lithosperic formation possibly acts here as well. This additional stress is crucial for the rise of rock burst events.

The origin of a rock burst event itself can be explained in a following way: the loading of rock around the periphery of an underground work is similar to the uniaxial compressive test performed in laboratories. The diagram of stress distribution by the uniaxial compressive text is shown in fig. 2. According to the stress distribution the sample divides to the compressive stress concentration zone and the tensile

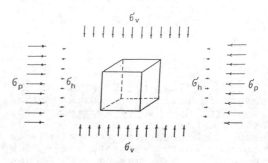

Fig. 1

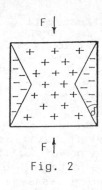

F ↓

F ↑

Fig. 2

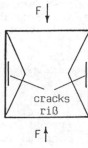

F ↓

cracks
riß

F ↑

Fig. 3

stress concentration zone [Bordia 1971]. The first zone mentioned is indicated by the "+" sign, the latter by the "-" sign. The size and shape of the zones depend on the internal friction angle of the rock:

$$\delta = \frac{\pi}{4} - \frac{\varphi}{2} \qquad (2)$$

where φ .... internal friction angle

Since the tensile strength of rocks is of about one order lower than their compressive strength, the tensile strength is first to be reached by loading. In homogenous rocks, the cracks parallel to the side walls of a sample start to open first. This means, the rock begins to unsettle. The stress acting in the beginning of that unsettling is called stress on an initiation boundary of unsettling. This stress can be traced from the volume deformation curve. For solid rocks, the value of unsettling initiation boundary stress often ranges around one half the

uniaxial compressive strength. This has been confirmed also by different measurements. The creation of cracks is accompanied by the sound effect audible in underground works as cracking. When the loading goes on, the cracks spread and the rock part located between the crack and the side wall of a sample gets under strut loading. After reaching the strength limit of that element, it unsettles - flakes off the sample. The similar events are witnessed in underground works. Sometimes, however, the two rock burst phenomena can be integrated into one - the newly created crack can immediately flake off. This is why cracking needn't always be followed by off-flaking. That's how the unsettling of tensile zones takes place.

If we apply this knowledge on an underground work under high horizontal stress, we can plot the situation as in fig. 4. If the maximum stress acts in the horizontal direction, concentration of tensile stress rises at the top of the underground work, and this part of the underground work periphery is first to unsettle, The top then acquires the triangle like shape.

The fact the unsettling starts to occur under stresses equal to approximately one half the uniaxial compressive strength allows us to assess the magnitude of confining stresses. There is a difference between the circumstances influencing the value of stress measured in compressive stress measurements in situ and in laboratory, but,

triangle shape unsettling
dreieckausbruch

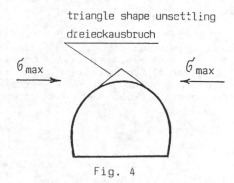

σ_{max} → ← σ_{max}

Fig. 4

for common types of rocks, the difference is not very significant. The main difference lies in the loading velocity. However, by the sufficient velocity of drifting an underground work, the differences will be lower. If rock burst events rise in rocks with strength of about 150 MPa, then, providing that unsettling initiation boundary is one half of that value, the confining stresses acting in those rocks would have to be approximately 75 MPa. After long-term stress acting, however, the strength decreases. Therefore, it can be supposed that horizontal stress acting in place of a rock burst event is at least 50 MPa.

The horizontal stress need not to be distributed uniformly in a rock mass. If both stiff and more deformable rocks occur in a rock mass, the horizontal stress will be higher in stiff rock and lower in more deformable rocks. The rock stifness is given by their deformability modulus. The higher the modulus of a chosen place in a rock mass is, the higher is the additional stress acting in that place. In the rock containing cracks that decrease the deformability modulus, the additional stress will be lower. Since the deformability modulus usually increases along with the rock strength, we can apply the stress distribution rule to the strength conditions.

The way of rock unsettling in an underground work depends on how a rock is able to accumulate the elastic energy and transmit it later to the surrounding rocks. This rock property is represented by the rock power-balance. The laboratories of our university are equipped with a special appliance to evaluate the power-balance by the compressive loading of rocks. The relative accumulated energy of rocks is given by

$$w = \frac{\sigma^2}{2E} \qquad (3)$$

where w ...... relative accumulated energy (MJ . m⁻³)

σ ...stress acting in the rock (MPa)
E ...... Young modulus (MPa)

If we suppose the stress to be 50 MPa and Young modulus 65.000 MPa, then the relative accumulated energy is 19 kJ . m⁻³ according to (3). How much of it is transmitted to the surrounding area depends on so-called power coefficient that is to be obtained by the rock power-balancing assessments.

The rock power-balance is calculated on the base of the power-criterion value [Petros 1974]:

$$\chi_e = \frac{W_v}{W_z + W_L} 100 \qquad (4)$$

where χ_e ...... power-criterion (%)
W_v ..... energy transmitted back in the moment of rock of sample unsettling (J)
W_z ...... energy of a rock sample possibly accumulated prior to unsettling (J)
W_L ...... energy accumulated by press-machine in the moment of unsettling (J)

The energy transmitted back in the moment of sample unsettling can be obtained by the force impulse sensor located under the rock sample.

The energy possibly accumulated prior to unsettling can be calculated from

$$W_z = \frac{V \sigma_r^2}{2 E_d} \qquad (5)$$

where σ_r ..... unsettling stress in a sample
V ..... volume
E_d .....deformability modulus

The energy accumulated by press-machine in the moment of unsettling can be calculated from the deformability characteristics of

208

a press and from the force acting in the moment of unsettling.

3. SOME WAYS OF ENSURING THE WORKERS´ SAFETY IN AREAS ENDANGERED BY ROCK BURST PHENOMENA

The measures that could prevent the underground workers from the danger of rock off-flaking can be divided into two groups: active and passive measures. While active measures can prevent off-flaking itself, the passive measures only limit its dangerous consequences. The passive ways of prevention include various kinds of rock support, safety barriers, administrative entrance and stay limitations to cendangered places etc. The anchors, for instance, can partially prevent the top of an uderground work from destruction. By that, they partially serve as an active measure. To prevent the off-flaking in the zones between the anchors, it is suitable to equip the underground work also with a wire-netting that set would serve as a combination of active and passive measures.

However, no kind of support can be installed up to the face of an underground work. That is why those measures don't fully eliminate the danger of accidents caused by

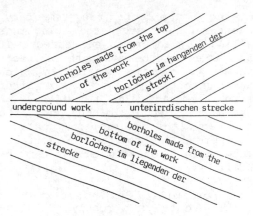

Fig. 6

the off-flaking. To eliminate the influence of rock burst events substantially, the causes of their rising must be eliminated. As the present analyses conclude, the principal cause of rock burst phenomena in uderground works is high additional horizontal stress. That additional stress can be substantially limited by creating larger free area, which allows deformation of surrounding rocks. That can be done by blastings at the top and the bottom of an uderground work. This situation is shown in fig. 5.

The side view is presented in fig. 6.

The unsettling of rock at the top and the bottom of an underground work will cause a substantial decrease of additional horizontal stress and thus destressing of a rock mass in the zones of future underground construction. The size of thus unsettled area depends on the size of performed blastings.

The effect of suggested blastings is similar to the effects of overmining and undermining of horizontal deposits, e.g. coal seams. The maximum stress direction of those seams is vertical, the situation is 90° turned in the considered case. The diference between the influence of over - or undermining and the unsettling blastings lies in the fact that the

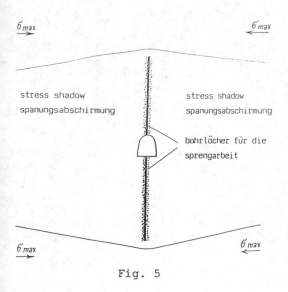

Fig. 5

mining effects are temporary, whereas the suggested blastings will be of permanent effect. It is given by the strength of rocks in the granite massif, in which the unsettling will stop after some time and the situation will get stabilized. By mining, on the contrary, the deformations continue to spread up to the surface. Then, the geostatic stress is restored.

The possibility of rise of rock burst events is strongly influenced by the corelation between the direction of underground works and the direction of maximum additional horizontal stress. If an udeground work is drifted in the direction of maximum additional horizontal stress, the risk of rock burst phenomena will be minimalized.

4. CONCLUSION

As it follows from the analysis of underground work deformations and the analysis of rock burst phenomena by drifting of those works in solid rocks, the dangerous rock burst events in depths of about 1.000 m are caused by high additional horizontal stresses. To propose effective rock burst preventive measures, it is necessary to find out direction and magnitude of additional stresses. Moreover, it is suitable to get information about the rock properties that are directly connected with the rock burst events. They are, first of all, compressive strength, unsettling initiation boundary, and power-balance of rocks. The best way of determining the stress distribution in a rock mass is the hydrofracturing method. The quicker and more approximate way of determining the state of stress is core boring with assessment of entirety and disc-shapeness of the core. If there is a system of underground works, the best preventive measure is the blastings made from the top and the bottom of an underground work. It would be suitable to assess the effectivity and range of that measure by measuring the stress before and after the blastings. The best preventive measure for the underground works made under given conditions seems to be the use of anchors combined with the wire-netting.

REFERENCES

Bordia,S.K.-Havíř,J.-Petroš,V.: The possibility of application of volume deformation of rock samples obtained by the uniaxial compressive tests to the assessment of quality of rock mass. Proceedings from the International conference on mining under heavy overburden. VVUÚ Ostrava, 1971 (in Czech)

Petroš,V.: The elastic energy of press-machine and it´s influence on measurements of rock power-coefficient. Proceedings from the Conference on newest knowledge in rock mechanics.ČSVTS Ostrava, 1974 (in Czech)

Anker in Theorie und Praxis, Widmann (Herausgeber) © 1995 Balkema, Rotterdam. ISBN 90 5410 577 1

Die Sicherung der Edertalsperre – Bericht über eine außergewöhnliche Felsanker-Anwendung

The stabilization of the Eder dam – A report on an extraordinary application of rock anchoring

H. Schwarz
Stump Spezialtiefbau GmbH, Hannover, Germany

ZUSAMMENFASSUNG: Zur dauerhaften Sicherung der 47 m hohen Schwergewichtsmauer waren dringende Verstärkungsmaßnahmen erforderlich. Die Sanierungsmöglichkeiten werden diskutiert und die Entscheidung zugunsten der sogenannten Verankerungslösung begründet. Diese sieht eine Vorspannung des Mauerkörpers gegen den Untergrund mit 104 Stück, 80 m langen Schwerlastankern vor. Die besonderen Anforderungen bestanden in den außergewöhnlich hohen Gebrauchslasten von 4500 kN und extrem hohen Anforderungen an die Bohrgenauigkeit. Die erforderlichen Voruntersuchungen, Einzelheiten zur Konstruktion der Anker und die begleitende Meßtechnik werden beschrieben. Bisher einmalige Ankerprüflasten von 12500 kN wurden erreicht. Die Arbeiten wurden in technischer und terminlicher Hinsicht erfolgreich abgeschlossen.

ABSTRACT: The permanent protection of the 47 m high gravity dam required urgent strengthening measures. The possibilities for such a renovation will be discussed and the decision for the anchoring solution explained. This solution designates a prestress of the dam body against the underground with 104, 80 m long gravity anchors. The special conditions include the extraordinarily high working loads of 4500 kN and the extremely high demands on the drilling precision. The necessary preliminary exploration, details of the construction of the anchors and the accomanying measurement technology are described. Thus far, unique anchor test loads of 12500 kN have been achieved. The work has been completed on schedule and with technical success.

1. DAS BAUWERK

Die Edertalsperre gehört mit einem Stauraum von mehr als 200 Mio m³ zu den größten Talsperren Deutschlands. Sie wurde in den Jahren 1908 - 1914 als gekrümmte Schwergewichtsmauer unter Verwendung einheimischer Materialien hergestellt. Der Mauerkörper wurde aus Grauwacke-Bruchstein in Trasszement-Mörtel errichtet.

Wichtige Daten:
Mauerhöhe 47 m,
Kronenlänge 400 m,
Überlauflänge 240 m,
Basislänge 270 m,
Kronenbreite 6 m,
Basisbreite 36 m.

Als Untergrund stehen zu je 50 % Grauwacke- und Tonsteinformationen des Karbon an.

2. VERANLASSUNG FÜR DIE BAUMASSNAHME

Langjährige Bauwerksuntersuchungen hatten zum Ergebnis, daß im Mauerfußbereich erhebliche Porenwasserdrücke wirksam wurden, die bei der um die Jahrhundertwende durchgeführten Planung der Baumaßnahme nicht berücksichtigt worden sind (Abb. 1).

Statische Untersuchungen nach heutigen Standards unter Berücksichtigung der gemessenen Sohlwasserdrücke und der inneren Wasserdrücke führten zu der Auflage, den höchsten zulässigen Talsperrenwasserstand gegenüber dem Vollstau um 1,50 m abzusenken. Damit konnte die Überlaufkrone der Staumauer für die Hochwasserentlastung nicht mehr genutzt werden.

Hinzu kam die Forderung gemäß DIN 19702, die eine Abführung des sogenannten Jahrtausend-Hochwassers in Höhe von 1100 m³/sek. verlangt. Die bisherigen Abmessungen der Überläufe waren

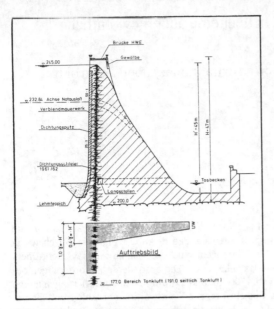

Abb. 1: Mauerquerschnitt Ist-Zustand
Cross-section of the dam, initial condition

für das Jahrhunderthochwasser mit 590 m³/sek. ausgelegt.

Schließlich soll über diese statischen und hydrologischen Anforderungen hinaus durch weitere Restaurierungsarbeiten die Lebensdauer des Bauwerks um weitere 80 - 100 Jahre verlängert werden.

3. DAS SANIERUNGSKONZEPT

3.1 Entwicklung des Sanierungskonzepts

Eingehende statische Voruntersuchungen am Gesamtsystem von Mauerkörper und Untergrund hatten zum Ergebnis, daß zur Erzielung der erforderlichen Standsicherheit ein Gewichtsdefizit von ca. 2000 kN je lfm Mauer durch äquivalente Zusatzmaßnahmen auszugleichen war.

Zur Diskussion standen 4 Varianten:

1. Verankerung der Mauer im Grauwacke-Tonstein-Untergrund durch vorgespannte Daueranker.

2. Aufbringen zusätzlicher Auflasten im Kopfbereich der Mauer.

3. Dichtung der Mauer durch wasserseitige Vorsatzschale

4. Rückwärtige Abstützung der Mauer.

Für die Beurteilung der Sanierungsvorschläge waren insbesondere folgende Gesichtspunkte maßgeblich:

• Eine Vollabsenkung des Stausees für die Dauer der Bauzeit war im Hinblick auf die Interessen der Anlieger und die wirtschaftliche Bedeutung des Tourismus für das Waldecker Land ausgeschlossen.

• Das äußere Erscheinungsbild der Mauer als Baudenkmal und dominierende Sehenswürdigkeit der Ederlandschaft sollte soweit wie möglich erhalten bleiben.

• Wirtschaftliche Gesichtspunkte

• Kürze und Zuverlässigkeit der Ausführungstermine.

Nach Abwägung aller Gesichtspunkte fiel die Entscheidung der Wasser- und Schiffahrtsverwaltung des Bundes als zuständige Baubehörde zugunsten der sogenannten Verankerungslösung.

3.2 Die außergewöhnlichen Anforderungen an die Sanierung

Die kennzeichnenden Maßnahmen für die Sanierung der Staumauer gehen aus Abb. 2 hervor und gliedern sich wie folgt:

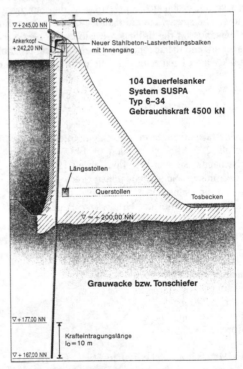

Abb. 2: Ausführungskonzept
Plan of execution

- Abbruch von Überbau und Mauerkrone.
- Herstellen eines neuen Überlauf-Bauwerks mit Kontrollgang als Lastverteilungsbalken für die Aufnahme der hohen Ankerkräfte.
- Gestaltung des neuen, hydraulisch optimierten Überlaufs und Wiederherstellung des Überführungsbauwerks.
- Herstellen der Daueranker,
Abstand i. M. 2,25 m,
Gebrauchslast 4500 kN,
Ankerlänge ca. 75 m,
Ausführung vom Überführungsbauwerk aus.
- Begleitende Injektionen im unteren Drittel des Grauwacke-Mauerwerks sowie unterhalb der Maueraufstandsfläche im bereichsweise durchlässigen Grauwacke- bzw. Tonschiefergebirge.

Insbesondere an die Bohr- und Verankerungstechnik werden mit dieser Aufgabenstellung Anforderungen gestellt, wie sie bisher zumindest in Europa einmalig sind:

1. Aus den hohen Gebrauchslasten resultieren außergewöhnliche Abmessungen für die Ankerzugglieder und damit besondere Anforderungen für die Herstellung, das Handling und die besonderen Belastungszustände beim Einbau der fast 80 m langen Anker.

2. Extrem hohe Anforderungen an die Richtungstreue der Bohrungen und ihre Geradlinigkeit im Hinblick auf die Bauwerksgeometrie, insbesondere im Bereich des unteren Kontrollgangs.

3. Fehlbohrungen oder unzureichende Tragfähigkeit einzelner Anker müssen durch geeignete Vorsorgemaßnahmen zuverlässig ausgeschlossen werden, da wegen der beengten geometrischen Verhältnisse der Einbau etwaiger Ersatzanker nicht möglich ist.

Die technischen Daten im einzelnen:
- Insgesamt 104 Stück Dauerfelsanker System STUMP-SUSPA, Typ 6-34, 34 Litzen x 0,6", St 1570/1770.
- Mittlerer Ankerabstand a = 2,25 m, minimaler Abstand im Bereich der Notablässe 1,85 m.
- Ankerlänge i. M. 75 m, Krafteintragungsstrecke 10 m.
- Bohrtiefen ab OK-Überbau ca. 83 m, davon Anteil der Mauerwerksbohrung im Grauwacke-Bruchstein-Material ca. 42 m.
- Ausführung der Bohrungen als Kernbohrungen mit durchgehender Kernentnahme.
- Bohrlochneigung 3,2 ° zur Lotrechten, in der Grundrißprojektion senkrecht zur Mauerlängsachse.
- Kritische Abstände im unteren Kontrollgangbereich: 1,25 m zur wasserseitigen Dichtung, 1,25 m zum innenliegenden Längsstollen.

Um insbesondere eine Beschädigung der wasserseitigen Dichtung zu vermeiden, war eine zulässige Bohrtoleranz von max. 1 % im Bauvertrag zugestanden worden. Dies entspricht im Mauerfußbereich einer zulässigen Abweichung von 40 cm.

Außerdem war in Mauerlängsrichtung eine größere Bohrlochabweichung wegen der geringen gegenseitigen Ankerabstände auszuschließen. Andernfalls hätten sich die Bohrungen im Ankerfußpunktbereich bei gegenläufiger Abweichung berühren oder gar überschneiden können.

Durch geeignete Bohrtechnik war sicherzustellen, daß die Bohrungen praktisch geradlinig verlaufen. Der minimal zulässige Krümmungsradius von 500 m durfte nicht unterschritten werden, um Umlenkkräfte mit ungünstiger Wirkung auf das wasserseitige Mauerwerk und unerwünschte Querbeanspruchung des gespannten Ankerzuggliedes zu vermeiden.

Für die Herstellung der Bohrungen wurden von uns folgende Verfahrensschritte vorgesehen:

1. Vorauslaufende Kernbohrung mit durchgehendem Kernaufschluß im Seilkernbohrverfahren, Ø 146 mm.

2. Laufende Bohrlochvermessung mit computergestütztem, neuartigen Meßverfahren.

3. Zwischenzeitliche Abdichtungs- bzw. Vergütungsinjektion.

4. Durchführung der Aufweitungsbohrung auf einen Bohrloch-Ø von 273 mm.

4. DIE BAUAUSFÜHRUNG

4.1 Besondere Voruntersuchungen

4.1.1 Konstruktive Ankerdetails:
Bekanntlich werden in Deutschland gemäß DIN 4125 allgemeine bauaufsichtliche Zulassungen für die Ausführung von Dauerankern verlangt. Deren Spektrum reicht bis 2000 kN Gebrauchslast. Für den hier vorliegenden Fall mit Gebrauchslasten bis 4500 kN mußte also eine sogenannte Zustimmung im Einzelfall eingeholt werden. Unter Einbeziehung maßgeblicher Fachinstitutionen wie Bundesanstalt für Wasserbau, Institut für Bautechnik, Ingenieurbüro WBI und anderer Fachgutachter wurden sämtliche konstruktiven Bestandteile des Ankers überprüft bzgl.
- Festigkeit
- Maßhaltigkeit
- Dauerbeständigkeit.
Hierbei waren projektspezifische Randbedingungen zu berücksichtigen, die sowohl bei Bauzuständen z. B. im Zuge des Ankereinbaus als auch im End-

zustand auftreten, wie z. B.

- Aufnahme der hohen hydrostatischen Drücke aus der Einwirkung der 70 m hohen Zementsäule beim Einführen des Ankers in das Bohrloch.
- Beanspruchung der Zugglieder infolge Eigengewichts (3,5 t) aus der Einwirkung von Umlenkkräften beim Ankereinbau und
- Berücksichtigung der großen Abmessungen von Anker-Ø (180 mm) und Ankerlänge (i. M. 75 m) bei den Herstellvorgängen im Baustellenwerk und beim Handling im Zuge des Einbaus der Anker in das Bohrloch.
- Konstruktive Maßnahmen zur Kompensierung der hohen Dehnwege (ca. 45 mm) beim Anspannen der Anker.

Die grundlegenden konstruktiven Einzelheiten des Schwerlastankers mit seinen kennzeichnenden Komponenten, Materialien für mechanischen Schutz und Korrosionsschutz, Kopfausbildung (Abb. 3) entsprechen weitestgehend der grundsätzlichen Gestaltung der STUMP-SUSPA Daueranker, für die allgemeine bauaufsichtliche Zulassungen für den Normalfall bis 2000 kN Gebrauchslast vorliegen.

Die für den Schwerlastanker in bezug auf seine Abmessungen und besonderen Beanspruchungen erforderlichen Ergänzungen und Festlegungen wurden im Rahmen einer Dokumentation zusammen-

gefaßt, die Grundlage für das detaillierte Qualitätssicherungssystem für die Ankerherstellung wurde. Die Arbeiten wurden sowohl im Baustellenwerk als auch beim Ankereinbau Schritt für Schritt durch Eigen- und Fremdüberwachung begleitet und dokumentiert.

Hierüber wird ausführlich in dem Beitrag von GAITZSCH (1995) berichtet.

4.1.2 Vorauslaufende Feldversuche:

Gemäß DIN 4125 sind grundsätzlich für Daueranker vorauslaufende Eignungsversuche durchzuführen. Damit soll die Eignung des anstehenden Baugrunds für die Aufnahme der planmäßigen Gebrauchslasten überprüft und der Nachweis für das geeignete Einbauverfahren geführt werden.

Im Hinblick auf die außergewöhnlichen Anforderungen an die Schwerlastanker der Edertalsperre kam den Vorversuchen besondere Bedeutung zu.

Ihre Planung erfolgte mit folgender Zielsetzung:

1. Gewinn von Aufschlüssen über die erreichbare Haftfestigkeit der Ankerverpreßkörper im anstehenden Gebirge als Grundlage für die Dimensionierung der Krafteintragungslängen, sowohl für die Grauwacke-Formationen als auch für die Tonschieferbereiche.

2. Nachweis eines geeigneten Bohrverfahrens für die geforderte richtungsgenaue Ausführung der Kernbohrungen Ø 146 mm.

3. Nachweis eines zuverlässigen Verfahrens für die laufende Bohrlochvermessung.

4. Optimierung der Vergütungsinjektion bzgl. der erforderlichen Rezepturen und Injektionsschritte, insbesondere im Bezug auf gegenseitige Beeinflussung benachbarter Bohrungen bei Umläufigkeiten usw.

5. Nachweis eines geeigneten Verfahrens für die Bohrlochaufweitung auf 273 mm.

Die zukünftigen Krafteintragungsstrecken der Schwerlastanker waren ca. zu je 50 % in Grauwacke- und Tonschieferformation herzustellen. Dazu wurden jeweils 3 Stück Versuchsanker mit gestaffelten Eintragungslängen von 3 m, 5 m und 7 m vorgesehen.

Um die Zuverlässigkeit des Verankerungssystems auch für weit höhere Lasten nachweisen zu können, wurden von der Firma STUMP BOHR GmbH auf eigene Kosten zwei zusätzliche Schwerlastanker mit 55 Litzen x 0,6" unter sonst gleichen Bedingungen in der Tonsteinformation eingebaut. Sie sollten bis zum Pullout beansprucht werden.

Die Anker wurden mit neuartigen Lichtwellenleitern bestückt, um den Dehnungsverlauf innerhalb der Krafteintragungsstrecke im Zuge der Lastaufbringung registrieren zu können.

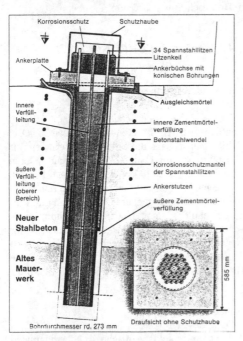

Abb. 3: Ankerkopfausbildung
Construction of the anchor head

Das Ergebnis der Probebelastungen übertraf alle Erwartungen:

Es wurden keine kennzeichnenden Unterschiede im Verformungsverhalten bei den insgesamt 6 Stück 4500 kN-Ankern registriert, und zwar unabhängig davon, ob die Krafteintragungsstrecke in Tonstein oder Grauwacke eingebunden waren und 3 m, 5 m oder 7 m betrugen.

Darüber hinaus konnte auch eine problemlose Lastaufbringung bei den "55-Litzern" mit insgesamt 12500 kN ohne etwaige Anzeichen von Kriechtendenzen oder beginnendem Bruch registriert werden.

Die Auswertung der Lichtwellenleitermessungen ergab ferner, daß die Prüflast von 12500 kN bereits nach 3 m voll in den umgebenden Tonstein eingeleitet worden ist (Abb. 4, 5).

Daraus wurde erkennbar, welche großen Sicherheitsreserven bei dem für die spätere Ausführung vorgesehenen Verankerungssystem vorhanden sind. Schließlich wurden die eigentlichen Bauwerksanker mit 10 m Krafteintragungsstrecke ausgeführt.

Bzgl. der Bohrgenauigkeit ergaben die Vorversuche Abweichungen von weniger als 1 % und erfüllten damit die Anforderungen beim Durchbohren der Gebirgsformationen.

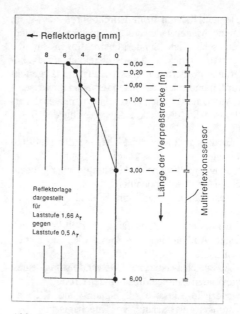

Abb. 5: Dehnungsverteilung in der Verankerungslänge des Stahlzugglieds
Distribution of elasticity in the anchoring length of a steel traction segment

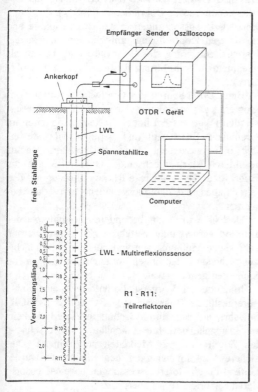

Abb.4: Schema des SICOM-Multireflexionssensors
Model of the SICOM multirefector sensor

Zusammenfassend bleibt festzustellen, daß die vorauslaufenden Probebelastungen außerordentlich wertvolle Aufschlüsse für die spätere Ausführungsplanung lieferten und derartige Tests für die Ausführung von Schwerlastankern allgemein als unverzichtbar gefordert werden sollten:

1. Nachweis der grundsätzlichen Eignung des anstehenden Baugrundes

2. Nachweis der Eignung des vorgesehenen Ankersystems

3. Nachweis der erzielbaren Bohrgenauigkeit

4. Nachweis der geeigneten Meß- und Kontrolleinrichtungen

5. Hinweise für die Planung der Arbeitsabläufe, insbesondere für die Injektionsschritte

6. ergänzende Hinweise für die Arbeitsvorbereitung, insbesondere für die Gerätekonstruktion

7. Zusätzliche Sicherheiten für die Termingestaltung.

4.2 Geräteplanung und Geräteeinsatz:

Insbesondere die hohen Anforderungen an die Bohrgenauigkeit machten im Zuge der Arbeitsvorbereitung besondere gerätetechnische Überlegungen erforderlich. Unter Zugrundelegung der Erfahrungen aus den Feldversuchen und im Hinblick auf die zu erwartenden Risiken beim Durchbohren des 46 m mächtigen Mauerwerks aus Grauwacke-Bruchstein wurden schließlich Sonderkonstruktionen für die Bohrgeräte entwickelt, die ausschließlich auf die besonderen Anforderungen dieser Baustelle ausgerichtet wurden und folgende Besonderheiten aufweisen:

• Ausrüstung mit einem horizontalen Verschiebetisch für die exakte Ansteuerung der Bohrposition
• Exakte geometrische Sicherung der Lafetteneinstellung
• Schwingungsarme Lagerung aller beanspruchten Gerätekomponenten
• Beschränkung auf eine Mindestanzahl an Gelenken
• Besonders leistungsfähiger Bohrkopf mit spezieller Aufhängung zur Gewährleistung eines ruhigen Laufs
• Ausgleichswaage zur Gewährleistung eines gleichmäßigen Bohrandrucks an der Krone
• spezielle Einsätze für Spann- und Klemmeinrichtungen als Schutz vor mechanischen Beschädigungen der Präzisionsrohre
• spezielle Führungen für die Bohrrohre zur Vermeidung von Schwingungen.

Für die Kernbohrungen waren zwei vorauslaufende Präzisionsbohrgeräte im Einsatz (Abb. 6).

Abb. 6: Kernbohranlage im Einsatz
Core drilling equipment in operation

Sie wurden ergänzt durch ein spezielles Injektionsbohrgerät, das im Pilgerschrittverfahren vor- und zurücksetzen mußte, um die wechselseitig erforderlichen Injektionsarbeiten kontinuierlich durchführen zu können.

Die abschließenden Aufweitungsbohrungen erfolgten als Imlochhammerbohrung mit einer speziellen Pilotkrone Ø 273 mm. Die hierfür erforderlichen schweren Bohrrohre ergaben für den Bohrstrang ein Eigengewicht von 7 t. Für den sicheren Austrag des Bohrgutes aus 80 m Tiefe mußte im Hinblick auf den unterschiedlichen Zerteilungsgrad des Bohrkleins ein Hochleistungskompressor eingesetzt werden, der den nötigen Spülstrom von mehr als 30 m/sek. im Bohrlochringraum sicherstellte. Das anfallende Bohrgut wurde gesammelt und zu einem besonderen Baustellenklärwerk geführt. Hier wurden Feststoffe und Wasser separiert. Nach Durchlaufen einer Neutralisationsanlage konnte das restliche Spülwasser in den Unterlauf der Eder eingeleitet werden.

4.3 Ablauf der Bohr- und Injektionsarbeiten

Das neu erstellte Brückenbauwerk diente als Arbeitsplanum für die Bohrarbeiten (Abb. 7).
Besondere Aussparungen in der Brückenplatte und im unteren Kontrollgang ermöglichten eine zuverlässige Zentrierung der Bohrrohre und dienten als geeignete Anbohrführung für die oberen 8 m Bohrstrecke.

Beim Durchbohren des Grauwacke-Mörtel-Konglomerats der Staumauer war mit einer häufig wechselnden Folge extrem hoher Gesteinsfestigkeiten im Grauwacke-Bruchstein und geringer Mörtelfestigkeit im Trasszement-Mörtel zu rechnen. Daraus resultieren stark wechselnde Querkräfte auf die Bohrkrone, die zusätzliche Risiken in bezug auf den geradlinigen und richtungsgenauen Bohrverlauf bedeuten.

Deshalb wurde mit besonderer Sorgfalt bei der Bohrlochvermessung verfahren. Zunächst wurden sehr kleine Meßintervalle von nur 3 m gewählt, bis ausreichende Erfahrungen vorlagen und zunehmend Sicherheit gegeben war.

Im Zuge der Vorversuche hatte sich eine computergestützte Bohrlochsonde System MAXIBOR bewährt, die sich durch besondere Zuverlässigkeit und baustellengerechte Handhabung sowie schnellen Zugriff auf die Meßdaten auszeichnete. Die sichere Funktionsfähigkeit des Meßgeräts wurde laufend durch Referenzmessungen an einer besonderen Kalibrierstrecke überprüft.

Darüber hinaus wurde selbstverständlich besonderer Wert auf die Schulung des Fachpersonals und

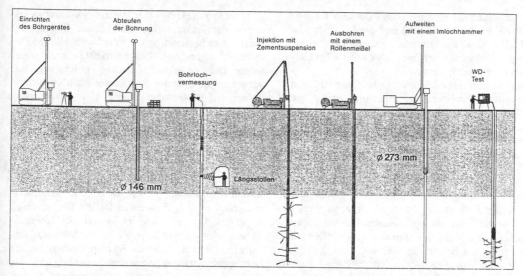

Einrichten des Bohrgerätes · Abteufen der Bohrung · Bohrlochvermessung · Injektion mit Zementsuspension · Ausbohren mit einem Rollenmeißel · Aufweiten mit einem Imlochhammer · WD-Test

Ø 146 mm · Längsstollen · Ø 273 mm

Abb. 7: Schematische Darstellung der Bohr- und Injektionsarbeiten

Schematic illustration of drilling and injection operations

Abb. 8: Einbau des ersten Ankers und Einsatz der Umlenkkonstruktion

Installation of the first anchor and insertion of the diverting device

die Einhaltung der vorgegebenen Parameter für den Andruck an der Krone (2 t) und die Drehzahl (ca. 170 -250 Umdrehungen/min.) gelegt. Eine unzulässige Bohrlochabweichung mußte unter allen Umständen vermieden werden.

Im Mauerfußbereich konnte aus dem unteren Kontrollgang heraus eine Überprüfung der Sollage der Bohrungen durchgeführt werden. Dies geschah durch Ortungsmessungen auf induktiver Basis. Außerdem wurden direkte Suchbohrungen mit Kernaufschluß in Einzelfällen durchgeführt, um die Ist-Lage der jeweiligen Bohrung durch Augenschein überprüfen zu können. Die sorgfältige Arbeitsvorbereitung hat sich ausgezahlt. Nach Abschluß der Bohrarbeiten kann ein Ergebnis vorgezeigt werden, das alle optimistischen Erwartungen übertraf. Die 104 Stück Kernbohrungen wurden mit einer mittleren Bohrlochabweichung von 0,45 % auf 80 m Solltiefe niedergebracht, bei einer max. Abweichung von 1 %. Über Einzelheiten zur Qualitätsüberwachung und zur Bohrlochvermessung wird von GAITZSCH (1995) berichtet.

4.4 Ankereinbau:

Nach erfolgter Aufweitung des Bohrlochs auf 273 mm wurde ein abschließender WD-Test als Nachweis für den erreichten Injektionserfolg als wichtige Voraussetzung für den Ankereinbau durchgeführt. Wurden die vorgegebenen Kriterien bzgl. der Dichtigkeit des vergüteten Gebirges nicht erreicht, wurde die Injektion wiederholt. Nach Freigabe des Bohrlochs für den Ankereinbau wurde das im Baustellenwerk vorbereitete ca. 75 m lange und 3,5 t schwere Ankerzugglied zur Einbaustelle befördert. Durch eine geeignete Rollenlagerung wurde sichergestellt, daß die Transportmanöver zwängungsfrei unter Ausschluß schädigender mechanischer Einwirkung erfolgte.

Um den Anker aus der horizontalen Förderrichtung in die vertikale Einbauposition zu bringen, wurde eine spezielle Umlenkkonstruktion als Einbauhilfe entwickelt (Abb. 8). Hierbei wurde der Anker über eine Umlaufbahn geführt, deren Krümmungsradius dem zulässigen Biegeradius des Zuggliedes von 2,5 m entsprach. Die Umlenkkonstruktion wurde über einen Einbaurahmen mittels Autokran aufgenommen und das über beidseitige Windenhalterung gesicherte Ankerzugglied sukzessive zur Einbaustelle gefördert. Hierbei war für eine jederzeit stabile und kontrollierte Führung von Gehänge und Ankerzugglied Sorge zu tragen. Nach Absetzen des Einbaurahmens über der Einbaustelle konnte das Zugglied Zug um Zug in das Bohrloch abgelassen werden.

Innerhalb des ca. 8 m tieferen Kontrollgangs wurde der Ankerkopf schließlich auf Sollage justiert. Anschließend wurde die vorbereitete Zementsuspension vom Bohrlochtiefsten nach oben aufsteigend eingebracht, wobei gleichzeitig im Ringraum anstehendes Restwasser nach oben verdrängt und ausgetragen wurde. Der Ankereinbau war damit abgeschlossen.

4.5 Spannarbeiten

Im Zuge der Prüflast in Höhe von 6750 kN treten Zuggliederdehnungen bis 45 cm auf, die einen entsprechend großen Pressenhub für die Spannpressen verlangen. Da außerdem im Inneren des Kontrollganges sehr beengte räumliche Verhältnisse vorlagen, mußten auch für die Spannpressen kompakte Sonderanfertigungen entwickelt werden (Abb. 9). Auch in Kompaktbauweise (Durchmesser 1,0 m, Höhe 1,0 m) verblieb ein Pressengewicht von 3 t. Diese Masse mußte zwängungsfrei über die Ankerüberstände verfahren und die 34 Litzen jederzeit kontrolliert eingefädelt werden. Hierzu wurde eine spezielle Hubwagenkonstruktion mit kardanisch angeschlossenem Rahmen eingesetzt, die die Presse sorgfältig in jede Position mit Feinstabstimmung einfahren konnte. Die mobile Einsatzfähigkeit der beiden Spannpressen war von besonderer Bedeutung, weil die Ankerkräfte nur stufenweise aufgebracht werden durften, um eine gleichmäßige Eintragung in den Mauerkörper sicherzustellen:

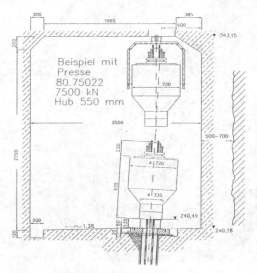

Abb. 9: Einsatz der Spannpressen
Installation of tension press

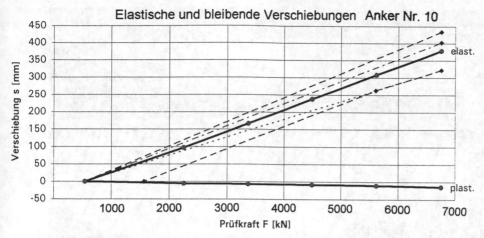

Abb. 10: Abnahmeprüfung (6750 kN)   Acceptance testing (6750 kN)

1. Durchführung einer Abnahmeprüfung mit einer Prüflast in Höhe der 1,5fachen Gebrauchslast, entsprechend 6750 kN.

2. Entlastung und Festlegung nach Abschluß der Prüfung auf 50 % der Gebrauchslast (2250 kN).

3. Nach Abschluß sämtlicher Abnahmeprüfungen 2. Belastungsgang mit Lastaufbringung auf nunmehr 80 % der Gebrauchslast.

4. Letzter Arbeitsgang mit endgültiger Festlegung auf 100 % = 4500 kN.

4.6   Ergebnis der Abnahmeprüfungen

Die bei den Spannvorgängen registrierten elastischen und plastischen Verformungen der Ankerzugglieder zeigen fast modellhafte Übereinstimmung. Sämtliche 104 Anker haben die Abnahmeprüfungen problemlos bestanden, Unregelmäßigkeiten sind nicht aufgetreten (Abb. 10).

Aus den bleibenden Verformungen und den geringen Kriechwerten ist zu folgern, daß der Krafteintrag in den Tonstein- bzw. Grauwacke-Untergrund eine sehr hohe Sicherheit beinhaltet.

Die an 10 Meßankern installierten Lichtwellenleiter zeigten Verformungen im Krafteintragungsbereich nur auf den ersten 2 - 3 m. Die unteren verbleibenden 7 m der Krafteintragungsstrecke sind also im Zuge der Abnahmeprüfungen noch gar nicht in Anspruch genommen worden und bieten dementsprechende Tragkraftreserven.

Die in den vorauslaufenden Feldversuchen bereits nachgewiesenen großen Sicherheiten wurden in der Ausführung eindrucksvoll bestätigt. Die durchgeführten intensiven Injektionsarbeiten und die sorg-

fältige Behandlung der Bohrlochwandung im Zuge der Bohrarbeiten waren dabei wichtige Voraussetzung für das einheitlich sichere Tragverhalten.

5.   SCHLUSSBEMERKUNG

Die insgesamt 104 Stück Anker wurden planmäßig innerhalb der vorgesehenen Bauzeit von einem Jahr eingebaut. Alle Termine konnten exakt eingehalten werden, obwohl der Ausführung ein streng überwachtes Qualitätssicherungssystem zugrunde lag. Qualität und Wirtschaftlichkeit müssen sich also nicht ausschließen.

Am 06. Mai 1994 wurde die sanierte Mauer im Rahmen eines Festaktes übergeben und die Sanierungsmaßnahme als hervorragende Ingenieurleistung gewürdigt. Nach mehr als 10 Jahren Teilstau konnten Besucher und Anwohner wieder das eindrucksvolle Bild der gefüllten Talsperre mit überströmter Krone genießen (Abb. 11).

Die Sanierung der Edertalsperre stellte in bezug auf die Ankerarbeiten außergewöhnlich hohe Anforderungen. Durch besonders eingehende Voruntersuchungen, die für anspruchsvolle Arbeiten dieser Art unabdingbar sind, konnten für Bohrtechnik und Ankertechnik gesicherte Grundlagen geschaffen werden. Sie waren die Voraussetzung dafür, daß die qualitativ hochwertige Ausführung planmäßig und termingerecht abgeschlossen werden konnte. Hierzu trug ein strenges Qualitätsmanagement bei, das sowohl dem hohen Sicherheitsbedürfnis gerecht wurde, aber auch Grundlage für einen geordneten Arbeitsablauf war.

Die außergewöhnliche Aufgabenstellung machte

Abb. 11: Die sanierte Staumauer am Tage der feier-
lichen Einweihung

The renovated dam on the day of opening
ceremony

bei den Verankerungsarbeiten in fast allen Arbeits-
bereichen Neuentwicklungen erforderlich, für die
kurzfristig gesicherte Brauchbarkeitsnachweise zu
erbringen war. Hier kam es ganz besonders darauf
an, daß Auftraggeber und beteiligte Fachingenieure
sachkundig und kooperativ mitwirkten.

Nicht zuletzt soll zum Ausdruck kommen, daß
der Erfolg derart anspruchsvoller Arbeiten von der
langjährigen Erfahrung ausgesuchter Fachfirmen und
der Zuverlässigkeit ihres Personals abhängt.

Mit der vorgestellten Maßnahme zur Sanierung
der Edertalsperre ist verdeutlicht worden, daß die
Ankertechnik in bezug auf die Größe der Gebrauchslasten und die Abmessungen der Anker nach
oben keine Grenzen kennt und sich auch unter
schwierigen Herstellbedingungen als zuverlässige
und dauerhafte Bauweise für zukünftige Einsätze
empfiehlt.

LITERATURHINWEISE:

Gaitzsch, H., 1995. Besondere technische
Aspekte bei Überwachung und meßtechnischer

Begleitung der Felsankerarbeiten an der Edertalsperre. Tagungsband Int. Symposium - Anker in
Theorie und Praxis - Rotterdam: Balkema
Edertalsperre 1994. Wasser- und Schiffahrtsverwaltung des Bundes, Wasser- und Schiffahrtsdirek-
tion Mitte. Hannover: Eigenverlag
Wittke, W., Schröder, D., 1994. Upgrading the
stability of the Eder masonry dam. In Hydropower & Dams, volume one, Issue five, Surrey:
Aqua ~ Media International

Improvement of the slope stability on the volcanic-sedimentary rocks by means of anchors

Sicherung einer aus vulkanischen Sedimentgesteinen bestehenden Böschung

E. Marchidanu
Technical University of Civil Engineering, Bucharest, Romania

N. Sima
S.C. GEOSOND S.A., Bucharest, Romania

ABSTRACT: The paper submits the solution of improvement of the slope stability on the volcanic-sedimentary rocks by means of anchors in embed zone of local materials dam. There are presented the main geological and geomechanical characteristics of rock slope, the scheme of anchoring network adopted, type of anchors used, construction method of works as well as the findings concerning the behaviour of slope eight years after commissioning of the works.

ZUSAMMENFASSUNG: Der Beitrag beschreibt die Sicherung einer Böschung aus vulkanischen Sedimentgesteinen zur Herstellung eines Aushubes für einen Schüttdamm.
Mit geologischen und geophysikalischen Erkundungen wurde in der Böschung eine zerklüftete Zone abgegrenzt. Die Klüfte sind Zentimeter bzw. Dezimeter weit und verlaufen im allgemeinen parallel zur Böschung.
Für den Sperrenaushub mußte auch teilweise die zerklüftete Zone angeschnitten werden, wodurch die Neigung der Böschung von 30 - 35° auf im Mittel 45° versteilt wurde. Um die Böschung für die Aushubarbeiten und den späteren Bestand zu sichern, erfolgten der Aushub, die Zementinjektionen und der Einbau der Anker stufenweise, von oben beginnend. Insgesamt wurden 70 Vorspannanker mit einer Länge von je 26,5 m in Bohrungen mit einem Durchmesser von 108 mm, mit einer Haftstrecke von 5,5 m in Zementmörtel versetzt. Ausreißversuche ergaben eine Tragkraft von 800 - 1 000 kN je Anker, sodaß man für die Stabilitätsberechnung eine Tragkraft von 800 kN angenommen hat.
Die acht Jahre dauernden Beobachtungen seit Ende der Bauarbeiten zeigen ein stabiles Verhalten der Böschung.

1 GENERAL DATA

Within the general hydrotechnical management scheme for the Târnava Mare river basin, on the western frame of the Oriental Carpathians at 5 km upstream of Zetea locality (Fig.1) a local material dam with clay core was built.

It has a height of 48 m and the storage reservoir created has a volume of 43 million c.m. The placement of the whole work-dam and lake bed is situated in the very development zone of the volcanic rocks represented by the volcanogen-sedimentary formation which, in the strict storage zone, consists exclusively of volcanic tuffogene brecciae and agglomerates.

Within the dam placement the left slope of the valley has a mamelon shape with rises 120 m above the flood plane, the slope surface having an inclination of 30-35°.

At the stage of geological technical studies when the first pits were executed in the fixing dam

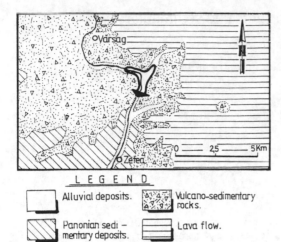

Alluvial deposits.

Vulcano-sedimentary rocks.

Panonian sedi‑ mentary deposits.

Lava flow.

Fig.1. Geological map for the
Zetea reservoir area
Abb.1. Geologische Karte des
Zetea Speicherbeckens

zone in the left slope, many gaps
were found, most of them being
cracks with widths of some centime-
tres and decimeters sometimes exceeding one meter. The direction of
the cracks, generally speaking, was
parallel to the slope surface.
These first findings led to the e-
laboration of an ample geological

and technical program which was con
tinued long time after the begining
of dam construction.

The conclusions of the fulfilled
research allowed the adaptation of
the execution project to the land
peculiarities for dam foundation in
that area and the adaptation of
some efficient solutions concerning
the foundation conditions, the tigh
tening of the rock massifs against
infiltrations and the ensuring of
the slope stability during works
and after commissioning the storage
reservoir.

2 GEOLOGICAL TECHNICAL STUDIES MADE IN THE LEFT SLOPE

The initial program of geological -
technical was completed and diversi
fied step by step according information from the research activity.

The geological-technical research
program, wich was completely
achieved, comprised the following
work categories:
- geophysical prospectings with
electrometrical and seismical methods;
- 10 wells with depths between 5
and 12 m;

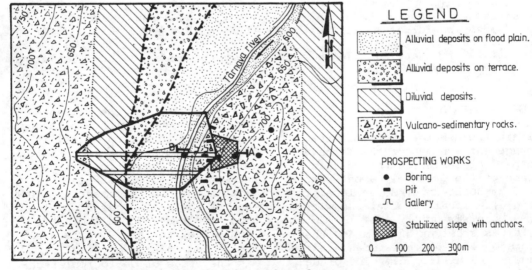

L E G E N D

Alluvial deposits on flood plain.

Alluvial deposits on terrace.

Diluvial deposits.

Vulcano-sedimentary rocks.

PROSPECTING WORKS

• Boring

− Pit

⌐ Gallery

Stabilized slope with anchors.

0   100   200   300 m

Fig.2. Geological map for the Zetea dam placement
Abb.2. Geologische Karte für die Baustelle der Zetea Talsperre

- 3 mining galleries with lenght up to 80 m, where some compressibility and rock-rock shearing experimental attempts were made;
- 5 drillings executed in mechanical system, with continuing recovery and depths up to 80 m, in some of them Lugeon tests an experimental injections with cement suspensions being made;
- 3 experiments consisting of pouring water marked with radioactive tracers into the wells and following the water circulation through the rock massif;
- 2 experiments made in situ through groups of injection drillings and observation drillings in order to test the fine, tuffogene materials to be eroded and hydro-dynamically carried to big flowing gradients of the underground water;
- standard analyses and complex experiments made in the laboratory on samples taken from the rock massif with regard to the establishment of the mineralogical composition, structure, texture, the rate chemical alteration, physical cha-

racteristics, mechanical resistance, elasticity, stability against infiltration curents etc.

The geological-technical map of the dam location and the placement of wells, galleries and drills executed in the left slope are shown in Figure 2.

3 MINERALOGICAL, PETROGRAPHICAL AND GEOMECHANICAL CHARACTERISTICS OF THE SLOPE ROCKS

The slope consists of a chaotic deposit of agglomerates, volcanic breccias and cinerits with different grades of granulation, cracking and alteration in which there are dissipated andesite stones, with angular forms, whose dimensions can reach one cubic metre.

The horizons mostly cineritic have reduced thickness, are of grey colour sometime violet-reddish and they have a crypto-crystalline structure and a porous texture. In the cinerite mass there are frequently met crystals of horn-

Table 1. Main phisical and mechanical characteristics of rocks slope
Tabelle 1. Physikalische und mechanische Haupteigenschaften der Böschung

Characteristic		Symbol	Unit measure UM	Number of the tested samples $n°$	Quadratic mean value deviation S	Mean value A^n
Apparent density		ρ_a	g/cm^3	192	0.263	2.01
Porosity		n	%	192	8.69	24.04
Compressive dry strength		σ_{cd}	MPa	91	27.6	18.81
Compressive saturation strength		σ_{cs}	MPa	113	33.05	16.90
Modulus of elasticity		E	MPa	3	–	4,870
Modulus of deformation		D	MPa	3	–	3,385
Angle of internal friction	At the pick	Φ	Degrees	3	–	$35°$
	Residual	Φ_r	Degrees	3	–	$30°$
Cohesion	At the pick	c	MPa	3	–	0.9
	Residual	c_r	MPa	3	–	0.3

blende and phenocrystals of feldspars and pyroxenes.

Breccias and volcanic agglomerates predominate from the surface
point of view and they are made of
fragments and stones of andesite
with angular or rolling forms fixed
in a tuff matrix of white colour.

From the granulation point of
view the rock is heterogeneous the
zones mostly cineritic alternating
with zone made of agglomerations of
fragments and big stones of
andesite with its interspaces filled with tuffogene material.

The fragments and stones of andesite have a black-grey or a reddish
-grey colour, a structure generally
speaking porphyrical and a massive
texture and when they are in advanced stage of alteration the
texture becomes vacuolar.

From the mineralogical point of
view, the andesite fragments con-
tains feldspars, hornblendes and
volcanic glass.

The main geomechanical characteristics of the volcanic-sedimentary rocks in the slope are presented in Table 1.

4 SOLUTIONS ADDOPTED FOR ENSURING THE SLOPE STABILITY DURING AND AFTER THE DAM BUILDING

The geological-technical research
allowed the clear demarcation of
that zone affected by cracks (Fig.
3) and the adopting the optimal
solutions for dam foundation and
the ensuring the slope stability
during and after the building of
some works auxiliary to the dam.

Supposing that the most probable sliding zone of the slope is
situated at the base of that zone
affected by cracks, the stability
calculations using the following
geotechnical values: γ = 20 kN/m^3;
Φ = 30° and c = 0 kPa, confirmed
that at the natural inclination
the slope is stable.

In the project for fixing the
dam in the left slope (Fig.4), the
adopted solution was from the
foundation land of underground of
dam, affected by cracks, to remove
by excavation about 50% following
that under the excavation level a
concrete screen of Bioge type to
be built, with a depth of 8.5- 13
m which is to be fixed in that
rock unaffected by cracks. Along
after the Bioge screen an

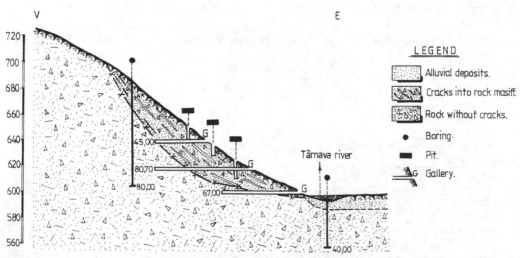

Fig.3. Geological section showing the left slope affected by cracks
Abb.3. Geologischer Längsschnitt durch die von Rissen geschädigte Zone

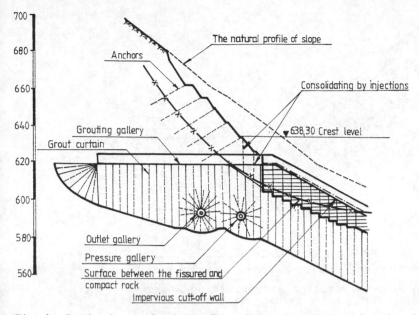

Fig.4. Designing solutions for slope management in the dam fixing zone
Abb.4. Ausführungsentwurf für die Böschungssicherung im Dammbereich

injection certain was de signed up
to 30-35 m depth.

Under the elevation of dam top
the project provided enforcement
works with mortar and cement injections for that rock affected by
cracks remained under the clay
core.

On to the excavations executed
for dam foundation and their exten
sion on the slope up to 680.00 ele
vation, the slope inclination was
modified up to a general inclination angle of about 45°.

For the slope resulted from excavation the calculations show a
diminishing of stability factor up
to the equilibrium limit (Table 2)

In order to ensure the slope sta
bility it was necessary to increase the resistence forces to
sliding with at least 10-15 %.

According to the alternative a-
nalysis the best solution proved
the anchoring of the excavated
slope with pre-tensioned anchors
and the enforcing of the rock volume affected by cracks through
mortar injections and cement suspensions.

Table 2. Factor of safety for st-
bility of slope at the slide
Tabelle 2. Werte des Instabili-
tätsfaktors beim Böschungsa
-brutschen

Probable slip surface	Slope		
		Excavated	
	Natu-ral	Before ancho-ring	After ancho-ring
ABCD	1.25	–	–
ABC	–	0.92	1.23
ABE	–	1.03	1.36

The execution order of excavation, enforcing by cementing and
anchoring work is shown in Fig.5.

5 DESIGNING AND EXECUTION OF
  ANCHORING WORKS

In order to achieve a stability
factor of about 1.25 an anchor net
work was designed as you can see
in Figure 6.

The distance between the anchor

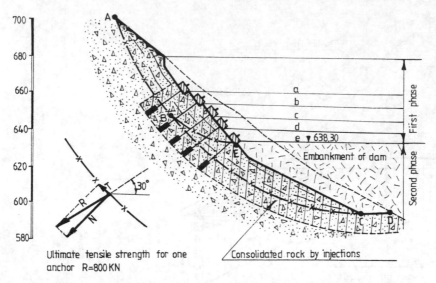

Fig.5.Phasing the slope excavation and enforcing works
Abb.5.Bauetappen für den Aushub und die Sicherung mit Anker

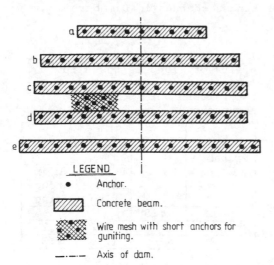

LEGEND

•   Anchor.

▨   Concrete beam.

▩   Wire mesh with short anchors for guniting.

—·—·—   Axis of dam.

Fig.6. Principle scheme for the distribution of anchors over the slope
Abb.6. Schema der Verteilung der Anker auf der Böschung

strings, measured on the slope sur face, was of 10 m and between the anchors on the same string a distance of 5 m was kept.

The dimensioning of the anchoring system was made according to the results of some attempts to pull three test anchors out.

At the first anchoring drillings there were many dificulties caused by the instability of the dril led holes, on the depth space corresponding to the zone affected by cracks. Due to this situation the solution was the enforcement of rocks by cement injections in descending system, made in those holes executed for introducing the anchors.

The anchors were introduced in the drilled holes with diametre $\Phi=108$ mm, executed with an inclination on $30°$ to the horizontal line, being fixed and cemented in the healthy rock on a length of 5.5 m.

At the pulling ont attempts of three anchors executed on the slope there were obtained forces resistant to pulling of 800-1000kN at every anchor.

At the checking calculations for the anchored slope stability there was taken into account a resistant force due to an anchor of 800 kN - - value considered to be covering enough if we take into account the effect of geomechanical improve-

Table 3. Data about the enforcing works for the slope rocks through cementing and anchoring
Tabelle 3. Daten der Arbeiten zur Böschungssicherung mittels Betonbalken und Anker

Row of anchors Elevation	Boring Φ=108 mm			Quantity of cement used for injection	Anchors type SPB (fascicle wire of steel Φ = 7 mm)				
	Number of boring	Deepth	Total length		Number of anchors	Number wires for one anchor	Length of anchor	Cemented length	Force of strain
-	-	m	m	tons	-	-	m	m	kN
a̲ 664.00	11	14-25	194	12,671	9	12	15.5-20.5	5.5	383-535
b̲ 658.00	14	25	350	97,740	14	24	26.5	5.5	975-1,000
c̲ 651.00	15	25	375	72,050	15	24	26.5	5.5	800-875
d̲ 643.00	16	25	400	63,960	15	24	26.5	5.5	800-1,000
e̲ 636.00	17	25	425	76,950	17	24	26.5	5.5	750-1,000
Total	73	-	1,744	323,371	70	-	-	-	-

ment for the rock massif with the help of cement injections.

According to the adopted anchoring a compression force of 800 kN distributed to a 50m^2 surface wich corresponds to a unitary effort of 1.6×10^{-2} MPa.

The main data regarding the technical parameters and the anchor characteristics used for slope enforcement are presented in Table3.

The anchor endings on every string were passed through a steel concrete beam wich had the role of alloving the execution in good conditions of tensioning every anchor being at the some time an element of resistance and redistribution of the slope pushing effort for all the anchors on the string.

After ending the enforcement works through cement injections and anchoring works the whole excavated surface executed above the dam top elevation, was protected with concrete gunite over a wire net fixed on the rock with short anchors.

6 CONCLUSIONS

6.1 The volcanic-sedimentary rocks situated between the cliff type rocks and semi-cliff type rocks, are, generally speaking, very difficult from the geomechanical point of view, being characterized as very heterogenous, anisotrope and susceptible of instability due to the hydrodynamic carring of the fine tuffogene matrix in wich hard rock fragments and stone of magmatic origine are included.

6.2 Within the geological and geomorphological conditions which characterizes the left slope of the Târnava Mare river in the Zetea dam placement the slope stability solution by previous cement injec- tions and pretensioned anchors, proved efficient from the technical as well as economical point of view.

6.3 Eight years after the ending of works the results of geodesical and inclinometrical measurements

made in drilling holes with flexible tubes, confirm a very good
behaviour of the anchor system,
the slope instability tendency
unbeing noticed.

New anchor design for a hydropower station enlargement on the Danube river
Neuartiges Ankerkonzept für den Ausbau des Kraftwerkes Ybbs-Persenbeug

S. Strohhäusl
D2-Consult, Linz, Austria

ABSTRACT: System anchoring with 6-12 m long Swellex expansion anchors and self boring anchors with a shotcrete lining was used to support two up to 40 m deep excavation pits and a 220 m² tunnel in between, for the project "7th Machine" for the Ybbs-Persenbeug Danube power station. Rock excavation and support of the pits and the tunnels had to be performed under highly fractured and partially disintegrated rock conditions. Tunnel top heading in gravel was driven beneath a double jet-grouting screen, which provided stable conditions during excavation under cohesionless soil conditions. Use of large scale prestressed anchoring for this temporary construction phase could be greatly reduced.

ZUSAMMENFASSUNG: Durch die Verwendung eines Felssicherungskonzeptes, basierend auf einer Systemankerung mit 6 - 12 m langen Swellex-Expansionsankern und Selbstbohrankern mit einer Spritzbetonverkleidung, konnte die unter großem Termindruck und schwierigen geologischen Bedingungen herzustellende, bis zu 40 m tiefe Hauptbaugrube, des Bauvorhabens "7. Maschine", für das Donaukraftwerk Ybbs-Persenbeug, erfolgreich ausgeführt werden. Aufwendige Vorspannankerungen konnten damit für diese temporäre Bauphase stark reduziert werden. Die hohe Flexibilität, die Einbauzeit eines 12 m langen Swellex - Expansionsankers betrug unter günstigen Bedingungen, inklusive Bohren nur 15 min, erlaubte eine rasche Reaktion bei sich verschlechternden Gebirgsverhältnissen bzw. rasch ansteigenden Verformungen.

Der Felsausbruch erfolgte, nach anfänglichen Sprengversuchen, durch Meißeln. Dadurch konnten Erschütterungen minimiert und Ausbruch und Sicherung besser aufeinander abgestimmt werden. Trotz schwieriger geologischer Bedingungen und teilweise starkem Wasserzutritt wurden damit innerhalb von 3 Monaten 45.000 m³ Fels ausgebrochen und 4.800 m² Baugrubenwand gesichert. Für die Sicherung wurden insgesamt 800 m³ Spritzbeton, 24.000 lfm Swellex Anker, 8.000 lfm Selbstbohranker und 2.000 lfm Vorspannanker eingebaut.

Der Vortrieb der Kallotte des 220 m² großen Tunnels unter der Montagehalle des bestehenden Kraftwerkes, erfolgte in der Kies- Schotterüberlagerung im Schutze eines doppelten HDBV-Schirmes mit einer Sicherung bestehend aus Spritzbeton und Ausbaubögen. Der Vortrieb und die Sicherung in den Strossenabschnitten und der Sohle erfolgte analog zur Hauptbaugrube mittels Meißeln in Kombination mit einer Spritzbetonauskleidung und einer Systemankerung.

Im Zuge der Planung der Baugrubensicherung wurden neben konventionellen Gleitkeil- und Gleitkreisberechnungen erstmals Berechnungen mit einem neuen Finite Elemente Programm durchgeführt. Dieses Programm ist in der Lage jeden einzelnen Anker entsprechend seiner Wirkungsweise zu erfassen, und verformungsbedingte Änderungen der Ankerkräfte, sowie die Interaktion Anker-Gebirge wirklichkeitsnah zu simulieren. Während des Ausbruchs wurde über eine Rückrechnung und eine Neuberechnung der zu erwartenden Endverformungen mittels einer neuentwickelten Finite Elemente Programmes ein wertvolles Instrument für die Bauüberwachung geschaffen.

1. INTRODUCTION

In the course of equipping the Ybbs-Persenbeug Danube power station with an additional turbine, an inlet channel with turbine building and a downstream outlet was erected on the orographic right side of the existing power station. On account of the surface conditions, an excavation had to be made over a length of 180 m, a width of on average 20 m and up to 40 m in depth. This open excavation pit was interrupted in the area of the power plant axis by a 32 m long tunnel that had to be mined, with a cross section of approx. 220 m² (Fig.1).

Underneath an overburden of up to 20 m high, consisting of gravel and filled material from the former power station excavation pit, there are the crystalline rocks of the Bohemian mass which had to be cut into to a depth of up to 20 m in the course of the construction work. The support of the main excavation pit in the gravel overburden was effected using a vertical structure of bored pile walls which were anchored several times, and also partly by slurry walls (Fig.2). In addition to this, two horizons of struttings were installed in the inlet channel excavation pit. (Fig.3)

In the course of the preliminary studies, the construction technology problems anticipated for the rock zone, caused by the strong jointing, the tectonic fracturation and partly by the weathering of the rock became evident (Fig.4). Furthermore, the tight construction schedule and the proximity of the Danube as well as the machines and buildings that are sensitive to shock and settling had to be taken into consideration when drawing up the excavation and tunnelling concept.

2. STABILITY ANALYSES

For the dimensioning of the support media as well as for determining the stability of the parts of the main excavation pit for the project "7th machine", that are situated in the rock, rock-mechanical analyses were performed using the sliding wedge and slip circle method. Because of the very non-uniform foliation and stratification as well as the constantly changing degree of separation and weathering there was no point in taking the rock structures into consideration. The rock characteristic values (friction angle $\varphi=35°$, cohesion $c=90$ kN/m³) that were assumed from the results of the geological studies and prognoses, formed the basis for these analysiss. This assessment also took into consideration, amongst other things, the results of shearing tests on strongly separated and jointed rock samples with and without joint filling, as stated in the literature.

For the pore water pressure in the slip planes investigated, a triangular distribution with the maximum value at the rock surface and achieving zero at the deepest point in the excavation pit, was assumed. The value at the upper edge of the rock was established in every analysis cross section in line with the ground water lowering prevailing there. The assumption of the reversed triangular load was effected presupposing a destressing of the pore water by the arrangement of drainage bore holes and/or via existing anchor bore holes.

In all 3 analysis cross sections situated on the water side, because of the afore-mentioned drainage through bore holes and ground water lowering, half the pressure was used respectively from the water level of the Danube. On the land side, in the 3 analysis cross sections investigated in the tailwater, at the top edge of the rock, half of the originally effective ground water pressure was applied (Fig.5). This assumption was effected on the basis of the ground water lowering performed behind the bored diaphragm walls. In the upstream section, the full ground water level at the top edge of the rock was reckoned with because of the absence of ground water lowering.

The horizontal water force affecting on the bank wall, caused by the level of the Danube water was applied, in the 3 analysis cross sections situated at the water side, at only 50% as a result of the influence of the bank wall (gravity retaining wall Fig.6).

This assumption was verified by the proof that at least half of these horizontal forces were diverted via friction in the bottom joint into the rock formation outwith the critical sliding wedge. In addition to the sliding wedge analyses, slip circle analyses were performed in the deepest cross section. These proved to be of no significance for the anchor calculation.

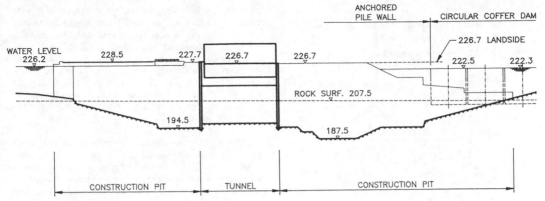

ANCHORED PILE WALL   CIRCULAR COFFER DAM

WATER LEVEL
226.2   228.5   227.7   226.7   226.7   226.7 LANDSIDE   222.5   222.3

ROCK SURF. 207.5

194.5   187.5

CONSTRUCTION PIT   TUNNEL   CONSTRUCTION PIT

Fig.1 Construction Pit - Longitudinal Section
Baugrube - Längsschnitt

Fig.2 Construction Pit - Outlet Tunnel
Baugrube - Auslauftunnel

Fig.3 Construction Pit - Inlet Tunnel
with two horizons of struttings
Baugrube - Einlauftunnel mit 2 horizontalen Aussteifungen

Fig.4 Highly jointed and fractured rock
in the Construction Pit
Stark zerklüfteter und zerlegter Fels in
der Baugrube

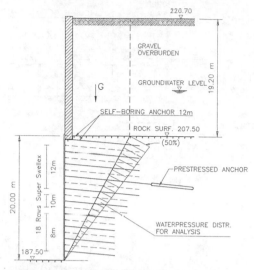

Fig.5 Landside Cross Section for analysis showing load conditions and anchor distribution
Landseitiger Berechnungsquerschnitt mit Belastung u. Ankeraufteilungen

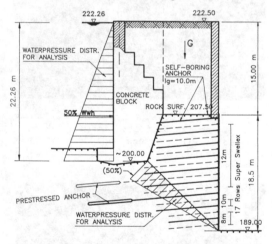

Fig.6 Waterside Cross Section showing load conditions and anchor distribution
Wasserseitiger Berechnungsquerschnitt mit Belastung und Ankeraufteilungen

3 ROCK SUPPORT CONCEPT FOR THE MAIN EXCAVATION PIT

On account of the tight constructional stipulations on the one hand, and the difficult, constantly changing underground conditions on the other hand, the attempt was made to develop as flexible a concept as possible to ensure the temporary constructional measures for the sup-

port of the excavation pit. This support concept had to be quickly modifiable in event of any deterioration in the safety level, caused by changed geological conditions or increasing deformations, without involving any delay to the construction time. The measures of support to be used for this had to be installed within a short time and had to be effective quickly.

These requirements were complied with by a planning based on a nail wall concept. The most significant element here was a system anchoring made from immediately effective expanding anchors (Swellex anchors) with a minimum load bearing capacity of 200 kN. Because of the great depth of the excavation pit, for the first time, lengths for this type of anchor of up to 12 m were used (Fig.7). The anchor pattern designed was between a 1m x 1m division in the deep excavation pit sectors and a 1.5 m x 1.5 m division in the shallow sections. In addition to this anchoring, a reinforced shotcrete shell was made for surface protection and to prevent loosening immediately after excavation.

Fig. 7 Installation of 12 m SWELLEX Anchor
Einbau eines 12 m SWELLEX Anker

In areas with insufficient bore hole stability, as for example in the area at the top edge of the rock, the Swellex expanding anchors were replaced by self boring anchors of the same length and with a load bearing capacity of 300 kN. In the case of rock slope heights exceeding 13 m, additional 1 to 2 rows of prestressed anchors were installed.

These anchors were designed for 1000 kN and tensioned to 600 kN. These prestressed anchors were adjustable and relievable. With their load bearing capacity reserves they provided additional safety. The constant checks performed on the anchor forces were a valuable contribution to the measuring programme for construction.

The originally planned excavation by blasting, involving a depth of round of max. 10 m and a depth of 3 m was changed over to bit chiselling after the first unsatisfactory experiences. The reasons for this were the too great shocks and the unsatisfactory performance resulting from extensive safety measures that had to be undertaken. Another reason was the continuous interruptions in the production of the support.

To relieve the ground water pressure, attention was paid to maintaining the drainage possibility over the longitudinal folds of the installed Swellex anchors. In event of greater accumulations of ground water, additional drainages were sunk. The setting up of a nominal ground water pressure directly behind the shotcrete shell and the adjacent rock zones could be prevented, in line with the analysis assumptions. The planned anchor pattern was compacted during execution upon encountering particularly disturbed rock formations.

The prerequisite behind the implementation of such a concept is the constant supervision of the excavation surfaces and the state of the rock with verification of the assumed rock parameters and a measuring programme with continuous interpretation of the deformations and forces measured.

4 ANALYSIS OF DEFORMATION AND ANCHOR FORCE WITH THE FINITE ELEMENT METHOD

In order to predict the deformations occuring during the individual excavation phases as well as the supervision on the rock characteristic values assumed for the rock areas, analysiss were performed in line with the finite element method in the deepest excavation pit cross section (3). The simulation of the anchor was effected for the first time using new anchor elements which permit the determination of the mode of action of every individual anchor while considering the changes in the anchor forces as brought about by deformations (4.5).

The load history is simulated in a total of 15 load cases (Table 1, Fig.8).

Table 1. Load cases for Finite Element Analysis
Lastfälle für die Finite Element Berechnung

LOAD CASE	DESCRIPTION
0	Primary stresses
1	Increasing the bank wall
2	Producing the bored piles
3	1st excavation section, prestressing the 1st anchor row
4	2nd excavation section, prestressing the 2nd and 3rd anchor row
5	3rd excavation section, prestressing the 4th anchor row
6	4th excavation section, prestressing the 5th anchor row
7	5th excavation section, placing the self boring anchors
8	6th excavation section, placing the Swellex anchor
9	7th exc. sec., prestr. the 6th anchor row, placing Swellex anchors
10	8th exc. sec., prestr. the 7th anchor row, placing Swellex anchors
11	9th exc. sec., prestr. the 8th anchor row, placing Swellex anchors
12	10th excavation section, placing the Swellex anchors
13	11th excavation section, placing the Swellex anchors
14	12th excavation section, placing the Swellex anchors

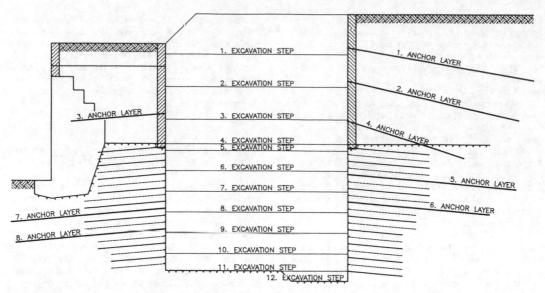

Fig.8 Excavation steps considered in FE Analysis
Aushubreihenfolge in der FE Berechnung berücksichtigt

5 BACK-ANALYSIS

With the results of the rock deformations measured during excavation, a back-analysis to the actual rock characteristic values was performed upon reaching half the excavation depth. With these characteristic values, a new analysis was then performed to get the total deformations to be expected.

The back-analysis was effected by iterative finite element analysis. In the new analysis the anchoring was simulated analagous to the first analysis by individual anchor elements. The load history was simulated again in a total of 15 load cases.

The rock parameters from the back-analysis and used for the new analysis are (in brackets - the values used for the first analysis):

unit weight γ 25kN/m3(25kn/m3)
friction angle φ: 35°(35°)
cohesion c: 90 kN/m2 (90 kN/m2)
elasticity module E:1400 MN/m2 (2000 MN/m2)
Poissons´s ratio ν: 0.2(0.2)

The following characteristic values were determined and used for the superposing layers of soil:

unit weight γ 21kN/m3(21kn/m3)
friction angle φ: 32.5°(32.5°)
cohesion c: 4 kN/m2 (4 kN/m2)
elasticity module E:50 MN/m2 (100 MN/m2)
Poissons´s ratio ν 0.33 (0.33)

On the land side, in the area of the bored pile wall, the deformations determined reached 26.0 mm (first analysis 15.9 mm) and 20.4 mm (first analysis 14.8 mm) at the top edge of the rock. On the water side, the maximum deformations from the analysis in the bored pile area amount to 30.8 mm (first analysis 9.5 mm) and in the rock area 15.4 mm (first analysis 6.0 mm) (Fig.9).

The anchor forces determined in the analyses and shown in Fig.10 for the Swellex anchors are under the load bearing capacity except for the lowest anchor layer on the land side in the area of the pump sump. This local excess can be traced back to the two-dimensional consideration of the pump sump arranged there. Presupposing a pump sump length of approx. 15 m, an additional redistribution of stresses in the longitudinal direction, that was not taken into consideration in the analysis, can be assumed. Three additional rows of self boring anchors, with lengths between 8,0 m and 12 m were installed.

234

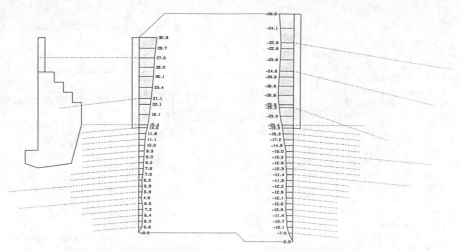

Fig.9 Pit Wall deformation from Finite Elemente-Analysis
Baugrubenwand Verformung nach Finite Element Berechnung

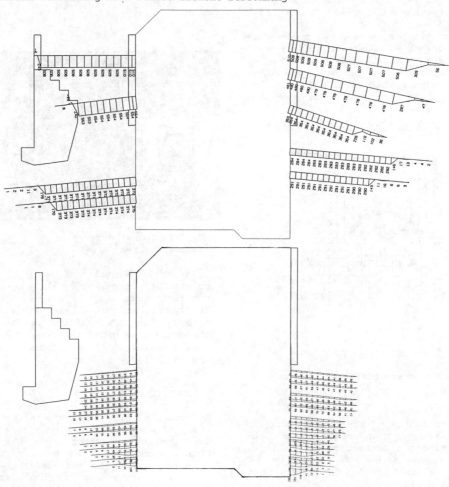

Fig. 10 Anchor force distribution from Finite Elemente Analysis
Auftretende Ankerkräfte laut Finite Element Berechnung

6 TUNNELLING

The 33 m long tunnel construction, to be erected for the channel race, with an excavation cross section of 220 m², lies with its top heading at a height of approx. 4 m in the gravel overburden. The remaining 12 m are again situated in the partly strongly disturbed formations of the Bohemian mass.

The low overburden of approx. 8 m to the foundations of the plant building, situated over the tunnel, called for tunnelling that had as low a settlement as possible. Therefore, the tunnelling concept was based on a three-division top heading tunnelling with a supporting earth core. To provide a save tunnelling procedure a double jet-grouting screen with foot widening in the area of the contact surface at the top edge of the rock was performed.

The jet-grouting screen was produced in three sections with section lengths of 12 to 15 m and 75 to 81 individual piles per double screen. For further support, a shotcrete lining and steel rips were installed. These rips were fitted approx. 1 m in the rock.

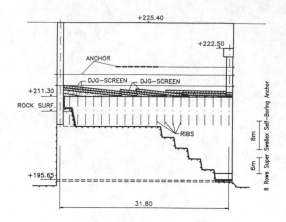

Fig.11b Tunnel excavation sequence
Longitudinal section
Tunnel Ausbruchsreihenfolge -Längsschnitt

Fig. 12 Tunnel excavation
Tunnelarbeiten

The maximum settlement, originally stipulated by the client of the construction contract, amounting to a total of 10 mm at the surface, was almost achieved with this concept. The maximum settlement at the surface amounted to 17 mm with a tangent gradient in the settlement trough of approx. 1:1000.

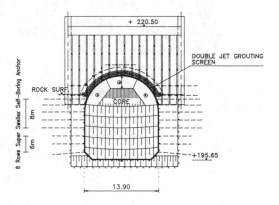

Fig.11a Tunnel excavation sequence
Cross section
Tunnel Ausbruchsreihenfolge - Querschnitt

The protection of the benches and the invert was effected, analagous to the construction pit protection in the rock, using system anchoring consisting of Swellex or self boring anchors and a shotcrete shell.

7 SUMMARY

By using a rock support concept consisting of a system anchoring with 6 - 12 m long Swellex expansion anchors and self boring anchors with a shotcrete lining, it was possible to successfully execute the up to 40 m deep excavation pit for the project "7th Machine" for the Ybbs-Persenbeug Danube power station, even under a very great pressure of time and difficult geological conditions. As a result, it was possible to greatly reduce prestressed anchors for this temporary construction phase. The high degree of flexibility, the installation time for a 12 m expansion anchor amounted under favourable conditions including boring to partly only 15 min., permitted quick reaction in the cace of deteriorating rock conditions and quickly increasing deformations.

After initial blasting attempts, rock excavation was effected by bit chiselling. As a result shocks were minimized. Inspite of difficult geological conditions and also, to some extent, pronounced water inflow, some 45.000m³ of rock were excavated in three months and 4.800m² of excavation pit wall was supported. For the support, a total of 800 m³ of shotcrete, 24.000 running metres of Swellex anchors, 8.000 running metres of self boring anchors and 2000 running metres of prestressed anchors were installed.

The tunnelling of the top heading for the 220 m² tunnel under the hydropower plant building was effected in the gravel-overburden under the protection of a double jet-grouting screen, supported by shotcrete and steel rips. Tunnelling and the support in the bench sections and in the invert were performed analagous to the main excavation pit, using bit chisels in combination with a shotcrete lining and system anchoring.

LITERATURE

Brandl, H. 1992. Konstruktive Hangsicherungen, Grundbautaschenbuch, 4th Edition Part 3. Ernst&Sohn, Berlin.

Hoek E., Bray J.W. 1981. Rock Slope Engineering. The Institution of Mining and Metallurgy, London.

Swoboda, 1989. Programmsystem FINAL. Finite Elemente Analyse linearer und nicht linearer Strukturen, Version 6.2. Druck Universität Innsbruck.

Swoboda G., Marence M., 1991. FEM modelling of rock bolds,Proc. of 7th Int. Conf. Computer Methods and Advances in Geomechanics, p. 1515-1520, Cairns.

Swoboda, Marence M., Strohhäusl S., Erten H. 1993. Design of rock bolts with numerical models. Proc. Seminar on import of computational mechanics on engineering problems, p. 147-155, Sydney.

Anchors in Theory and Practice, Widmann (ed.) © 1995 Balkema, Rotterdam. ISBN 90 5410 577 1

Two examples of transmission of anchor loads of hydraulic structures without anchor heads

Zwei Beispiele aus dem Wasserbau für die Anwendung von Vorspannankern ohne Ankerköpfe

V. N. Zhukov
'Hydroproject' Institute, Moscow, Russia

S. V. Ternavski, Y. O. Zal'tzman & A. A. Lyoubomirov
'GEOTECHNICA' Co. Ltd, Moscow, Russia

ABSTRACT: Such method of load transmission was required by specific features of anchored structures. At the same time the method produced an economic effect. All anchors were cable, of a bonded type. In both examples permanent achors were provided with temporary anchor heads. The heads were removed after tensioning of tendons, grouting of their free lengths and strength gaining. Transmission of load from the prestressed tendon to the anchored structure was effected through the 2nd bond length lying on the free length. The cables consisted of 7-wire strands. The breaking load of each cable was 1 600 kN and the working load was 800 - 900 kN. More than 1 800 pieces of such anchors were installed. The service life of the anchors was designed for more than 50 years.

ZUSAMMENFASSUNG: Üblicherweise werden die Kräfte von Vorspannankern während der gesamten Nutzungsdauer der Anker über die Ankerköpfe in das Bauwerk übertragen.
In zwei besonderen Fällen mußte von dieser Bauweise abgegangen werden. Die Vorspannung erfolgte über provisorische bzw. mehrmals benutzbare Ankerköpfe. Nach Injektion einer zweiten Verankerungsstrecke im Bereich des Ankerkopfes und ausreichender Festigkeit des Injektionsgutes wurden die temporären Ankerköpfe entfernt. In beiden Fällen wurden vertikale, blockierte Litzenbündelanker von 20 m bis 21 m verwendet. Die Vorspannkraft betrug 105 - 115 Mp.
Das erste Mal wurde diese Bauweise von Zhukov im Jahre 1989 beim Sajano-Schuschenskaja-Wasserkraftwerk angewendet. Dort wurde die Standsicherheit der Bodenplatte vom Tosbecken nachträglich mit 300 Ankern erhöht. Es bestand die Gefahr, daß durch Schwindrisse die Ankerköpfe korrodieren. Außerdem konnten durch die Verwendung von temporären Ankerköpfen Schalungsarbeit und Bauzeit eingespart werden.
Das zweite Mal wurde die Methode bei der Reparatur einer Schleuse vom Moskau-Kanal angewandt. Nach 60 Betriebsjahren mußten die Schleusenmauern infolge Korrosion der Bewehrung verstärkt werden. Aus Platzgründen war die Ausführung eines permanenten Ankerkopfes auf der Schleusenmauer nicht möglich.

1 GENERAL DESCRIPTION OF "ZHUKOV" METHOD

Normally the load of prestressed anchors is transmitted to the anchored structures through anchor heads left to stand for the whole period of operation of anchors (Fig. 1 - a). We had to violate this tradition with installation of a great number of permanent anchors (1300 pieces). These were sufficiently large anchors charcterized by the following features: length - 20 m; tendon - 7 seven-wire strands; breaking load of tendon - 1600.8 kN; calculated working load - 784-882 kN; type of anchor - bonded; service life - permanent, more than 50 years. The principle feature of these anchors consisted in the fact that temporary anchor heads which were removed from the anchor within 2-3 weeks after prestressing and grouting of the free length were installed on permanent anchors. The anchoring by such method was called by us as "Zhukov". We had to do this because of specific features of anchored structures which created problems associated with transmission of anchor load to them. The problems came into existence because these structures had been already in operation and we had only to adapt to them but not to dictate our will. In this case we were deprived of advantages that are available with builders. They may provide beforehand for the suitable traditional method of load transmission in the form of anchor heads of any type. At two our projects for various reasons we had one common problem - impossibility to make anchors with heads left for the whole period of operation. Thus at the Sayano-Shushensk HPS it was impossible to accommodate heads on the surface of massive slabs of the stilling basin (Fig. 1 - a) and on the lock walls of the Moscow Canal there was no room for installation of jacks on the heads. The problem could be solved by a "surgical" method, i.e. to destroy everything that impending installation of permanent heads and jacks and then to restore everything that was destroyed. For instance, it was possible to make recesses in a concrete surface at the Sayano-Shushensk HPS (Fig. 1 - b) and at locks of the Moscow Canal to remove a massive parapet. However the cost of these activities would be very high: in the first case a functional reliability of the spillway will be reduced and in the second case exorbitant expenses and a threat to navigation will take place. This price will be unacceptable.

It has been possible to solve this problem at both structures by a simpler method rejecting permanent anchor heads and changing over to their temporary use during a short interval of time when the cement mix was setting at the free length section. These temporary

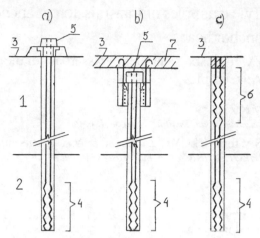

Figure 1. Methods of transmission of loads from cable anchor to structure: a - conventional through anchor head on the surface; b - through anchor head into recess (Tarbela dam); c - without anchor head through 2nd (upper) bond length; 1 - anchored structure; 2 - zone of location of bond length; 3 - suface of structure; 4 - 1st (lower) bond length; 5 - anchor head; 6 - 2nd (upper) bond length; 7 - fibre reinforced concrete; 8 - plastoconcrete

Fig. 1. Methoden der Lastübertragung vom Anker auf das Bauwerk: a - konventionell mit Ankerkopf auf der Oberfläche; b - mit versenktem Ankerkopf (Tarbela-Sperre); c - ohne Ankerkopf, mit zweiter (oberer) Haftstrecke; 1 - zu ankernder Bauteil; 2 - Zone für die Verankerung; 3 - Oberfläche des Bauteiles; 4 - erste (untere) Haftstrecke; 5 - Ankerkopf; 6 - Zweite (obere) Haftstrecke; 7 - Glasfaserverstärkter Beton; 8 - plastischer Beton

heads were installed on temporary steel stops and when their necessity disappeared they were removed and reused. The removal was realized only after formation of 2nd bond length on the free length. The end of the tendon protruded from the bore hole was removed as well. The prestressing load was transmitted to anchored structures through the 2nd bond length. This is the main idea of our method of anchoring. The completed anchor is shown in Fig. 1 - c. It appeared possible to realize this method because the body of anchored structures was so lengthy (massive) that the 2nd bond length was housed entirely in it. This bond length was grouted simultaneously with grouting of the free length. Its strength was sufficient to take the prestressing load. Conducted experiments and observations confirmed this. The removed heads

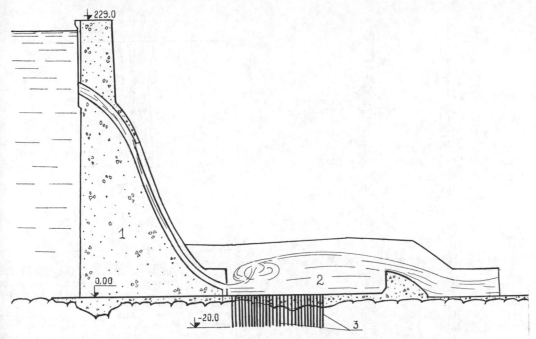

Figure 2. Anchoring of slabs of stilling basin bottom at Sayano-Shushensk HPS on the Yenisei river by prestressed anchors: 1 - Dam; 2 - Stilling basin; 3 - Cable anchors.
Fig. 2. Verankerung der Bodenplatte für das Tosbecken bei der Sayano-Shushensk-Wasserkraftanlage am Yenisei-Fluß durch vorgespannte Anker: 1 - Sperre; 2 - Tosbecken; 3 - Anker.

were reused not less than five times and frequently more than that. Besides embedded parts in the concrete for the heads at bore hole mouths were not necessary.

In the long run the problem was solved together with a produced economic effect.

2 ANCHORING AT THE SAYANO-SHUSHENSK HPS

Here about 350 anchors (Fig. 2) were installed to fix massive blocks of the stilling basin bottom to the rock foundation of about 10 m in thickness. The necessity of anchoring was generated by attempts to increase the stability of slabs experiencing very high loads when passing the flood of the Yenisei river through spillways of the concrete dam. The stilling basin was designed for dissipation of water energy at discharge 13600 m³/s with a drop of 220 m. Specific energy of the discharged flow reaches 200 MW/m and is one of the largest in modern hydraulic practice. Velocity of the flow at the entrance into the stilling basin is about 60 m/s. Such a high energy produces high static and dynamic loads and oscillating pressure in interblock joints even at very small openings of about 0.2 mm. High velocity of the water flow caused abrasion and cavitation damage of the concrete surface. This was the specific feature of the place of the structure where anchoring is required.

The principle problem was how to transmit the load from anchors to the slabs of the stilling basin and at the same time to avoid their damage. Of course, first the traditional method of transmission of loads through anchor heads was considered. For this purpose the variant of arrangement of heads in rectangular recesses 60x60 cm in size and up to 70 cm in depth was considered. Their walls should be of a reverse slope. After completion of anchoring these recesses should have been concreted in flush with the surface. The similar engineering solution

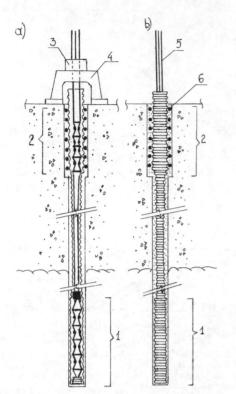

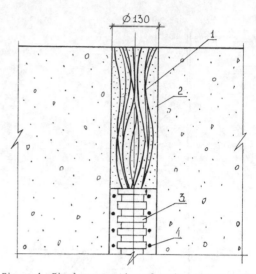

Figure 4. Final preparation of anchor top on surface of stilling basin bottom: 1 - Wires of unlaid strands 2 - Plastoconcrete; 3 - Tandon in polyethylene corrugated tube; 4 - Spiral reinforcing.
Fig. 4. Ausbildung der Verankerung in der Bodenplatte des Tosbeckens: 1 - Einzeldrähte; 2 - Plastischer Beton; 3 - Geripptes Hüllrohr aus Polyäthylen; 4 - Spiralbewehrung.

Figure 3. Main stage of anchoring: a - Tensioning on temporary stops with temporary head; b - View of anchor after removal of temporary head and temporary stop; 2 - Upper bond length; 3 - Temporary tension head; 4 - Temporary head; 5 - Tendon; 6 - Spiral reinforcing.
Fig. 3. Vorgehen bei der Ankerung: a - Vorspannen mit temporärem Ankerkopf; b - Anker nach Entfernung der temporären Einrichtung; 1 - Untere Haftstrecke; 2 - Obere Haftstrecke; 3 - Temporärer Ankerkopf; 4 - Temporäre Vorrichtung zur Vorspannung; 5 - Vorspannkabel; 6 - Spiralbewehrung.

was already employed by the VSL company at the spillway of the Tarbela dam in Pakistan in 1977.

However in our case such solution seemed to be unreliable and inconvenient in use. The reasons for this idea were the following. The concrete filling such large recesses could have been destroyed by oscillating pressure in contraction cracks between new and old concrete. Quality of cleaning of recess bottoms would have not always been appropriate because they were always filled with water. It was impossible to remove water from the surface of the stilling basin bottom. The recesses would have been so close to

each other (upto 1.5 m) that it would have been very difficult to perform cleaning operations on such surfaces.

And finally it would have been difficult to provide the required quality of all anchors with abundance of manual operations in half-submerged and repeatedly drying pits.

Consequently the second variant was used by the "Zhukov" method. It allowed the area of recesses on the surface to be reduced by several times. Thus, if in variant with recesses 60x60 cm in size the area of each recess was 3600 m², then with the "Zhukov" method the size of the recess was decreased down to 13 cm in diametre with area 133 cm², i.e. by 27 times. After applicable in-situ tests this method was adopted. Sequence of operations according to this method and preparation of the anchor top after completion of work are shown in Fig. 3 and 4.

After all protruding parts of the anchor were removed, the mouth of the bore hole was cleaned down to 30 cm mechanically, was dried, primed and all wires of the strand were unlaid and this section was

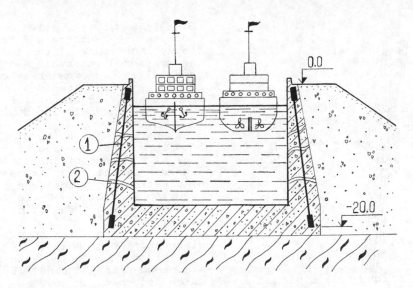

Figure 5. Repair of old reinforced concrete lock walls of the Moscow Canal
1 - Cracks in walls; 2 - Cable prestressed anchor; 3 - Lower bond length; 4 - Upper bond length.
Fig. 5. Reparatur der Schleusenmauern beim Moskau-Kanal: 1 - Risse; 2 - Vorspannanker; 3 - Untere Haftstrecke; 4 - Obere Haftstrecke.

filled with plastoconcrete in flush with the surface. In a small hole a sound filling was produced which was adequately bonded with the tendon. It was the firm belief that such filling will be not broken away by the water flow. The 1992 flood and floods of the following years did not disclose any damages of the surface of the stilling basin bottom.

Check of strength and upper bond lengths was carried out immediately on the bottom of the stilling basin.

The lower bond length was located in hard crystalline paraschists and orthoschists. The check was performed by the conventional FIP procedure and during 10 months with the use of compression dynamometer which was installed under the temporary head. The results demonstrated the absence of creeping. In September 1990 it was 1031 kN and in July 1991 it was less by 2% which in our opinion points to stabilization.

The upper bond length was tested and checked at two stages. At the 1st stage before the beginning of work 9 bond lengths of the tendon of three types in concrete slabs of the stilling basin bottom were tested. By the results of strength tests of these bond lengths all design and

process parameters of the upper bond length were specified. At the 2nd stage the behaviour of the upper bond length was checked by selective measurements of tendon movements in downward direction at disconnection of temporary heads from tendons. If these movements of tendons in downward direction at the level of the bore hole mouth were more than 5 mm, then frequently the decision was made to install a new anchor.

3 ACNHORING OF LOCK WALLS OF THE MOSCOW CANAL

Old reinforced concrete structures in the world appear in increasing numbers. Many structures require repairing. Prestressed anchoring is one of the suitable engineering means of such repair. From 1989 we have been using this method for repair of old reinforced concrete lock walls of the Moscow Canal. The Canal was constructed in the early thirties. For half a century of operation many cracks appeared in the walls, a part of reinforcement was damaged by corrosion. It was necessary to restore this portion of reinforcement and to raise up shear resistance on the surface of cracks.

243

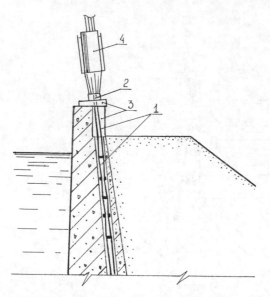

Figure 6. Tensioning of anchor on lock walls: 1 - Tendon; 2 - Temporary head; 3 - Temporary Stop; 4-Jack.
Fig. 6. Vorspannen der Anker auf der Schleusenmauer: 1 - Vorspannkabeln; 2 - Temporärer Ankerkopf; 3 - Temporäre Fixierung; 4 - Vorspannpresse

It was accepted that the most suitable and simple method of solution of this problem is installation of cable prestressed anchors (Fig. 5).

Practically this engineering problem does not differ from anchoring at the Sayano-Shushensk HPS and from the known anchoring of rock masses. In the given case anchors should piece through cracked concrete and should be fixed in a stable zone which was a foundation slab here. Its thickness of about 5 m was sufficient for formation of the lower bond length.

Anchor heads were installed on welded special-form stops which transmits the load mainly to the top of the walls and partly to the massive parapet (Fig. 6). The second bond length was formed in the upper portion of bore holes.

Strength and creeping of the lower bond length were checked in the process of tensioning (by the FIP method) and after tensioning - by instruments. Duration of observations of forces of tensioned but ungrouted tendons was 10, 2, 1 and 0.5 months. Forces were determined by compression dynamometers, IPS-160 of

"DIGES" Company installed on 10 anchors between anchors and temporary stops. A decrease of the force below the mentioned intervals did not exceed 1-2%. In the majority of cases the force left in tendons by the moment of grouting of the free length was within the calculated working load.

Strength and creeping of the upper bond length was checked by measurement of movement of the tendon in downward direction at transmission of prestressing loads from temporary heads to the tendon. There were the following three types of observations:

1. Short-term - in the process of disconnection of heads;
2. Mid-term - during 9 days;
3. Long-term - during 2.5 years which are going on now.

Short-term observations were carried out from the immovable base with the use of indicators with a value of division 0.01 mm mounted on 7 tendons and with the use of a conventional measuring rule. Farther only measuring rules were used for measuring at all anchors. In most cases (90%) the tendon moved into the bore hole for not more than 5 mm and in 5% of cases - for 7-8 mm; and only in rare cases - for 9-12 mm. These values were comparable with the values of drawing of strands into concrete at making the prestressed structures at the plant-manufacturer according to the process of tensioning "on stops". There this value at releasing the tensioning from the stops to the concrete is about 2.5 mm. We thought that the excess of the plant value by 2-3 times is not a bad result for the conditions of the construction site. And consequently we took a value of 7 mm as a reference one for checking the quality of the upper bond length at all anchors.

Mid-term observations were also carried out from the immovable base and by the similar indicators. At three anchors the end of the tendon for the first day moved for 0.02 mm and for the rest of 8 days it was immovable. At one anchor the end of the tendon did not move at all. At the latter anchor the end moved for the first day for 1.1 mm and for the rest 8 days it did not move at all (Fig. 7).

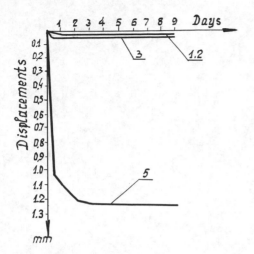

Figure 7. Movement of tendon downward after removal of temporary heads during mid-term observations: 1,2,3,4 - number of observed anchors.
Fig. 7. Bewegung des oberen Endes der Spannglieder nach Entfernung der temporären Vorspanneinrichtungen mittelfristige Beobachtungsergebnisse. 1,2,3,4 - Beobachtete Anker

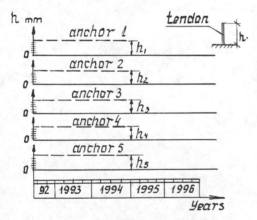

Figure 8. Height of tendon above mouth of bore hole after removal of head.
Fig. 8. Länge des aus dem Beton herausragenden Teiles des Spanngliedes bei 5 Ankern, langfristige Beobachtungsergebnisse.

Long-term observations are shown in Fig. 8. As may be seen from the graph that there are not downward movements of five tendons for 2.5 years. All this points to a reliable operation of the second bond lenght.

For 5 years of anchoring about 1000 anchors were isntalled at locks. In this case 1200 walls were consolidated. Anchors were installed spacing 1-2 m apart.

By the results of observations of horizontal deformations it was observed that seasonal deformations of the top of the walls across the axis of the lock after consolidation by anchors were decreased by 2-3 times.

It provides reason to state the completion of successful anchoring of locks generally and the use of the "Zhukov" advantageous method in particular.

4 CONCLUSIONS

The "Zhukov" method allows the prestressed anchoring of massive structures to be carried out with transmission of anchor loads on them not through anchor heads but through the second bond head.

This method may appear indispensable if the following is required:

1. To minimize the damage of the concrete surface and adjacent structures;

2. To eliminate completely the protruding parts on the surface of the structure;

3. To decrease the cost of anchoring through the reduction of anchor heads by several times (through their reuse) and eliminate completely the embedded parts for them.

This method is the most advisable where:

1. Anchoring is to be carried out from the surface on which the water will be flowing with high velocity during operation, e.g. on the bottom of the stilling basin, spillways, chutes;

2. At very small spaces where it is impossible to install a jack on the anchor head right on the mouth of the bore hole;

3. Where higher architectural requiements for invariability of the view of the surface of the structure exist.

This method may be used for anchoring to the foundation not only structures in operation but structures under construction as well. It may be also useful for repair of old concrete and stone structures with cracks.

This method does not exclude the possibility of connection of tendons with reinforced concrete beams

(walls, buttresses, etc.) on the surface of rock masses and structures.
For this purpose after removal of
the head the end of the tendon protruding from the bore hole and having the length which is required for
connection may be left.

A separate conclusion pertaining
to conventional anchors. The capacity of bond prestressed anchors to
transmit the load to the massive anchored structure only through the
2nd bond length even without anchor
head demonstrates that conventional
anchors with anchor head if they
are bonded are characterized by reliability higher than that of unbonded ones. If after many tens of years
corrosion (invisible from outside)
damages the bond between the tendon
and the head, then the anchor will
still transmit the load through the
2nd bond length and no failure will
take place. It goes without saying
that for this purpose the bond
length should be made properly.

REFERENCES

Use of rock anchors at Tarbela. Wa-
    ter Power & Dam Construction, Feb-
    ruary, 1978: 44-47.

Some aspects of current ground anchor design and construction in the United Kingdom

Einige Gesichtspunkte zum derzeitigen Stand bei Entwurf und Ausführung von Ankern in Großbritannien

Michael J.Turner
Applied Geotechnical Engineering, UK

ABSTRACT: The design and construction requirements for ground anchors in the UK are generally required to comply with the British Standards Code of Practice for Ground Ancorages, BS8081, published in 1989. Aspects and requirements of the British Code are reviewed particularly those dealing with design responsibility, design parameters and corrosion protection. Current UK practice is reviewed in comparison with BS8081 particular reference to major ground anchoring contracts in Wales.

ZUSAMMENFASSUNG: Entwurf und Ausführung von Ankern im Untergrund müssen in Großbritannien im allgemeinen der Britischen Norm "Code of Practice for Ground Anchorages BS8o81" entsprechen, die 1989 veröffentlicht wurde.
Ein wesentlicher Gesichtspunkt dieser Norm ist der Versuch, die verschiedenen Zuordnungen der Verantwortlichkeit sowohl für den Entwurf als auch für die Ausführung von Ankerungen im Untergrund zu klären und zu definieren. Weiters wird gefordert, daß insbesondere bei Freispiel-Ankern aus Stahl auf den Korrosionsschutz besonders geachtet wird. Es werden drei Stufen beschrieben: kein Schutz, einfacher oder doppelter Korrosionsschutz. Daher müssen Daueranker einen einfachen oder doppelten Korrosionsschutz erhalten.
Der vorliegende Beitrag beschreibt den gegenwärtigen Stand der Praxis in Großbritannien in Übereinstimmung mit den britischen Normen an Hand von zwei größeren Projekten mit Ankern: das eine betrifft die Stützwände vor den Portalen des Pen-Y-Clip Tunnels im Zuge der Autobahn A 55 in Nord Wales, wo etwa 1600 m Daueranker bei bis zu 30 m hohen Stützmauern eingebaut wurden; das andere Projekt betrifft die Sicherung einer Straße in Süd-Wales, die durch eine Hanggleitung gefährdet ist, wo sowohl Anker mit hoher Tragfähigkeit als auch Boden-Nägel mit sehr niedriger Tragfähigkeit einen Teil des Stabilisierungsprojektes bilden.
Ein wesentlicher Teil dieser Arbeiten waren die Last- und Verformungs-Messungen.
Einige weitere Themen der BS8081 werden in Hinblick auf ihre Auswirkungen auf die gegenwärtige britische Praxis behandelt. Insbesondere wird der Einfluß auf die Kosten von doppelt korrosionsgeschützten Litzen-Ankern erörtert.

1.0 INTRODUCTION

The requirements for the design, installation and testing of ground anchors in the UK are generally extremely stringent and governed by the British Standards Code of Practice for Ground Anchorages, BS8081, published in 1989.

One feature of BS8081 is that it attempts to clarify and define the various issues of responsibility for both the design and construction of ground anchorages. The Code makes clear that it is most important that this aspect of the work is addressed within the philosophy of the contract and in its specification.

BS8081 also pays careful attention to the requirements for the corrosion protection of the steel tendon. It identifies two categories of anchorages: either temporary or permanent. Temporary anchorages are regarded as having a design life of no more than two years and are typically required to function only for the duration of some construction activity. Permanent anchorages are generally required to provide support to the structure for its full design life, but the category is also deemed to include "temporary" anchorages if they have a service life beyond two years.

In addition the Code identifies three classes of protection: no protection (where no specific measures are taken to ensure the corrosion protection of the tendon), single protection and double protection. Typically, permanent anchorages must be either single or double corrosion protected. Temporary anchors often have no protection, but depending upon circumstances and risk can also be single or double protected.

Great importance is also placed upon both preliminary load testing and routine testing of working anchors.

2.0 DESIGN ASPECTS OF BS8081.

2.1 Responsibilities.

The British Code of Practice identifies various areas of responsibility that have to be addressed within the contract . This emphasises that the "designer" of the anchors might be employed by the client, the main contractor, a consulting engineer retained by the client, or a specialist contractor, and that some aspects of ground anchor design, particularly the individual anchor loads, spacings and orientation and corresponding free anchor lengths might, depending upon the particular contract, be the responsibility of an engineering team working for either the Client or the contractor. Thus BS8081 attempts to make clear the various issues that have to be clarified within a particular contract so that individual responsibilities are known, defined and understood.

2.2 Corrosion protection.

Table 1 summarises the categories and classes of protection defined by BS8081. It is proposed to centre discussion here upon the protection classes for permanent anchors: i.e single and double protection.

TABLE 1: Categories and classes of protection (from Table 11 of BS8081).
TABELLE 1: Schutzkategorien und -klassen (gemäß Tabelle 11 von BS8081).

PROPOSED CLASSES OF PROTECTION FOR GROUND ANCHORAGES

Anchorage category	Class of protection
Temporary	With no protection
	With single protection
	With double protection
Permanent	With single protection
	With double protection

The British Code defines single protection as implying that one physical barrier against corrosion is provided to the steel tendon prior to installation. Double protection on the other hand implies the provision of two barriers, where the purpose of the outer, second, barrier is to protect the inner barrier against the possibility of damage during tendon handling and placement. The second barrier therefore provides additional insurance, and the Code emphasises the difference between the degree of protection of the tendon once installed in the ground, and that of the tendon as supplied. The implication of this distinction is that it would appear to be permissible for the second, outer, barrier to be damaged during the installation stage, provided the inner, primary, barrier remains intact. Quite how much damage is acceptable has not, so far as the author is aware, been formally addressed on any specific contract.

The Code also draws attention to the fact that the least protected zone of an anchorage defines the class of protection provided. Thus attention to the anchor head and to joints and boundaries within the system are important, if not paramount, to the designer of the anchor tendon.

BS8081 also pays attention to, and attempts to codify, the nomenclature of ground anchorages and techniques. In this regard, for example, Figure 1 illustrates the nomenclature and major features of a typical corrosion protected ground anchor. A typical single or double protected anchor tendon will consist of an encapsulated tendon bond length (which transfers the anchorage force into the fixed anchor length), a free tendon length, and an anchor head. It will be noted that the fixed anchor length is not necessarily the same as either the tendon bond or encapsulation bond length. Correspondingly, the free anchor length is not necessarily the same as the free tendon length. This differentiation is intentional within the Code and allows flexibility of design by the designer.

Permanent strand anchors. In the case of ground anchorage tendons fabricated from prestressing strands, the double protection requirement is usually provided by two concentric plastic sheaths.

Figure 2 illustrates a typical double protected strand anchor complying with BS8081 (after Turner, Selley and Cooper, 1992). The reason for this design approach is that the cement or resin grout used to fill the inner sheath and thus encapsulate the steel strands is not considered to be capable of transferring the anchorage forces without uncontrolled cracking.

Strand anchorages probably comprise 90% of the UK market.

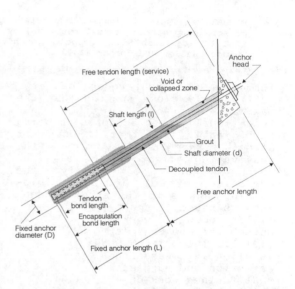

FIGURE 1: Nomenclature for corrosion protected ground anchorage (from BS8081/1989
BILD 1: Bezeichnunen für korrosions-geschützte Untergrundanker (gemäß BS8081/1989)

Permanent bar anchors. Where permanent anchors are fabricated from threaded or deformed bar, such as the Dywidag or Macalloy prestressing quality steels, the encapsulation grout is accepted as forming part of the double protection system. Thus only a single corrugated plastic sleeve is required to complete the double protection, as illustrated in Figure 3, overleaf.

The reasoning behind this is that the ribbed configuration of the steel bars is deemed to act as a crack inducer to the encapsulation grout. This causes cracks to be

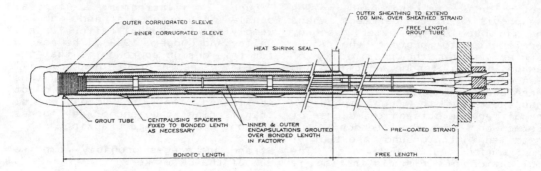

FIGURE 2: Typical double protected ground anchorage constructed from prestressing strand (from Turner, Selley and Cooper (1992).
BILD 2: Regelausbildung doppelt geschützter vorgespannter Litzenanker (von Turner, Selley und Cooper (1992).

formed in the grout under load at a sufficiently close spacing that the maximum crack width has traditionally been accepted as being sufficiently fine that it does not affect the ability of the passivating effect of the alkaline cement grout to nullify the corrosive effect of any fluids gaining access to the steel tendon. This is based, in particular, upon the work of Ostermeyer and Scheele (1978). The British Code is slightly towards this assertion and suggests that further work on this subject would be beneficial. Barley (1994) has made similar points in this regard.

Bar tendons are widely thought of as being unwieldy and difficult to manage. Joint designs are available, however, that allow the continuity of corrosion protection to be completed in the same operation as the joint is made. The anchors can then be installed in a similar manner to that of running a drill string down the borehole. Bar tendons account for perhaps 10% of ground anchors in the UK. They are also widely used for soil nailing.

2.3 Design of the anchor to ground bond.

The load transfer between the anchor tendon and the soil or rock within which it is embedded depends upon the design of three boundaries.
* Anchorage grout to ground
* Encapsulation to anchorage grout
* Tendon to encapsulation grout.

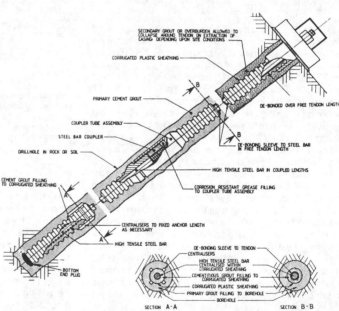

FIGURE 3: Typical double protected ground anchorage constructed from prestressing bar.
BILD 3: Regelausbildung doppelt geschützter Untergrundanker bei vorgespannten Stäben.

Bond between anchorage grout and the ground.

The relationship between the design anchorage force, Tw, and the required load transfer length, L, is usually determined by a skin friction equation of the form

$$Tw = \frac{\pi \; d \; L \; S_{ult}}{F}$$

where d is the diameter of the borehole

S_{ult} is the ultimate bond or skin friction at the grout/ground interface, and

F is the required global factor of safety

Values for S_{ult} are derived from a wide variety of sources: such as pile-related bearing capacity formulae, supplemented by documented pull-out tests on specially constructed test anchors and the like.

The load carrying capacity of anchors in soils and weak rocks can also be enhanced by multi under-reaming or by post-grouting techniques, where additonal grout is injected into the fixed anchor length at high pressure after the initial anchorage grout has set.

In addition, some success has been reported on single bore multiple anchor tendons - a technique whereby a number of essentially single strand anchors are placed within the same borehole. Each strand anchor is designed to transfer load over a separate section of the anchor hole (e.g. Barley, Eve and Twine, 1992)

Encapsulation to anchorage grout bond.

The design of the corrugated plastic sheath forming this interface is typically required to be such that a mechanical interlock will be formed between the external and internal grout on either side of the boundary. The British Code recommends a maximum ultimate bond stress of 3.0 N/mm² at this interface where cement grouts are used, assuming that the bond is uniform over the surface. Higher values can be adopted if adequately proven. Bond tests confirm this is a reasonable but conservative low bound, and in most cases this is not a critical interface in design.

It has to be remembered that, where two concentric sheaths are used, the inner sheath will be the critical interface.

Tendon to encapsulation grout bond.

For temporary anchors, where the steel tendon is embedded directly into the anchorage grout and the tendon bond length is usually the same as the fixed anchor length, the length necessary for the grout/ground bond is usually more than adequate to ensure against failure of the tendon/ground bond. In the case of permanent anchors, the encapsulation length is often shorter than the fixed anchor length (because a shorter encapsulation length makes for easier handling of the tendon at installation). In such a case the tendon/grout interface can become critical.

The British Code suggests that a value for the ultimate bond stress at this interface when using cement grout should be limited to 2.0 N/mm² for clean prestressing strand or deformed bar, rising to 3.0 N/mm² for strands which have been locally noded (where the outer wires of each strand are unravelled and then rewound around small collars placed onto the kingwire at intervals to form small nodes). Tests undertaken by the writer suggest that a value of 2.0 N/mm² for plain strand is reasonable, but that deformed bars can develop in excess of 5.0 N/mm² ultimate bond stress. The validity of the 50% increase in bond value for noded strands to 3.0 N/mm² is more questionable. Our tests indicated that such nodes would have to be at close centres to achieve the stated 50% increase in bond.

3.0 CONSTRUCTION EXPERIENCES WITH BS8081.

3.1 Pen-y-Clip Tunnel.

The Pen-y-Clip tunnel scheme forms

part of the A55 expressway, and takes the trunk road through the rocky headland of Pen-y-Clip, just west of Conwy in North Wales.

The tunnel project included two major anchored panel walls, also known as "Brazil walls" up to 30 metres high, to support steep, unconsolidated scree and quarry-related deposits lying at or close to their angle of repose on top of a steeply inclined bedrock surface. A typical cross-section through one of these walls is shown on Figure 4. Three further anchored walls were required outside the tunnel approaches, and, in all, some 1600 permanent double corrosion protected anchors were required for the anchored retaining walls.

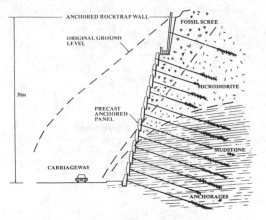

FIGURE 4: Pen-y-Clip: typical section through high panel "Brazil" walls.
BILD 4: Pen-y-Clip: Regelquerschnitt durch die "Brazil"-Stützmauer.

Although the project design predated the official publication of the code, the anchors were designed and constructed in general accordance with the requirements of BS8081. They were installed into a wide range of geotechnical materials, varying from strong to extremely strong microdiorites, through sheared and altered mudstones to glacial till and rock slope screes containing boulders and rock debris from upslope quarrying operations.
The anchoring contract included an extensive permanent monitoring system using hydraulic load cells to monitor the load behaviour of the

ground anchorages in the retaining walls. Movement gauges were also installed. Additionally, a series of research anchorages were installed to monitor the long term performance of such double protected anchors.

The project is more fully described by Turner, Selley and Cooper (1992).

Design. The design of the anchored retaining walls was undertaken by the Engineer retained by the Client. This included the development of the design philosophy, the overall design of the anchored structures, and the definition of the safety factors to be employed. This also included the specification of the corrosion protection requirements of the anchorage system, the anchorage spacing and orientation, minimum free anchor lengths and individual anchorage loads.

The specialist anchoring contractor was then responsible for the design of the individual fixed anchor lengths to safely provide the required anchorage force, together with the design and construction of the double protected tendons which would provide the required force. Thus the Engineer undertook responsibility for the philosophy, design and specification of the global scheme, whilst the Contractor was essentially responsible for the design and performance of each individual anchor. The anchorage loads varied up to a maximum of 600 kN. The design fixed anchor lengths varied between 3.0 m and 12.5 m depending upon loads and anchoring stratum.

A separate, pre-contract test programme was undertaken to investigate the construction and performance of double protected anchorages typical of the type that would be installed in the contract works. This work enabled definitive design values to be obtained for ground anchorages installed within the varied strata, as outlined in Table 2, overleaf.

Installation: The anchor holes were constructed using an overburden drilling system (ODS) with air flush techniques, producing a 220 mm diameter reamed and drilled hole

TABLE 2: Pen-y-Clip: designed bond values for grout to ground interface (after Turner, Selley and Cooper, 1992) Factor of safety on ultimate bond stress (design) was 3.0 to give working bond stress.
TABELLE 2: Pen-y-Clip: Entwurfswerte für die Haftfestigkeit des Injektionsgutes (nach Turner, Selley und Cooper, 1992) Sicherheitsfaktor gegen die Bruchfestigkeit 3,0 (Entwurf).

Ground Type	Borehole diameter (mm)	Ult. Bond Stress, s_{ult} (measured) (kN/m²)	Max. Bond Stress, s_{max} (no failure) (kN/m²)	Ult. Bond Stress, s_{ult} (design) (kN/m²)	Wkg. bond Stress, s_w (design) (kN/m²)	Remarks
Glacial till	160 -220	113*-417		140*	46.7	* Lowest result assumed to be affected by softened material due to length of time hole was maintained open.
Mudstone	115-160	416		400	133	
Microdiorite	160-220	-	371-547	540	180	
Fossil scree	160	477		450	150	

with a 203 mm O.D. temporary drill casing.

The drilling system also proved very successful with regard to alignment of the drillholes, which were allowed a maximum deviation of 1 in 30 from their specified alignment. Such tolerances are more onerous than proposed in BS8081, which suggests that an angle tolerance of $\pm2.5°$ should be allowed in initial setting up, together with an acceptable deviation of 1 in 30. The deviation of the holes from their specified alignments was checked using a Reflex-Fotobor dip and direction indicator and demonstrated that a high degree of accuracy could be achieved in installation.

Three grout mixes were developed for the site: a neat cement grout, a nominal 1.4:1 cement/sand grout and a nominal 0.8:1 cement/sand grout. The latter utilised a super-plasticiser to limit the water demand of the mix. The quality of grout was constantly checked by site tests consisting of flow cone, wet density and bleed measurements for each day's grouting, and cube testing. It is considered that the testing of grout cubes is a useful check on the quality and strength of the grout, while flow cones and wet density measurements are of value for checks on the consistency of the grout. Bleed tests for grout to be used for ground anchors may in practice be of little significance, since a fully corrosion-protected ground anchorage in accordance with BS8081 does not rely on the anchorage grout for any contribution to its corrosion protection.

All anchor tendons were fabricated in factory facilities off-site and delivered to site in batches on purpose-made transportation frames, to minimise handling and damage.

Stressing and testing: Stressing was carried out using multi-strand jacks fitted with load cells and digital read out units. Jacks, pressure gauges and load cells were regularly calibrated in general accordance with the recommendations of BS8081.

The load-time performance of a proportion of anchors on each wall is monitored using the permanent load cells, which are read on a regular basis. In addition, on the high panel walls, overall wall movement and deformation are monitored using targets mounted on the panels. Movements of these targets were monitored from remote survey stations both during the construction contract and after the completion of the works. Load cell measurements on the anchored cantilever wall in fact monitored marked load fluctuations in the anchors as the construction works progressed.

3.2 A4061 Rhigos Road.

The A4061 connects the Rhondda Fawr valley with Hirwaun in the South Wales coalfield area of Mid-Glamorgan. Just north of the town of Treherbert the road is located along the eastern slope of the valley. In this locality the road crosses an ancient landslip area, slow movement of which has affected the road almost from the date of its opening some sixty years ago.

The scheme was designed to stabilise the road where it crosses the landslip area. The remedial work consisted of two parallel rows of anchored concrete strongpoints which provided the main stabilising and restraining force across the slip plane. In addition, soil nails were installed at the head and toe of the anchored strongpoints. The scheme is illustrated on Figure 5.

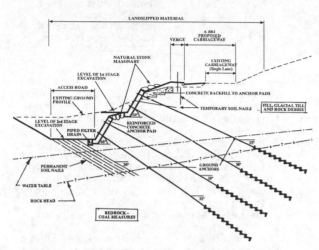

FIGURE 5: Typical cross-section through nailed and anchored slope stabilisation works: A4061 Rhigos Road.
BILD 5: Regelquerschnitt der Vernagelung und Verankerung im Zuge der Böschungs-sicherung bei der A4061 Rhigos Road.

The restraint was formed by a precast concrete panel wall, with up to four rows of rock anchors inclined at an angle of 25° to 35° and founded in bedrock. Some 150 rock anchors were required, each with a design working load of 1100 kN and fabricated from 8 No 15.2 mm drawn strands, in a similar configuration to Figure 2, to form a double protected ground anchor.

Almost 200 temporary nails were installed at the head of the slope to assure the stability of the temporary cut face that was formed to allow the installation of the upslope anchor pads and their associated ground anchorages. A further 500 permanent soil nails were installed to provide soil support below the toe of the lower anchor pad, as illustrated in Figure 5. These were installed at an angle of 30 degrees below the horizontal and were intended to reinforce the

zone of soil and weathered bedrock that could remove support to the anchored wall if it were to continue slipping. The works also included extensive toe and slope drainage work to remove and channel any excess water pressures.

The bedrock consisted of siltstones and sandstones of Coal Measure age, with subordinate mudstones and thin coals, overlain by made ground from the road construction and glacial till deposits. The overburden materials typically consisting of firm to stiff sandy clays with gravel and cobble size fragments of sandstone.

Design: Again, for this contract, the design of the anchored wall and the determination of the anchor loads and inclinations were all undertaken by the Engineer.

The specialist contractor was responsible for the design of the individual anchor, including both fixed and free anchor lengths to sustain the required designated loads, together with the design of the double protected tendons.

To confirm the design parameters, three proving tests were installed and tested at the site prior to installing the contract anchors. Design bond values derived from these tests are outlined on Table 3, overleaf.

Drilling and installation: Drilling and installation techniques were similar to those used at Pen-y-Clip, with nominal hole diameters varying between 160 and 220 mm. Again both neat cement and sand/cement grouts were used for the anchorage grout.

Stressing and testing: Stressing was carried out using 350 tonne multi-strand jacks. Generally the anchors within each panel were stressed simultaneously to limit eccentric loading upon the panels. The major problem in the contract stressing work was that large horizontal movements of the pads occurred under the imposed loads. In excess of 150

TABLE 3: Rhigos Road: Ground anchors: design bond values for grout to ground interface.
TABELLE 3: Rhigos Road, Bodenanker: Haftfestigkeiten für die Wahl des Injektionsmaterials.

Ground Type	TCR %	RQD %	Borehole diameter (mm)	Ult. Bond Stress, s_{ult} (measured) (kN/m^2)	Max. Bond Stress, s_{max} (no failure) (kN/m^2)	Wkg. bond Stress, s_w (design) (kN/m^2)
Grey, highly to slightly weathered MUDSTONE, generally weak and with claybands, but with approx 20% of stronger siltsone.	100	0 -7	220	552	-	183
Grey, thinly to medium bedded, fresh to slightly weathered SILTSTONE, with subordinate mudstone and occ. clayey bands	100	40-50	220	No failure	868	267
Grey, thinly to medium bedded, slightly weathered SILTSTONE, moderately strong to strong.	100	70-80	220	No failure	868	267

mm of movement was not uncommon. Although permanent load cells were not utilised on this contract, all the anchors were designed to allow check loading and load adjustment at a future date if required by the engineer.

Soil nails: The temporary nails were installed within very heterogeneous ground conditions: made ground overlying and intermixed with glacial till.

The permanent nails were installed within glacial till and landslip debris and weathered bedrock.

Nail loads varied from 35 to 175 kN. The tendon size varied from 25 to 40 mm diameter HY rebars.

The design of the permanent nails required that they be double protected in a similar manner BS8081 ground anchors. The protective scheme adopted was a single grout-filled corugated plastic sleeve.

4.0 SUMMARY

The Pen-y-Clip and Rhigos Road projects were undertaken with the recommendations and requirements of BS8081 in mind. The ground anchorage tendons are double protected and comply with all the requirements contained in BS8081 with respect to the corrosion protection of the system.

Some salient features of BS8081, which, in the writer's opinion, have been highlighted by both these and other recent ground anchorage contracts, may be summarised as follows:-

(a) The role, responsibilities and duties of the designer and the specialist contractor have to be addressed. BS8081 highlights that the specialist contractor can have design duties and responsibilities under the contract that effectively make him a part of the design team in some aspects of his role and a part of the contracting team in others.

(b) The Code intended to make the options of single or double protection open to suit the particular conditions and the evaluation of the designer.

The effect, however, has been that almost every permanent anchorage in the UK is now required to be double protected. The perception is that "double" is not much more expensive than "single". In reality, double protected strand anchorages require bigger holes. In the 1970's the writer was associated with the installation of what BS8081 would term as single protected anchorages for the major Thames flood prevention bank-raising scheme with a notional design life of 60+ years. These were most often installed within a drilling system utilising a 90 mm diameter borehole (115 cased diameter). The Pen-y-Clip double protected anchorages are installed within a 160 mm minimum diameter borehole with a 195 mm cased diameter. Typically the minimum borehole diameter for any double protected strand anchorage would be 125-140 mm. The implication in terms of drilling costs is that a double-protected anchorage could be twice as expensive to install as a single protected anchorage.

(c) BS8081 is quite clear that encapsulation length is not the same as fixed anchor length. The great bulk of ground anchorages with which the writer is concerned do not take account of this fundamental difference. The effect is that longer and more cumbersome encapsulation lengths are being installed than are often necessary.

ACKNOWLEDGEMENTS

The Pen-y-Clip scheme was designed and supervised by Travers Morgan on behalf of the Welsh Office. The main contractor for the project was John Laing Civil Engineering Ltd. The consulting engineers for the Rhigos Road scheme were Halcrows on behalf of Mid Glamorgan County Council, and the main contractors were Taylor Woodrow Civil Engineering Ltd.

The ground anchoring contractor for both projects was PSC Freyssinet Ltd. The author was retained by PSCF for the contractor-related aspects of the anchor designs.

REFERENCES

BARLEY, A. D. (1994)
"A question of experience"
Ground Engineering,
June/July, pp 13-16

BARLEY, A. D., EVE, R and TWINE, D. (1992)
"Design and construction of temporary ground anchorages at Castle Mall Development, Norwich"
Int. Conf. on Retaining Structures, Cambridge, 20-23 July.

BS8081:1989
British Standards Code of Practice for Ground Anchorages.
British Standards Institution, London.

OSTERMEYER, H. and SCHEELE, F. (1978)
"Research and Ground Anchors in non-cohesive soils"
Revue Francais de Geotechnique.
No 3, 92-97.

TURNER, M. J., SELLEY, P. J. and COOPER, G. M. (1992)
"Ground anchorages to the retaining walls at Pen-y-Clip"
Int. Conf. on Retaining Structures, Cambridge, 20-23 July.

Anker in Theorie und Praxis, Widmann (Herausgeber) © 1995 Balkema, Rotterdam. ISBN 90 5410 577 1

Die Auswirkung der geologischen Unstetigkeiten auf die Deformationen der verankerten Baugrubenwände der U-Bahnstationen Istanbul

Effect of the discontinuities on the deformations developed in the supporting systems of Istanbul Subway Stations

Mahir Vardar & Ismail Eriş

Abteilung für Ingenieurgeologie und Felsmechanik, Fakultät für Bergbau, TU Istanbul, Turkey

ZUSAMMENFASSUNG: Die Istanbuler U-Bahn ist seit 1992 im Bau. Die geplanten 10 Stationen, die in offener Bauweise hergestellt werden, sind 11 bis 35 m tief. Zur Zeit werden vier Baugruben ausgehoben und mit verankarten Stahlbetonwänden gesichert. Ein ausgedehntes Meßprogramm dient zur Überwachung der U-Bahnröhren und der Baugruben, somit auch der beeinflußten schon bestehenden Bauwerke. Die Arbeit versucht die Abhängigkeit der Deformationen von den geologischen Unstetigkeiten darzustellen.

Bei der Entwicklung der Deformationen an der Baugrubenwand sind unter anderen drei Faktoren maßgebend. Diese sind das Grad der Werwitterung, die Raumstellung der geologischen Unstetigkeiten und die lithologische Beschaffenheit der Felsmassen. Die gemessenen Deformationen waren nicht senkrecht zu der Baugrubenwand. Sie wurden hauptsächlich von der Schichtung und Klüftung kontrolliert.

ABSTRACT: Construction of Istanbul Subway has began in 1992. The Subway route is located at the European side of Bosphorus along the north-south direction. It connects the historic peninsula of Istanbul to Pera. The route is 16.3 km long and it passes through the metropolitan area of Istanbul having a population of about 13 million. Construction methods provide a detailed control of deformations and settlements.

Total cross section area of the tunnel, which is designed as parallel double tubes, is 36 m². Tunnels, which are a depth of 17 to 40 m from the ground surface, connects the stations. The depths of the stations vary from 11 m to 35 m. 7.5 km length of route was constructed by using NATM. Excavation and sheet wall construction of 4 stations out of 10 were completed. Deformations developed due to excavation processes were affected by discontinuities.

A formation of carboniferous age underlies a 1-4m thich artificial or residual-transported soil fill. It consists of conglamera, sandstone, and siltstone, and it is extremely weathered in some location. The formation is folded and very fractured. Faults and joints have caused the rock mass being crumbled. Magmatic dykes (diabase and andesite) cuts the rock mass frequently. Ground water table has not been found through the route of the subway. However, a transient leakage due to the discontinuities has been observed.

In this paper, in situ rock mechanics measurements of deep excavations supported by anchors were represented and it was depicted how deformations were affected by discontinuities in rock mass. In the evaluation of the deformations along the excavation walls three factors have been examined. They are the degree of weatering, orientation of discontinuities and lithology of rock mass. According to the results deformations were not perpendiculer to the excavation wall and their direction was controlled by discontinuities.

1 EINLEITUNG

Die Istanbuler U-Bahn ist seit 1992 im Bau. Die erste Bauphase beinhaltet auf der europäischen Seite dieser gigantischen Metropole mit ca. 13 Millionen Einwohnern eine 16.3 km lange Route in der Nord-Süd Richtung, welche die Altstadthalbinsel über eine eigene Bahnbrücke mit dem historischen Pera-Viertel verbindet.Zwei bergmännisch aufgefahrene Doppelröhren von je 36 qm verbinden die 11 bis 33m tief liegenden Stationen, die in offener Bauweise hergestellt werden. Die Überlagerung über den Röhren beträgt 17 bis 40 m. Bis jetzt sind 7.5 km mit der NÖT fertiggestellt. Von den geplanten 10 Stationen wurden bereits vier Baugruben ausgehoben und mit verankarten Stahlbetonwänden gesichert. Die Deformationen und Setzungen werden mit entsprechenden Baumaßnamen kontrolliert. Ein ausgedehntes Meßprogramm dient zur Überwachung der U-Bahnröhren und der Baugruben, somit auch der beeinflußten schon bestehenden Bauwerke.

Die U-Bahnstationen verdienen wegen ihrer Abmessungen mit 220m Länge, 23m Breite und bis zu 33 m Tiefe eine besondere Beachtung. Beim Aushub dieser tiefen Baugruben werden häufig Verformungen der Geländeoberfläche, Verschiebungen der Wände und auch Hebungen der Baugrubensohlen festgestellt. Es ist dabei von großer Bedeutung zu wissen, wie und warum diese Deformationen sich einstellen und wann und in welcher Größenordnung sie die Baugrube gefährden könnten. Die Beantwortung dieser Frage ist unter anderem von der Baugrubenabmessung und deren Geometrie, von der Aushubmethode und deren Geschwindigkeit, von der Verbauart, sowie von der Größe der Stützkräfte und vor allem aber auch von den Baugrundverhältnissen abhängig.

In dieser Arbeit wird gezielt nur die Auswirkung der ingenieurgeologischen Beschaffenheit des Baugrundes auf die Deformationen der verankerten Baugrubenwände untersucht.

1.1 *Der Baugrund als inhomogenes Medium*

Für ingenieurmäßige Bauzwecke ist die allgemeine geologische Benennung und Betrachtung des Baugrundes nicht alleine ausreichend. Es ist dabei auch von Bedeutung zu wissen und zu definieren, welche Faktoren das Gesamtverhalten dieses bestehenden Naturmaterials gestalten und ändern. Deshalb betrachtet die Ingenieurgeologie den Baugrund unter den folgenden vier unterschiedlichen Aspekten:

a) Der Felsgrund
b) Der Boden
c) Das Übergangsgestein
d) Die (oberflächliche) Auffüllung

Verknüpft man aber die ingenieurgeologischen Daten mit den Felsmechanischen entsprechend der Systemgrösse, so kommt man in demselben geologischen Vorkommen mit denselben Gebirgsverhältnissen und -bedingungen zu den unterschiedlichsten mechanischen Verhaltensweisen der jeweils zu berücksichtigenden Felsmassen.

In der Abbildung 1 ist diese Beziehung graphisch und tabellarisch dargestellt. Von Links nach Rechts sind zuerst die Benennung des in Systemgröße zu behandelnden Felskörpers und dessen vereinfachtes Aussehen, dann seine entsprechende Lage in σ-ε bzw. ($\Delta\sigma$-$\Delta V/V$) -Verhalten auf der Pre- und Post Failure-Kurve angezeigt. Letzlich wird noch angemerkt, mit welcher mechanischer Annahme das in-situ Verhalten dieser Gebirgsmasse angeschätzt und simuliert werden kann. Es ist hieraus ersichtlich, daß dasselbe Gebirge, je nach der Systemgröße monolithisch, annäherd mit den Gesteineigenschaften und -parameter, oder aber auch als Wenig-, Viel- oder Mehrkörpersystem mit der jeweiligen Verbandfestigkeit einem technischen Eingriff entgegenwirken kann. Die Aufgabe des Felsbaumechanikers scheint deshalb jene zu sein, diese Umwandlung der Gesteinsparameter zu den wahren Gebirgsparametern zu erklären und sie möglichst zutreffend zu bestimmen. Für diesen Zweck sind systematische Labor- und in situ Versuche und -messungen am istanbuler Grauwackengebirge durchgeführt worden. Die Abbildung 2 faßt einen Teil dieser Ergebnisse zusammen und schildert, wie verschieden die Bruchfestigkeiten und E-Moduli im Labor und im Gelände sein können.

Bis zu 33 m tiefe Baugruben der U-Bahnstationen in Istanbul werden in den oben genannten vier verschiedenen Medien ausgehoben. In den ersten Metern liegt die künstliche Auffüllung oder der angewachsene Boden vor. Darunter bis zu 20m Tiefe befindet sich das Übergangsgestein als mehr oder minder verwittertes Gebirge. Als Muttergestein besteht der Baugrund hauptsaechlich aus Karbonalten Konglomeraten, Sandsteinen, Grauwacken und Schluff- und Tonsteinen. Diese Felsmassen sind stark geklüftet, gefaltet und häufig durch Störungszonen und Verwerfungen zerrüttet.

Magmatische Gänge (Diabase und Andesite) durchziehen vielerorts diese Gesteine. Deshalb können die gemessenen Verformungen an den

Baugrubenwäden nicht allein mit den Annahmen der Kontinuumsmechanik erklärt und berechnet werden, da diese nicht nur einem einzigen elastischen Entlastungsprozeß unterliegen.

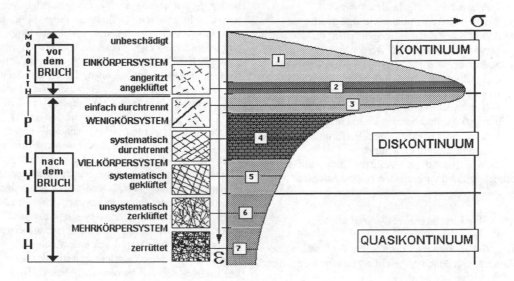

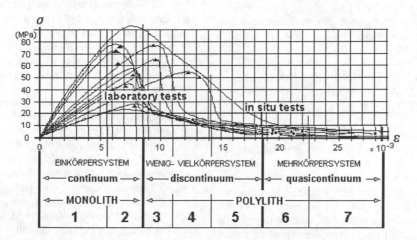

1.2 Aufgaben der Verankerung im Felsbau

Bevor Anker geplant und eingebaut werden, sollten folgende Fragen beantwortet sein:

1) Was ist die Aufgabe des Ankers?
- Soll er äussere Kräfte ins Gebirge übertragen?
- Soll er den Kräften entgegenstehen, die durch den Aushub im Gebirge frei werden?
- Soll er ein Wenig-, Viel- oder Mehrkörpersystem durch Zusammenbinden der einzelnen Kluftkörper zu einem Monolith umwandeln?
- Soll er in situ eine armierte Felsmasse herstellen, die denZug- und Scherkräften standhalten kann?
- Soll er die zeitabhängigen Verformungsprozesse kontrollieren und berichtigen?

2) Wie ist diese Aufgabe am einfachsten, schnellsten und wirksamsten zu gewährleisten?

1.3 Wie kann man dem Gebirge beistehen?

Die Verbandfestigkeit des Gebirges soll während des Aushubes möglichst geschont bleiben oder wenn nötig, verbessert werden.

Die Kluftkörperbeweglichkeit des Gebirges darf durch den Aushub nicht aktiviert werden.

Die Umwandlung der Spannungszustände soll dem primären Spannungszustand angepaßt sein.

Die zeitabhängigen Verformungen sollen durch die Schnelligkeit und Effektivität des Bauablaufes verhindert werden.

Das Entstehen des Kluftwasserdruckes sowie des Porendruckes soll anhand der Drainagebohrungen unter Kontrolle gebracht sein.

2. DIE BEISPIELHAFTEN VERFORMUNGEN DER BAUGRUBENWAENDE VON ZWEI U-BAHNSTATIONEN

Zur Kontrolle der Stabilität der Baugrubenwände wurden in allen U-Bahnstationen in erster Linie die räumlichen Verformungsraten mit Hilfe der verschiedenen Meßverfahren systematisch gemessen und ausgewertet. Vor dem Baubeginn wurden für diesen Zweck hinter den geplanten Grubenwänden tiefe Inklinometer eingesetzt und während des Aushubes wurden noch planmäßig, synchron zu den Aushubsphasen, an die verankerten Baugrubenwände Vermessungspunkte fixiert und auch Meßanker eingebaut. Die Setzungen der Geländeoberfläche sowie die Bewegungen der Nachbarbauten wurden opto-trigonometrisch aufgenommen.

2.1 U-Bahnstation Osmanbey

Die Osmanbey-Station befindet sich unter einer sehr wichtigen, belebten Hauptstraße der Stadt. An den beiden Seiten stehen alte, bis achtstöckige Wohn- und Geschäftsbauten. Deshalb ist diese 11 m tiefe Baugrube weitgehend mit Meßpunkten und Kontrollmessgeräten bestückt. Wegen der geologischen Verhältnisse (Lithologie, Schichtung, Kluftstellungen, Störungszonen, Heterogenität u.a.) war die westliche Wand empfindlicher. Deshalb wurde ein 14 Meter langer Inklinometer ca. 4 m vor der Grubenwand in den Baugrund eingesetzt. Die Verformungen wurden systematisch abgelesen. In der Abbildung 3 sind die zeitlichen Horizontalkomponenten dieser Verschiebungen parallel und normal zur verankerten Baugrubenwand in Abhängigkeit der Tiefe und der geologischen Beschaffenheit des Baugrundes graphisch dargestellt.

Es sind dabei drei Bewegungsbereiche zu unterscheiden. Der erste Bereich belegt die ersten 5 Meter und zeichnet sich durch einem von oben nach unten abnehmenden Verschiebungscharakter aus. Die horizontale Gesamtverformung dieses Bereiches beträgt in der obersten Zone 38.3 mm und ihre Komponenten senkrecht und parallel zur Baugrubenwand sind jeweils 34 mm und 17.5 mm. Sie nehmen aber mit der Tiefe fast linear ab und bekommen Werte von 18 mm und 9 mm. Der zweite Bereich durchzieht die folgenden 2.5 Meter und zeigt gleichbleibende Verschiebungsraten von jeweils ca. 18 und 8 Milimetern, welche hauptsächlich nur mit dem kinematischen Verhalten eines starren Körpers vergleichbar sind. Dann kommt zwischen 7.5 und 11 Metern ein Bereich mit unterschiedlichen Verschiebungsformen, die ein Aufeinandergleiten verschiedener Partien eines Schichtenpaketes vermuten laeßt.

Um diese Phänomene näher zu untersuchen, wurden diese gewonnenen Meßergebnisse mit den ingeniergeologisch-geomechanischen Auswertungen und Daten in Zusammenhang gebracht. In der Abbildung 4 ist von oben nach unten das Folgende dargestellt; der Lageplan der Baugrube mit der vereinfachten Geologie (4a), der geologische Schnitt der westlichen Baugrubenwand (4b), dieAusbreitungszonen der Verwitterungsklassen (W_2 bis W_4 nach), die Lagen der für verschiedene Verfahren angesetzten Meßpunkte und die Aushubstiefen an derselben Wand (4c) und letzlich die zeitliche Entwicklung der totalen Horizontalverschiebungen in Abhängigkeit von den etappenweisen Aushubsvorgängen (4d).

Die Feststellungen und Interpretation dieser Abbildung dürfen folgendermaßen zusammengefaßt werden:
Die Baugrube ist nn einer geologisch so interessanten Stelle ausgehoben, daß das Grauwackengebirge hier fast mit all seinen Eigenschaften vortritt. Die Schichten fallen mit 57^0-70^0 aus der Grubenwand hinaus. Die Schichtenfolge wechselt in einigen Metern ab und sie werden von einer 6-7,5 m dicken Störungszone und zwei stark verwitterten Diabasegängen durchtrennt. Die Korngrössen und sogar auch die mineralogischen Zusammensetzungen der gesteinsbildenden Partien, die unter anderen hier das Ausmaß und die Ausbreitung der Verwitterung bestimmen, ändern sich in kurzen Abständen. Die Verwitterungskarte der Grubenwand stimmt mit dem geologischen Schnitt weitgehend überein und läßt den Baugrund ausreichend und zweckentsprechend modellieren.

Demgemäß befindet sich der eingesetzte Inklinometer in den ersten 3 Metern in einem künstlich aufgefülltem Material und dann in geschichteten und geküfteten Schluff-, Ton- und Sandstein. Die Störungszone ist von der Inklinometerbohrung nicht betroffen.Keine Diabasegänge durchörtern diesen Inklinometermeßbereich.
Die gemessenen Horizontalkomponenten des Verschiebungsvektors ergeben rechnerisch einen durchschnittlichen Winkel von 63^0 (62^0 46'). Da hier die Schichtstellungen um N(18^0-21^0)W, (60^0-67^0)NE wechseln, soll der Fallwinkel mit dem Norden 69^0 bis 70^0 betragen. Dies ist ein unumstrittener Beweis dafür, daß hier die Verformungen hauptsächlich auf den Schichtungsflächen aktiviert werden. Daß aber noch mit der Tiefe kein kontinuirliches Verformungsbild entsteht, läßt erkennen, daß bestimmte, gesündere Schichtpakete sich als Starrkörper auf den wiederstandsärmeren bewegen.

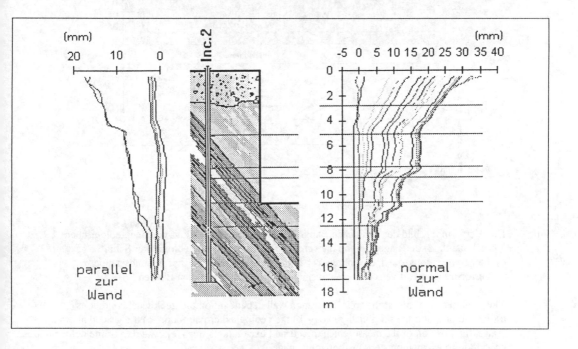

Abb.3   Geologischer Schnitt und der Verlauf der horizontalen Verschiebungen der Inklinometerachse vor der westlichen Grubenwand. U-Bahnstation Osmanbey/Ist.
(I bis III die unterschiedlichen Beweglichkeitstypen)

Fig.3   The geological cross section and the development of the lateral displacements along the inclinometer axis in front of the west excavation wall. Osmanbey Metro station /Ist.
(I to III : the different Types of the Deformations)

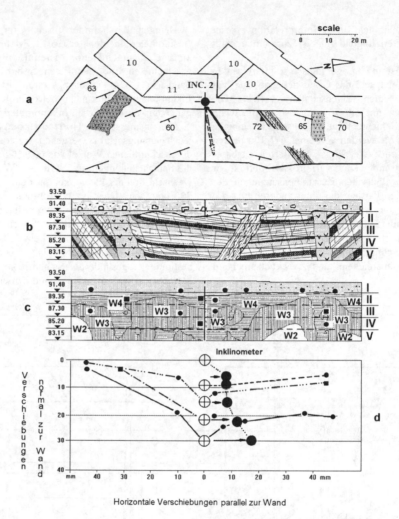

Horizontale Verschiebungen parallel zur Wand

Abb.4   Das Verformungsbild der westlichen Baugrubenwand in Abhängigkeit von den geologischen
Verhältnissen a) der Lageplan und die vereinfachte Geologie b) geologischer Schnitt
c) dieVerwitterungszonen, d) dieLage der angesetzten Meßpunkte und Aushubsetappen
e) Verformungverlauf verschiedener Aushubsetappen in horizontalen Ebenen

Fig.4   The total deformations at the west excavation wall dependent on the geological conditions
a) the site plan and the simplified geology b) the geological cross section c) the graded alteration
zones, the position of the measurement points and excavations steps e) evalution of the deformations
according to excavation steps in ahorizontal plain

2.2 *U-Bahnstation Gayrettepe*

Ein anderes Beispiel bildet das Verformungbild des
Inklinometers hinter der südöstlischen Wand der
28 m tiefen Baugrube der U-Bahnstation Gayrettepe,
wo die geologischen Bedingungen günstiger waren.
Die Schichten fallen hier durchschnittlich mit ca. 15^0

in die Baugrube. Aber da durch diesen kleinen
Neigungswinkel die Reibungskraefte nicht
überwunden werden, kommen auch die vorhandenen
Diskontinuitäten nicht in Bewegung, solange die
Anker das Felsgefüge zusammenbinden und den
nötigen Ausbauwiederstand leisten. Trotz der
Ähnlichkeit mit den ersten 5 Metern verformt sich

262

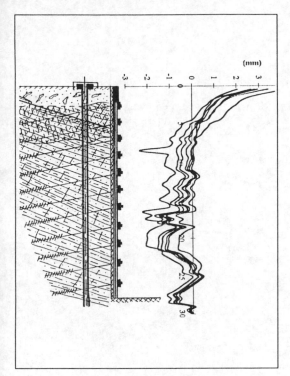

gegeben wird und deren mechanische Festigkeit überschritten ist. Deshalb sollten die Baugrundverhältnisse durch Kartierung der Homogenbereiche und der Unstetigkeiten gut verstanden und für die Projektanten genaustens geschildert sein, damit die verankerten Grubenwände sicher und ökonomisch dimensioniert werden können.

LITERATUR

Kovacevic M.S., I.Jaserevic, Y. Kulic,1993. Plane equilibrum of rock slopes with anchors. Symp. on Assesment and Prevention of Failure Phenomena in Rock Engineering, Proc., p.p.653-660, Istanbul

Ulrichs, K.R.,1981. Untersuchungen über das Trag- und Verformungsverhalten verankerter Schlitz- wände in rolligen Böden. Die Bautechnik,4/1981, S.124-132.

Vardar, M., Eris, I., 1994. Derin ve genis kazilarda saglamlastirma-destekleme önlemleri üzerine bir arastirma. ZM.5 Ulusal kongresi, cilt 2, S.314-327,Ankara

Abb.5 Interessanter Verlauf der Verschiebungen mit negativen Raten an der südöstlichen Wand der U-Bahnstation Gayrettepe.

Fig.5 Interesting course of the deformations curve with negative rates by the SE-excavation wall of the Gayrettepe Metro station.

die Grubenwand sogar ins Gebirge hinein, weil die Ankerkräfte zu hoch angeschätzt und berechnet sind. Das Zusammenschließen der Klüfte spielt dabei die größte Rolle. Die Teilbeweglichkeit der geklüfteten Felsmasse führt unter anderen zu diesem unruhigen Verformungsbild.

3 SCHLUSSFOLGERUNGEN

In geklüftetem und/oder geschichtetem Gebirge spielen die geometrische Größen und die Raumstellungen der Unstetigkeiten auch bei der Baugrubensicherung die maßgebende Rolle. Außer der elastischen Entlastung bestimmter Homogenberiche sind auch zeitabhängige plastische Verformungen auf den aktivierten Kluft- und Schichtflächen zu erwarten, solange durch den Aushub den Kluftkörpern eine kinematische Freiheit

Anker in Theorie und Praxis, Widmann (Herausgeber) © 1995 Balkema, Rotterdam. ISBN 90 5410 577 1

Meßtechnik im Zuge der Ankerherstellung

Measuring techniques in the process of anchor construction

H.-Walter Gross
Bilfinger + Berger Bauaktiengesellschaft, Mannheim, Germany

Reinhold Völkner
Grün + Bilfinger GmbH, Wien, Austria

ZUSAMMENFASSUNG

Einige wenige Sonderfälle der Ankertechnik machen es erforderlich, die genaue räumliche Lage der Injektionsstrecke zu vermessen. Grund dafür ist z.B. die Gefahr von Schäden an benachbarten Bauwerken durch den Verpreßvorgang oder die ungewollte Lasteinleitung.
Diese Ankerbohrungen können einfach und schnell mit einer Spezialsonde vermessen und CAD-unterstützt in räumlichen Bezug zum Nachbarbauwerk gebracht werden.
Diese Technik wurde 1994 bei BILFINGER + BERGER Bauaktiengesellschaft auf einer Baustelle in München eingesetzt. Hier mußten die Verpreßstrecken einer Stirnwandverankerung im Hinblick auf den Mindestabstand zu bestehenden Tunnelröhren ermittelt werden. Entsprechend den gemessenen Abstandswerten zwischen Tunnelaußenschale und Verpreßstrecke wurden dann die Verpreßparameter festgelegt.

ABSTRACT

Only some applications of the anchoring technique make it necessary to measure the exact three-dimensional orientation of the grouting area to neighbouring constructions.
The reason for this is the risk of damage by the grouting pressure or the unintended loading.
These anchor holes can be measured relatively uncomplicated and quickly by a special probe and they can be brought in three dimensional connection with the neighbouring construction by aid of CAD.
This technique was used in 1994 by BILFINGER + BERGER Bauaktiengesellschaft on a construction site in Munich. Here the grouting sections of an anchored sheet pile wall in regard to the radial minimum distance to existing tunnel tubes had to be calculated. According to distance values by steps the grouting parameters were fixed.
The measuring equipment which was used disposes of inclination probes for the verticality with correction of torsion. The lateral deviations are calculated with an angular movement probe, which is also situated in the equipment.
With the CAD-system used the values of the three systems of coordinate axes of the tunnel tubes, the building pit and the orientation of the anchor are computed to one system and the absolute distances are calculated. For that purpose the software algorithms are used.
The described system is a fast and elegant solution in the case of the above-mentioned problems.

EINLEITUNG

Die Herstellung der Ankerbohrungen selbst bedarf im Grunde genommen selten besonderer Meßtechniken. Eher spielt Meßtechnik hierbei im Zusammenhang mit Lasten nach der Fertigstellung des Ankers eine Rolle. Im Falle, daß die Ankerbohrung aber mit anderen Bauwerken in ungünstige Wechselwirkung zu geraten droht, muß ein steuerbares Bohrverfahren oder eine nachträgliche Orientierungsmessung der nicht gesteuerten Bohrung eingesetzt werden. Ebenso kann die Lagevermessung der Verpreßstrecken bei sich kreuzenden Ankern bei Baugrubenwänden erforderlich sein. Die Festlegung von Verpreßdrücken und spätere Funktion kann gegebenenfalls dann erst im Zusammenhang mit den benachbarten Bauelementen erfolgen.

Steuerbares Bohren bedeutet hier einen oft nicht tragbaren technischen und wirtschaftlichen Aufwand.

Die zweite angesprochene Lösung dagegen, als reine Orientierungsmessung, kann schnell über die Verwendungsfähigkeit der Bohrung Auskunft geben. Die Istlage muß dann nach Erfordernis mehrmals während der Bohrung oder einmalig nach Erreichen der Endtiefe geprüft werden.

VERMESSUNG KLEINLOCHBOHRUNGEN

Bei Ankerbohrungen müssen aufgrund deren Abmessungen und Geometrie Ausrüstungen für Kleinlochbohrungen zur Orientierungsmessung eingesetzt werden. Mit vorwiegend vertikalem Verlauf sind solche Bohrungen relativ einfach, z.B. als inkrementale Inklinometermessungen durchführbar.

Sobald wie bei Ankerbohrungen üblich, die horizontale Komponente überwiegt müssen hier aufwendigere Techniken eingesetzt werden.

Während die Neigung zur Vertikalen weiterhin mittels Inklinometersensoren sichtbar wird, ist für den Azimut, also den seitlichen Verlauf, die Messung zusätzlich zu instrumentieren. Eine teure Variante hierzu ist der Kompaßeinsatz mit Kreiseltechnik, da ein magnetisches System im Bohrrohr, sofern nicht antimagnetischer Stahl verwendet wird, versagt. Einfacher ist hier der Einsatz eines eingebauten Gelenksensors für das inkrementelle Aufsummieren der Azimutwerte. Die gesamte Meßstrecke wird in Schritten entsprechend der Sondenlänge eingeteilt und Schritt für Schritt werden die vertikalen Inklinometerwerte und die horizontalen Gelenkwerte

(=Azimutkomponeneten) aufsummiert. Dadurch entsteht eine Gesamtstrecke, bestehend aus einzel vermessenen Kleinstrecken, die in ihrer Summe zuverlässig das Ergebnis darstellen.

EINSATZ BAUSTELLE „HASENBERGL"

Diese Technik mittels Inklinometer und Gelenksensor wurde bei BILFINGER + BERGER Bauaktiengesellschaft auf der Baustelle Hasenbergl in München erfolgreich eingesetzt.

Auf dieser Baustelle wurde die Anschlußwand zu einem Nachbarlos in mehreren Lagen verankert. Nach den bestehenden unkritischen ersten beiden Ankerlagen A und B war die Ankerlage C dicht über der Tunnelschale des Nachbarloses auszuführen. Zehn Anker dieser Ankerlage C waren auf ihren Bohrverlauf hin zu vermessen, und durften nur unter Einhaltung eines Mindestabstandes zur Tunnelaußenschale eingebaut und verpreßt werden. Wesentliches Know-how war hier bei der Weiterverarbeitung der gemessenen Werte zum Ergebnis gefordert. Hier mußten die räumlichen Koordinaten der dreidimensional gekrümmten Körper von Tunnelaußenschalen und Ankerbohrungen zueinander in Bezug gebracht werden.

DAS VERMESSUNGSSYSTEM

In Zusammenarbeit mit der Firma Glötzl, Karlsruhe, wurde eine Multifunktionssonde eingesetzt. Aus-

Eingebautes Führungsrohr / Installed guidance tube

Meßwerteingabe / Taking data

wahlkriterien waren eine im Versuch nachgewiesene Genauigkeit, ein gutes Handling der Gerätschaft und positive Erfahrungen mit ähnlichen Geräten dieses Fabrikats. Die Multifunktionssonde ist im wesentlichen mit Gravitationssensoren und einem Knickgelenk mit Lasertechnik ausgerüstet. Die Gravitationssensoren liefern den Verlauf der Bohrung zur Vertikalen und eine Verrollungskorrektur der Sonde. Der seitliche, also horizontale Verlauf, ergibt sich aus dem Signal eines optischen Sensors, auf den ein Laserstrahl im Knickgelenk auftrifft.

Der Sondenkörper ist mit Radsätzen bestückt und wird in Meterschritten durch ein vorher ins Bohrloch eingebautes Führungsrohr gezogen.

Dabei werden die jeweiligen Meßwerte der Sensoren protokolliert und per Tabellenkalkulation verarbeitet. Die Messung wird ein zweites Mal als Umschlagmessung durchgeführt, wodurch eine Fehlerkompensation erfolgt. Die Startposition, also Nullagenmessung der Sonde erfolgt durch die geodätische Vermessung des Startrohres, in dem sich die Sonde vor und nach der Messung sowie beim ersten Meßschritt befindet.

MESSTECHNISCHER ABLAUF

Nach dem Abbohren der gesamten Ankerlänge wird das Innengestänge der Verrohrung gezogen und ein Nutrohrgestänge mit Distanzhaltern zum Außenrohr eingebaut. Vor dem ersten Rohrschuß dieses Gestänges befindet sich ein pneumatischer Packer, der nach Erreichen der Endtiefe aufgeblasen wird. So

wird ein nennenswerter Sandeintrieb während der Meß- und der Auswertephase verhindert. Die Lafette des Bohrgerätes wird nun so ausgerichtet, daß sie als Stativ zur Sondenaufnahme dient. Auf das eingebaute Nutrohrgestänge wird das letzte Rohrstück, welches die Sonde beinhaltet, aufgebaut. Dieses Rohrstück wird geodätisch auf seine Position als Nullage der Sonde vor dem Einfahren eingemessen. Die Sonde wird nun mit dem Meßkabel verbunden, aus ihrer Arretierung gelöst und an den vordersten Punkt der Bohrung geschoben. Bei zu geringer Neigung oder gar Steigung wird ein Spezialgestänge eingesetzt. Nach dem Anschließen des Anzeigegerätes an das Meßkabel wird die Sonde nun Meter um Meter zurück gezogen und die Werte des jeweiligen Schrittes in eine PC-Tabellenkalkulation eingegeben. Ist so die gesamte Strecke einmal abgefahren, wird die Sonde zusammen mit dem Startrohr axial um 180 Grad gedreht und der Vorgang als Umschlagsmessung wiederholt.

DIE AUSWERTETECHNIK

Durch Koordinatentransformation werden hierbei die gegebenen Daten der Tunnelaußenschalen, die geodätischen Daten am Bohransatzpunkt und die Messungen der Multifunktionssonde als das eigentlich geforderte Ergebnis aufgezeigt. Der räumliche Verlauf der beiden Tunnelröhren bzw. der der Außenschalen wird als Datensatz im 3-D-Modus als CAD-Datei erstellt. Alle weiteren Meßdaten werden in der gleichen Weise dieser Datei zugeführt. Mit Hilfe dieser Technik ist es möglich, die Ergebnisse über CAD-interne Algorithmen gestützt zu errechnen

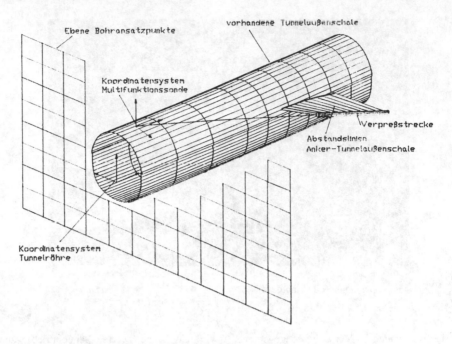

Ebene Bohransatzpunkte

vorhandene Tunnelaußenschale

Koordinatensystem
Multifunktionssonde

Verpreßstrecke

Abstandslinien
Anker-Tunnelaußenschale

Koordinatensystem
Tunnelröhre

Dreidimensionale Ansicht einer Tunnelröhre mit dem Verlauf der Ankerbohrung
Three dimensional drawing of tunnel tube and anchor hole

und darüber hinaus übersichtlich räumlich darzustellen. Die verschiedenen gemessenen Bohrungs-
verläufe der jeweiligen Anker und deren Verpreßstrecken werden per Layertechnik einzeln abgelegt.
So kann das Einzelergebnis und auch die Gesamt-
ansicht eingeblendet und geplottet werden. Als
abschließendes Ergebnis der Messungen werden hier
die Abstände der jeweiligen Ankerverpreßstrecken
zur entsprechenden Tunnelaußenschale aufgezeigt.
Neben der räumlichen Ansicht der Ankerlage werden
die tiefenabhängigen Abstände zwischen Verpreßstrecke und Tunnelaußenschale als Tabelle aus-
gegeben. Festgelegte Abstände zur Tunnelaußenschale bestimmen die Kriterien für die Verpreßparameter des jeweiligen Ankers. Bei Unterschreiten
eines Mindestabstandes muß die Verpreßstrecke axial
in Richtung Bohrlochmund soweit verschoben
werden bis zulässige Werte erreicht sind.

Die beschriebene Vorgehensweise ist eine elegante
Lösung, die Sollagen von Ankern nicht nur bei
geometrisch komplizierten Baugruben zu überwachen. Hierbei können Mindestabstände zwischen
den einzelnen Ankern festgelegt, und die räumliche
Situation per CAD anschaulich betrachtet werden.

Anker in Theorie und Praxis, Widmann (Herausgeber) © 1995 Balkema, Rotterdam. ISBN 90 5410 577 1

Erfahrungen mit Ankerausbau im deutschen Steinkohlenbergbau

Rock bolting experience in German coal mining

H.Witthaus
DMT-Gesellschaft für Forschung und Prüfung mbH, Essen, Germany

ZUSAMMENFASSUNG: Seit 1950 werden Anker im deutschen Steinkohlenbergbau als Gruben-ausbau und zur Gebirgsvergütung verwendet. Unter den geologischen Randbedingungen und be-trieblichen Zwängen hat sich eine differenzierte Verfahrensweise zur Bemessung von Ankerausbau entwickelt, die sich aus den sicherheitlichen Anforderungen ableitet. Sowohl die Verformung der Anker als auch ein kritischer Kräftenachweis werden betrachtet, um Anker in unterschiedlichen An-wendungsgebieten einzusetzen. Aus verschiedenen Anwendungen werden zwei Einsatzfälle be-schrieben, um die Erfahrungen beispielhaft darzustellen.

ABSTRACT: Coal-mining in Germany is practised as longwall mining and mostly influenced by geo-logical and geomechanic parameters of the deposit. More than 40 years of experience in rockbol-ting techniques led to a special way of proceeding. Deformation of roadway cross-section results from high rock pressure caused by the weight of overlaying strata and additional influences by for-mer and actual mining operations. Therefore the max. height of deformation to take over by rock-bolting must be defined for each drifting project. Geomechanic parameters of depth, structure, and resistance of roof and underground of the seam are basis for the prediction of convergence in roadways. This prediction has to consider various influences of winning for showing the height of rockbolting deformation during the lifetime of support. Especially working boundaries of former mining horizons above or under the seam horizon must be considered. The max. convergence to take over by rock bolting has to be compared with this amount of prediction to estimate the possi-bility of application for each drifting-project. Different constructions of bolt design and grouting allow different amounts of deformation. Generally simple designed bolt rods with different kinds of surface structures are used. As a rule not preloaded rods with lengths of about 2.20 to 3.00 m and diame-ters between 22 and 31 mm are grouted by using glueing cartridges filled with resin adhesive (cartridged polyester-grout) in boreholes of 28 to 48 mm. Mostly the strength of tension varies between 250 and 500 kN. High deformation in roadway cross-sections requires high yielding capa-city of bolt rods. To extend this capacity constructive elements are used in construction of rock bolt design. New rock bolts e.g. the combibolt show optimal suitability for the compensation of deforma-tion exceeding the elasticity range of simple designed bolt types. In this lecture roadway support by rockbolting is shown to give an abstract of using this elements for support. There is also rockbolting in German coal-mining as additional support e.g. in cases of combined support systems and additional yielding support.

1 EINLEITUNG

Anker werden etwa seit dem Jahre 1950 im deutschen Steinkohlenbergbau verwendet. Der Einsatz als Grubenausbau und Gebirgsvergütungsinstrument ist geprägt von den Randbedingungen und Zwängen geologischer Besonderheiten und betrieblicher Abläufe. Aus negativen Erfahrungen in den 60er und 70er Jahren entwickelte sich eine differenzierte Verfahrensweise für die Dimensionierung und Durchführung. Insbesondere muß berücksichtigt werden, daß Verformungen des Hohlraumquerschnitts und Auflockerungen im Gebirge selbst mit Ausbaumitteln nur selten mit wirtschaftlich vertretbarem Aufwand zu verhindern sind.

Als Maß für die Querschnittsverminderung in Grubenbauen wird die Konvergenz herangezogen. Sie kann im deutschen Steinkohlenbergbau bis zu 100 % der Streckenhöhe betragen. Aufgrund langjähriger Erfahrungen ist es möglich, die Konvergenz im Vorfeld abzuschätzen und geeignete Ausbaumittel auszuwählen und anzupassen.

Sofern die äußeren Bedingungen es zulassen, kommt Ankerausbau zum Einsatz, der sich durch ergonomische und wirtschaftliche Vorteile auszeichnet. Nachteilig wirkt sich jedoch das begrenzte Verformungsvermögen bisher üblicher Ankerbauformen aus. Neben ausschließlichem Ankerausbau haben sich Anker jedoch als Zusatzausbau in verschiedenen Anwendungen bewährt.

2 BETRIEBLICHE RANDBEDINGUNGEN DES DEUTSCHEN STEINKOHLENBERGBAUS

Der Abbau im Ruhrrevier, der größten deutschen Lagerstätte, geht derzeit in Teufen zwischen 600 und 1300 Metern um. Es werden verschiedene Flöze mit Mächtigkeiten zwischen 0,8 und 3,5 m im Strebbau abgebaut. Für diese Bauweise ist es erforderlich, Strecken im Nebengestein und in der Flözebene aufzufahren und zur Förderung, Fahrung, Wetterführung und zum Materialtransport zu nutzen.

Teilweise werden diese Strecken nicht nur durch das Gewicht der überlagernden Schichten, sondern zusätzlich auch durch dynamische Einflüsse des Abbaus belastet[1]. Weiterhin wirken sich Abbaubegrenzungen früherer Abbaubetriebe in anderen Flözebenen teilweise negativ auf die Druckverteilung im Gebirge aus.

Bei hohen Streckenkonvergenzen kommt nachgiebiger Unterstützungsausbau aus mehrteiligen Stahlbögen zum Einsatz. Dieser wird stellenweise durch eine Baustoffschale ergänzt, um hohen Belastungen durch Ausbaustützdruck zu begegnen. Instandhaltungsmaßnahmen und Senkarbeiten sind aber keine Ausnahme sondern die Regel.

3 ANWENDUNGSGEBIETE UND DIMENSIONIERUNG

Anker dienen im deutschen Steinkohlenbergbau als:
- alleiniger Ausbau in der Form Anker-Maschendraht-Verbundausbau
- Abfanganker mit temporärer Ausbaufunktion am Streb-Strecken-Übergang
- Schubsicherungsanker mit der Aufgabe abgleitende Gebirgsschichten in geneigter Lagerstätte zu sichern
- Vergütungsanker mit der Funktion der Gebirgsstabilisierung in aufgelockerten Bereichen
- Aufhängungsinstrument zur Verdübelung der Lasten von Transporteinrichtungen und sonstigen Betriebsmitteln

Im Rahmen dieses Vortrages kann auf die speziellen Belange einzelner Anwendungen nicht eingegangen werden. Daher wird beispielhaft die Anwendung als Anker-Maschendraht-Verbundausbau im folgenden dargestellt, um die Dimensionierung einer Ankerung zu verdeutlichen:

Die geologischen und betrieblich bedingten Verhältnisse der Gebirgsbeanspruchung im deutschen Steinkohlenbergbau sind großen lokalen Schwankungen unterworfen. Es hat sich in der Vergangenheit gezeigt, daß einheitliche Bemessungsverfahren unter rein statischen und festigkeitstechnischen Gesichtspunkten nicht anwendbar sind. Insbesondere die Eigenschaften des Gebirges lassen sich während der Planungsphase nur grob abgeschätzen.

---

[1] Der Überlagerungsdruck ergibt sich aus der mittleren Wichte des Deckgebirges und der Teufe.

Aus Beobachtungen unter Tage und Grundsatzuntersuchungen an Modellprüfständen resultierte daher ein Bemessungsverfahren für Ankerausbau, welches nicht an der ausbautechnischen Nutzung der Anker, sondern an deren sicherheitlicher Funktion orientiert ist.

Dieses Verfahren beruht auf der Kluftkörpermethode nach GÖTZE, welche maximal mögliche Belastungskörper definiert, die in den Hohlraum fallen können. Aus Untersuchungen des Auflockerungsverhaltens im Streckenmantel geht danach hervor, daß der kritische innere Reibungswinkel des Gebirgsverbandes geeignet ist, sogenannte Kluftkörper abhängig von der Querschnittsform zu beschreiben. Es kann damit die Form und auch das Gewicht entsprechender Belastungen quantifiziert werden, was die Grundlage einer Ausbaubemessung liefert.

Die Dimensionierung nach dieser Methode muß jedoch berücksichtigen, daß Kluftkörper unabhängig vom Auflockerungsverhalten plötzlich gelöst werden und den Ausbau schlag- oder impulsartig belasten. In der Praxis der Ankerdimensionierung wird daher der 'Fall aus der Höhe Null' berücksichtigt. Damit ist gemeint, daß der zwar formschlüssig aber nicht aktiv gesetzte Anker auf eine zusätzliche dynamische Lastkomponente ausgelegt sein muß.

Die Gesamtlast ergibt sich dann nach den Grundsätzen der mechanischen Energieerhaltung zum Doppelten der statischen Gewichtskraft des Belastungskörpers, wenn keine dämpfenden Eigenschaften des Ausbaus berücksichtigt werden.

In Bild 1 ist beispielhaft dargestellt, welches Ergebnis die Dimensionierung einer Ankerung im deutschen Steinkohlenbergbau hat, und welche Belastungsannahmen für eine bogenförmige Strecke getroffen werden.

Neben dem oben beschriebenen Nachweis ausreichender Ausbaukräfte des Ankerausbaus muß die Verformung der Anker bei auftretender Konvergenz beachtet werden. Die kritische Verformbarkeit der Ankerstangen wird daher mit den Bewegungen im Gebirge verglichen, um die Anwendbarkeit der Ankertechnik unter den örtlichen Randbedingungen abzuschätzen.

Hierzu wird die Konvergenz anhand empirisch gefundener Gleichungen prognostiziert und eine zulässige Streckenkonvergenz aus den technischen Eigenschaften des Ankers und den geometrischen Randbedingungen definiert.

In Bild 2 sind die Berechnungsgrundlagen des Konvergenzvergleiches zusammenfassend dargestellt.

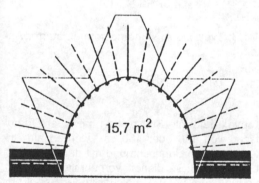

15,7 m^2

Endgültiges Ankerschema
(Betriebsoptimum)
bei der derzeitigen Bohrtechnik

Bild 1: Ankerschema für eine bogenförmige Strecke im deutschen Steinkohlenbergbau

Fig. 1: Boltscheme for Arch-Type Support in German Coal-Mining

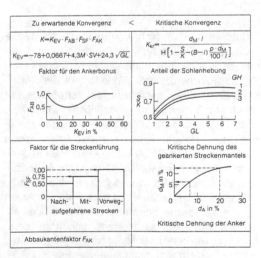

Bild 2: Verformungsnachweis für die Anwendung von Ankerausbau

Fig. 2: Deformation Reference for Bolt Support

Kennwerte für den Gebirgsaufbau, die Belastung aus überlagernden Schichten und Abbau sind im wesentlichen maßgebend für die Konvergenzprognose.

Auf der anderen Seite ergibt sich die zulässige Konvergenz aus der Ankerlänge und der Verformbarkeit der Anker, die in eine zulässige Dehnung des Streckenmantels umgerechnet wird. Weiterhin wird aus den Festigkeiten der Nebengesteinsschichten berechnet, welcher Anteil der Konvergenz auf die Ankerung wirkt.

In neueren Anwendungen wird angestrebt, verschiedene Ausbausysteme zu kombinieren, um den stetig steigenden Belastungen zu begegnen, die mit zunehmender Gewinnungsteufe auftreten.

So eignet sich Ankerausbau unter bestimmten Bedingungen als Sicherung Vor-Ort und kann im nachgeschalteten Bereich einer Streckenauffahrung durch Unterstützungsausbau verstärkt werden. Besonders wirkungsvoll ist diese Verfahrensweise, da zum Zeitpunkt der Erstellung die Belastungen im Streckenmantel nicht voll ausgebildet sind. Maximale Spannungen treten in Abbaubegleitstrecken erst in der Nähe des Strebes auf und bewirken dann hohe Auflockerungen des Gebirges.

4 ANKERBAUFORMEN UND ANKERANORDNUNG

Anker werden als nicht vorgespannte Elemente radial zum Querschnitt in Bohrlöchern auf gesamter Länge mit mineralischem Mörtel oder Kunstharz gesetzt. Als Werkstoff für die Ankerstangen wird hochwertiger Stahl verarbeitet, der Mindestanforderungen bezüglich Festigkeit, Dehnung und Kerbschlagbiegevermögen genügen muß. Übliche Tragkräfte der Anker liegen zwischen 240 und 500 kN je Anker, bei Außendurchmessern von 22 bis 31 mm und Ankerlängen zwischen 2,20 m und 3,00 m. Hauptsächlich werden einfache Bauformen starrer Anker verwendet, deren Schaft durchgehend oder abschnittsweise mit einer Haftprofilierung versehen ist. Neuere Entwicklungen verfügen über zusätzliche Nachgiebigkeitselemente.

In Bild 3 finden sich Ausführungsbeispiele typischer Ankerstangen. Am Ankerfuß läuft der Schaft als Mischspitze scharfkantig aus, um in Verbindung mit Kunstharzpatronen die Durchmischung der Mörtelkomponenten im Bohrloch

zu optimieren. Die Profilierungslängen des Ankerschaftes entsprechen mindestens dem Maß der minimalen Verbundlänge des Ankermörtels, um die maximale Tragkraft des Ankers aufzunehmen. Ein Zweikant am Ankerkopf dient als Setzhilfe zum Eindrehen der Anker.

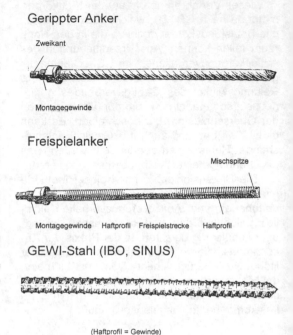

Gerippter Anker

Zweikant

Montagegewinde

Freispielanker

Mischspitze

Montagegewinde   Haftprofil   Freispielstrecke   Haftprofil

GEWI-Stahl (IBO, SINUS)

(Haftprofil = Gewinde)

Bild 3: Typische Ausführungsformen starrer Anker

Fig. 3: Typical Rigid Bolts

Konvex geformte Stahlplatten mit einer Grundfläche von 120 oder 450 cm² werden als Ankerplatten am Bohrlochmund mit der Ankermutter befestigt. Sie dienen vornehmlich der Verbindung zum Verzug, welcher in der Regel aus einem Drahtgeflecht mit Drahtstärken zwischen 3 und 4 mm und Maschenweiten zwischen 50 und 75 mm besteht.

Das Einbringen der Anker ist über Ankerbohr- und Setzeinrichtungen mechanisiert und bietet vor allem bei Streckenauffahrungen mit Teil- oder Vollschnittmaschinen die Möglichkeit optimierter Betriebsabläufe. Bild 4 zeigt das Einbringen der Anker und des Verzuges. Der Maschendraht wird von der Firstmitte ausgehend über den Umfang abgerollt und mit Ankern über den Umfang arretiert.

Bild 4: Einbringen der Anker und des Verzuges

Fig. 4: Support with Bolt and Lagging

5 WEITERENTWICKLUNGEN

5.1 Nachgiebige Ankersysteme

Besondere Bauformen und differenzierte Verfahrensweisen zeigen im Einsatz erfolgreiche Möglichkeiten, die Anwendungen des Ankerausbaus zu erweitern.

So haben neu entwickelte Kombianker ein sehr hohes Verformungsvermögen bei Zug- und Scherbelastungen. Es führt in der Anwendung dazu, daß von diesem Anker weit mehr Streckenkonvergenz zerstörungsfrei aufgenommen werden kann als von starren Ankern. Bild 5 zeigt den konstruktiven Aufbau eines solchen Ankers und Kennlinien der Kraftaufnahme und Verformung im eingebauten Zustand. Ein Nachgiebigkeitselement am Ankerkopf sorgt dafür, daß zusätzliche Verformbarkeit konstruktiv erreicht wird.

Der Ankerschaft besteht bei dieser Bauform aus einem Kernstab, welcher von einem Rohr umhüllt wird. Der Kernstab entspricht im Durchmesser und Kraftaufnahmevermögen einem Ankerstab der Tragkraft 250 kN. Das zusätzliche Hüllrohr erhöht den tragenden Querschnitt und führt zu Ankertragkräften von ca. 470 kN.

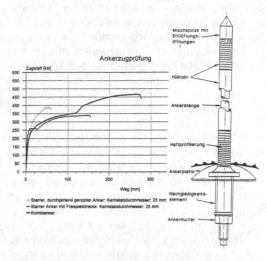

Bild 5: Kombigleitanker und zugehörige Kennlinien des eingebauten Ankers bei Zugbelastung

Fig. 5: Combi-Slip-Bolt and Accessory Characteristics of The Bolt in The Case of Tension Load Stress

Dehnungen des eingebauten Ankers von maximal 270 mm unter Zugbeanspruchung und Scherwege bei zusammengesetzten Belastungen auf Zug und Scherung von 240 mm können von einem derartigen Anker aufgenommen werden. Starre Bauformen sind im Vergleich hierzu in der Lage nur ca. die Hälfte des Weges bei Zug- und 1/6 des Weges bei zusammengesetzter Beanspruchung schadlos zu überstehen. Gerade die Verformbarkeit bei zusammengesetzter Belastung ist jedoch für Gebirgsanker relevant, da Auflockerungen in der Regel an den vorhandenen Schwachflächen im Gebirge orientiert sind und so nicht-axiale Belastungen der Anker bewirken.

Kritische Verformungen treten innerhalb des Streckenmantels einer bogenförmigen Strecke vor allem in den First- und Oberstoßbereichen auf. In der Anwendung von Ankerausbau resultieren daraus sogenannte 'Modulare Ankersysteme', mit unterschiedlichen Ankerbauformen in Stoß- und Firstbereichen. So kann ausschließlicher Ankerausbau bis zu Konvergenzen von ca. 40 % ausgelegt werden, während starre Anker (Bild 3) höchstens 20 % Streckenkonvergenz überstehen können. Um derartige

Systeme wirkungsvoll einsetzen zu können, werden angepaßte Verfahren zur Untersuchung der Verformungen angewendet. Numerische Modelle bieten neue Ansatzpunkte, die jedoch an dieser Stelle nicht erörtert werden können.

5.2 Flexible Anker

In untertägigen Hohlräumen mit üblichen Streckenquerschnitten bis 26 m² sind die Platzverhältnisse oft sehr begrenzt. Dies führt in der Praxis dazu, daß die gebirgsmechanisch erforderlichen Ankerlängen mit starren Ankern oft nicht realisierbar sind. Zu diesem Zweck wurden Seil- und Bündelanker entwickelt, deren Flexibilität größere Ankerlängen erlauben. Alternativ vorstellbare und außerhalb des Bergbaus übliche, gemuffte Ankerstangen verbieten sich, da die Verformbarkeit des Ankers an Muffenverbindungen sehr gering ist und eintretende Konvergenzen zum frühzeitigen Versagen der Anker führen. In Verbindung mit neu konstruierten Ankerbohr- und Setzeinrichtungen, dem Flex-drill-System, zeigen Betriebsversuche mit flexiblen Ankern, daß besonders am Übergang von Streb zu Abbaubegleitstrecke eine Gebirgsvergütung erzielt wird. So kann die Leistungsfähigkeit des Gewinnungsbetriebes erhöht und die Wirtschaftlichkeit gesteigert werden.

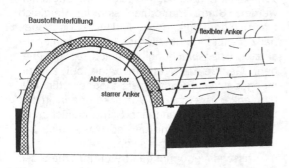

Bild 6: Gebirgsvergütung im Streckensaum durch Bündelanker
Fig. 6: Strata Consolidation in Roadsides by Flexbolts

In Bild 6 ist abschließend ein Einsatzbeispiel gezeigt: 4,0 m lange Bündelanker im Saumbe-

reich einer Abbaubegleitstrecke verwendet, um einen hohen Sicherheitsstandard und eine Leistungssteigerung des Gewinnungsbetriebes zu erreichen.

Die Anker werden unter einem Winkel von 50 bis 70 gon zur Horizontalen aus einer Strecke mit 5,0 m Breite in den Stößen gesetzt und somit die angeschnittenen instabilen Flözhangendschichten verfestigt. Geringerer Nachfall und optimierte Ausbautechnik aus Streb-Strecken-Übergang sind die Folge. Ein Ankeransatzpunkt in ca. 1,20 m Höhe wäre mit keiner anderen vorhandenen Technik realisierbar.

6. ZUSAMMENFASSUNG UND AUSBLICK

Ankertechnik wird im deutschen Steinkohlenbergbau als Ausbau und Hilfsmittel der Gebirgsvergütung eingesetzt. Die geologischen und geomechanischen Einflüsse der Lagerstätte und des Strebbaus bilden ungünstige Voraussetzungen, um statische Bemessungsverfahren anzuwenden. Aus der über 40jährigen Erfahrung mit Ankerausbau resultiert eine Methode der Dimensionierung, in der sicherheitliche Aspekte im Vordergrund stehen. Sie hat sich seit nunmehr 20 Jahren bewährt, bildet jedoch gleichzeitig eine Begrenzung der Anwendbarkeit auf Gebirgsverhältnisse mit geringer Konvergenzerwartung. Dies liegt begründet in den bisher üblichen einfachen Ankerbauarten starrer Ankerstangen mit geringem Verformungsvermögen.

Weiterentwicklungen von Ankerbauarten zielen daher auf eine größere zulässige Verformbarkeit des eingebauten Ankers. Der neuentwickelte Kombianker zeigt in den bisherigen praktischen Anwendungen, daß auch Streckenkonvergenzen bis zu ca. 40 % mit Ankerausbau beherrschbar sind.

In der letzten Zeit werden zudem Anwendungen angestrebt, in denen verschiedene Ausbausysteme zusammenwirken, um den stetig steigenden Belastungen wachsender Gewinnungsteufe und zunehmendem Durchbauungsgrad der Lagerstätte zu begegnen. Auch hieraus ergeben sich Möglichkeiten, die ergonomischen und wirtschaftlichen Vorteile dieser Ausbaumethode im deutschen Steinkohlenbergbau in größerem Umfang als bisher zu nutzen.

LITERATUR

Götze, W.; Stephan, P.; Wiegand, H.-A.: Anwendungsgrenzen, Einsatzbereiche und künftige Entwicklung der Ankertechnik, Glückauf 118 (1982), Nr. 21

Götze, W.: Bruchvorgänge und Verformungen um Strecken im geschichteten Gebirge. Ergebnisse aus Modellversuchen und Untertagebeobachtungen, Diss. Berlin 1968.

Wittenberg, D.: Entwicklung und Prüfung von Ankersystemen für Anwendungen im hochbeanspruchten Gebirge, 2. intern. Kolloquium 'Ankerausbau im Bergbau' der RWTH Aachen, Aachener Beiträge zur Rohstofftechnik und -Wirtschaft, März 1995.

Krahe, J.: Der Ausbau am Übergang Streb-Strecke in flacher Lagerung, Glückauf 123 (1987, Nr. 19, S. 1205-1211. Der Aufwand für das Abfangen der Streckenbögen am Übergang Streb-Strecke, Glückauf 115 (1979), Nr. 24, S. 1177-1179.

NN: Rösler Drahtwerke: Rollbare Verzugmatte für Ankerausbau, Firmen-Information

3 Types of anchors
Ankertypen

Internal unbonded cables/anchors for temporary protection of buildings

Innere Spannglieder/Anker ohne Verbund für temporäre Sicherung von Gebäuden

A.B.Ajdukiewicz & A.T.Kliszczewicz
Silesian Technical University, Gliwice, Poland

ABSTRACT: Many cases of different building structures are known in which some actions are temporary and/or occur once only or few times during the lifespan of a building. Such cases are connected usually with the new construction works beneath existing buildings or excavations in the vicinity. Similar situations are also typical for buildings in regions where influences of deep underground mining works cause specific combinations of serious transient deformations of subsoil. Sometimes, these actions are considered from the beginning at design of building, while in other cases, probably more often, the necessity of strengthening regards existing structures for which such actions have not been expected at design.

Since the additional strengthening members, particularly in underground parts of buildings, should carry the large forces the strong, preferably prestressed anchors must be used. On the other side the period of special influences is relatively short and introduction of permanent strengthening would not be the proper way. Therefore, the external unbonded tendons should be considered, with the possibility of their removal. Some special problems usually appear when such a protection of the structure must be done from inside of the building only. The construction of temporary protection for multi-storey buildings has been considered in design work and some tests. The results of these works and examples of constructional details are the main essence of the report presented.

ZUSAMMENFASSUNG: Es gibt viele Beispiele für Baukonstruktionen, bei denen gewisse Beanspruchungen vorübergehend sind und nur einmal oder selten während ihres Bestandes auftreten. Solche Fälle treten meist bei der Herstellung von Stollen unter bestehenden Gebäuden oder in der Nähe befindlichen Baugruben auf. Ähnliche Situationen treten auch bei speziellen Untergrundverhältnissen in Bergbaugebieten auf.

Manchmal wird dies schon bei der Planung von Bauwerken berücksichtigt, meist aber werden Verstärkungen der Fundamente oder tragender Teile der Gebäude erst nachträglich eingebaut, wobei die Vermeidung von Spannankern große Bedeutung hat. Durch die temporäre Beanspruchung ist jedoch eine dauerhafte Verstärkung nicht erforderlich. Daher ist in vielen Fällen der Einsatz von außen liegenden Spanngliedern mit der Möglichkeit, diese wieder abzubauen, zweckmäßig. Die dabei auftretenden Probleme sind Gegenstand vieler Projekte und Forschungen. Ergebnisse dieser Arbeiten und Ausführungsbeispiele sind Inhalt dieses Beitrages.

1. INTRODUCTION

A strengthening of buildings as a protection against transient impedency from a variety of underground works - beneath and in vicinity alike - is nowadays necessary more and more often.

Typical stuations are the following (Fig.1):
- new road or railway tunnels uder built over areas,
- new pipelines in urban areas,
- deep foundations of new buildings or other excavations near existing buildings,
- ground subsidence due to changes in underground water conditions,
- influences of underground human activity resulting in subsoil deformation, particularly mining works.

These actions result sometimes in significant internal forces in buildings, particularly in their lowest parts.

2. STRUCTURAL ANALYSIS ASPECTS

Recently, the simplified analysis of endangered buildings was based on the expected displacements, separated into three components:
- unequal vertical subsidence, resulting in subsoil curvature,
- horizontal deformation of subsoil,
- inclination of ground surface causing inclination of the whole building.

These components were considered separately, and results of calculations were finally summed. Such an approach was characteristic for the mining subsidence analysis, since the combination of actions was too complex for simultaneous taking into account.

Other common approximation was the limitation of the analysis to linearly-elastic stage. It led to extremly large forces in foundations, which were hardly covered by reinforcing members.

Presently, the rough approximations may be replaced by much more adequate analysis.

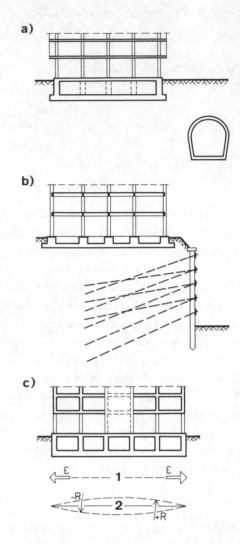

Fig.1. Examples of transient impedency of buildings caused by underground activity:

a) tunnel driving, b) excavations in neighbourhood, c) influences of subsidence from deep mining works

1 - horizontal tensile/compression deformations,

2 - changes of subsoil curvatures

Bild 1. Beispiele für vorübergehende durch unterirdische Arbeiten verursachte Gefährdung eines Gebäudes: a) Aushöhlen eines Tunnels, b) tiefe Gruben in der Nähe, c) Einflüsse der bergbaulichen Senkung:

1 - horizontale Deformation von Druck und Zug,

2 - Veränderungen in der Untergrundkrümmung

Common consideration of the system consisting of building and subsoil, subject to various influences of underground actions is possible now. The advanced nonlinear models for structures and ground can be analyzed by incremental-iterative numerical methods.

Different paths of strains and stresses together with taking into account many variants of time-dependent phenomena allow to estimate the range of realistic internal forces in structural members quite accurate (Ajdukiewicz & Majewski 1994). This is a good basis for the proper design and construction of strengthening members.

There are two basic cases in practice. The first one is the design of new buildings, which should be protected against incidental actions. This is typical problem of design and construction if only the values of expected subsoil deformations are determined.

The second case is much more popular today. It concerns existing buildings which should be strengthened to carry previously not expected actions from subsoil deformation. The constructional possibilities of reinforcing are usually limited, so the advanced materials and strengthening methods should be taken into account.

Safety of structures and resistance against cracks at transient actions must be considered individually, according to the variety of structures. In general, the forces to be carried are relatively large. Therefore, strong reinforcing anchors in form of prestressed cables have to be usually taken into account.

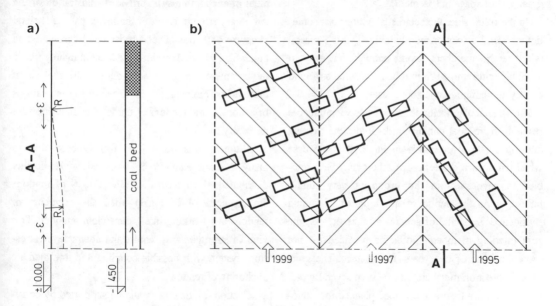

Fig.2. Example of the group of similar buildings in the housing estate endangered with mining subsidence in different periods;
a) simplified cross-section of the variable subsidence trough,
b) layout of the planned longwall mining activity under the estate in successive periods
Bild 2. Beispiel für eine ganze Gruppe ähnlichen Gebäude innerhalb einer Wohnsiedlung,
die durch Bodensenkung gefährdet werden;
a) vereinfacher Schnitt durch veränderliche Senkungsmulde,
b) Skizze der geplannten bergbauliche Wirkung auf die Gebäudewände in folgenden zeitarbeitschnitte

For example, in the building 27m long and 13m wide, which may be subject to moderate mining influences in longitudinal direction (expected values: $+\varepsilon = 0.3\%$, $R_{min} = 12$km), the total tensile force in this direction has been assessed as large as about 4.2MN (420 t) in the middle of building length. This value was calculated from the condition of cracks in existing R.C. basement limited to 0.1mm. More than 85% of the force should be carried by additional external reinforcement - anchored bars or prestressing cables. In spite of the fact that transient actions in full range may be considered as accidental and the load factor is usually taken as equal to 1.1 or even 1.0 only at reinforcement design, the resulting number of anchors is very large and they may fill a great part of space in basement.

On the other side, the accidental actions last sometimes several weeks or months only. This is a reason why the the members strong enough, but easy to assembly and removal should be considered as a kind of recycled equipment in some cases.

The spectacular example of successive influences resulting from coal mining works under the existing housing estate is presented in Fig.2. It is the case when temporary protection members may be assembled in succession in the group of very similar buildings. As the period of development of mining subsidence is usually of several months or less, the strengthening members should be provided for not more than one or two years, for instance. The same set of protection members may be used several times. If there is more than one coal bed planned for exploitation under the same area (after the period of several years necessary for re-balance of the ground) the set of anchors may be reused in the same nests in walls.

3. CONSTRUCTIONAL EXAMPLES

The individual situations do not allow to introduce the general rules for temporary protection.

In mid-rise buildings the actions transmitted from subsoil concentrate mainly in foundation strips and basements (70-80%), but part of them turns over to the higher structural members, e.g. wall or slab ties. The distribution of forces depends on the longitudinal and flexural rigidity of basement structure (Ajdukiewicz & Majewski 1993).

Most often, the structure of basement does not allow to introduce too big prestressing, but demands the large load-carrying capacity of strengthening tendons and their anchorages in case of accidental action. Sometimes, the protection must be ensured against the action of both signs, e.g. hogging and sagging moments resulting from ground unequal subsidence. In such a case more complex arrangement of anchors must be provided. Illustration of the temporary strengthening of the lower part of skeletal structure with box-shaped basement is presented in Fig.3. The external tendons are located mainly inside the basement space, but according to the analysis of the cracked building subject to various combinations of actions, some tendons at the level of first floor are necessary too.

As the actions transmitted from subsoil due to many reasons cannot be assumed as uniformly distributed, the intermediate anchorages of tendons are required (4 in Fig.3); then the segments of tendons are working like independent anchors. This kind of strengthening require the adequate prestressing system with possible coupling and intermediate anchoring of tendons.

Selection of anchors should be supported by consideration of the ductility of tendons and the probability of unexpected local actions in particular parts of building. From both these reasons in the building presented in Fig.3 the tendons (1) in basement were designed to be stressed up to 40% of their strength f_{py}, while the smaller tendons (2) in slab level - up to 55% of f_{py}.

The example in Fig.3 shows the detached building with full access from outside (5) to fasten the external anchorages (3). Very often the access to building basement from outside is not possible due to adjacent buildings. Then the strengthening should be provided from inside only. In such a case some kinds of ten- dons are useful only, particularly those which could be joined from relatively short segments by means of simple couplers. The simplified exemplary situation is presented in Fig.4.

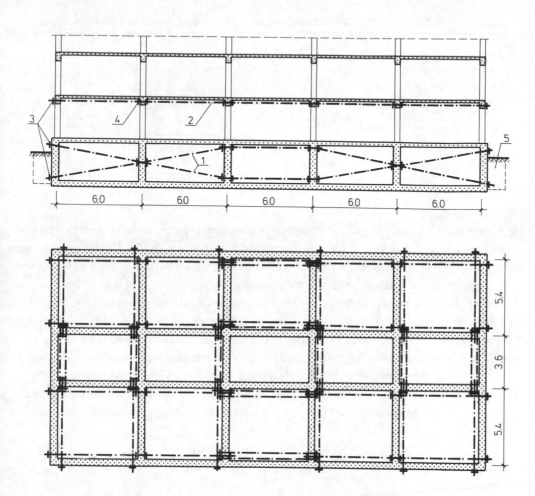

Fig.3. Unbonded tendons for temporary anchoring of the box-shaped structure of basement in skeletal building endangered by mining influences: 1 - anchors for protection against horizontal tensions and curvatures of both signs, 2 - anchoring of structure in slab levels, 3 - external anchorage, 4 - intermediate anchorage, 5 - possible access from outside

Bild.3. Freispielanker für temporäre Verankerung "Kasten" Kellerkonstruktion in einen durch die bergbauliche Einflüsse gefährdeten Skelettgebäude: 1 - Sicherungsankern gegen horizontale streckungen und kümmungen beder Zeichen, 2 - Verankerung der Konstruktionen im Deckestand, 3 - äußere Verankerung, 4 - mittelbare Verankerung , 5 - möglicher Zugang von Außen

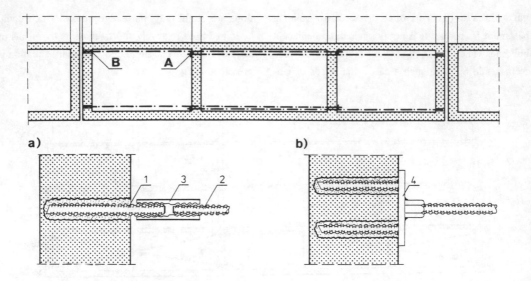

a)   b)

Fig.4. Example of temporary longitudinal strengthening of the bottom-part of the building existing between adjacent buildings (no access from outside): A - stressing anchorage of bars, B - fixed anchorage : a - single anchorage up to 120kN load-capacity in R.C. wall of about 0.30m thickness, b - combined anchorage with several fasteners in wall; 1 - threadbar (e.g. $\varnothing$32mm) anchored in resin,
2 - main removable bar (e.g. $\varnothing$25mm), 3 - transition coupler (32/25mm), 4 - anchor plate with nut
Bild.4. Beispiele für temporäre Längestärkung des unteres Teils eines Gebäudes das zwieschen zwei anllegenden Gebäuden gelegen ist (kein Zugang von Außen): A - Spannankerung des Stabglied, B -Blindverankerung
a - Einzele Verankerung von Tragfähigkeit bis 120kN in einer Stahlbetonwand ungefähr 0.30m dick,
b - Komplexverankerung mit einigen Einzelverankerungen in einer Wand;
1 - Rippenstab (z.B.$\varnothing$32mm) Verankert im Harz, 2 - Hauptrippenstab zum wegscheiben (z.B.$\varnothing$25mm),
3 - Muffekuplung (32/25mm), 4 - verankernde Festplatte mit den Mutter

At least three requirements are important in the situation presented in Fig.4:
- possibility of introduction of tendons in segments,
- stressing anchorages (A) inside the building,
- fixed anchorages (B) in external walls.
The example in Fig.4 shows the proposal for the real building. Threadbar anchors $\varnothing$25mm have been designed with fixed anchorages of two types: (a) for strong concrete in the wall, and (b) for uncertain strength of concrete.

4. SELECTION OF ANCHORS

The choice of proper anchoring system for the temporary strengthening of buildings requires consideration of many aspects. The following are usually the basic:
- design for an existing building or newly erected,
- straight or broken lines necessary for tendons,
- strength of building members for possible stressing and fastening of anchorages,

- accessibility to the building from outside,
- period of necessary remaining of protection elements in the building,
- humidity of the environment inside and outside the building,
- corrosion aggressiveness expected,
- re-use of the strengthening members.

In newly designed buildings the location of straight or curved tendons inside walls, beams or ties is possible in cored ducts as well as in ducts formed by sheath left in place, while in existing buildings the possibility is reduced in practice to external tendons, with local contacts with a structure.

The important aspect of anchors selection is the strength of material in the members where tendons have to be fastened. In existing buildings there are usually no elements massive enough to fasten the strong anchors. Other limitations are connected with the access of prestressing jacks, as the space available is usually small. These reasons essentially differentiate the systems useful at strengthening of buildings from those used at temporary prestressing of bridges or other large structures. The necessary tests may also vary, according to the recommendations of FIP Commission 2 - 1991, 1992. For instance, to strengthen the building shown in Fig.3, where external access was possible and inclined cables were required, two well known systems were recommended (according to *Post-Tensioning Manual* - 1990): monostrands Dywidag-MS or VSL-MS (Fig.5a) and threadbars Dywidag TS (Fig.5b). Both of them are useful for small and medium forces, in the environment of moderate aggressiveness. For cases of severe or very severe exposure conditions the systems without metal parts can be borrowed from rockbolt technology (e.g. Weidmann FRP, 1992) - Fig.5c.

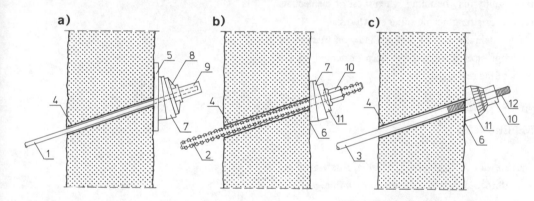

Fig.5. Proposals for anchorage practice of inclined unbonded tendons for moderate prestressing forces:
a) Dywidag Monostrand System - effective force up to 170 kN, b) Dywidag Threadbar System, up to 500kN, c) metalless Weidmann FRP- Rockbolt , up to 100kN; 1 - strand ⌀15mm (0.6") in polyethylene tube filled with grease, 2 - threadbar ⌀25mm, 3 - fiberglass/resin bar ⌀22mm, 4 - drill-hole throughout concrete wall, 5 - anchor plate, 6 - smoothened concrete surface, 7 - wedge washers, 8 - typical Dywidag anchorage with wedge jaw, 9 - grease cap, 10 - anchor nut, 11 - inclined pad with spherical bearing, 12 - thread sleeve

Bild 5. Praktische Beispiele der Verankerungen für geneigte Freispielanker von der durchschnittlichen Spannkraft: a) Dywidag MS - Effektivkraft bis 170kN, b) Dywidag TS - bis 500kN, c) Weidmann GFK Ankertechnik - bis 100kN; 1 - Seil ⌀15mm mit dem Schmiermittel ausgefüllen Polyäthylenmantel, 2 - Rippenstab ⌀25mm, 3 - Glasfaser/Harz Stab ⌀22mm, 4 - Durchbohre durch Stahlbetonwand, 5 - Druckplatte, 6 - geschliftene Betonfläche, 7 - Keilunterlagscheibe, 8 - typisch Verankerung Dywidag mit der Kegelbacke, 9 - Schutzaufsatz mit dem Schmiermittel, 10 - Mutter, 11 - Keilkugellager, 12 - Gewindemuffe

5. CONCLUSIONS

The temporary strengthening of buildings to carry transient actions transmitted from subsoil became the more and more frequent necessity in last time. Although the most spectacular examples can be found in mining regions, similar problems may result from changes in subsoil caused by other reasons.

Particular structural situations should be considered individually, but usually the values of additional forces in structures could be estimated rather roughly as well in newly designed as in existing buildings. Therefore, the strengthening members should be provided with significant margin of safety at considering ultimate limit state or resistance against cracking.

From functional and economical points of view these relatively strong and costly protection members should be removed when not necessary. The best opportunity may be found when most of elements is reused for strengthening other structures.

The main aspects of design and construction of buildings strengthening have been presented on the basis of the real situations.

REFERENCES

Ajdukiewicz A. & Majewski S. 1993. Analysis of reinforced-concrete wall structures with regard to nonlinearity of structure-subsoil system. Proceedings of International Conference on *"Analytical Models and New Concepts in Mechanics of Structural Concrete"* - May 1993; Bialystok University of Technology, Poland: 9-26.

Ajdukiewicz A. & Majewski S. 1994. Nonlinear analysis of building-subsoil system subjected to deformations due to mining subsidence. *Archives of Civil Engineering*. Polish Academy of Sciences, Vol.XL, No 3-4/1994: 319-344.

FIP Commission 2. 1991. State-of-the-Art-Report. *Materials and Systems for External Prestressing*. 7th revised draft, July 1991.

FIP Commission 2/4. 1992. Prestressing Materials and Systems - Practical Construction. *Recommendations for Acceptance of Post-Tensioning Systems*. Version 10th, March 1992.

Post-Tensioning Manual. 1990. 5th edition. Post-Tensioning Institute, Phoenix, USA.

Weidmann - FRP (Fiberglass Reinforced Plastic). 1992. Presentation in Proceedings of the 2nd International Conference on *"Fracture and Damage of Concrete and Rock"*. Vienna, November 1992.

Anwendung von Ankern aus Glas-Epoxid-Verbundstoff im polnischen Salzbergbau

The application of glass-epoxide bolts in Polish salt mining

B. Barchanski
Berg- und Hüttenakademie Krakow, Poland

KURZFASSUNG: Das Salzbergwerk "Wieliczka" ist der älteste in der Welt tätige Bergbaubetrieb. Infolge des im Laufe von 700 Jahren geführten Abbaus der Salzlagerstätte entstanden der aktuelle Stand und die aktuelle Struktur des Bergwerkes. Seit 1989 werden - neben anderen Sicherungsmaßnahmen - in den Kammern Anker aus Glas-Epoxid-Verbundstoff verwendet. Diese Anker haben eine Länge von 10 m; der Außendurchmesser beträgt ca. 27 mm und das im Kern des Ankers befindliche Rohr zur Entlüftung und Dränagierung hat einen Durchmesser von ca. 14 mm.
Die Anker werden in Bohrlöchern mit Zement-Magnesit-Mörtel befestigt.

ABSTRACT: The salt mine in Wieliczka is the oldest mine known in the world. Its present shape and structure are the result of the over 700-year-exploitation of the salt deposit. The workings, being over 300 km long, are located on 9 levels at the depth ranging from 50 to 300 m. There are 2 040 openings whose total cubic capacity is approx. 7,0 - 7,5 million m³. Because of the unique character of the mine (the "Wieliczka" mine is on the UNESCO list of places of the world cultural heritage) for a long time some of the openings have been undergoing a time- and labour-consuming process of saving from destruction (compression). Since 1989, apart from other preservation means, bolts made of glass-epoxide composite have been applied in the openings. Their length is up to 10 m, the outside diameter equals 27 mm and the hole in the rod is approx. 14 mm in diameter. The bolts are set in a cement-magnesit grout.
A systematic monitoring of the technical condition of the openings has confirmed that the applied method produces positive results. Up till now approx. 50 000 running meters of this type of bolts have been introduced. The basic argument for applying them in salt mining, apart from their strength being comparable to that of the steel bolts, is their incomparably higher durability and rust resistance as well as an easy way of inserting.

1.0 Einleitung

Das in der Nähe der historischen Hauptstadt Polens Krakow gelegene Salzbergwerk in Wieliczka wurde von den polnischen Königen gern ihren Gästen als Beispiel präsentiert, das den Reichtum des Landes beweisen sollte. Dieses Bergwerk ist ein Zeuge der Geschichte.

Bis zum heutigen Tag ist aus jener Zeit ein Teil der Grubenbaue erhalten geblieben. Es wird geschätzt, daß das Volumen aller Grubenbaue in Wieliczka 7,5 Mio. Kubikmeter Hohlräume beträgt.

Diese Einmaligkeit des Salzbergwerkes in Wieliczka ist offiziell von der UNESCO anerkannt und im Jahre 1978 ist das Salzbergwerk in die Liste des Weltkulturerbes eingetragen worden. Die Begründung von

der UNESCO für die Wieliczka-Grube lautete wie folgt: „Das Salzbergwerk in Wieliczka ist das einzige Bergbauobjekt in der Welt, welches ununterbrochen seit dem Mittelalter bis heute in Betrieb ist. Seine originellen Abbauhöhlräume (Gehsteige, Rampen, Abbaukammern, Schächte) haben eine Gesamtlänge von etwa 300 km. Sie wurden auf 9 Grubensohlen bis zur Teufe von etwa 327 m untergebracht; sie stellen alle Etappen der Entwicklung der Bergmannstechnik in verschiedenen Epochen dar. Die einmalige Sammlung von Einrichtungen und Bergmannswerkzeugen im Inneren der Grube stellt in der natürlichen Umwelt die Systeme des Salzabbaus sowie die Entwässerung, Bewetterung und Beleutung der Grube vor. Hier befinden sich auch Denkmäler, die mit Salzsiedereien verbunden sind, deren Beginn bis in die prähistorischen Zeiten reicht (Neolithikum). Zu denen, die man heute sogar in der ganzen Welt nicht mehr trifft, gehören Bergmannskunstschätze, nämlich Bildhauereien im Salz und ganze reich geschmückte untertägige Kapellen".

Im Jahre 1992 fand ein dramatischer Kampf um die Rettung des Salzbergwerkes in Wieliczka vor einer Überschwemmung statt. Die Wassergefährdung wurde zum Glück beherrscht.

Hinsichtlich der Einmaligkeit des Salzbergwerkes und des äußerst komplizierten Prozesses der Rettung der Salzgrubenbaue werden in dem folgenden Aufsatz die Erörterungen über die Anwendung von neuen Ausbauarten in polnischen Salzbergwerken ausschliesslich auf den in Wieliczka angewendeten Ausbau beschränkt.

2.0 Bergtechnisch-geologische Bedingungen

Den geologischen Schnitt über die Wieliczka-Salzlagerstätte stellt die Abb.1 dar.

Die in der Salzserie auftretenden Gesteine bilden eine plattenförmige und salzstöckige Lagerstätte. Die plattenförmige Lagerstätte wird durch Sandsteine und Ton mit Anhydrit und Gips begleitet.

Die Mächtigkeit dieser Schichten beträgt ca. 50 m. Auf den unteren Schichten lagert eine Serie von 3 bis 5 Schichten von Grünsalz, die mit Durchwachsenem (Ton und Anhydrit) unterteilt sind.Die Mächtigkeit dieses Grünsalzes schwankt von 0,5 bis 4,0 m; und die der Zwischenlagerungen von 0,2 bis 1,5 m. Das Grünsalz besteht aus dicken Salzkörnern von Durchschnitt bis zu einigen Zentimetern mit kleinen Körnern von Gips und Anhydrit. Über dem schichtenförmigen Grünsalz lagert das Flöz von sogenantem „Schibiker" Salz, das von einer Schicht von Ton und Anhydrit geteilt wird. Das Schibiker- Salz wird von einem umfangreichen Komplex des des sog. spissum sal (lat.) von einer tonigen oder schlammigen Schicht geteilt, die die Mächtigkeit von 0,2 bis 0,5 m hat. Der unterschiedliche petrographische Bau des Salzes wurde durch die Änderungen in der miozänen Salzserie verursacht.

Unter einzelnen Salzschichten befindet sich dünnes Verwachsene aus Ton, Sandstein und Anhydrit, dessen Mächtigkeit von wenigen Millimetern bis zu einigen Zentimetern beträgt. In dem mittleren Teil des Komplexes tritt das Zwischenlagerungen von der Mächtigkeit von 2 bis 5 Metern.

Die gesamte Mächtigkeit des Salzkomplexes beträgt bis zu 30 Metern.

Die plattenförmige Lagerstätte lagert über der stockigen Lagerstätte und alle ihren Schichten sind durcheinander gemischt ohne die Folge der einzel-

nen Schichten einzuhalten. In der unmittelbaren Nachbarschaft des Salzes (Kontaktzone) kommt am haufigsten der Zuber vor. Das Bergemittel bildet vor allem graues Ton ohne deutliche Schichtung. Außerdem treten in kleinen Mengen Bruchstücke von Ton-Kalk-Dolomit-Gestein mit Stücken von Anhydrit und Flysch.

Diese sehr schweren geologischen Bedingungen sind noch dadurch erschwert, daß das Bergwerk zu der 1. Kategorie der Wassergefährdung zählt, und manche Reviere zu der 2., und sogar der 3. Kategorie.

In einem so unter dem geologischen Gesichtspunkt komplizierten Gebirge verursachte der Mensch (durch den über 700 Jahre geführten Abbau) eine Gefährdung für die im Salz hergestellten Grubenbaue. In vielen Kammern (z.B. in der Kapelle der Seligen Kinga , dieser Grubenbau hat die Ausmaße 12m x 18m x 55m) entstanden Risse und Klüfte aufgrund des komplizierten Standes der Gebirgsspannungen. Die zuständigen Behörden unternahmen die Maßnahmen, die o.g. Grubenbaue vor der Zerstörung zu retten. Zu den wichtgen Entscheidungen in dieser Hinsicht gehörte die Wahl der Ausbauart. die dem Prozess der Zerstörung entgegenwirken könnte. Infolge der komplexen Untersuchungen hat man sich im Jahre 1989 entschlossen, den Ankerausbau aus Glas-Epoxid-Verbundstoff einzuführen.

3.0. Ankerausbau aus Glas-Epoxid-Verbundstoff

3.1. Untersuchungen zu der Eignung des Ankerausbaus
3.1.1. Einführung

Die unten angeführten Untersuchungen wurden zu dem Zweck durchgeführt, die Genehmigung des Bergamtes für die Zulassung der neuen Ankerausbauart in dem Salzbergwerk Wieliczka zu erhalten. Die Untersuchungen umfaßten die Bestimmung (2):

- der Zugfestigkeit und der Scherfestigkeit der Ankerstange aus Glas-Epoxid-Verbundstoff,
- der Festigkeit der o.g. Ankerstange in der Umhüllung aus dem Magnesit-Mörtel.

3.1.2. Die Grundparameter des zu untersuchenden Ankerausbaus

Es wurden die Ankerstangen untersucht (sog. Tragelement), die aus dem Glas-Epoxid-Verbundstoff hergestellt wurden, auf der Basis der Glsafasern und des Epoxidharzes als Verbundstoff.

Der Bau der Stange stellt die Abb. 2 dar. Der innere Drain - ein Röhrchen von dem Durchmesser 12/8

288

mm , aus hartem Polyäthylen- bildet ein Kern, auf dem Glasfasernstränge,getränkt mit Epoxidverbundstoff, verseilt werden. Das Formen der Glasfasernstränge erfolgt durch ihre Verlegung parallel zu der Drainachse mit der Verdrehung bis zu zwei Umdrehungen auf 1 m der Drainlänge. Im Prozess des Formens der Stange werden die einzelnen Fasernbündel verpresst, damit die äußere Oberfläche mit deutlichen Eintiefungen und Wölbungen entsteht, die das Haftvermögen der Stange an den Magnesitbindestoff erhöht. Die spirale Form der Bündel im Verhältnis zu den parallel zu der Stangenachse verlegten Bündel erhöht die Fähigkeit zur Energie-Akkumulation und die Zugfestigkeit. Das ist analog wie in den Stahlseilen.

Der Drain ist nicht kohärent mit der Stange und läßt sich ungehindert ausziehen.

Die Untersuchungen der Zugfestigkeit der Stangen von gemitteltem Durchschnitt 24,6 mm (siehe Abb. 2) haben erwiesen, daß die Zerstörung von der Kraft $F_1 \approx 200$ kN hervorgerufen wird. Die Untersuchungen, die die Scherkraft derselben Stange bestimmten, haben gezeigt,daß die Scherkraft $F_2 \approx 36$ kN beträgt.

Die Scheruntersuchungen derselben Stange in der Umhüllung aus Magnesit-Mörtel (siehe Abb.) haben gezeigt, daß bei der Kraft:

- $F_3 \approx 26$ kN erfolgt die Abscherung der Umhüllung aus dem Magnesit-Mörtel ,

- $F_4 \approx 36$ kN erfolgt die Abscherung der Stange.

Bei dem Auftreten des komplizierten Spannungsstandes, z.B. Abscheren mit der Biegung, beträgt die Zerstörungskraft $F_5 \approx 60$ kN.

3.2. Praktische Anwendung des neuen Ankerausbaus im Salzbergwerk "Wieliczka".

3.2.1. Einführung

Nach einer Reihe von komplexen Untersuchungen, die teilweise oben dargestellt wurden, hat das Bergamt seine Genehmigung erteilt, die neuen Anker in dem Bergwerk Wieliczka anzuwenden.

Die Technologie der Einführung der neuen Anker, die Einführung selbst und Anwendung bis zum heutigen Tag hat die Firma "Przedsiębiorstwo Budowy Kopalń" aus Lubin (Unternehmen für Schachtanlagenbau Lüben) übernommen (Legnica-Głogów-Kupferrevier - LGOM).

3.2.2. Die Grundparameter des neuen Ankerausbaus, der im Salzbergwerk Wieliczka angewendet wird.

Nach den erfolgreichen Versuchen in situ, die durch das Unternehmen "Przedsiębiorstwo Budowy Kopalń" Lubin nach der von dieser Firma entwickelten Technologie durchgeführt wurden, hat man begonnen, die technischen Probleme zu lösen, die mit der Einführung der Anker im industriellen Ausmaß verbunden sind.

In diesem Zusammenhang hat man zwei Themenbereiche untersucht:

A. Die mit den Ankern selbst verbundenen Fragen,

B. Die mit der Technik des Einsetzens der neuen Anker im Salzgebirge verbundenen Probleme.

A. Die zur praktischen Anwendung in den Grubenbauen der Grube Wieliczka zugelassenen Anker sollen folgende Voraussetzungen erfüllen:

- Die Ankerstange hat den äußeren Durchmesser = 27 mm +1 mm; mit dem Loch im Innern (mit glatter Oberfläche) von Durchmesser = 14 mm,

- Die Länge der Ankers (je nach Bedarf) beträgt von 2,2 m bis 10,0 m.

- Die äußere Oberfläche des Ankers soll auf der ganzen Länge "gekerbt, geriffelt" sein, um gute Zusammenbindung mit dem Magnesit-Verbundstoff zu sichern .

- Die Zerstörung des Ankers infolge der Zugkraft kann nicht auftreten, wenn die Kraft kleiner als 100 kN ist.

B. Das Einsetzen der Anker im Salzgebirge soll durch folgende Maßnahmen gesichert werden:

- jeweils soll individuell für jeden Grubenbau, hinsichtlich der sehr unterschiedlichen geologischen Bedingungen, die Einteilung der Ankerlöcher geplant werden.

- Die Bohrlöcher sollen auf der ganzen Länge mit dem Durchmesser = 42 mm gebohrt werden.

- Der Bohrlocheintritt soll auf den Durchmesser = 65 mm erweitert werden.

- Die Anwendung, wenn es notwendig sein soll, der Verzugsmatten aus verschiedenartigem korrosionsfestem Material.

- Die Anwendung des Verbundstoffes von entsprechenden Qualität.

Die Parameter des Verbundstoffes (in diesem Fall - Magnesitmörtel) sollen während des Kontaktes mit der Salzumgebung konstant bleiben. In der flüssigen Phase wird der Verbundstoff in das Bohrloch eingepresst (Druck bis 0,5 mPa) und darf die Struktur der Ankerlochwände nicht verletzen. Das Binden und Erhärten/Verfestigung des Verbundstoffes darf keine Schrumpfungen und Risse hervorrufen. Die Risse könnten in Zukunft den eventuellen Weg zur Wassermigration bilden.

Am günstigsten als Verbundstoff erwies sich der Magnesitmortel. Dieser Mörtel wurde auf der Basis folgender Bestandteile hergestellt:

1. Magnesitbindemittel in Form von kaustischem Magnesit, kaustischem Dolomit oder Magnesiumoxid. Das Bindemittel soll min. 45% MgO und max. 4,5% CaO enthalten.

2. Wässrige Lösung der Chloride oder Sulfide von zweiwertigen Metallen ($MgCl_2$ x H_2O, $FeSO_4$ usw.),

3. Der minerale Füllstoff in Form von Quarzsand oder gemahlener Schlacke.

4. Plastifizierungsmittel (z.B. Bentonit)

5. Die kapillaraktiven Mittel (z.B. Salze der Resinosäuren, Fettsäuren).

6. Süßes Anmachwasser.

Die Art der Einpressung des Verbundstoffes - des Magnesitmörtels - in das Loch mit dem eingeführten Anker ist schematisch auf der Abb. 4 dargestellt. Wenn sich das Verbundmittel in der Entlüftungsleitung zeigt, wird in diesem Moment das Einpressen eingestellt (siehe Abb. 4.9).

4.0. Zusammenfassung und Schlußfolgerungen

Der seit 1989 sukzessiv in dem Salzbergwerk Wieliczka eingeführte neue Ankertyp hat sich vollständig bewährt. Auf Grund der Beobachtungen der über 5000 m eingesetzten neuen Anker kann folgendes festgestellt werden:

1. Die Anker aus Glas-Epoxid-Verbundstoff sowie der Magnesit-Mörtel sind korrosionsfest in der chemisch stark aggressiven Umgebung.

2. Expansion (Ausdehnungsvermögen) des Epoxid-Stoffes im Laufe des Abbindungsprozesses im Ankerloch sichert sehr gute Befestigung des Ankers im Gebirge. Im Fall eines stark rissigen Salzmantels um die Kammer verstärkt ihn der in das Gebirge eindringende Magnesitmörtel in solchem Maß, daß die weitere Destruktion des Mantels gebremst wird. Neben dem Prozess der Verfestigung erfolgt auch sehr gute Abdichtung des Gebirges, was die freie Wassermigration verhindert.

Quellennachweis

Barchanski, B.,*Das Salzbergwerk Wieliczka zur Zeit Agricolas und jetzt,* EMC '94, Agricola-Ehrung 1994, Freiberg 1994

Informationen der Firma Przedsiebiorstwo Budowy Kopaln Lüben 1991-1994

Gruppenarbeit, *Geschichte der Krakauer Saupen,* Ministerium für Kultur und Kunst, 1988

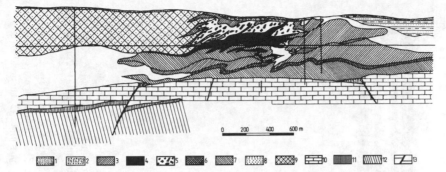

S   N

Abb. 1. Der geologische Schnitt durch die Steinsalzlagerstätte Wieliczka nach J. Poborski
      1. Quartär, 2. Ton - Gips - Hut, 3. „Chodenickie" - Schichten, 4. Salzlagerstätte,
      5. Taubesgestein, 6. Evaporit - Fazies (schwetelig), 7. „Skawina" - Schichten,
      8. Oberkreide (Paleogen), 9. Karpatenflysch (Außenrand), 10. Oberejura,
      11. Mittlere - u. Unterejura, 12. Perm, 13. - Verwerfung.
Fig. 1. Geological cross-section of rock-salt deposit in Wieliczka (acc. to Poborski)
      1. Quaternary system, 2. Silit-gypsum overlay, 3. The Chodenice beds, 4. Rock-salt
      deposit (evaporates of chloride facies), 5 Waste formations surrouding lumps of the
      green salt, 6. Evaporates of sulphate facies (gypsum and anhydrite), 7. The Skawina
      beds, 8. Upper Cretaceous system - Older Tertiary system, 9, Flysch of the Carpathians
      edge, 10. Upper Jurassic system, 12. Permian system, 13. Faults.

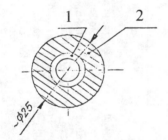

Abb. 2. Der Querschnitt einer Verbundstoffstange
      1. Rohr (Drain) aus Polyäthylen Ø12/8,
      2. Verbundstoff.
Fig. 2. Cross-section of a composite rod
      1. Polyethylene Ø 12/8 pipe (drain),
      2. Composite.

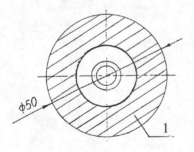

Abb. 3. Querschnitt einer Verbundstoffstange
      in der Umhüllung aus Magnesitmörtel
      1. Magnesitmörtel.
Fig. 3. Cross-section of a composite rod set
      in magnesite grout
      1. Magnesite grout.

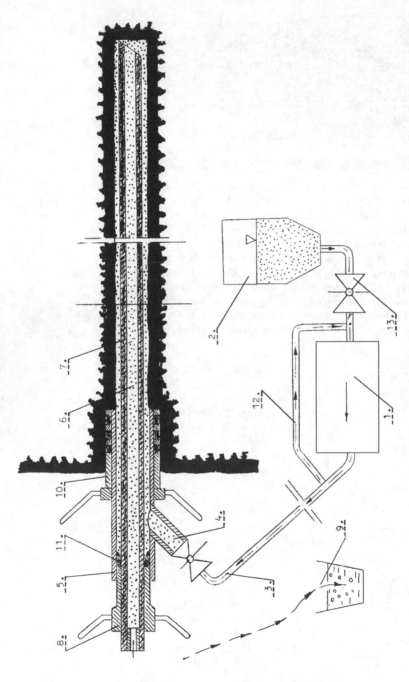

Abb. 4. Schema des verankerten Gebirges samt der Anlage zur Einpressung des Magnesitmörtels (ohne Einhalten des Maßstabes) 1. Zementationspumpe, 2. Behälter mit Verbundstoff / Bindemittel, 3. Einpressungsleitung, 4. Druckstutzen mit Ventil, 5. Kopfstück (Packer), 6. Verbundstoff / Bindemittel, 7. Anker aus Glas-Epoxid, 8. Entlüftungsendstück, 9. Entlüftungskontrolle, 10. Äußere Büchse des Kopfstückes, 11. Gummiringe, 12. Umführungsleitung der Pumpe, 13. Saugventil.

Fig. 4. Scheme of bolted rockmass together with the facilities to pump the magnesite grout (not to scale) 1. Cementation pump, 2. Tank with binding agent, 3. Pumping pipe, 4. Connector pipe with valve, 5. Head, 6. Binding agent, 7. Glass-epcxide bolt, 8. Venting nozzle, 9. Venting check, 10. Head outer sleeve, 11. rubber rings, 12. pump by-pass pipe, 13. suction valve.

292

Anchors in Theory and Practice, Widmann (ed.)© 1995 Balkema, Rotterdam. ISBN 90 5410 577 1

Theory and practice of the Single Bore Multiple Anchor System

Theorie und Anwendung des Single Bore Multiple Anchor Systems

A. D. Barley
Keller Colcrete, Wetherby, UK

ABSTRACT: The new concept of installing a multiple of unit anchors in single bore hole has allowed a considerable increase in anchor capacities in soils and weak rocks. Fixed anchor lengths as much as 30m long can now be efficiently utilised to achieve failure loads of 2000 to 3000kN in clays. Where circumstances demand, the anchors can be double protected for permanent works, or for temporary works the steel tendons can be fully withdrawn after use.

ZUSAMMENFASSUNG: Das neue Ankerkonzept des "SBMA-Ankers" beinhaltet den Einbau mehrerer Einzelanker in ein Bohrloch und ermöglicht dadurch eine beträchtliche Erhöhung der Ankerkräfte in Lockerböden und verwittertem Fels. Es können Haftstreckenlängen bis zu 30m hergestellt werden und es besteht die Möglichkeit Ausziehkräfte von 2000 bis 3000kN in tonigen Böden zu erzielen. Der "SBMA-Anker" kann sowohl als Daueranker mit zweifachem Korrosionsschutz oder als Temporäranker mit kompletter Rückgewinnung der Stahlteile ausgeführt werden.

1.0 INTRODUCTION

A typical ground anchor tendon with a 6m fixed length will, at test load, need to extend some 15 to 20mm at the proximal end of the fixed length prior to any load being transferred to the distal end of the tendon. It is unusual for the elastic behaviour of the grouted soil around the anchor tendon to be compatible with the elasticity of the tendon and allow a uniform distribution of load along the fixed length. Thus, it is widely acknowledged that, in the majority of circumstances, debonding at the tendon/grout or the grout/ground interface must occur as anchor load increases and prior to any load being transferred to the distal end of the fixed length. This phenomenon is commonly known as progressive debonding and is associated with grossly non-uniform distribution of bond stress along the fixed length at all stages of loading. Information has been published by a multitude of researchers on this topic.

Progressive debonding generally results in a highly inefficient use of the in situ ground strength; in the load condition where the ground strength deep in the fixed length is being utilised, the ground strength above has been exceeded and only a residual strength is available there at the anchor soil interface (Fig 1). However, a system that can transfer the load simultaneously to a number of short lengths in the fixed anchor bore without the occurrence of progressive debonding, will mobilise the in situ ground strength efficiently and result in a considerable increase in anchor capacity (Fig 1). This is the principle of the single bore multiple anchor.

2.0 THE SINGLE BORE MULTIPLE ANCHOR CONCEPT

The system involves the installation of a multiple of unit anchors into a single borehole. Each unit anchor has its own individual tendon, its own unit fixed length of borehole, and is loaded with its own unit stressing jack. The loading of all the unit anchors is carried out simultaneously by a multiple of hydraulically synchronised jacks which ensures that the load in all unit anchors is always identical.

In a situation where the load transfer mechanism from tendon to grout eliminates progressive debonding, or where the unit fixed lengths are short enough to be unaffected by the progressive debonding, then in a homogenous stratum the maximum ground strength can be mobilised (by bond) uniformly and simultaneously over the entire fixed length. Furthermore, with such a system there is no theoretical limit to the total overall fixed length utilised whilst in normal anchors little or no increase in load capacity is expected with fixed lengths greater than 8 to 10m.

In the case of non-homogenous soil conditions in the fixed length, each unit fixed length can be designed for the appropriate condition. If the soil is

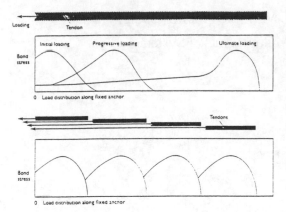

Fig.1

a) Progressive debonding along a normal anchor fixed length
a) Progressiver Spannungsverlauf entlang einer Haftstrecke

b) Single bore multiple anchor simultaneously loads a number of short fixed lengths
b) Single Bore Multiple Anchor Spannungsverlauf mehrerer kurzer Haftstrecken

weaker in the upper fixed length, then the proximal unit anchors will have longer unit fixed lengths than those at greater depth such that when equal load is applied to each unit anchor, each one is mobilising the same percentage of the ultimate grout/ground bond capacity or such that each failure occurs simultaneously. Albeit, if the unit anchors are founded in soil conditions with different creep characteristics, the unit fixed lengths would be designed such that each unit anchor design complies with the appropriate creep criterion in the working condition.

The SBMA system can also be designed for the encounter of soil with strength reducing with depth or with strength varying throughout the fixed length, or even for the encounter of very weak bands of soil at irregular depths. In the latter case the number of unit anchors is designed to allow for a potential failure of one or two unit anchors whilst the remaining intact unit anchors will still sustain the total anchor working load with an appropriate factor of safety.

3.0 PRACTICAL CONSTRAINTS

3.1 General

Theoretically the multiple anchor system would work to its maximum efficiency when utilising a large number of low load capacity unit anchors each with relatively short unit fixed lengths over which no progressive debonding exists. However,

the following constraints control the actual number of unit anchors and the unit anchor capacities:
1. The bond length or bond mechanism used at the tendon/grout interface of each unit to allow safe use of the full tendon capacity;
2. The diameter and type of the corrosion protection of the fixed anchor (encapsulation);
3. The influence of passage of the "free" length tendons from the deep unit anchors (distal) on the bond capacity of the shallower unit anchors (proximal) and their
congestion in the borehole;
4. The arrangement at the anchor head of the multiple of individual jacks in the hydraulically synchronised stressing system. (All unit anchors have different free lengths and hence require different amounts of extension and ram travel).

3.2 The unit anchor tendon

The difficulties in handling and coupling rigid bars, and the extremely low capacity of a single wire tendon immediately exclude both types of tendon from consideration.

Strand is readily available in three sizes, 12mm, 15mm and 18mm with a type variety in each group (normal, superstabilised, and dyform or compact). Extensive research information from strand and encapsulation pull out tests has allowed a number of options of bond mechanism to be considered (Ref 1). These range from non-deformed strand to deformed strand, or deformed tendon, or to mechanical locking devices. For permanent works requiring the encapsulation of the strand within a double plastic corrugated duct system (developed to comply with the corrosion protection requirement of BS8081 (Ref 2)) the deformed strand system was chosen, whilst for temporary works either deformed strand or a mechanical device for removable anchors is available (Ref 3).

Although research has established that the full capacity of the entire range of strands could be achieved within encapsulation lengths of 1 to 1.5m, in practice the unit encapsulation lengths have been standardised in the 2 to 3m range as a general safeguard.

Further research has determined that encapsulation size complete with a double plastic layer, could be as little as 22mm, but the common diameter now in use is 50mm for ease of fabrication.

Initially unit anchors contained only single strands but the demand for higher unit tendon capacity to ensure failure at the ground/grout interface in preliminary trial anchors necessitated incorporation of two strands. Subsequent development has confirmed that a multiple of strands may be satisfactorily incorporated into the double protected encapsulations of individual unit anchors to allow mobilisation of even higher unit anchor loads.

3.3 Multiple stressing jack and load measurement arrangement

The initial choice of the unit anchor tendon system determined that the range of test load required in production unit anchors was between 200 to 300kN (75% characteristic strength of strands). In order to demonstrate factors of safety in the range of 2 to 3 in the test loading of preliminary trial anchors, or to achieve failure at the grout/ground interface, unit anchor test loads up to 600kN have accordingly been required.

In utilising a multiple of hydraulically synchronised jacks, the arrangement which maintains the unit anchor tendons on the minimum pitch circle diameter has been found to be most appropriate. This allows use of normal 150mm to 200mm diameter ducts at the head of the anchor with only nominal deviation of the strand alignment through the jacks. The five unit jack arrangement shown in photo 1 has been most appropriate but an alternative using seven jacks is also feasible.

Each of the jacks is coupled via a central manifold to a single hydraulic power-pack. Thus, during load application the load in each unit anchor is always the same. The hydraulic pressure is measured by a pair of matching calibrated gauges and, based on the ram area of the identical jacks, the applied load is known. Any error in measurement of pressure is identified immediately by observation of discrepancy between the two gauge readings and by checking with the gauge pressure on the powerpack itself. Any friction within the "system" can be established by carrying out load and unloading cycles. Owing to continual difficulties over a 15 year period in achieving compatibility between loads established from pressure gauge readings with those recorded by load cells (strain gauged, vibrating wire or hydraulic), more emphasis has now been placed on determining loads by accurate reading of hydraulic pressure gauges alone.

In the case of preliminary trial anchors, each individual jack also has its own pressure gauge and lock off valve. If, from the load/extension data, the failure or onset of failure of a unit anchor is suspected then its valve is closed and the load in that unit can be observed independently while further testing of the other unit anchors is continued.

3.4 Unit anchor fixed lengths

Having established from the multiple jack arrangement the optimum number of unit anchors, and from the tendon system the range of working and test capacities of unit anchors, then the design of unit anchor lengths can be made. However it should always be borne in mind that in the vast majority of conditions the shorter unit fixed lengths (2 to 4m) are more efficient than longer unit fixed lengths (4 to 8m) and thus an appropriate choice of borehole diameter is also relevant; typically 200mm dia in stiff clays, 140mm dia in stronger glacial clays and weak rocks, and 140mm dia cased holes in medium dense fine sands.

3.5 Effect of adjacent tendons on proximal unit fixed anchor lengths

All mechanisms which transfer load from tendon to grout or encapsulation to grout subject the grout to bursting stresses (Ref 4). Owing to the very limited tensile strength of cementitious grout it is, in the majority of cases, the surrounding soil or rock which effectively confines the grout and prevents the grout column bursting at low loads. The presence of a number of strands in close proximity and within a compressible sheaths, adjacent to the bond system of the proximal anchors, provides a considerable weakness in the grout column and reduces the effective confinement. Research has been carried out to investigate the influence of the presence of the adjacent strands on the bond capacity of both encapsulations and mechanical devices (Ref 5). In soil conditions where confining stresses are limited a system of surrounding the adjacent strands in non-compressible sleeves and reinforcing the grout has been developed to ensure these problems do not result in low capacity pull-out failure.

From the testing of the numerous anchors it has been established that friction within the free length of the strands of distal anchors can, due to their passage of upper encapsulations, be greater than that in proximal anchors. For this reason it is recommended that the lower limit of the apparent tendon free length acceptance criterion is 80% (or strand extensions are not less than 80% theoretical). This limit is consistent with current European practice, but somewhat less than 90% recommended in BS8081. It should be borne in mind that via the nominal friction the load is still transferred into the overall fixed anchor length.

3.6 Effect of load change in a production SBM anchor

It has been normal practice in the U.K. for over a twenty year period to apply a preload of 110% of working load to production anchors.
This generally provides more than a reasonable overload to ensure that, within the life of the anchor, load loss due to soil creep or tendon relaxation does not cause the load to fall below working load. This procedure complies with BS8081 and as such is applied to more than 95% of installed anchors. However, there are occasions in which the full working load is not applied to an anchor and subsequent load change results entirely from the amount of movement of the anchor head in the axial direction.

When SBM anchors are installed for use in the normal applications where full working load is applied, then no special considerations are necessary. However, where the anchors are intended to be partially or fully loaded by structural movement of the anchor head, then consideration must be given to the designed variations in the unit anchor free lengths. When the anchor head moves, the load increase in proximal unit anchor will be greater than that in a distal unit anchor due to its shorter elastic length; thus the load locked into each unit anchor at a datum or an intermediate level must be varied such that when the calculated amount of movement occurs, necessary to load the anchor, then after this movement the unit loads will be equal, and no individual unit anchor overloaded.

4.0 TEST ANCHOR PROGRAMMES

One of the major benefits accruing from the installation and testing of preliminary trial anchors using the multiple anchor system is that each unit anchor provides a full and comprehensive set of data with regard to its own elastic and non-elastic behaviour and bond capacity; i.e. a five unit anchor provides five times as much data as a normal anchor. Attempts have not been made to fully isolate the grout column associated with each unit anchor, and it is accepted that some upward transfer of load may exist between unit anchors during normal loading. However, in the trials carried out to date, the determination of failure capacity of some middle or lower unit anchors has not been prevented by this phenomenon. Furthermore, after reaching a general stage of failure, subsequent tests have been carried out to substantiate the information from individual unit anchors. The proximal anchor is loaded to failure first, and the associated grout column pulled away remote from the one below. This is repeated working progressively towards the distal anchor. In addition to the trials carried out to establish ultimate capacities, in the majority of cases, load holding tests have been carried out at locked off loads of 1.1 x working load to ensure load loss does not exceed 1% load per unit time over 8 time periods (5, 15, 50, 150, mins; 8, 24 hours; 3, 10 day) in order to comply with the requirements of BS8081. No SBM unit anchors tested to date have failed this criterion and generally losses have been well within these limits.

5.0 DESIGN APPROACH AND DESIGN DEVELOPMENT FOR ANCHORS IN CLAY

The current design rules applied to non-postgrouted shaft anchors in clay generally follow those developed for bored piles in equation:

$$Tf = \pi . D . L . \alpha . Cu \qquad (1)$$
Tf = ultimate load in kN
D = bore diameter (m)
L = fixed anchor length (m)
α = adhesion factor
Cu = average undrained shear strength over the fixed anchor length kN/m²

Recommended values of α established from piling are in the 0.2 to 0.5 range whilst a range of 0.3 to 0.6 has been achieved in normal anchoring. For anchors in London Clay a value of 0.45 is often considered appropriate. Although it is acknowledged in anchoring that the proportion of the clay shear strength mobilised reduces with increase in fixed anchor length, the above formula makes no allowance for such a phenomenon. It is accepted in the piling and anchoring industries that the adhesion factor, α, allows for variation of founding stratum and variation in drilling and construction techniques. To accommodate another variable within this acknowledged factor, and for anchors only, would be confusing. Thus it is appropriate to reintroduce an "efficiency factor", "fs", in place of "α" for this purpose (originally recommended by Bassett (Ref. 6) but not specifically for relating efficiency to length of fixed anchor)

i.e. $Tf = \pi . D . L . fs . Cu.$ \qquad (2)

The majority of SBM anchors installed and tested were drilled with open hole, water flush techniques, but some in the London Clay were augered. The unit fixed lengths tested have ranged from 2.5 to 7m, whilst fixed length of normal anchors has ranged from 10m to as much as 23m grouted length in clay.

Unfortunately, it is not always economic for the site investigation to provide full and comprehensive data on the clay shear strength over the full depth range. In an increasing number of situations, particularly in boulder clays and glacial tills, only standard penetration test data is available. Such data can be used in two ways to design the fixed length of the anchor:
i) Make use of the relationship and factors recommended by Stroud (Ref 7) to allow clay shear strength to be estimated:

$Cu = f_1 . N$
where f_1 = factor ranging from 4.4 to 6.0(3)
N = Standard penetration test value

Thus make use of the derived clay shear strength value in the previous equation 2.

ii) On the basis of failure loads exhibited in the trial anchor, determine a direct
relationship between bond stress and N for anchors in clays and;

$Tf = f_{10} . N$ \qquad (4)
where Tf = ultimate bond stress
$Tf = \pi . D . L . f_{10} . N$ \qquad (5)
where f_{10} = factor.

Such relationships have previously been proposed by Littlejohn for anchors in chalk (Ref 8), and Barley for anchors in chalk, mudstone and sandstone (Ref 9). Consistent with design approach above (equation 2), it should be possible to incorporate the same or similar efficiency factor, fs, related to the choice of fixed length.

i.e. Tf = π. D. L. fs. f10. N   (6)

Attempts have been made in the analysis of test data from 2 normal anchors and 61 unit anchors to substantiate the above design criteria and establish values of the recommended factors fs and f10 (Clause 7.0)

6.0 TEST ANCHORS IN CLAYS

6.1 Hampton, Surrey

Two 20m deep 105mm diameter holes were augered 13.5m into stiff to very stiff London Clay. A normal 5 strand encapsulated anchor tendon was installed in one hole whilst 5 unit anchor tendons were installed in the second (Fig 7). Identical tremie grouting techniques were utilised. The normal anchor achieved a failure load of 370kN, whilst the SBMA achieved 660kN during multiple loading. During subsequent tests which progressively loaded each individual unit anchor from top down, the summation of unit capacities was 905kN, and failure bond stresses ranged from 146kN/m² in the proximal anchor to 303kN/m² in the distal anchor. Two longer anchor holes were augered to 30m (23m into the London Clay (Fig 7)). The normal anchor achieved 470kN, whilst the SBMA achieved 980kN. The summation of the unit anchor capacities when failed individually was 1280kN, and failure bond stresses ranged from 78kN/m² in the longer proximal anchor, to 227kN/m² without failure in the lower anchors. Thus, during direct loading against a load cell, the multiple anchors achieved capacities of 178% and 208% of the capacities of normal type anchors in 20m and 30m deep anchor holes, whilst the actual summation of unit anchor capacities (which would have been achieved using the currently available synchronised jacking system) achieve in excess of 240% that of normal anchors. Soil strength information from an adjacent borehole allows direct comparison of the failure bond stresses with the range of undrained shear strengths of the clay.

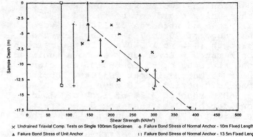

Fig.8 Exhibited failure bond stresses of SBMA and normal 20m anchors
Fig.8 Dargestellte Bruchspannungen eines "SBM-Ankers" und eines normalen Ankers von 20m

Eight out of the ten unit anchors were failed as were both normal anchors. Fig 8 indicates that all of the short 2.5m unit anchors mobilised bond stresses equivalent to the full clay shear strength. Considering the full 13.5m of normal anchor grouted in the clay as being effective, then some 40% of the average clay strength was mobilised. Fig 9 suggests that the 4m unit anchors utilised 70% of the clay strength at failure, whilst the full 23m grouted length of the normal anchor attained a bond stress of 22% of average clay strength. Considering only 10m of the 23m being effective, this rose to 36% (neglecting any contribution from the upper 13m at failure).

6.2 Chingford, North London

A single inclined trial SBMA anchor was open bore water flushed to 40m through overlying materials and into London Clay. A total fixed length of 20m in a 190mm dia bore was utilised by 6 No. unit anchors, each within 3.33m fixed length. The anchor achieved a maximum load of 1450kN recorded on a single anchor load cell when the synchronised jacking system was not available. Unit anchor capacities during induced progressive failure were 220 and

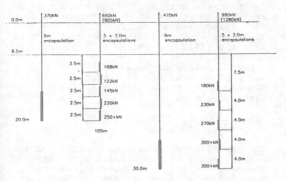

Fig.7 Normal and SBM anchors in London Clay
Fig.7 Normale und "SBM-Anker" in London Clay

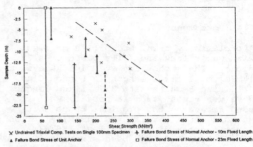

Fig.9 Exhibited failure bond stresses of SBMA and normal 30m anchors
Fig.9 Dargestellte Bruchspannungen eines "SBM-Ankers" und eines normalen Ankers von 30m

230kN on the two proximal units, but the other units could not be failed at 250kN. Failure bond stresses of 110kN/m² were in the range of 70 to 78% of the clay shear strength (140 to 160kN/m²) in the relevant depth range. Lower unit anchors achieved bond stresses of 45 to 65% of the clay shear strength without failure. During a load hold period of 10 days, load losses at 660kN reached 3.1% of lock-off load, considerably less than the 8% acceptable in BS8081. The 28 production SBM anchors were subsequently proof loaded to between 675 and 900kN without difficulty, and no problems occurred during the load test periods.

6.3 Heathrow, West London

A single inclined SBM anchor was open bore water flushed through overlying gravels into stiff to very stiff London Clay. A total fixed length of 28m in a 190mm bore was mobilised by 5 unit anchors. Unit fixed lengths were varied, 7, 5, 7, 5, 4m to investigate efficiency of unit fixed lengths and influence of increasing clay shear strength. Using the synchronised jacking system, a total load of 2144kN was achieved. The two proximal anchors failed during this operation at 375 and 463kN. Subsequent controlled loading, which progressed failure top down, achieved a failure load of 446kN on the third unit anchor, and no failure at 539kN and 590kN on the distal units. Mobilised bond stresses were 90, 155, 107kN/m² at failure and 181 and 247kN/m on the non-failed units.

The information on clay shear strength in the SI report is limited but standard penetration tests have been carried out throughout the depth of the London Clay. This has allowed the investigation of a direct relationship between ultimate bond stress of the unit anchors in the clay and the N value at the relevant depth (fio factor).

The failure bond stresses of the two 7m unit anchors equated to 3.7 and 3.1 times the average "N" value at the relevant depth, whilst that of the 5m unit anchor equated 5.3N. The two lower anchors with unit lengths of 5 and 4m did not fail at bond stress values of 4.9N and 6.1N.

6.4 Portsmouth

A preliminary trial anchor was drilled with waterflush and fully cased to a 28m depth. The anchor achieved 1056kN over a 17.5m total fixed length in London Clay utilising 5 units with 3.5m fixed lengths. N values ranged from 28 to 45. Failure bond stress on the proximal three unit anchors were 116, 116 and 140kN/m² whilst lower anchors did not fail at 156kN/m². fio factors relating bond stress to N ranged from 3.6 to 4.1 whilst efficiency factors (fs) calculated from Cu = 4.4N ranged from 0.72 to 0.83.

6.5 Southampton   (Ref. 10)

A single inclined test anchor was installed using water flush and full length casing with a total fixed length of 17.5m in the Bracklesham Beds (firm to stiff silty clay with bands of silty clayey e.g. fine sand). Over the depth of the five 3.5m long unit anchors the clay shear strength ranged from 152 to 200kN/m² and SPT values from 29 to 44. (This indicates fi factors from 4.5 to 5.2). End of casing grouting was carried out with pressures up to 7 bar. At 1337kN, failure of the proximal anchor occurred (168kN) but lower anchors could not be failed at 284 to 300kN. Failure bond stress of 155kN/m² and non failure bond stresses of 205kN/m² equated to the clay shear strength with fs = 1.0 and fio value at failure of 5.3.

6.6 Boston, Lincolnshire

Two trial anchors were installed for testing to failure in very stiff clay with chalk gravel and occasional cobbles, using fully cased water flush system. The 16m fixed lengths consisted of 5 unit anchors with 3.2m unit lengths. The anchors achieved 1400 and 1480kN with only four of the total 10 unit anchors indicating failure. Failure bond stresses ranged from 188 to 210kN/m² and six unit anchors did not fail at 216kN/m². Only limited standard penetration test results were available indicating 20 in upper firm/stiff clay and 38 in very stiff clay zone. With interpolation of this data, (N = 34 at level of failed unit anchors) an fio factor of 5.5 is calculated at failure, and fio of 5.7 in the non-failed anchors.

6.7 Newcastle Stage 1

Two trial anchor holes were open hole drilled using waterflush to construct a 20m total fixed length in very stiff sandy gravelly clay (Glacial Till). Soil strength was considered to be represented by the increase in N values with depth. This relationship was N = (15 + 3.1Z) where Z = depth in metres. The unit fixed lengths were designed on the basis that anchor bond stress is proportional to N. Unit fixed lengths of 7.0, 5.2, 4.2 and 3.6m were installed, and the trial anchors tested to 1923 and 2037kN without any unit failures.

Bond stresses of between 180 and 322kN/m² were achieved without failure and these equated to fio factors of 6 and 8 without failure.

6.8 Newcastle Stage 2

The two trial anchors were constructed at a lower depth in the Glacial Till than those in Stage 1, and this involved the penetration of granular glacial deposits for the distal anchors (dense sand and gravels). Thus it was necessary to fully case the borehole to depth albeit with the same finished hole diameter (140mm) as

Stage 1. Inspection of plotting of N value with depth indicated strength development relationship of N = 25 + 3.9Z. Design unit fixed anchor lengths were 6.5, 4.5, 4.0, 3.0, 3.0, with the last two lengths founded in the granular stratum. Trial anchors achieved 2915 and 3062 kN without failure and mobilised bond stresses ranged from 204 to 467 kN/m².

One trial anchor, identical to those taken to test load, was subjected to load hold tests over a 10 day period. The two proximal unit anchors lost 1% and 2% load over 6 time periods (to 24 hours) as compared with the tolerable loss of 6%. The distal anchor had lost 2% load at 3 days and no further loss after 10 days. The behaviour was very satisfactory.

7.0 DATA ANALYSIS AND DESIGN RECOMMENDATIONS

From the 61 unit anchors tested, 21 unit anchors and 2 normal anchors were failed which allowed calculation of failure bond stresses. Of the 21 unit failures site investigation data presented clay strength in terms of Cu values (or reasonable interpolation of such) in the depth range of 11 units. Use of recommended values of fi (4.4 in London Clay) allowed reasonable estimate of clay shear strength at depth of 3 other units. This data allows the presentation of Fig 11 showing the values of efficiency factor fs against fixed length. This indicates that the full clay shear strength can frequently be mobilised in bond when short length unit anchors (2.5 to 3.5m) are installed (fs = 0.95 to 1.0). However, in the 3.5m to 4.0m range

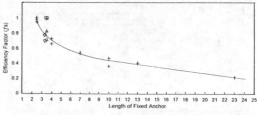

Fig 11 - Efficiency (fs) vs Length of Fixed Anchor

+ Hampton (Augered)   ◇ Chingford (Open hole/water flush)
▲ Portsmouth (Full cased/water flush)   □ Southampton (Full cased/water flush)

fs can vary from 0.66 to 1.0. With fixed lengths greater than 4m there is a continual fall-off in efficiency. The use of Equation 2 along with efficiency factor values presented in Fig 11 allows a more accurate estimate of the ultimate capacity of straight shaft anchors founded in clay than those obtained from Equation 1 recommended in BS8081. Furthermore use of Equation 2 is particularly appropriate for the design of SBM anchors where optimisation of the bore diameter and the unit fixed length may now be made with confidence with due consideration of bond efficiency.

Figure 12 provides a full worked example of such a design, where the relationship between Cu and depth in London Clay is established from S.I. data, and the example is represented schematically for clarification.

Of the 21 induced failures of unit anchors, 11 took place in founding strata where ground strength was represented by a

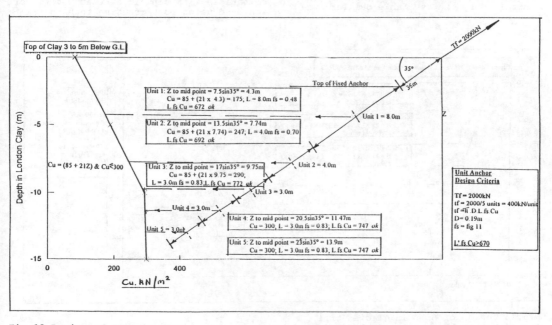

Fig.12 Design of SBMA in clay
Fig.12 Konstruktion eines SBMA in Clay

299

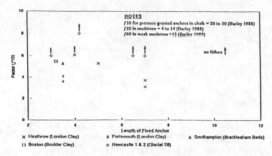

Fig 13 - Design of Bond Stress/N Factor (f10) vs Length of Fixed Anchor

range of N values at the relevant depth. Values of f10 ranged from 3.2 to 5.0 in London Clays, and 4.1 to 5.0 in the Boulder Clay. In the Glacial Till in Newcastle f10 ranged from 6 to 8 without anchor failure. In trial anchors where unit fixed lengths were varied, only a few failures occurred, thus it has not been possible to produce an efficiency factor (fs) relationship specifically for anchor designs based on standard penetration test values. However, in the absence of such direct data the use of Fig 11 should still be appropriate since the efficiency factor is relatable to the soil, the construction technique and to the fixed length used, and not to the mode of representation of soil strength. The limited number of f10 values are illustrated in Fig 13.

Fig 14 provides a guide to the design method that could be utilised at Newcastle Stage 2 to determine the length and distribution of the unit fixed anchors when controlled by the available SPT information.

8.0 ANCHORS IN FINE SANDS

A number of trial SBM anchors and normal anchors have been installed using end of casing grouting techniques into medium to fine sand with variable content of silts and clays. The test data has not yet been fully analysed to allow presentation of reasonably precise efficiency factor versus fixed length relationship as portrayed in Fig 11. However, the presence of a similar trend is apparent and accepted.

BS8081 recommends the use of the following equation:

$$Tf = L \, n \, \tan \phi \qquad (7)$$

where
L = fixed anchor length
n = load capacity factor ranging from 135 to 165kN/m in fine to medium sand
ϕ = shearing angle of soil

This equation makes no allowance for the occurrence of progressive debonding or fall off in load/m length as fixed length increases. In the absence of a Fig 11 applicable to fine sands, initial studies have arrived at an appropriate mathematical expression for the efficiency factor to be applied to equation 7:

$$Tf = L \, n \, \tan \phi \, . \, fs \qquad (8)$$

where $fs = (0.91)^{L.\tan \phi}$

Graphical representation of this relationship presents curves very similar to those of Ostermayer. In the author's experience the capacities of anchors achieved by end of casing grouting techniques in fine sands are normally quite compatible with capacities that Ostermeyer achieved using post grouting techniques.

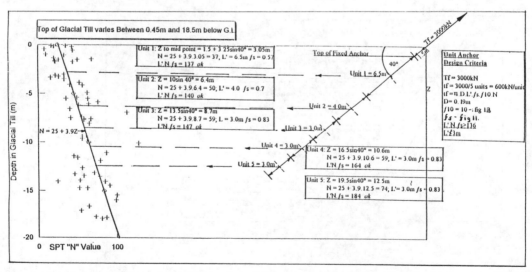

Fig.14 Design of SBMA in glacial till
Fig.14 Konstruktion eines SBMA in glazialem Ton

300

It must be fully appreciated however, that in equation 8 the load capacity value 'n' is heavily influenced by construction technique, and grout pressures must at least be in the 5 to 10 bar range. Specialist contractors may have their own appropriate proven values of n but the application of an efficiency factor is still appropriate.

9.0 OVERALL SUMMARY

During the 7 years since inception, the single bore multiple anchor system has attained loads in straight shafted anchor holes in soils, mixed ground and weak rocks that could not have previously been considered attainable:
London Clay – 2150kN
Glacial Till – 3000kN without failure
Medium dense medium to fine sand – 2400kN without failure
Highly weathered Chalk – 1280kN without failure.

Working loads in the range of 750kN to 1500kN or possibly more, are now available in soils. Some 10,000 unit anchors have now been installed and tested, the majority being for permanent usage, and a small number with special fully removable tendon facilities. Failure of only a handful of unit anchors has occurred due to ground conditions, but with the sound performance of other unit anchors in the bore, no redrilling or remedial works have been necessary as a result of the inbuilt safety system.
The considerable amount of data provided in trial SBM anchors with differing unit fixed lengths has allowed the development of efficiency factors for anchors founded in clays, and, although not finalised, also in fine silty clayey sands. This information provides a better understanding of the influence of progressive debonding on anchor capacity and for the first time accurately quantifies its effect.

10.0 ACKNOWLEDGEMENT

The Author would like to thank Keller Colcrete of the Keller Group for continual support in the research and development of this completely new (now patented) anchor system. Combined research and studies in conjunction with the University of Surrey has allowed development of spread sheets for portraying the distribution of bond stress along the length of the tendon (Mr R Woods) and the development of a mathematical expression for the efficiency factor applied to sand anchors (Mr K Barkhordari).

REFERENCES

Ref 1 Barley, A.D. 1989. Gunbarrel pull out tests on strand and encapsulations 1974 to 1988. Confidential.
Ref 2 B.S.I. Publication 1989. British Standard Code of Practice for ground anchorages.
Ref 3 Trummer, F. Barley A.D. 1995. Removable multiple anchors. Conf. Salzburg.
Ref 4 Yuen-Cheong Lo 1979. Investigation of the effect of lateral restraint on ground anchor failure at the grout/tendon interface. Thesis, University of Sheffield.
Ref 5 Barley, A.D. 1991. Effect of adjacent strands on encapsulation pull-out capacity. Confidential.
Ref 6 Bassett, R.H. 1970. Discussion to paper on soil anchors. ICE Conf. on ground engineering, London.
Ref 7 Stroud, M.A. 1974. The standard penetration test in insensitive clays and soft rocks. Proc. European seminar on penetration testing, Stockholm.
Ref 8 Littlejohn, G.S. 1970. Soil anchors, ICE conf. on ground engineering, London.
Ref 9 Barley, A.D. 1988. Ten thousand anchorages in rock. Ground Engineering Sept, Oct, Nov 1988.
Ref10 Barley, A.D. 1991. Slope stabilization ground anchor system in rocks and soil. ICE Conf. on slope stability engineering, Isle of Wight.

Kunstharze als Ankermörtel: Anwendungstechnik, mechanisches Verhalten und Korrosionsschutz

Resin grouts for bolting and anchoring: Application, mechanical performance and corrosion protection

W.Cornely & M.Fischer

CarboTech, Berg- und Tunnelbausysteme GmbH, Essen, Germany

ZUSAMMENFASSUNG: Seit nahezu vierzig Jahren werden Polyestermörtel-Patronen zur Vermörtelung von Ankern im Bergbau eingesetzt. Neben diesem Verfahren sind in der Zwischenzeit eine Reihe anderer kunstharzgebundener Ankermörtel entwickelt worden. Meist handelt es sich dabei um Epoxid- oder Polyurethanharze.
Gemeinsam ist diesen Bindemitteln das schnelle Aushärten und damit die schnelle Belastbarkeit des Ankers. Die erreichten Druckfestigkeiten und insbesondere Zugfestigkeiten sind deutlich höher als bei zementösen Mörteln. Der E-Modul ist, je nach verwendeten Harz und vor allem dem Füllgrad, unterschiedlich; bei vergleichbar hohen Festigkeiten gibt es Bindemittel spröde wie Zement oder so elastisch, daß diese Mörtel Punktbelastungen ausgleichen können.
Die Resistenz gegenüber angreifenden Medien ist allgemein gut bis sehr gut. Die spezifischen Charakteristiken dieser Stoffe erfordern eine spezielle Verfahrenstechnik, die ein sicheres und schnelles Einbringen der Stoffe erlauben. Ergänzt wird diese Technik durch besondere Injektionsanker wie das „Injektionsrohr mit Ankerwirkung", das als Besonderheit einen integrierten Packer aufweist. Die Nutzung dieser spezifischen Eigenschaften ermöglicht in einer Vielzahl von Fällen eine effektive Verfahrenstechnik und gleichzeitig eine wirtschaftliche Problemlösung.

ABSTRACT: Since almost fourty years, polyester resin cartridges for roof bolting are used in mining. In addition to this process, a range of other resin based grouts for bolting and anchoring have been developed, in most cases epoxy or polyurethane resins.
All these grouts have in common fast curing and, thus, fast loading of the anchor. The compressive and especially the tensile strengths which are achieved are considerably higher than with cementitious grouts. The modulus of elasticity differs according to the resin and, in the first place, to the filler content. There are grouts, both with similar high strength, on the one hand as stiff and brittle as cement and, on the other hand those with a certain yield capacity, able to equialize high point load.
The resistance against aggressive media is good to very good for epoxies and polyurethanes, only fair for polyesters. The specific characteristics of these materials require a special processing, enabling a safe and quick placement. This technique can be further improved by using special injection bolts as the „injection pipe with bolting effect" incorporating an integrated packer. With special fabric hose packers, even large soil anchors can be installed very fast and economically. By making advantage of the specific characteristics, resin grouts offer an effective processing and simultaneously an economic solution in a multiplicity of bolting and anchoring applications.

1. Einführung

Das Setzen von Ankern ist eine Technik mit einem weitgespannten Anwendungsbereich, es reicht von kleinen Dübeln, wie sie in der Befestigungstechnik benutzt werden, bis zu Grundankern, die mehrere Dutzend Meter Länge haben und weit über 1.000 kN Traglast aufweisen. Man unterscheidet zwischen vorgespannten und schlaffen, starren und nachgiebigen, mechanisch verspannten und vermörtelten Ankern; das heißt im Vordergrund der Diskussion steht im allgemeinen der Anker selbst hinsichtlich seiner Werkstoffeigenschaften, seiner Oberflächengestaltung, seiner konstruktiven

Ausbildung und viele Dinge mehr. Weniger Aufmerksamkeit wird dem Ankermörtel geschenkt. Dem Ankermörtel kommt die Aufgabe zu,
◆ die Spannungen des Gebirges auf den Anker zu übertragen (Ausbauanker) bzw. am Anker angreifende Kräfte ins Gebirge zu übertragen (Lastanker) oder
◆ ggf. den Anker vor Korrosion zu schützen.
Doch nicht nur der Endzustand des Einbaus ist wichtig, es kann auch erforderlich sein,
◆ die Spannungen möglichst frühzeitig zu übertragen.
Darüber hinaus spielen für die Auswahl eines Ausbaumittels auch bauverfahrenstechnische Aspekte eine Rolle, nämlich
◆ möglichst einfach und sicher appliziert werden zu können und darüber hinaus natürlich
◆ preiswert zu sein.

Sehr häufig werden als Ankermörtel mineralische Baustoffe eingesetzt. Solche mit Kunstharzen als Bindemittel finden ebenfalls vielfache Einsatzgebiete. Es können folgende Typen unterschieden werden:
• gefüllte ungesättigte Polyester- oder Vinylester-Mörtel in der Befestigungstechnik (UP-Harze)
• gefüllte Epoxid- und Polyurethanmörtel (EP- und PU-Harze)
• ungefüllte, nichtschäumende Polyurethan-Harze (PU-Harze)
• ungefüllte, nichtschäumende Silikatharze (OM)
• ungefüllte, schäumende Polyurethan-Harze (PU-Harze).

2.   Anforderungen an Ankerbindemittel

Die einschlägigen Normen zum Ankern [1] verlangen vom Hersteller genaue Angaben hinsichtlich der Bauweise der Anker und ihres Einbaus, zu den erforderlichen Eigenschaften des Bindemittels wird erstaunlich wenig ausgesagt. Die DIN 4125 verweist auf die Zementnorm EN 196 bzw. die DIN 4227, die sich mit dem Verpressen von Spannkanälen beschäftigt. Die DIN 21521 beschreibt ein Verfahren zur Ermittlung der minimalen Verbundlänge und macht dabei die Vorgabe, daß die Druckfestigkeit der zementösen Baustoffe 5 MPa überschreiten soll, ehe der Anker geprüft wird; kunstharzverklebte Anker, die zum soforttragenden Ausbau eingesetzt werden, sollen bereits nach 30 min geprüft werden.D.h. letztlich wird nicht der Baustoff, sondern die Wirkung der Kombination Gebirge/Bindemittel/Zugglied geprüft.
Welche Kriterien als signifikant angesehen werden können, soll im folgenden diskutiert werden.

2.1   Mechanische Eigenschaften

Der Parameter, der meist zur vergleichenden Bewertung von Baustoffen herangezogen wird, ist die einachsige Druckfestigkeit, eine Eigenschaft, die vergleichsweise leicht zu messen ist (wenngleich hier verschiedene Meßvorschriften herangezogen werden können [2]). Weiters werden oft Biegezugfestigkeit [3] und die daraus auch rechnerisch zugängliche Zugfestigkeit [4] angegeben. Die Belastung eines Zuggliedes gegenüber dem Gebirge führt zu Scherspannungen. Deshalb wurde für UP-Harz-Mörtel als Parameter die Stanzfestigkeit, d.h. eine Scherfestigkeit mit begrenzter Querdehnung vorgeschlagen [5] [6]. Der Kraftübergang zwischen den Oberflächen von Anker/Mörtel/Gebirge wird durch Haftscherfestigkeiten bestimmt, die jedoch unseres Wissens selten erfaßt wurden. Angaben der Werte müssen natürlich die Verschiedenheit der Gebirgs- und Ankeroberflächen berücksichtigen und sind somit nur begrenzt in die Realität übertragbar. Die Angabe dieser Größe setzt natürlich voraus, daß eine echte Haftung tatsächlich existiert, was aber nicht immer der Fall ist (s. 3.1).
Eine große Bedeutung wird von einigen Fachleuten dem E-Modul beigemessen. Die australische Schule besteht auf einem möglichst hohen Wert (> 10 GPa), da man so ein Nachgeben des Mörtels und damit eine Entfestigung des Gebirges verhindern will. Ziel dieses Ankerausbaus ist die Herstellung eines steifen Balkens in der Firste von rechteckigen Strecken des Bergbaus [7]. Als Referenzversuch mißt man das Nachgeben eines Ankers im zu ankernden Gebirge auf eine Einklebelänge von 250 - 300 mm.
Andere Autoren befürworten insbesondere im druckhaften Gebirge einen niedrigen E-Modul [8]. Demnach kommt dem Baustoff die Aufgabe zu, die Spannungen aus dem Gebirge auf das eigentliche Zugglied, den Anker mit seinem E-Modul von 200 GPa (Stahl) oder 45 GPa (GfK) zu übertragen. Besondere Belastungen entstehen dabei an auf Zug oder Schub beanspruchte Trennflächen (Bild 1), an denen sich die Spannungen sprunghaft ändern. Diese Spannungsspitzen können durch nachgiebige Stoffe besser abgefangen werden als durch spröde, die an dieser Stelle eher reißen. Hierdurch entsteht eine lokale Lastspitze im Zugglied; darüber hinaus wird die Korrosionsschutzschicht des Baustoffes geöffnet.
Auch bei Stoffen mit niedrigerem E-Modul wird eine lokale Spannungsspitze dazu führen, daß in der unmittelbaren Nachbarschaft der Trennfläche der Haftverbund zwischen Gebirge und Baustoff versagt, jedoch keinesfalls reißt der Injektionskörper; die Haftscherfestigkeit ist bei Kunstharzen in der Regel geringer als die Zugfestigkeit.

Tabelle 1: Wichtige Daten verschiedener Kunstharzbindemittel
Selected Data of Different Resin Binders

		UP-Harz-Patronen / SiS-Patronen	EP-Mörtel / Carbolith EP	PU-Mörtel / Carbolith PU	ansteifendes Harz / Carbothix H	Silikatharz / Geodur OM	Schaumharz / Bevedol S-Bevedan	Schaumharz / Bevedol WF-Bevedan	Zement-mörtel / Ankermörtel CT85
Einachsige Druckfestigkeit nach 5 h	MPa	70	46	78	85	25	25	70-90	2.5
gemäß EN 196 nach 24h		80	52	86	95	65	25	70-90	50
nach 7 d		90	n.b.	94	95	85	25	70-90	85*
Biegezugfestigkeit nach 24 h gemäß EN 196	MPa	25	10	51	60	5	18	30	n.b.
Zugfestigkeit nach 24 h ber.	MPa	17	7	34	40	3.5	12	20	n.b.
E-Modul gemäß ASTM C215-91	GPa	16	11	11	2-3	1.5	0.8	3	45
Kugeldruckfestigkeit von 64 MPa	h	n.b.	3	1	24	n.b.	-	0.5	n.b.
Topfzeit bei 10°C	min	0.5-6	4-5	4-5	4-5	4-5	2-3	1	n.b.
Kriechverhalten		++	++	++	++	++	++	++	++
Viskosität bei 10°C vor dem Mischen nach dem Mischen	Pa.s	pastös 150 -	pastös 260/220 250	pastös 275/235 500	flüssig 0.9/0.8 500	flüssig 0.8/0.9 2-3	flüssig 0.8/0.9 0.8	flüssig 0.9/0.9 0.9	pastös - -
Korrosionsschutz		-	+	+	++	+	+	++	+
Beständigkeit gegen Sulfate		++	++	++	++	++	++	++	---o
Säuren		o-+	++	++	++	++	++	++	---o
Alkalien		o	++	++	++	++	++	++	++

* 28d

-	schlecht	poor
o	genügend	fair
+	gut	good
++	sehr gut	very good

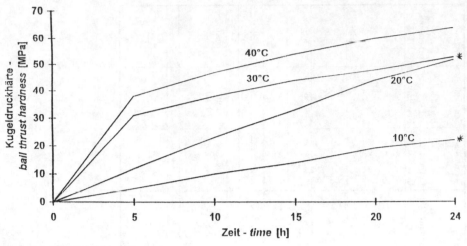

* Endwert 60 MPa nach 2 bis 5 Tagen
  Final value 60 MPa after 2-5 days

Bild 1 :   Festigkeitsentwicklung von Carbothix H durch Messung der Kugeldruckhärte
            in 4-mm-Schichtdicke bei unterschiedlichen Temperaturen
            Development of ball thrust hardness of CARBOTHIX H in a 4-mm-layer at
            different temperatures

Entscheidend für diesen Haftverbund ist das
Schwindverhalten des Baustoffes. Zementöse
Baustoffe schwinden im allgemeinen. Deshalb
werden in der Regel spezielle Quellmittel zugesetzt,
wobei allerdings die Festigkeiten deutlich reduziert
werden. Schwinden der Ringsäule des Bindemittels
führt zur Schwächung, wenn nicht zum Versagen des
Haftverbundes zum Gebirge. In diesem Fall werden
die Kräfte durch Reibung und vor allem Formschluß
übertragen. Riefen im Bohrloch, aber auch eine
Profilierung des Ankers in der Haftstrecke fördern
die Ankerwirkung und sind in vielen Fällen sogar
Voraussetzung für ein Funktionieren des Ankers. Bei
Felsankern werden nicht profilierte, glatte Bereiche
als Freispielstrecken betrachtet.
Kunstharze weisen wie Kunststoffe allgemein zwei
Zustandsbereiche auf; einen spröden und einen verformbaren. Der Übergang ist durch die Glastempe-
ratur gekennzeichnet, oberhalb der die Festigkeiten
und Moduln erheblich abfallen. Bei den hier betrachteten Kunstharzen liegt diese Temperatur bei mind.
60 °C, also außerhalb der normalen
Gebrauchstemperaturen, so daß diese Eigenschaft
hier nicht weiter diskutiert werden soll.
Eine weitere für Kunststoffe typische Eigenschaft ist
das Kriechen (Fließen). Eine Meßvorschrift bietet die
DIN 4093 [9], die die zeitliche Verformung unter
einer konstanten Last beschreibt. Die Temperatur-
abhängigkeit ist zu beachten. Bei den hier

besprochenen Kunstharzen der CarboTech werden
diese Spannungszustände nicht erreicht.

2.2.   Zeitliche Entwicklung der mechanischen
        Eigenschaften

Wenn man über Vorteile von Kunstharzen beim
Ankern spricht, wird in erster Linie das schnelle
Aushärten genannt. Erwünscht ist in der Regel, daß
der Anker sofort trägt, daß sich aber andererseits der
Anker über eine längere Zeitspanne problemlos
einbauen läßt, zwei Anforderungen, die gegenläufig
sind, zwischen denen man den richtigen Kompromiß
finden muß. Oft spielt dabei das Einbauverfahren
eine wichtige Rolle. Oft wird unterschätzt, welche
Bedeutung die Umgebungsbedingungen auf diesen
Prozeß haben. Hier spielen vor allem die
Gebirgstemperatur und die Schichtdicke des Harzes
bzw. die Dicke des Ringspaltes eine Rolle. Nach
einer Faustregel verdoppelt bzw. halbiert sich die
Geschwindigkeit organischer Reaktionen mit 10 °C
Temperaturunterschied. In dünnen Ringspalten wird
die Reaktionswärme sehr gut an das umgebende
Gebirge abgeführt. Es hat sich deshalb bewährt, den
Ankermörtel in Schichtstärken, die der Dicke des
Ringspaltes entsprechen, temperiert aushärten zu
lassen und die Festigkeitsentwicklung durch
Messung der Kugeldruckhärte [10] zu verfolgen
(Bild 1).

306

2.3.   Chemische Beständigkeit und Korrosion

Zementmörtel gelten wegen ihrer hohen Alkalität als Korrosionsschutz für Stahl. Treten allerdings, verursacht durch Spannungen, makroskopische Risse auf, haben zusitzende Wässer freien Zugang.
Ein weiteres Problem der schnellen Zementmörtel ist, daß sich bei Zutritt von Sulfaten Ettringit bildet und so das Gefüge des Zementsteins zerstört wird. Diese Reaktion kann naturgemäß bei Kunstharzen nicht eintreten, sie sind sulfatresistent.
Eine Passivierung des Stahls wie beim Zementmörtel kommt - mit Ausnahme des GEODUR OM - bei Kunststoffen nicht in Frage. Wegen ihrer hohen chemischen Beständigkeit und geringen Durchlässigkeit werden sie oft im Korrosionsschutz eingesetzt.

2.4.   Verfahrenstechnische Aspekte,
        Wechselwirkungen mit der Umgebung

Je nach Stoff werden die Kunstharze unterschiedlich verarbeitet. Eine der entscheidenden Größen für die Pumpbarkeit ist die Viskosität der Harze bzw. Harzmörtel. Gefüllte Harze sollten bei der Lagerung nicht separieren. Bei Ankern, die über Kopf eingesetzt werden, kann eine hohe Fließgrenze des Bindemittels wichtig sein. Dies ist nicht nur aus verfahrenstechnischen Gründen, sondern auch wegen der Arbeitshygiene wichtig.
Während man die ätzenden Eigenschaften von Zementmörteln scheinbar als gegeben hinnimmt, stehen die hygienischen und brandtechnischen Eigenschaften der Kunstharze oft unter Kritik. Im Bergbau wird der Umgang deshalb durch Zulassungen geregelt.
Ökologischen Kriterien kommt eine wachsende Bedeutung zu. Großflächiger Kontakt mit dem Grundwasser dürfte jedoch die Ausnahme beim Ankern sein, zumindest solange es sich um schnellreagierende Kunstharze handelt, die im Gegensatz zu Zementen nur begrenzt in die Kluftsysteme eindringen. Zahlreiche, an der Praxis orientierte Gutachten haben in jüngerer Zeit belegen können, daß CarboTech-Kunstharze auch bei Kontakt mit Grundwasser ohne Schaden eingesetzt werden können.

3.   Stoffe und Verfahren

Im folgenden werden die obengenannten Aspekte für die wichtigsten Kunstharze und Verfahren abgehandelt.

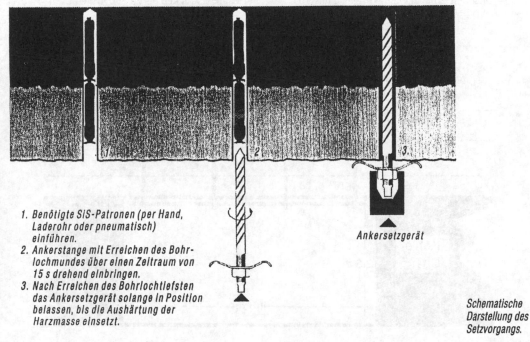

1. Benötigte SIS-Patronen (per Hand,
   Laderohr oder pneumatisch)
   einführen.
2. Ankerstange mit Erreichen des Bohr-
   lochmundes über einen Zeitraum von
   15 s drehend einbringen.
3. Nach Erreichen des Bohrlochtiefsten
   das Ankersetzgerät solange in Position
   belassen, bis die Aushärtung der
   Harzmasse einsetzt.

Ankersetzgerät

Schematische
Darstellung des
Setzvorgangs.

Bild 2 :   Das Ankern mit Harzpatronen
               Roof bolting with resin cartridges

3.1. UP-Harz-Mörtel

Ungesättigte Polyesterharze (UP-Harze) bestehen im Lieferzustand aus doppelbindungshaltigen (daher ungesättigt genannten) Polyestern mittleren Molekulargewichts, dem Styrol als ebenfalls doppelbindungshaltiger Reaktivverdünner beigefügt wird. Angeregt durch Peroxide als Härter, polymerisieren die Verbindungen zu einem dreidimensionalen Netzwerk aus. In der Befestigungstechnik werden statt der Polyester zuweilen die gegen alkalische Verseifung beständigeren Vinylester verwendet.

Die Polymerisationsreaktion führt zu einem Schwund. Durch Füllen mit Zuschlägen aus Calciumcarbonat und/oder Quarz läßt sich das Schwinden somit reduzieren, daß es selbst bei 2 mm Ringspalt den Anker sicher hält, im wesentlichen durch Formschluß wie Zementankermörtel.

Werden so geankerte Bereiche mit niedrigviskosen PU-Harzen nachinjiziert, beobachtet man deshalb in aller Regel, daß das Injektionsharz an den Ankerköpfen austritt.

UP-Harze selbst werden nicht injiziert. Versuche während der letzten Jahre in Großbritannien haben nicht zur Akzeptanz durch die Praxis geführt.

Bereits in den Fünfziger Jahren wurde bei der Bergbau-Forschung das Patronenverfahren entwickelt [11]. Während in der Befestigungstechnik Glaspatronen mit grobem Quarzkorn als Füllstoff eingesetzt werden, hat sich als Anwendungsform im Berg- und Tunnelbau die Harzpatrone in Folie mit feinem Füller weltweit durchgesetzt. In einer kontinuierlich arbeitenden Fertigungsmaschine wird ein Flachfolienband so gefaltet und verschweißt, daß eine Harz- und eine Härterkammer entstehen, die gleichzeitig gefüllt werden. Üblich sind Patronenlängen zwischen 300 und 750 mm, zuweilen auch länger; die Durchmesser liegen in Europa zwischen 23 und 32 mm. Die Abbindezeiten sind extrem kurz; bei den normalen Einstellungen liegen sie zwischen 2 und 3 min.; schnelle Patronen binden in 30 s und weniger ab.

Zur Anwendung:

Das standfeste Bohrloch wird mit der benötigten Menge an Patronen bestückt. Eine Ankerstange, meist Rippentorstahl, wird eingeführt und mit ca. 450 Upm und einem Vorschub von 6 m/min eingetrieben. Die Ankerspitze zerstört die Kammern der Patrone; die Ankerprofilierung vermischt die Komponenten. Bei Verwendung schneller Patronen im Bohrlochtiefsten hält der Anker nach wenigen Sekunden sich selbst und in der Folge das Gebirge (Bild 2). Bei Bedarf können sie vorgespannt werden. Dieses Verfahren ist zu einem Standardverfahren im Kohlenbergbau, neben Deutschland vor allem in USA, Südafrika, Australien und neuerdings Großbritannien geworden. Ergänzt durch angeankerte Profilbleche („W-straps") stellt dies den Regelausbau in Rechteckstrecken bis zu 500 m Teufe dar (Bild 3).

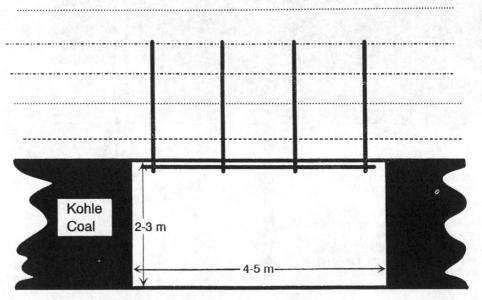

Bild 3 :   Profil einer geankerten Rechteckstrecke
            Profile of rectangular road with roof bolt support

Die Etablierung als Standardverfahren hat teilweise auch zur Mechanisierung des Verfahrens geführt. Große „Roofbolter" im Tunnelbau verfügen über die Möglichkeit, die Patronen pneumatisch ins Bohrloch zu fördern. Die Ankerlängen liegen zwischen 1 und 3 m, wesentlich längere Anker lassen sich wegen des hohen Eindringwiderstandes nicht setzen.

Wichtig für eine ordnungsgemäße Durchführung des Verfahrens ist die richtige Abstimmung zwischen den Durchmessern von Bohrloch, Patrone und Anker, ebenso eine gute Zerreißbarkeit der Folie und eine günstige geometrische Anordnung der Härterkammer. Nur so läßt sich eine gute Mischung erreichen. Nicht vermeiden läßt sich, daß die Plastikfolie als Trennfläche im Mörtel wirkt und daß Luft oder Wasser beim Mischen mit eingeschlossen werden. Dies führt zu Bildung von Lunkern. Ein Korrosionsschutz ist somit nicht gegeben (dasselbe gilt für Zementpatronen).

Wasser führt zu einer leichten Verzögerung des Aushärtens bei gleicher Endfestigkeit.

Obwohl UP-Harzpatronen organische Bestandteile enthalten, sind aus der vierzigjährigen Geschichte des Verfahrens keine Brände bekannt.

Ihre Dauerhaftigkeit haben UP-Patronen-Verklebungen in einem Langzeitversuch auf der Versuchsgrube Tremonia in Dortmund bewiesen. Nach 10 Jahren regelmäßiger Prüfung bis an die Fließgrenze des Stahls bzw. einer Dauerbelastung mit 35 KN zeigten sich keinerlei Kriecherscheinungen oder ein sonstiges Nachlassen der Ankerwirkung [12]. Auch Erschütterungen durch Sprengungen beim Tunnelvortrieb führen nicht zu meßbaren Setzungen der Anker [13].

Weltweit werden pro Jahr ca. 100 Millionen Klebeanker installiert [14]. Das Ankern mit UP-Harzpatronen hat sich somit als ein Verfahren durchgesetzt, mit dem einerseits im Vortrieb sofort nach dem Abschlag gesichert werden kann und andererseits kurze Felsanker auch dauerhaft als Ausbau genutzt werden können.

3.2. Epoxid- und Polyurethan-Mörtel

Anders als beim ungesättigten Polyester verbinden sich bei EP- und PU-Harzen zwei Komponenten in einer Polyaddition zu einem hochmolekularen Netzwerk. Bei Epoxidharzen sind das die eigentliche Epoxidkomponente und ein Aminhärter, bei Polyurethan sind dies Polyole und Isocyanate. Polyaddukte schwinden beim Aushärten nur wenig oder gar nicht, Polyurethane quellen sogar meist leicht durch Kohlendioxidbildung, die durch Wasserspuren in der Polyolkomponente verursacht wird. Typisch für Polyaddukte ist, daß ihre

Bild 4: Ankermörtelpumpe für gefüllte 2-K-Harze
Bolting grout pump for two component resin mortars

Viskosität auch lange nach dem Mischen niedrig bleibt und sie, insbesondere EP-Harze, binnen kürzester Zeit erstarren.

Die Applikation erfolgt über eine zweikomponentige Injektion. Durch mineralische Füllstoffe erhalten die Harze eine hohe Viskosität, so daß der Inhalt der Liefergebinde über einen Auspreßstempel der Dosierpumpe zugeführt werden muß (Bild 4). Getrennt fließen die beiden Komponenten zum Bohrloch, wo sie in einem Statikmischer innig miteinander vermischt werden. Insbesondere bei Polyaddukten ist eine homogene Vermischung Voraussetzung für ein einwandfreies Aushärten. Normalerweise erfolgt die Ankerung im Injektionsverfahren über einen Rohranker aus Stahl oder vorzugsweise glasfaserverstärktem Polyester. Genauso können natürlich Seil- oder Bündelanker mit Injektionsschlauch verpreßt werden (Bild 5). Es ist aber auch möglich, das Bohrloch nach dem Füllmörtelverfahren zu füllen und dann einen konventionellen Stabanker einzuführen. Während der Injektion füllt der Mörtel nicht nur lunkerfrei das Bohrloch, sondern auch größere Klüfte, die vom Bohrloch durchschnitten werden, d.h. es findet auch eine gewisse Gebirgsvergütung statt. Durch die Injektion wird so - im Vergleich zum Patronenverfahren - in gestörten Gebirgszonen eine Krafteinleitung in das Gebirge erst ermöglicht.

Insbesondere EP-Harze sind als gute Kleber bekannt. EP-Harze sind anders als PU-Harze jedoch sehr empfindlich gegenüber Mischfehlern. 10 % Mindermenge einer Komponente kann zu einer deutlichen Reduzierung der Festigkeiten führen. Die hohe Haftfestigkeit in Verbindung mit dem schwundfreien Aushärten führt zu einem noch besseren Verbund als bei UP-Harzen und somit zu einem

309

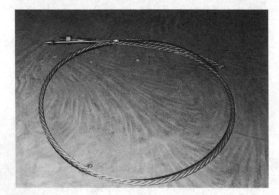

Bild 5: a. Seilanker mit Injektionskanal
         Cable bolt with injection pipe
      b. GfK-Litzenanker
         Fiberglass cable bolt with injection pipe

höheren Sicherheitsfaktor.
Das Bild 5 zeigt die flexiblen Injektionsanker der
CarboTech. Im oberen Teil ist der Stahlseilanker mit
23 mm Durchmesser und 30 t Bruchlast zu sehen, im
unteren der korrosionsbeständige GFK-Litzenanker
mit 35 t Bruchlast.
Wasser im Bohrloch wird verdrängt und durch die
hochviskose Konsistenz des Mörtels nicht
eingemischt. EP-Harze kleben auch an feuchten
Oberflächen gut, bei PU-Harzen wird die Haftung
etwas verringert. EP-Harze brauchen, je nach
Schichtdicke und Temperatur, einige Stunden zum
Abbinden, PU-Harze können beliebig schnell
eingestellt werden.
Beide Harze gelten als außerordentlich beständig
gegenüber chemischem Angriff, sie sind deshalb auch
als Korrosionsschutz wirksam. Zu diesem Zweck
werden diese Harze ja auch als Beschichtungen
eingesetzt.

3.3.   Ungefüllte, nichtschäumende
        Polyurethanharze

In den beiden vorhergehenden Abschnitten war stets
von gefüllten Harzen die Rede. Nichtgefüllte Harze
lassen sich demgegenüber wesentlich einfacher
verarbeiten. In Carbothix H steht ein Polyurethan-
harz zur Verfügung, dessen Einzelkomponenten sich
wie normale Flüssigkeiten pumpen lassen. Sofort
nach dem Mischen nimmt das Harz jedoch eine
pastenartige Konsistenz an, so daß es auch aus
Überkopf-Bohrlöchern nicht mehr ausfließt [15].
Vorzugsweise werden Injektionsanker eingesetzt.
Bei der Applikation wird wie im vorigen Abschnitt
verfahren; die Pumptechnik ist allerdings wesentlich
einfacher. Die selbstansaugenden Pumpen sind
relativ klein, so daß sie von Hand versetzt werden
können (Bild 6).

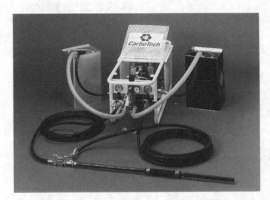

Bild 6: Verpreßpumpe für nichtgefüllte 2-K-Harze
        Resin grout pump for two component resin
        grouts

Bild 7: Aus dem Gebirge geborgene Anker mit
        Carbothix-PU-Harz verklebt
        Bolts sealed with CARBOTHIX PU-resin,
        salvaged from the gob

Ebenso wie das gefüllte Material dringt Carbothix auch verfestigend in größere Spalten des Gebirges ein (Bild 7).
Carbothix ist sehr unempfindlich gegenüber Fehldosierungen. 10 % Unter- bzw. Überschuß mindern die Festigkeit des ausgehärteten Harzes nicht signifikant. Als ungefülltes Material verfügt es über einen E-Modul, der im Vergleich zu Zementmörteln um den Faktor 10 niedriger liegt, während die Biegezugfestigkeit um den Faktor 10 höher ist. Wie bereits im einleitenden Kapitel gesagt, kann das Harz dadurch einen wesentlich größeren Teil der Spannungen aus dem Gebirge auf den Anker übertragen, ohne daß es dabei reißt. Wie FE-Berechnungen an einem konkreten Beispiel gezeigt haben, bleiben sowohl Harz als auch Stahl in ihrem elastischen Arbeitsbereich wirksam. Statt im Bereich der Lastspitze zu fließen, nimmt das gesamte Stahlzugglied die Spannung reversibel auf. Der Korrosionsschutz für den Anker bleibt erhalten.

3.4. Silikatharze (Organomineralharze)

Eine Variante der Polyurethanharze stellen die Silikatharze dar. Sie sind Hybride zwischen organischen und mineralischen Harzen. Dem Isocyanat als einer Komponente steht ein Wasserglas, also eine Natriumsilikatlösung, mit speziellen Additiven gegenüber. Das Isocyanat reagiert mit dem Alkali-Polyharnstoff, während des Wasserglas durch das dabei entstehende Kohlendioxid ausgefällt wird. Die Verschiedenartigkeit der Komponenten erfordert ein besonders sorgfältiges und intensives Mischen im Statikmischer. Das Mischharz dickt im Verlaufe der wenige Minuten dauernden Reaktion kontinuierlich an, um schließlich zu erstarren. Die Endfestigkeiten stellen sich allerdings erst nach etwa sieben Tagen ein.
Zum Applizieren kann dieselbe Technik wie bei Carbothix verwendet werden. In normalem Gebirge ist allerdings die Verwendung eines Packers erforderlich, um ein Ausfließen aus dem Bohrloch zu verhindern, d.h. das Füllverfahren unter Verwendung von Stabankern ist nicht anwendbar.
Häufig werden Silikatharze in zerstörtem Gebirge eingesetzt, in dem standfeste Bohrlöcher nicht erstellt werden können. Deshalb werden hier Injektionsbohranker verwendet. Meist ist in diesen Fällen das Gebirge so beansprucht, daß der größere Teil des Harzes in die Risse und Klüfte fließt und es so durch Vergütung erst ankerbar macht. Die erreichbaren Druckfestigkeiten liegen in derselben Größenordnung wie die der anderen Harze. Der E-Modul und auch die Zugfestigkeit sind allerdings deutlich geringer als bei Carbothix. Silikatharze schwinden leicht.
Das ausgehärtete Harz hat einen pH-Wert > 9,5 und wirkt somit als passivierendes Korrosionsschutzmittel für den Stahl. Gegenüber chemischen Angriff durch Säuren sind Silikatharze erstaunlich beständig.

3.5. Schäumende PU-Harze

PU-Harze, die üblicherweise zu Verfestigungsinjektionen verwendet werden, können auch als Ankerbindemittel eingesetzt werden. Man verwendet dazu dieselbe Verfahrenstechnik wie in den beiden letztgenannten Fällen. Die Verwendung eines Packers ist jedoch unerläßlich, vor allem um das Rückfließen der niedrigviskosen Harze zu verhindern. Darüber hinaus dient der Packer dazu, die Expansion schäumend eingestellter Harze zu begrenzen. Ein zu starkes Aufschäumen schwächt

Tabelle 2: Mörtel/Anker-Typ-Kombination

	Litzenanker im Grundbau	Stabanker z.B.: CarboTech M27	Seilanker z.B.: CarboTech S 23	Injektionsrohr mit Ankerwirkung IRMA	Injektionsbohranker IBO/IBI
UP-Mörtel	-	X	-	-	-
EP- oder PU-Mörtel	X	X	X	-	X
Ansteifende PU-Harze	X	X	X	X	X
Silikatharze	X	-	X	X	X
PU-Injektionsharze	X	-	-	X	X

311

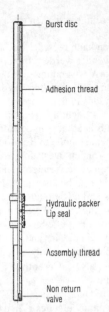

— Burst disc

— Adhesion thread

— Hydraulic packer
Lip seal

— Assembly thread

Non return
valve

Bild 8: Injektionsrohr mit Ankerwirkung (IRMA)
Injection bolt with bolting effect

die Festigkeiten des erhärteten Harzes stark.
Für diese Anforderungen hat CT-BT das sog.
Injektionsrohr mit Ankerwirkung (IRMA) entwickelt [16]. Das IRMA besteht aus einem durchgehenden Stahlrohr mit Gewinde an luft- wie
bergseitigen Ende (Bild 8).
Entscheidend ist der integrierte Gummipacker, der
durch das Verpreßmedium über zwei Löcher und
Lippenventile im Rohr auf ca. 40 bar vorgespannt
wird. Bei Überschreiten dieses Drucks gibt nämlich
eine Berstscheibe am Rohrausgang für das Harz den
Weg frei in den Ringraum des Ankerbohrloches und
ins Gebirge.
D.h. der Anker wird durch den Injektionsvorgang
selbsttätig verspannt; das IRMA ist sehr
anwenderfreundlich. Verschiedene Typen mit
Bruchlasten bis zu 250 kN sind erhältlich. Als
Bindemittel werden meist PU-Injektionsharze wie
Bevedol S- oder Bevedol-WF-Bevedan oder
Geodur-Silikatharz, aber auch zementöse Baustoffe
eingesetzt. Im Steinkohlenbergbau wird dieses
Verfahren regelmäßig zur Vergütung des
Streckensaumes in Abbaustrecken eingesetzt.
Bislang war ausschließlich von Gebirgsankern gemäß
DIN 21521 die Rede, wo die besprochenen Harze
auch im wesentlichen ihre Anwendung finden. Aber
auch beim Setzen von Bodenankern des Erd- und
Grundbaus gemäß DIN 4125 kann man sich die
vorteilhaften Eigenschaften der Kunstharze zunutze

machen. Damit ist in erste Linie das schnelle
Aushärten, d.h. die schnell verfügbare Tragwirkung
gemeint. Das schnelle Aushärten kann darüberhinaus
das Wegfließen des Ankermörtels aus der
eigentlichen Verpreßzone verhindern. Desweiteren
ergeben sich Vorteile bei der Logistik, der
Verfahrenstechnik und der erforderlichen Einbauzeit.
Nachgewiesen wurden diese Eigenschaften bei einer
Hangsicherung im Buntsandstein der ICE-Strecke
der Deutschen Bundesbahn zwischen Fulda und
Kassel [17]. Der Hang sollte mit Mehrstabankern von
25 - 45 m Länge und 6 m Verpreßstrecke gesichert
werden. Durch den klüftigen felsige Grund flossen
große Mengen des konventionellen Ankermörtels ab,
wobei Spitzenwerte von bis zu 35 t erreicht wurden.
Mit zwei verschiedenen Techniken konnten hier
Kunstharzverpressungen ihre Wirksamkeit
nachweisen.
a) Zunächst wurde der Mehrstabanker 6-7 m vom
bergseitigen Ende entfernt mit einem 1 m langen
Gewebeschlauch (Bullflex) überzogen, der als
Packer diente. Durch diesen Packer und in die
Verpreßstrecke hinein wurde ein PE-Rohr geführt,
das in beiden Bereichen Austrittslöcher aufwies. Als
Injektionsharz wurde ein schnellreagierendes PU-
System vom Typ Bevedol-Bevedan verwendet, das
auf ein dreifaches Aufschäumen eingestellt war. Das
Mischelement saß unmittelbar vor dem Packer. Nach
Injektion von knapp 500 kg Harz stieg der Druck an,
die Verpreßstrecke (theoretisches Volumen ca. 80 l)
und die angrenzenden offenen Klüfte waren verfüllt.
Nach dem Abkühlen des frischen Harzes, d.h. zwei
Stunden nach der Injektion wurden die Anker gemäß
DIN 4125 mit der 1,33fachen Gebrauchslast von
695 KN, also 925 KN, beaufschlagt, ohne daß die
Anker unzulässige Verformungen gezeigt hätten. Bei
einem Versuch wurde die Verpreßmenge auf 300 kg
reduziert mit dem gleichen Prüfergebnis.
b) Durch eine zusätzliche Maßnahme gelang es, die
Verpreßmenge und somit die Installationszeit noch
deutlich zu verringern. Dazu wurde nicht nur der
Packerbereich, sondern die gesamte 7 m lange
Verpreßstrecke des Ankers mit einem Gewebesack
überzogen. Das Gewebe ist teilweise durchlässig für
Polyurethan. Als Injektionsharz wurde das
vergleichsweise langsam reagierende Bevedol N-
Bevedan mit einer Reaktionszeit von 55 min bei
15 °C verwendet. Als Harzmenge reichten 55 kg und
10 min Verpreßzeit aus; zwei Stunden nach der
Injektion konnte kein Harz mehr nachinjiziert werden
(Bild 9).
Diese Anker wurden 15 Std. nach dem Verpressen
einer Abnahmeprüfung unterzogen, vier Tage später
einer Eignungsprüfung nach DIN 4125 unterzogen
(Bild 10, Bild 11).

a) Setzen des Ankers und Ziehen der Verrohrung

b) Verpressen des Packers

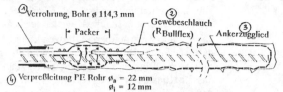

④ Verrohrung, Bohr ø 114,3 mm
|← Packer →|
② Gewebeschlauch (^{R}Bullflex)
③ Ankerzugglied
④ Verpreßleitung PE Rohr ϕ_a = 22 mm
ϕ_i = 12 mm

c) Verpressen der Haftstrecke

⑤ Fertiggestellter Anker

⑥ Gebirgsverzahnung

Verpreßkörper

a) Inserting the anchor and withdrawing the casing
b) Grouting of the packer
c) Grouting of the bond length

1. Casing, o.d. 114,3 mm
2. Fabric hose
3. Anchor rod
4. PE pipe
5. Finished anchor
6. Interlocking with stratum

Bild 9 : Schnitt durch einen Verpreßanker mit Gewebeschlauch; Injektionsverfahren
Sectional drawing of a soil anchor covered with fabric hose; injection process

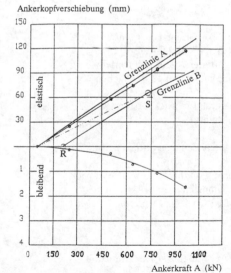

Ankerkopfverschiebung (mm)

Grenzlinie A
Grenzlinie B
elastisch
bleibend
S
R

Ankerkraft A (kN)

Bild 10: Verpreßanker mit Gewebeschlauch,
elastische und plastische Ankerkopfver-
schiebungen nach DIN 4125
Soil anchor with fabric hose, elastic and
plastic displacement of anchor head
acc. to DIN 4125

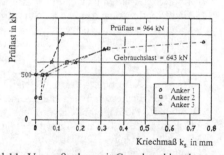

Prüflast in kN
Prüflast = 964 kN
Gebrauchslast = 643 kN
Anker 1
Anker 2
Anker 3
Kriechmaß k_s in mm

Bild 11: Verpreßanker mit Gewebeschlauch,
Kriechmaße nach DIN 4125
Soil anchor with fabric hose, creep. acc. to
DIN 4125

Die Bedingungen der Norm wurden voll erfüllt. Über
den Zeitraum eines Jahres wurden weitere
Kontrollmessungen durchgeführt, ohne daß negative
Erscheinungen wie Kriechen o.ä. aufgetreten wären.
Leider sind diese Ansätze zunächst in der
Anwendung nicht weiterverfolgt worden, da die
Behörden eine Kontamination des Bodens und des
Grundwassers befürchteten. Eine Vielzahl von
Gutachten, teilweise in Abstimmung mit dem

Deutschen Institut für Bautechnik, belegen jedoch mittlerweile, daß diese Gefahr nicht zu besorgen ist, so daß hiermit ein leistungsfähiges Verfahren zur Herstellung schnelltragender Grundanker zur Verfügung steht.

4.   Zusammenfassung

Kunstharze stehen in einer weitere Palette als Bindemittel für das Ankern zur Verfügung. Sie reicht von den Patronen auf Basis ungesättigter Polyesterharze, die gleichzeitig eine sofortige Sicherung im Vortrieb wie auch einen Systemausbau erlauben, über eine Vielzahl gefüllter und ungefüllter Harze auf Polyurethan und Epoxidharzbasis zum Fixieren aller Arten von Gebirgsankern bis hin zum wirtschaftlichen Verpressen von Grundankern mit Härtungszeiten von wenigen Stunden. Neben dem reinen Zeitvorteil, den die Kunstharze bis an die Grenze der Verarbeitbarkeit nutzen können, ist die gute chemische Beständigkeit, namentlich gegen Sulfate, Säuren und Laugen ein wesentliches Plus, das für Kunstharze spricht. Berücksichtigt man darüber hinaus die anwenderfreundliche Verfahrenstechnik und die geringen Mengen, die zur Erfüllen der Aufgaben benötigt werden, stellen Kunstharze in vielen Fällen eine technisch und wirtschaftlich attraktive Alternative zu zementösen Baustoffen dar.

Literatur

1a	DIN 4125, Verpreßanker
1b	DIN 21521, Gebirgsanker für den Bergbau und den Tunnelbau
2a	DIN EN196, Prüfverfahren für Zement
2b	DIN 18136, Untersuchung von Bodenproben: Best. der einachsigen Druckfestigkeit
2c	DIN 22025, Bestimmung der einachsigen Druckfestigkeit von Festgesteinen
2d	DIN 52105, Prüfung von Naturstein: Druckversuch
2e	DIN 53454, Prüfung von Kunststoffen: Druckversuch
3	DIN EN196, Prüfverfahren für Zement
4	Nach CEP/FIP-Richtlinien (1978)
5	Klaus Große, 1988 „Untersuchung zur Verbesserung der Haftung von kunstharzvermörtelten Ankern im Gebirge" Dissertation Clausthal-Zellerfeld.
6	SABS (South African Bureau for Standards), 1991, No 1534: Resin capsules for use with tendon based support systems
7	Chris Reynolds, 1995 „The Introduction of Australian Roofbolting Technology to U:K. Coal Mines" in: „Ankerausbau im Bergbau", Aachen (ISBN 3-86073-352-4)
8	C. Erichsen, W. Keddi, 1980 „Das Tragverhalten vermörtelter Anker und Entwicklung eines neuen Ankertyps", Vortrag 8. Nat. Felsmechanik Symposium 1990, Aachen.
9	DIN 4093, Einpressen in den Untergrund
10	DIN 53456, Prüfung von Kunststoffen: Härteprüfung durch Eindruckversuch.
11	J.C. Eaton, 1992 „Resin capsule development for roofbolting" Mining Technology, 1992, 301-303
12	Günther Nähring, 1987 „Die Versuchsgrube untersucht das Langzeitverhalten von Gebirgsankern", Kompaß, 1987, 183 + 186
13	G.S. Littlejohn, 1992 „Monitoring the influence of blasting on the performance of rock bolts at Panmaenbach Tunnel", zitiert in: Tunnels and Tunnelling, Sept. 1992, p. 36
14	James J. Scott, 1989 „Roof bolting is a sophisticated art", Coal, Vol. 26, No. 8, p. 59-69
15	EP 9302266
16	EP 0398839
17	W. Romberg, W. Cornely, 1991 „Temporäranker mit Verpreßkörpern aus Bevedol-Bevedan-Systemen" in: Geotechnik 14, 112-117.

Anker in Theorie und Praxis, Widmann (Herausgeber) © 1995 Balkema, Rotterdam. ISBN 90 5410 577 1

Betriebliche und technische Wirksamkeit des Rohrankers beim Auffahren eines Strebs in der bulgarischen Kohle

Operational and technical performance of the Tube-Anchor in the starting chamber of a longwall in a Bulgarian coal mine

H. Habenicht
Weißkirchen, Austria

N. Nikolaeff
Universität für Bergbau und Geologie, Sofia, Bulgaria

ZUSAMMENFASSUNG: Am Beispiel des bereits mehrfach beschriebenen Rohrankers der Type TFA werden ingenieurmäßige Berechnungen dargelegt, die es erlauben, unter Bezugnahme auf eine Gebisgsklassifizierung, eine Tragring-Mechanik zu formulieren, mit welcher ein konsequenter Berechnungsablauf unter Einbeziehung des Gebirges, der Anker, und der Ausbaubogen möglich wird. Ein Anwendungsbeispiel dafür und die praktischen Ergebnisse werden beschrieben.

ABSTRACT: The tube type friction anchor, symbolized by TFA, exhibits a cross section which has been formed to the shape of an ellipse. In a borehole of circular cross section the diameter of which is smaller than the large axis of the ellipse a friction fit is generated at the borehole wall which causes a reinforcement effect in the rock mass.

One of the eminent features of the TFA is its persistent resistance during large amounts of rock motion, due to the enduring frictional contact to the borehole wall, even though eventual excessive loads maybe developed by the rock.

This persistance offers a particular utilization in rock carrying rings of large deformations, such as in mild and squeezing rock. In cases where internal support is provided by steel arches the rock usually converges until the contact to the steel is completed. The TFA is suitable to provide resistance to the rock mass in the deformation range corresponding to the closure of the contact gap. This particular property makes the TFA suitable for the compound function of rock carrying rings that are not only reinforced by anchors but simultaneously supported by steel arches.

Nikolaeff has derived engineering formulas not only for the TFA as a reinforcing element but also for simulating the effect of the TFA in the rock carrying ring and also for the configurations involving steel arches. His design formulas allow the determination of the required support, and are given in the text of this paper.

The procedure starts from the calculation of the equivalent total magnitude of a support resistance that would be required to stabilize a rock carrying ring of certain quality at a certain selected amount of convergence. A subsequent step using the same mechanics is calculated for the equivalent support resistance that would be required from the arrangement of anchors if they would have to control the rock carrying ring at a certain allowed convergence. Finally, the difference between the originally determined total resistance for the rock carrying ring and the resistance that is provided by the selected anchor pattern forms the resistance magnitude to be provided by the steel arches.

The procedure is demonstrated on the example of a rectangular cross section of a starting chamber for a longwall. The practical drivage of the new cross section was completed successfully, and with a number of improvements.

1 EINLEITUNG

Nach rund 25 Jahren seit der
Erfindung des Rohrankers besteht in
Bulgarien umfangreiche Erfahrung
aus der Praxis. Der Rohranker,
bereits mehrfach in
Veröffentlichungen beschrieben, hat
sich in verschiedenen
Gebirgsklassen im bulgarischen
Bergbau bewährt. Neuerdings
abgeleitete technische Beziehungen
und Berechnungsschritte haben einen
weiteren Fortschritt ermöglicht,
der zur planmäßigen Verbesserung im
Auffahren eines ersten Aufhauens
neuer Form geführt hat. Dieser Fall
eines Einsatzes in einem
Braunkohlenbergwerk, welcher nach
eingehender Berechnung und
Vorabschätzung verwirklicht werden
konnte, wird nachstehend
beschrieben.

2 DER ROHRANKER

Die große Vielfalt der
Ankerbauarten (Habenicht 1976) wird
unter anderem auch bereichert durch
den Rohranker der Type TFA, zu
welchem neuerdings auch
eigenständige Berechnungen
entwickelt wurden, die einen
zusammenhängenden und schlüssigen
Entwurfsvorgang ermöglichen.
Beschreibungen des Rohrankers haben
bereits durch Nikolaeff (1987 und
1989) sowie durch Nikolaeff und
Paruschev (1983 und 1985/1)
stattgefunden. Durch die Erfindung
im Jahre 1971 in Bulgarien wurde er
vorerst besonders in jenem Land
eingesetzt. Dort wurde er auch
patentgeschützt, und die meisten
Erfahrungen werden aus dem
bulgarischen Bergbau berichtet (
Nikolaeff und Paruschev 1985/2
sowie dieselben 1989). Auch in
einigen anderen Ländern Europas
wird er mit Erfolg verwendet.

    Das Prinzip des Rohrankers beruht
auf der Reibwirkung, welche
entsteht, wenn ein Rohr, dessen
ursprünglicher Kreisquerschnitt zu
einer Ellipse umgeformt worden ist,
in ein kreisrundes Bohrloch
getrieben wird, dessen Durchmesser
kleiner ist als die große Achse der
Ellipse. Diese Konfiguration ist in
Abb. 1 dargestellt. Das Bohrloch
erzwingt die teilweise
Rückverformung des
Rohrquerschnittes im Bereich

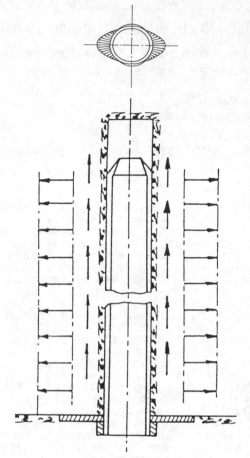

Abb. 1: Längsschnitt durch die
Bohrlochachse (unten) und
Querschnitt (oben) durch das
Bohrloch mit dem Rohranker.
Longitudinal section through the
borehole-axis (bottom), and cross
section (top) through the borehole
with the tube type friction anchor.

elastischer Formänderung, wodurch
Normalkräfte in radialer Richtung
an der Bohrlochwand entstehen.
    Wahlweise kann die Ellipsenform
auf das Bohrloch abgestimmt werden.
Die verhältnismäßig feinfühlige
Wahl des geeigneten
Bohrlochdurchmessers sichert den
Einrammwiderstand, die Haftung im
Gebirge, und somit die
Ausziehkraft. Abb. 2 zeigt ein
Diagramm der Haltekraft-Zunahme mit
größer werdender Haftlänge.

Bei technisch möglichen Ankerlängen
von bis zu 3,2 m (in Ausnahmefällen
bis zu 4,0 m) können demnach Kräfte
in der Größenordnung von 30 bis 90
kN, bei längerer Zeit sogar bis 120
erzielt werden.

Die Reißkraft am Anker erreicht
94 bis 134 kN, je nach der
Wandstärke des Rohres.

Mit zunehmender Zeit steigt nach
dem Einbauen der Haftwiderstand des
Ankers am Gestein. Bei Beträgen der
großen Achse des elliptischen
Querschnitts von 39 - 43 mm
erscheinen je nach Gebirgsart
Bohrdurchmesser von 36-42 mm
geeignet.

Eingebaut wird der Rohranker mit
einem Drehschlag-Hammer. Dazu
benutzt man einen Adapter, dessen
Schlagfläche auf der Seite des
Ankers so gestaltet ist, daß der
Anker beim Eintreiben aufgebördelt
wird. Somit wird durch die
Aufbördelung die Halterung der
Ankerplatte gesichert.

3 DIE MECHANIK

3.1 Der Anker

Die Haltekräfte des Ankers werden
durch den Anpreßdruck im Kontakt
des Rohres mit dem Bohrloch
bestimmt. Verschiedene Gesteine
bieten auch verschiedenen
Verformungswiderstand und
verschiedene Reibungskoeffizienten.
Willkürlich einstellbar sind die
Einflüsse der Wandstärke des
Rohres, der Exzentrizität der
Ellipse und die Abstimmung des
Bohrlochdurchmessers auf den
Rohrquerschnitt. Diese Mehrzahl von
Faktoren verlangt eine umsichtige
Wahl der Parameter unter
Berücksichtigung der Gegebenheiten
des Gebirges, nämlich der
Steifigkeit des Gesteins und der
Verformbarkeit des Tragrings. Der
Rohrwerkstoff aus einem Stahl der
Güte St-38 ändert seine Steifigkeit
beim Aufprägen der Ellipse nicht
bemerkenswert, so daß die
Charakteristik dieser Stahlgüte im
wesentlichen erhalten bleibt.

Eine Datenbasis aus
Ausziehversuchen in verschiedenen
Gesteinen und bei verschiedener
Geometrie der Ellipse und des
Bohrlochs, sowie bei abgestuften
Ankerlängen erlaubt eine gezielte
Zuordnung.

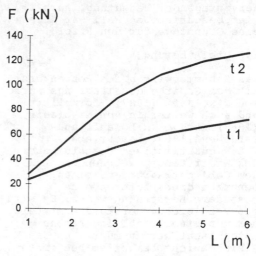

Abb. 2: Die Haltekraft F nimmt mit
der Haftlänge L aber auch mit der
Zeit zu. t1...sofort nach dem
Einbau; t2...nach 1 bis 2 Monaten.
The pullout force F increases with
increasing contact length L, and
also with increasing time.
t1...directly after mounting;
t2...after 1 to 2 months.

Unter den Eigenschaften dieses
Ankertyps ragt besonders seine
Beständigkeit der Grenzhaftspannung
hervor. Bei Überschreiten der
Grenzhaftspannung am Kontakt des
Ankers zum Bohrloch durch übergroße
Gebirgskräfte tritt lokales Gleiten
ein. Dabei ist auffallend, daß der
Gleitwiderstand nicht bei gößerer
Bewegung absinkt, sondern daß
durchwegs ein annähernd konstanter
Widerstand fortwirkt, der die
Erhaltung des Tragvermögens
sicherstellt.

Daraus resultiert eine
Ankerkennlinie, die bei
fortschreitender Verformung des
Gebirgstragrings, etwa bei
anhaltender Konvergenz, einen über
äußerst große Strecken konstanten
Widerstand aufweist (Nikolaeff und
Paruschev 1989). Die somit
überbrückbaren Verformmungen des
Gebirges können weit größer werden
als bei Spreizankern, Klebankern
oder vermörtelten Ankern, bei
welchen Defekte am Kontakt
eintreten können.

Diese Eigenschaft ist besonders
bei milden Gebirgsarten und bei
großer Konvergenz von

hervorragender Bedeutung. Auch im hier beschriebenen Fall war dies eine Grundlage für den Erfolg.

3.2 Die Ankerung

Das Zusammenwirken des Rohrankers mit dem Gebirge des Tragrings wurde von Nikolaeff (1987) formuliert und auch weiterhin von Nikolaeff (1989) und von Nikolaeff und Parushev (1989) beschrieben.

Zur quantitativen Feststellung des Widerstands, welchen die Anker dem Gebirgstragring zuleiten, benutzen diese Autoren eine besondere Berechnungsweise. Es wird ein Vergleichswiderstand ermittelt, der von einem Stahlring-Ausbau aufgebracht werden müßte, um das Gleichgewicht bei einem bestimmten Betrag der Konvergenz herzustellen.

Die Kräfte-Entwicklung des vergleichsweise ungeankerten Gebirgstragrings wird dabei ausgedrückt durch ein mechanisches Gesetz, in welchem die gebirgstechnologischen Parameter eingesetzt werden (Nikolaeff 1970). Dabei benutzt man auch eine Gebirgsgüte-Klassifikation.

Nikolaeff (1987) konnte in einer grundlegenden Arbeit diesen gerechneten Ausbauwiderstand pg herleiten aus einer Beziehung nach Paraschkewoff (1969), woraus hervorgeht:

$$pg \geq (2*w*H-SDM)*d/((1+d)*2**b)$$

worin b=(2+d)/d   Gl.1

Hierin bedeuten
w...spezifisches Gewicht der Überlagerung (kN/m3)
H...Teufe des Gebirgstragrings (m)
SDM...einachsige Druckfestigkeit des Gebirges (kPa)
d = m / (k-1), mit m und k Parameter aus den Mohr'schen Spannungskreisen zur Definition der Hüllgeraden nach Nikolaeff (1987).

Diese Form erhält man unter Zugrundelegung eines Gebirgstragrings, dessen Radius rg angeglichen ist an die Länge der Gebirgsanker. Aus praktischen Gründen und wegen der tatsächlichen Querschnittsgröße des Hohlraums wird der Radius gewählt in der Größe

$$rg = 2*rh,$$

worin rh den Radius des Hohraums darstellt.

Die Größe des Widerstands pg bildet die Basis des Bemessungsablaufes.

In einer weiteren hergeleiteten Gesetzmäßigkeit kann jener äquivalente Widerstand pe errechnet werden, den die Gebirgsanker in den Gebirgstragring einleiten. Die Berechnung von pe erfolgt ausgehend von der wahlweisen Festlegung der Ankerdichte und einer daraus resultierenden Konvergenz des geankerten Gebirgstragringes. Setzt man danach die ermittelte Konvergenz ein in die Gl. 1, so erhält man einen Widerstand, der dem ungeankerten Gebirgstragring entgegenwirken müßte, um das Gleichgewicht bei der angesetzten Konvergenz herzustellen.

Der somit bestimmte äquivalente Ausbauwiderstand (im realen Fall wird er einem geschlossenen Gleitbogen zugedacht) kann als Maß dafür eingesetzt werden, welchen Trageffekt die Anker erwirken. Die Größe pe berechnet Nikolaeff (1987) in folgender Form:

$$pe = A1 * A2 ** (1/(1+d))   Gl.2$$

mit
A1=(2*w*H-SDM)/(2*(1+d)*(k-1))
A2 = 3*(2*w*H+d*SDM)*(n+1)*rh / (4*pd*E*(d+1))

worin gilt:
n ... Poisson'scher Koeffizient
E ... E-Modul des ungestörten Gebirges (MPa)
pd... Haltekraft des Ankers bei Konvergenz (kN)

3.3 Kombination Anker und Stahlbogen

Bei Ankerung des Gebirgstragrings in mildem, druckhaftem Gebirge wird im Falle zu großer Konvergenz die Möglichkeit gewählt, durch zusätzliche Stahlbogen weitere Unterstützung einzubringen. Im Falle des Rohrankers wird die Berechnung unter Einsetzen einer jeweils gewählten Größe der Konvergenz durchgeführt.

Den Stützwiderstand pb des zusätzlich benötigten Bogenausbaus erhält man aus der Differenz des gesamten Stützdrucks pg der für den unausgebauten Gebirgstragring zu erbringen wäre, und dem Betrag des Stützkraftäquivalents pe, der den Ankern zugeschrieben wurde. So erhält man:

$$pb = pg - pe \qquad \text{Gl. 3}$$

Dieser Ausbaustützdruck pb wird praktisch erst verzögert wirksam, nämlich nach dem Schließen der Fuge am Kontakt des Bogens zur Hohlraumkontur. In manchen reellen Gebirgsarten und Hohlraumquerschnitten erreicht dieser Betrag 50 bis 100 mm. Der Rohranker ist geeignet, durch seine Gleitfähigkeit bei aufrechtem Widerstand, diesen Vorgang abzusichern. Wünschenswert ist daher der Einbau der Anker möglichst frühzeitig, insbesondere aber vor dem Setzen der Stahlbogen.

Der quantitative Aufwand an Stahlbogen NR nach der maßgebenden Größe pb läßt sich errechnen aus:

$$NR \geq B1 * (B2+B3) \qquad \text{Gl. 4}$$

mit
B1=(pb*bo*D**2)/fst
B2=0,0189*(1-M)/Wx
B3=(0,2125+0,2875*M)/(D*A)

Die Werte bedeuten:
NR ... Anzahl der Stahlbogen (Stk/lfm)
bo ... Breite des Hohlraums (m)
fst ...Zulässige Festigkeit des Stahls (kPa)
M ... Verhältnis von horizontaler zu vertikaler Komponente des generellen Spannungsfeldes
Wx ... Widerstandsmoment um die x -Achse des Stahlbogen -Profils (cm3)
A ... Metallquerschnitt im Stahlbogen-Profil (cm2)
D ... transformierter Durchmesser des Hohlraums (m)

4 EINSATZ IM BRAUNKOHLENBERGWERK PERNIK

4.1 Das Gebirge

Das hier betrachtete Flöz liegt in einer Teufe von 60 - 80 m und besitzt eine Mächtigkeit von 0,7 - 0,9 m. Es enthält eine Braunkohle der einachsigen Druckfestigkeit von 3 - 7 MPa. Liegendes und Hangendes bestehen aus einem milden Mergel mit einachsigen Druckfestigkeiten von Laborproben bei 6 - 9 MPa.

Besondere Bedeutung kommt dem hangenden Mergel zu, da die Kalotte der Kohlestrecken und der Aufhauen im Hangenden angeordnet ist.

Diese Kombination von Schichten verhält sich stark druckhaft und erlaubt in der Grube Pernik das Auffahren von Strecken und Aufhauen nur bis zu einem Durchmesser von 2,5 bis 3,0 m. Dabei kann es während der nutzbaren Standzeit allerdings zu Konvergenzen bis zu 80 cm kommen.

4.2 Das Aufhauen

Auch die für den Strebbau zur Vorrichtung hergestellten Aufhauen werden in Pernik nur in kreisförmigem Querschnitt bis maximal 3,0 m Durchmesser aufgefahren. Der Ausbau besteht in geschlossenen Gleitbogen-Ringen eines Glockenprofils verschiedener Querschnitts-Kennwerte. Ihr Einschubwiderstand liegt bei 150 kN für den gesamten Stahlbogen. Die herkömmlich übliche Ausbaudichte Ad beträgt ca. 3 Stk. je lfm.

Im Verlauf der Standzeit bildet sich mit einherschreitender Konvergenz eine Auflockerungszone um den Hohlraum, deren Gestalt elliptisch bis kreisrund sein kann.

Im vorliegenden Fall wurde eine nur geringe Exzentrizität dieser Zone beobachtet, so daß für ihre Definition ein einfacher Radius rg eingesetzt werden konnte. Für eine angenäherte Kreisform kann nach Paraschkewow (1969) die Beziehung gelten:

ro = rg / rh =

$$=((2*w*H+SDM)*d / (2*pg * (1+d)))**(d/2) \qquad \text{Gl. 5}$$

Darin sind w, H, d, m, k, pg, rg und rh wie in Gl. 1.

Die praktischen Arbeiten der Auffahrung mit Sprengbetrieb und manuellem Wegfüllen (Stojantschev et al. 1994) bedeuteten für die Mannschaft schwieriges Arbeiten auf beengtem Raum, wobei die

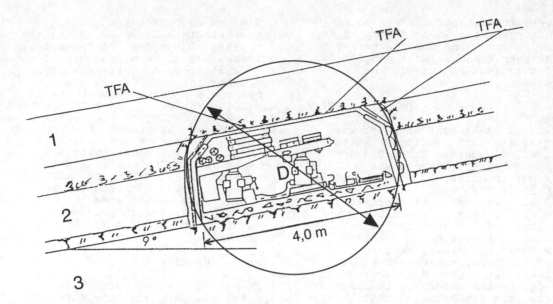

Abb. 3: Querschnitt des Aufhauens.
D...Durchmesser des äquivalenten
Kreisquerschnitts; 1...Hangendes;
2...Kohlenflöz; 3...Liegendes;
TFA...Position der Rohranker.

Cross section of the starting
chamber. D...diameter of equivalent
circle; 1...roof strata; 2...coal
seam; 3...footwall strata;
TFA...position of tube anchors.

anschließende Montage der Teile der
Ausbaugestelle in liegender Haltung
erfolgen mußten.

Die Längen solcher Aufhauen
betragen manchmal nur 30 m, meist
jedoch 60 m, was auch mit der
praktischen Streblänge von gleicher
Größe verbunden ist. Wegen der bis
zum Auffahrungsende bereits
eintretenden
Querschnittsverringerung gestaltet
sich danach auch der Antransport
der Strebausbau-Gestelle schwierig.
Gleiches gilt auch für die zu
installierende weitere
Strebausrüstung.

4.3 Der neue Querschnitt des
Aufhauens

Abgeleitet von der Gestalt und
Größe der Strebausrüstung,
insbesondere der in Pernik
verwendeten Ausbaugestelle, erweist
sich ein größerer Querschnitt des
Aufhauens erstrebenswert. Aufbauend
auf diesem Ansatz wurde ein
annähernd rechteckiger Querschnitt
geprüft, sowie eine

Querschnittsgröße, die ein
einfacheres Arbeiten und den
Einsatz einer Lademaschine
ermöglichen sollte.

Der betrachtete und auch in der
Grube erfolgreich ausgeführte
Querschnitt ist in Abb. 3
dargestellt.

4.4 Der Äquivalent-Querschnitt

In der Standsicherheitsuntersuchung
mußte der Rechteck-Querschnitt in
einen äquivalenten Kreisquerschnitt
umgerechnet werden, dessen
Durchmesser D von Nikolaeff aus der
Beziehung errechnet wird:

$$D = 0,5*h*(z**(1/2)+(z**2+1)**(1/2))$$

Gl. 6

Es gelten:
z = bo / h
bo... Breite des rechteckigen
      Aufhauens (m)
h ... Höhe   -- " --

Der Durchmesser des so ermittelten

äquivalenten Querschnitts wurde in Gl. 4 und in Gl. 5 anstelle des Durchmessers des Hohlraums (gegeben durch dessen Radius rh) eingeführt.

4.5 Kennlinien-Mechanik und Eingangsgrößen

In Betriebsversuchen der vorangegangenen Phase wurde bereits die Wirksamkeit der Rohranker in dem Verbundsystem in Erfahrung gebracht. Ein beachtenswerter Beitrag der Rohranker zum Stabilitätseffekt war festgestellt worden (Stojantschev et al. 1994).

Die Berechnung nach den voran aufgezeigten Beziehungen ergaben auch für den neuen äquivalenten Durchmesser bei erträglichen Größen der Konvergenz Größen des erforderlichen Ausbauwiderstands, den man mit dem Rohranker und den Stahlbogen erbringen konnte.

Die Größe der Konvergenz wurde praktischerweise als Betrag der Firstsenkung ausgedrückt, wobei nach dem Superpositionsprinzip die Gesamtsenkung mit Umax errechenbar ist aus den Teilsenkungen v für die Phase bis zum Kontakt der Hohlraumkontur mit dem Stahlbogen und U_o für die daran anschließende Senkung der Profilachse des Stahlbogens. Man erhält somit

$$U_{max} = U_o + v \qquad (m) \qquad Gl. 7$$

Hiefür ist jedoch U_o aus der Wirkung des Verbunds der Anker mit dem Gebirge zu ermitteln, wofür Nikolaeff folgende Form gefunden hat:

$$U_o = (1/(4+2*(f-b)))*((2*U_{max}/Na) -a*pd) \qquad (m) \qquad Gl. 8$$

Die Symbole bedeuten:
a ...Parameter der Arbeitscharakteristik des Rohrankers (m/N)
La ...Haftlänge des Rohrankers (m)
wa ...Neigung der außenliegenden Rohranker gegenüber der Horizontalen (°)
f ... La * (sin wa)/R
R ... = D/2 mit D nach Gl.6
b ... = (2+d)/d mit d nach Gl.1
Na ... Zahl der Anker je lfm der Strecke

Bei der Suche nach dem dafür passenden Ausbau kann man voraus

TABELLE 1: Einteilung der Gebirgsgüteklassen nach Bulitscheff, modifiziert durch Nikolaeff.
Rock mass classification after Bulitscheff, modified by Nikolaeff.

GEBIRGS-KATEGORIE	KOEFFIZIENT DER STAND-FESTIGKEIT S	STANDDAUER DES GEBIRGES
I	> 70	unbeschränkt
II	5 - 70	bis 6 Monate
III	1 - 5	240 - 360 Std.
IV a	0,55 - 1,00	96 - 240 Std.
IV b	0,25 - 0,55	24 - 96 Std.
IV c	0,03 - 0,25	10 - 24 Std.
V	< 0,03	< 10 Std.

die Zahl der Anker frei wählen, dafür den fiktiven Ausbauwiderstand bestimmen, und anschließend gemäß Gl. 3 den erforderlichen Widerstand des Bogenausbaus ermitteln.

Für die Benutzung der Gleichungen 1 und 2 benötigt man allerdings den Wert SDM, für dessen Bestimmung Nikolaeff (1987) ein eigenes Verfahren benutzt. Darin wird aus der Zugfestigkeit SZL und der Druckfestigkeit SDL von Laborproben deren Verhältnis T als

$$T = SZL / SDL \qquad Gl. 9$$

bestimmt. Weiter benötigt man den Standfähigkeits-Koeffizienten S nach Bulitscheff (1994), welcher einer Gebirgsklassifizierung entnommen wird. TABELLE 1 enthält diese Klassifizierung, welche für den vorliegenden Zweck von Nikolaeff in der Gebirgskategorie IV um die drei Unterstufen a,b,c erweitert worden ist.
Aus diesen Unterlagen wird die Berechnung von SDM nach der von Nikolaeff hergeleiteten Beziehung möglich:

$$SDM = c * SDL \qquad (MPa) \qquad Gl. 10 a$$

$$c = T + (1-T)*(S/(107*f_o))**(1/2) \qquad Gl. 10 b$$

Hierin stellt f_o die Festigkeitskennzahl nach Protodiakonov dar.

Dieser Rechenprozess wurde von Nikolaeff bereits an einer Mehrzahl

von praktischen Verhältnissen
angewendet und als aussagekräftig
erkannt.

4.6 Konkretes Beispiel des Aufhauens in Pernik

Im Fall des neuen
Rechteckquerschnitts für das
Aufhauen konnte unter Benutzung der
oben angeführten Mittel eine
Standfähigkeitsuntersuchung und ein
Unterstützungsprogramm berechnet
werden. Die freie Standdauer der
Firste konnte dafür mit 20 h
angesetzt werden, was den
Koeffizient der Standfestigkeit in
der Größe S=0,25 aus Tabelle 1
ergab. Die weiters eingesetzten
Zahlenwerte sind in Tabelle 2
gegeben.

4.7 Entwurfsresultate

Das Ergebnis des soeben
beschriebenen Berechnungsganges
erbrachte die in Tabelle 3
ersichtlichen technischen Daten des
geankerten und unterstützten
Gebirgstragrings.

4.8 Praktischer Erfolg

Die nach diesen Zahlen hergestellte
und ausgebaute Kammer des Aufhauens
erlaubte praktisch eine
störungsfreie Auffahrung in vollem
Querschnitt und den Einsatz einer
Lademaschine. Im fertigen Aufhauen
konnten die Ausbaugestelle für den
Streb als komplette Garnituren
eintransportiert und aufgestellt
werden, wogegen man für die kleinen
Kreisquerschnitte die Garnituren
hatte zerlegen und in Teilen
eintransportieren müssen.

Die strebseitigen Ulmensegmente
der Ausbaubogen wurden erst kurz
vor Beginn des Kohle-Schneidens
entfernt, während zur Unterstützung
der Firste im Aufhauen ein
Holzblock-Pfeiler auf den Kappen
jedes Gestells errichtet worden
war. Abb. 3 zeigt diese Anordnung
in der Phase vor dem Entfernen der
kohleseitigen Segmente.

Die abgeleiteten Vorteile
äußerten sich besonders in der
Arbeitserleichterung und der
Zeitersparnis. Deutlich verringert
wurden die Dauer der Auffahrung,
des Eintransportierens der

TABELLE 2: Eingangsgrößen für die
Berechnung des Aufhauens.
Given data for the design
calculations of the chamber.

bo = 4,5 m	m	= 1,0	
h = 2,0 m	k	= 1,7	
H = 80,0 m	M	= 0,8	
D = 4,47 m	Umax=	0,1	m
w = 20,0 kN/m3	a	= 5,5E-7 m/N	
SDL =12,0 MPa	pd	= 90,0	kN
f_0 = 1,2	d	= 1,43	
T = 0,125	wa	= 30	°
n = 0,38	La	= 2,4	m
E = 3,5 GPa			

TABELLE 3: Ergebnisse der
Berechnung für das Aufhauen.
Design results for the chamber.

S = 0,25	
c = 0,175	pg =123 kPa
Uo = 0,062 m	pe = 73,4 kPa
SDM= 2,1 MPa	pb = 49,4 kPa
NR = 0,94 Stk/lfm	
Praktischer Ansatz: Wähle NR = 1	

Strebgestelle und von deren Montage
und Aufstellung. Die Zeitersparnis
zufolge der Erleichterungen wird
seitens der Grube mit 2 Monaten
angegeben. Weiters wirkten sich die
verbesserten Arbeitsbedingungen und
der größere verfügbare Raum sehr
positiv bei der danach folgenden
Einrichtung der anderen Maschinen
für den Streb aus.

Durch Beobachtungen der Senkungen
an Firste, Ankern und Streckenbögen
konnten Hinweise gefunden werden,
nach welchen die Gebirgszone um das
Aufhauen eine Auflockerung
durchmachte, welche außerdem
druckmildernd gewirkt hat. Im
praktischen Ablauf fand man dafür
Indikationen, da sowohl der
rückwärtige Stoß wie auch vor allem
der Strebstoß sich als
verhältnismäßig druckfrei erwiesen,
wogegen bei den kreisförmigen
Aufhauen-Querschnitten eine gewisse

Druckhaftigkeit Erschwernisse gebracht hatte.

5 SCHLUSSWORT

Die eigenständige Entwicklung des Rohrankers, sowie seine Erprobung und seine praktischen Einsätze haben zu Erkenntnissen geführt, die sich auch in ingenieurmäßigen Berechnungen formulieren lassen. Die abgeleiteten Gesetzmäßigkeiten und quantitativen Beziehungen, die auch in ihrem Gedankengut und ihren Resultaten gewissermaßen den in Österreich eingetretenen Entwicklungen entsprechen (vgl. Seeber 1974, Habenicht 1983), haben sich dazu geeignet, eine Vorabschätzung für eine Pionierarbeit zu erbringen, deren Aussagen eine positive Entscheidung zugelassen haben, und deren Ergebnis auch in der Ausführung erfolgreich war.

LITERATUR

Bulitscheff, N. 1994. Mechanika podsemnich Sooruschenija. *Ver. Nedra,* Moskau.

Habenicht, H. 1976. Anker und Ankerungen zur Stabilisierung des Gebirges. Springer Verlag Wien New York, 192 pp.

Habenicht, H. 1983. The anchoring effects – our present knowledge and its shortcomings. *Internat. Symp. on Rock Bolting,* Abisko, Sweden, 1983, Vol.I, p.257–268.

Nikolaeff, N. 1970. Zur Vorauswahl von Untersuchungsmethoden und Anwendungen zur Parameterbestimmung des Ankerausbaus. *Tagungsberichte Internat. Bergbaukongress,* Madrid.

Nikolaeff, N. 1987. Eine Methode zur Bemessung des kombinierten Ausbaus aus Stahlringen und Ankern für Grubenbaue in weichen Gesteinen. *Minno delo,* Heft 11.

Nikolaeff, N. 1989. Rozwitie na Teoriata u Praktika na Ankernia Krepesch za podzemni Soraschenija. *Habilitationsschrift an der Universität für Bergbau und Geologie,* Sofia, Bulgarien.

Nikolaeff, N. & V. Paruschev 1983. Oval shaped friction pipe anchor. Theory and application in mining and underground construction. *Proc. Internat. Symp. on Rock Bolting,* Abisko, Sweden, p.425–428.

Nikolaeff, N. & V. Paruschev 1985. Possibilities for rock bolting application in deep mines. *9-th Plenary Scientific Session, Internat. Bureau of Strata Mechanics,* Varna, A.A. Balkema.

Nikolaeff, N. & V. Paruschev 1985. Ankeri u ankeren Krepesch. *Ver. Technika,* Sofia.

Nikolaeff, N. & V. Paruschev 1989. Ein neues Ausbausystem für schwere gebirgsmechanische Bedingungen. Grundlagen zur Bemessung eines kombinierten Streckenausbausystems aus Metallringen und Reibungs -Rohrankern. *Tagungsberichte 8. Internat. Gebirgsdrucktagung,* Varna, Beitrag B13/14.

Paraschkewoff, R. 1969. Mechanika na Skalite. *Ver. Technika,* Sofia.

Seeber, G. 1974. Problematik der Gebirgsklassifikation im druckhaften Gebirge. *Bundesministerium für Bauten und Technik, Straßenforschung.* Sonderdruck aus Heft 18, Wien, S. 29–36.

Stojantschev G., N. Nikolaeff, & V. Paruschev 1994. Effektives Einbauverfahren des mechanisierten Strebanlage für geringmächtige Kohlenflöze. *Proc. 16-th World Mining Congress,* Sofia, Beitrag C-11, S. 97–102.

Anker in Theorie und Praxis, Widmann (Herausgeber) © 1995 Balkema, Rotterdam. ISBN 90 5410 577 1

Selbstbohrende Injektionsanker, -nägel und Mikropfähle WIBOREX

Self-drilling injection bolts, nails and micro-piles WIBOREX

Jochen Kasselmann
F. Willich Berg- und Bautechnik GmbH + Co, Dortmund, Germany

ZUSAMMENFASSUNG: Die neuen Wege in der Ankertechnik begannen bereits 1983. Heute, 1995, kann man sagen, daß ein weiter Weg bei der Entwicklung der Injektionsbohranker zurückgelegt wurde. Dieser lange Entwicklungsweg zeigt seinen Niederschlag in einer Reihe von Patenten, die in den letzten Jahren erteilt wurden.

ABSTRACT: The report deals with a new bolting system on which development started in 1983. In the course of time, many important individual steps have been taken, culminating in the present very extensive use of injection drill bolts.
These bolts can be used for applications with tensile and compressive stresses ranging from 100 kN to more than 900 kN. Another milestone in the development was the passivation protection of drilling bolt rods, which now permits bolts with a free bolting length to be used even in contaminated soil. Efforts are currently under way to have stainless steel injection drill bolts approved as permanent bolts by the construction supervising authority in Berlin. This approval procedure is not yet concluded, but we assume that approval will be given in the near future. The report also shows how problems can be solved with the oppropriate process and application technology. The same applies to injection and bonding procedures.The field of application for injection drill bolts is not confined to the pure bolting area. This system also allows for nailing for slope protection. The larger systems can be used as piling. This application is certainly the most diverse and the most challenging to the structural engineer. However, the possibility of implementing targeted injection measures with self-drilling injection bolts should not be underestimated either. This also provides scope for new systems in injection technology. This is illustrated with some examples from the field of German mining, from extensive application in Austrian tunnelling, taking the Achberg Tunnel in Unken es an example, and from the clearance and sealing of an anhydrite gallery in demanding geological conditions.
Nor should those innovations which are still necessary be overlooked. The example of the development of sliding-head bolts shows how this innovative technology can be used not only in the field of self-drilling injection bolts but also in that of conventional bolt technology.

1 SYSTEMBESCHREIBUNG

Man kann heute eigentlich nicht nur von Injektions-bohrankern sprechen, sondern man muß deutlich fest-stellen, daß diese Technik in vielen Anwendungs-bereichen vertreten ist. Der Bohr- und Injektionsanker WIBOREX findet seinen Einsatz ebenso als Pfahl oder Nagel. WIBOREX Injektionsbohranker sind für An-ker, Pfähle, Injektionslanzen, Bodennägel sowie be-wehrte Erde im weitesten Sinne gleichermaßen gut geeignet.

Das Regelwerk DIN 4125 "Verpressanker" (Nov. 1990) gilt für Anker, die eine freie Ankerlänge aufwei-sen, vorgespannt werden und nur Zugkräfte aufneh-men.

Das Regelwerk DIN 4128 "Verpresspfähle" (April 1983) gilt für Kleinpfähle, die auf Zug und Druck be-lastet werden und auf ihrer gesamten Länge mit Zementstein ummantelt sind.

1.1 System

Mit den verschiedenen Entwicklungsstufen des WIBOREX-Systems wurde ein umfangreiches Zu-behörprogramm für die unterschiedlichsten An-wendungsfälle mit entwickelt. Alle Systeme basieren jedoch auf dem gleichen Grundprinzip:
"Stange - Bohrkrone - Muffe - Mutter"

1.2 Varianten

Das Prinzip des Bohr- und Injektionsankers

325

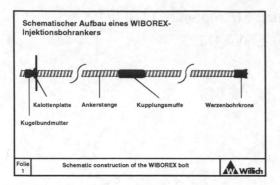
Schematischer Aufbau eines WIBOREX-Injektions-
bohrankers
Components of a WIBOREX injection drill anchor

"WIBOREX" beruht auf der Kombination von sieben
grundlegenden Ideen:
- Verwendung eines Rohres als Anker
- Schaffung einer preiswerten Einmal-Bohrstange
  und -Bohrkrone
- Benutzung des Ankerrohres für das Verfüllen
  des Bohrloches von Bohrlochtiefsten aus
- Verwendung eines Feinkornbaustahls mit hoher
  Kerbschlagzähigkeit
- Verwendung eines ferritisch-austenitischen rost-
  freien Stahles als Daueranker
- Ausbildung der Gewinderippen wie bei Betonstahl nach DIN 488
- Durchlaufende Gewinde gewährleisten indivi-
  duelle Einsatzmöglichkeiten
Um alle diese Vorteile auch in die Baupraxis umzusetzen
und wirtschaftlich nutzbar zu machen, waren umfangreiche Entwicklungen erforderlich. Dies zeigt sich
heute sehr deutlich an dem enormen Anwendungsspektrum, das durch die Produktpalette abgedeckt
wird. Folgendes Schaubild soll Ihnen einen Überblick
geben:

Ankertyp / Pfahltyp	Einheit	WIBOREX 30/16	WIBOREX 30/11	WIBOREX 40/16	WIBOREX 73/53	WIBOREX 103/78
Außendurchmesser	mm	30	30	40	73	103
Außendurchmesser für stat. Berechnung	mm	27,2	26,2	37,1	69,9	100,4
Innendurchmesser	mm	16	11	16	53	78
zul. Belastung auf Zug und Druck	kN	100	150	300	554	900
zul. Querkraft	kN	58	88	164	329	535
Höchstlast	kN	220	320	660	1160	1950
Gewicht	kg/m	3,0	3,5	6,9	12,8	24,7
kleinster Querschnitt	mm²	382	446	879	1631	3146
Kraft an der Fließgrenze	kN	180	260	490	970	1570
Folie 2	Technical Data of the WIBOREX system					▲▲ Willich

Technische Daten des WIBOREX-Systems
Technical data of WIBOREX-System

Anhand der Folie 2 können Sie erkennen, daß die An-
wendungsbreite von einer zulässigen Belastung auf
Zug und Druck von 100 kN bis 900 kN reicht. Diese
Werte werden erreicht unter folgenden Annahmen:
- Die zulässigen Lasten enthalten einen Sicherheits-
  faktor von S = 1,75 gegenüber der Fließlast.
- Die zulässige Querkraft wurde für Rohre - unter
  Berücksichtigung von Plastifizierung - errech-
  net.

1.3   Material

Wie bereits vorgängig stichpunktartig erwähnt, wird
das WIBOREX-System aus Feinkornbaustahl St E 355
bzw. St E 460 gemäß DIN 17124 als Ausgangsmaterial
hergestellt. Diese Feinkornbaustähle besitzen eine hohe
Kerbschlagzähigkeit (> 39 Joule) und sind unempfindlich gegen Spannungsrisskorrosion, Wärmebehandlung
und Querdruck. Der einfache Korrosionsschutz für
Kurzzeitanker, Verpresspfähle und Bodennägel wird
im Regelfall durch die Zementleimüberdeckung des
Stahls sichergestellt. Die Norm fordert eine Zementsteinüberdeckung des Ankerstahls von mindestens
- 20 mm im Boden   und
- 10 mm im Fels
gleichmäßig über den Umfang des Stahls. Eine
Passivierung des Stahls im Bereich der freien Ankerlänge ist bei Kurzzeitankern nicht erforderlich. Dies
aufgrund des gegen Spannungs-Riß-Korrosion un-
empfindlichen Feinkornbaustahl mit Restrohrwand-
stärken über 8 mm.
   Bei speziellen Anwendungsfällen, wie z.B. ver-
seuchten Böden, läßt sich der Korrosionsschutz durch
Passivierung des Stahls verbessern. Man spricht in
diesem Zusammenhang von Combicoating oder einer
Duplex-Beschichtung. Dabei handelt es sich um die
Kombination zweier rosthemmender Maßnahmen:
- Einer Feuerverzinkung nach DIN 50976, d.h.
  einer ca. 90 mm starken Schichtdicke. Dies
  entspricht einer flächenbezogenen Masse von
  mehr als 610 g Zink pro m²
- Einer zusätzlichen Epoxid-Pulver-Beschichtung
  nach DIN 55928 T5 mit einer Gesamtstärke von
  130 mm.
   Mit diesem Passivierungsschutz werden die
Korrosivitätsklassen bis 5 abgedeckt (DIN 55928 T1).
Zur Zeit ist dieser Schutz jedoch nur für die Systeme
WIBOREX 30/11 bis 40/16 verfügbar.
   Einen darüberhinausreichenden Korrosions-schutz
bietet der doppelte Korrosionsschutz für Daueranker,
dabei kommt ein ferritisch-austenitischer Stahl mit
22% Cr, 5,5% Ni, Werkstoff-Nr. 1.4462 zur Anwen-
dung. INOX-Stahl wird für die WIBOREX Injektion-
Bohrstange und die Kupplungsmutter verwendet. Die-
ser rostfreie Stahl INOX ist wissenschaftlich ausführlich untersucht durch die Bundesanstalt für
Materialforschung und -prüfung (BAM, Berlin,
Aktenzeichen 1.3/12279).

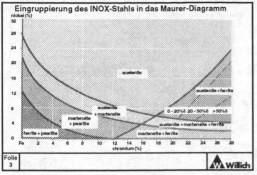

Eingruppierung der INOX-Stähle nach dem Maurer-Diagramm
Postition of INOX-steels in the Maurer-diagram

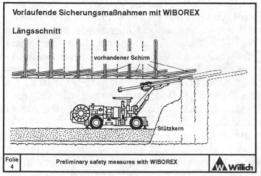

Vorlaufende Sicherungsmaßnahmen mit WIBOREX
Safeguards with WIBOREX

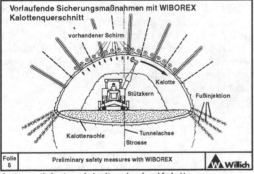

Automatisiertes Arbeiten in der Kalotte
Automatic operations at the roof section
of the tunnel

1.4 Verfahrens- und Anwendungstechnik

Umfangreiche Anwendungsbereiche liegen im Tunnelbau. WIBOREX Injektions-Bohranker werden z.B. eingesetzt bei der "Neuen österreichischen Tunnel-baumethode" für Anker über 3 m Länge im Lockergestein bei geringer Überlagerung und in Störzonen.

Die Injektionsmaßnahmen verbessern die Kohäsion der anstehenden Böden. Selbst in rolligen, schiefrigen oder sandigen Böden ist das Bohren ohne verrohrte Bohrung möglich. Arbeitstechnisch ermöglicht dieses System ein Erstellen von langen Ankern auf engstem Raum.

Auf Folie 4 und 5 sehen wir einen denkbaren Anwendungsfall mit mechanisiertem Bohren. Durch Beschränkung auf die geringen Rohrdurchmesser bis vielleich max. 40/16 ist ein händisches Einbringen der Verankerung möglich.

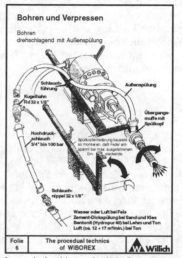

Bohren, drehschlagend mit Außenspülung
Rotary percussion drilling with flushing-installation

Bei Verwendung von Spülköpfen ist speziell darauf zu achten, daß mit Spülkopfanfederung gearbeitet wird.

Grundsätzlich sind jedoch lafettengeführte Bohrsysteme effektiver. Für die WIBOREX-Systeme 30/16 bis 40/16 sind linksdrehende, pneumatisch angetriebene Bohrhämmer mit einem Gewicht von ca. 30 kg sinnvoll. Der Luftverbrauch liegt je nach eingesetztem Bohrhammertyp bei ca. 6 m³/min, wobei der Arbeitsdruck ca. 6 bar betragen soll. Wird jedoch in bindigen Böden, z.B. Ton, mit Luft gespült, sind die o.g. Luftmengen nicht mehr ausreichend (ca. 17 m³/min). Für die WIBOREX-Systeme 73/53 und 103/78 sind hydraulisch betriebene, rechtsdrehende Bohrhämmer erforderlich. Das Gewicht des Hammers sollte idealerweise bei ca. 240 kg bzw. 400 kg liegen. Die erforderliche Vorschubkraft auf der Lafette ca. 50 kN. Das Bohrsystem sollte einen Drehmoment von 2500 Nm entwickeln. Bei diesen Werten handelt es sich um Zirka-Angaben, diese müssen den Gegebenheiten in Abhängigkeit von Bohrlänge und Geologie angepaßt werden.

Die bevorzugten Einsatzmöglichkeiten von WIBOREX Injektionsbohrankern liegen überall dort, wo bisher verrohrt gebohrt wurde; Z.B. in Kies, Kies Sand, Sanden, Geröll, Gehängeschutt, verwittertem Fels u.s.w..

Je größer die Gefahr ist, daß das Bohrloch einfällt, um so effektiver ist der Einsatz der Injektionsbohranker-Systeme. Das Ziehen des Bohrrohres entfällt, die Bohrungen werden im Durchmesser geringer, der Scherverbund und das Setzungsverhalten wird günstiger beeinflußt.

Ein bevorzugter Einsatz ist ferner bei beengten und schwer zugänglichen Baustellen. Durch die Möglichkeit den WIBOREX Injektionsbohranker beliebig zu koppeln und leichte Bohrgeräte zu verwenden, läßt sich dieser beispielsweise in Kellern, unter Brücken, auf Gerüsten, am Steilhang und in in Betrieb befindlichen Werkshallen zwischen den Maschinen einsetzen; dies gilt ebenfalls für Baustellen im Gebirge, die in schwer zugänglichen Zonen liegen.

1.5 Injizieren und Verpressen

Während bei den Systemen 30/16 bis 40/16 sowohl ein Injizieren mit Zement, Mörteln oder Kunststoffinjektionsharzen , wie z.B. WILKIT üblich ist, konzentriert man sich beim Verpressen der Systeme 73/53 und 103/78 auf Zemente und Mörtel. Hier wird bereits beim Bohren in nichtbindenden Böden Zementleim als Spül- und Stützflüssigkeit benutzt. Bei einem W/Z-Faktor von 0,4 bis 0,7 und einem Spüldruck von 5 bis 20 bar wird das Wasser der Stützflüssigkeit schnell durch den nichtbindenden Boden abgefiltert. Der so entstandene Filterkuchen stabilisiert das Bohrloch. Dieser Vorgang ist bekannt und vergleichbar mit der Herstellung von Bohrpfählen nach DIN 4014, auch hier stabilisiert der Filterkuchen aus der Stützflüssigkeit das Bohrloch.

Nach Beendigung dieses Vorgangs kann sich bei offenem Bohrloch ein Verpressdruck von bis zu 60 bar aufbauen, weil die Zementspülung bei einem W/Z von ca. 0,4 unter ständiger Rotation des Ankerrohres und weiter ansteigendem Druck gepumpt wird. Wie beim Nachverpressen wird der erstarrende Zementleim durch Rotation und steigendem Druck auseinandergerissen. Die Zementschollen verkeilen sich zwischen Ankerstange und der sich verfestigenden Bohrlochwandung. Der Vorgang ist vergleichbar dem des Nachverpressens und führt zu einem verbesserten Last- und Setzungsverhalten.

1.6 Überprüfen der Ankertragkraft

Falls der Anker mit Zement und nicht mit Organomineralharz (WILKIT) injiziert würde, kann er je nach Zementgüte frühestens 4 bis 7 Tage nach dem Verpressen belastet bzw. geprüft werden.

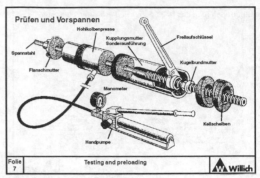

Vorspannen und Prüfen
Prestressing and testing

Bei dieser Abnahmeprüfung werden Verschiebungen in mehreren Zeitabständen ermittelt. Zur Prüfung wird eine Unterkonstruktion aus U-Profilen (Ankerstuhl) auf den Anker aufgesetzt. Hinter diesem Ankerstuhl eine Hohlkolbenpresse angebracht, die auf den Anker aufgeschraubt wird. Die Meßuhr, mit der die Bewegung des Ankerkopfes gemessen wird, ist auf einem Stativ befestigt. Wenn die Verschiebungsdifferenzen im nicht bindigen Boden zwischen der Messung nach 2 und 5 Minuten nicht größer als 0,2 mm sind, hat der Anker die Abnahmeprüfung erfüllt.

Beim bindigen Boden wird die Prüfung dahingehend abgewandelt, daß der Meßzeitraum vergrößert wird. Die Verschiebungsdifferenzen müssen zusätzlich nach 10 bis 15 Minuten festgehalten werden. Ein Anker hat im bindigen Boden die Abnahmeprüfung erfüllt, wenn die Verschiebungsdifferenz zwischen den Messungen nach 5 und 15 Minuten nicht größer als 0,25 mm ist.

Sollte die Differenz größer sein, so ist der Beobachtungszeitraum auszudehnen. Nähere Einzelheiten sind in der DIN 4125 nachzulesen.

Hat der Anker die Abnahmeprüfung erfüllt, wird er bei Berechnung mit aktivem Erddruck auf 80% und bei Berechnung mit Ruhedruck auf 100% seiner Gebrauchslast vorgespannt.

2. Anwendungsbereiche

Wie bereits unter Punkt 1. (Systembeschreibung) erwähnt ist die Bandbreite der möglichen Anwendungen enorm breit.

Die Unterschiede in statisch-konstruktiven Bereichen sollen die folgenden Systemskizzen aufzeigen:

2.1 WIBOREX als Anker

Nach DIN 4125 gilt für Anker, daß sie eine freie Ankerlänge aufweisen, vorgespannt werden und nur Zugkräfte aufnehmen.

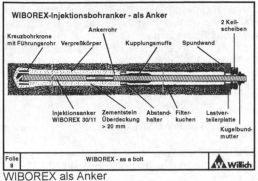

WIBOREX als Anker
WIBOREX as an anchor

Zur Sicherstellung der freien Ankerlänge sind konstruktive bzw. arbeitstechnische Vorkehrungen zu treffen.

Konstruktive Maßnahmen sind:

Schutz der freien Ankerlänge durch ein Hüllrohr aus Kunststoff oder der Einsatz von PUR-umschäumten Gewindestangen im Bereich der freien Ankerlänge. Hier sind von Willich Berg- und Bautechnik in den letzten Monaten umfangreiche Entwicklungen betrieben worden, die heute einen sicheren Einsatz möglich machen.

Arbeitstechnische Maßnahmen sind das Überbohren oder das Freispülen des erforderlichen Bereiches. Beide Maßnahmen beinhalten nicht unwesentliche Risiken. Unserer Auffassung nach ist daher eindeutig die PUR-Ummantelung im Bereich der freien Ankerstrecke zu bevorzugen.

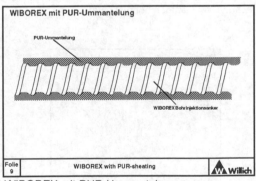

WIBOREX mit PUR-Ummantelung
WIBOREX with PUR-sheathing

2.2 Das WIBOREX-System als Nagel

Der Bodennagel unterscheidet sich im Prinzip nicht vom Kleinverpreßpfahl nach DIN 4128. Wenn jedoch der nach DIN erforderliche Mindestdurchmesser von 150 mm bei Ortbetonpfählen bzw. von 100 mm bei Verbundpfählen nicht erreicht wird, spricht man von

Bodennägeln. Bei Bodennägeln liegt der Stabdurchmesser bei bis zu 30 mm, somit deckt der WIBOREX 30/11 diesen Bereich voll ab.

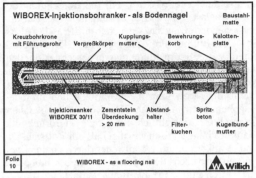

WIBOREX als Bodennagel
WIBOREX as a nail

Der vollvermörtelte Injektionsbohrstab wird kopfseitig in die Bauwerkskonstruktion eingebracht. Der Anwendungsbereich liegt in der Böschungssicherung und im Berg- und Tunnelbau.

2.3 Das WIBOREX-System als Pfahl

Die WIBOREX Injektionsbohranker eignen sich ausgezeichnet als Kleinverpresspfahl. Nach DIN 4128 werden diese auch Mikro-, Wurzel- oder auch Injektionspfahl genannt.

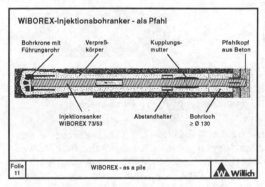

Pfahl nach DIN 4128/Pile according to DIN 4128

Der WIBOREX Kleinverpresspfahl besteht aus dem zentralen, selbstbohrenden Gewinderohr, das als Bewehrungsstab im Wesentlichen die Lasten auf Zug bzw. auf Druck in direkter Abhängigkeit vom Verpreßkörper über diesen in den Boden überträgt. Die hierbei aktivierte Mantelreibung ist ganz wesentlich abhängig von der angetroffenen Bodenart und der Pfahllänge. Wird auf eine Probebelastung verzichtet, sind die Grenzwerte nach DIN 4128 anzuwenden.

In der Regel werden jedoch Probebelastungen an mindestens zwei Pfählen oder mindestens 3% aller Pfähle durchgeführt. Die einfachste denkbare Versuchsanordnung benutzt zwei Nachbarpfähle zur Ableitung der Reaktionskraft. Für die Berechnung der Grenztraglast von Kleinverpresspfählen wird nach den vorliegenden Grundsatzversuchen von Prof. Dr. Floss, Prof. Dr. Blümel, Prof. Steinfeld bzw. Dr. Haag mit einem Verpreßkörperdurchmesser gerechnet, der dem doppeltem Bohrkronendurchmesser entspricht. Die Vorteile dieser Systeme liegen ganz eindeutig in ihrer Flexibilität, sich schnell und einfach den örtlichen Gegebenheiten anzupassen. Das Rohr ist bei gleichem Querschnitt hinsichtlich Biegung, Querdruck und Mantelreibung statisch günstiger als ein Vollstab. Die Endverankerung erfolgt entweder durch einen reinen Haftverbund nach DIN 1045, durch eine Gewindeplatte oder durch angeschweißte Betonstahlhaken.

2.4 Das WIBOREX-System als Injektionslanze

Das WIBOREX-System bietet speziell ab dem Rohrdurchmesser 73/53 mm ein neues System von Nachverpressventilen. Der Ansprechdruck liegt bei ca. 20 bis 50 bar.

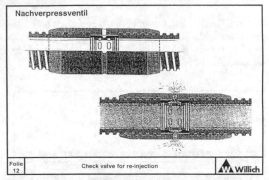

Nachverpressventil/Valve for re-injection

Die für den WIBOREX 73/53 üblicherweise verwendeten Kupplungsmuffen lassen sich durch die Verwendung einer verlängerten Muffenvariante in Injektionsmuffen umrüsten - das vorliegende Bild zeigt diese Möglichkeit auf.

Das so entstandene System kann man als selbstbohrendes-Manschetten-Rohr-Verfahren bezeichnen. Das Verpressen des Injektionsgutes erfolgt über eine Doppelpackereinheit, wie sie aus der Anwendungstechnik bestens bekannt ist. Nach dem Spülen der Packereinheit und des, in die Muffe eingebauten Kunststoff-Ventils, ist ein Nachverpressen möglich.

Gegenüber den herkömmlichen Kunststoff-Manschetten-Rohren spart dieses selbstbohrende-Manschetten-Rohr-Verfahren den Rückbau der verrohrten Bohrung und die Mantelmischung.

Außerdem wirkt das im Boden belassenen Rohr als Bewehrung, die konstruktiv oder statisch ihre Berücksichtigung finden kann.

Dimensionen kleiner, jedoch ebenso effektvoll, sind die WIBOREX-Systeme 30/16 und 30/11. In dem folgenden Kapitel "Anwendungsbeispiele" soll hierauf detailliert eingegangen werden.

3. ANWENDUNGSBEISPIELE

Erstellung eines Tragringes unter Verwendung von WIBOREX-Injektions-Bohrankern und WILKIT

Im Berg- und Tunnelbau kommt es recht häufig vor, daß Strecken und Tunnel in zu Konvergenzen neigendem Gebirge angefahren werden. An einem 1992 ausgeführten Projekt innerhalb des deutschen Bergbaus wird nachfolgend hier die Vorgehensweise erläutert.

Im stark zum Bladdern neigenden Tonstein wurden Strecken mit einem Querschnitt von ca. 30 m² mittels einer Teilschnittmaschine aufgefahren. Als Ausbau kamen Klebeanker und Spreizhülsenanker mit einem Drahtverzug zum Einsatz. Bereits bei den ersten Sicherungsmaßnahmen gab es erhebliche Probleme, da die für die Sicherung vorgesehenen Anker sich nicht mehr problemlos in die vorgebohrten Bohrlöcher schieben ließen. Die Folge war, daß die Ankerarbeiten recht langsam vorankamen.

Innerhalb weniger Monate nach Auffahrung der Strecke stellten sich Konvergenzen zwischen 50 und 70 cm ein. Man entschloß sich daher, einen selbsttragenden Gebirgsring durch eine kombinierte Anwendung von Injektionsbohrankern und gebirgsvergütenden Maßnahmen herzustellen.

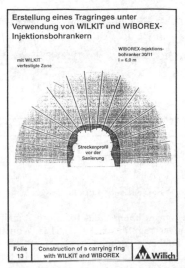

Phase 1

Trotz des geringen Ausbruchquerschnittes von rund 30 m² war es möglich, durch den Einsatz von WIBOREX Injektionsbohrankern 4 m bzw. 6 m lange Anker zu setzen. Die kuppelbaren Ankerstangen haben dies ohne Probleme ermöglicht. Zusammen mit den verlorenen Bohrkronen bietet dieses System diverse Anwendungsmöglichkeiten auf engstem Raum; kollabierende Bohrlöcher verlieren ihren Schrecken, da nicht mehr mit dem Verlust von Bohrstange und Bohrloch gerechnet werden muß.

Nach Beendigung des Bohrvorganges übernimmt die "Bohrstange" die Funktion einer Injektionsstange. Mittels einer Zweikomponenten-Kunstharz-Injektionsanlage, die für den Injektionsvorgang direkt mittels Adapter an den WIBOREX Injektionsbohranker angeschlossen wird, wurde das Organomineralharz WILKIT zur Erstellung des Tragringes injiziert. Durch diese Vorgehensweise wurde der Injektionsanker im Bohrloch festgesetzt und gleichzeitig das umgebende Gebirge so vergütet, daß unmittelbar nach Beendigung dieser Maßnahmen eine sehr deutliche Reduzierung der Konvergenzen eintrat. Nach Wiederherstellung des Sollprofils, wurden die WIBOREX Injektionsbohranker ihrer weiteren Funktion als Anker gerecht.

hohe Standfestigkeit und lange Lebensdauer auf. Außerdem erlaubt dieses Verfahren ein sicheres und kontinuierliches Arbeiten.

Problemlösung in anspruchsvollen geologischen Verhältnissen - Beispiel Achbergtunnel bei Unken

Seit Dezember 1993 rollt der Verkehr durch den knapp anderhalb Kilometer langen Achbergtunnel bei Unken in Österreich. Die zweispurige Röhre ist Teil der stark befahrenen österreichischen Ost-West-Verbindung, die von Salzburg aus über das sogenannte "Kleine Deutsche Eck" nach Tirol führt.

Das Ingenieurbüro Laabmayr + Partner war mit der Planung und technischen Betreuung des Gesamtprojektes von der Landesregierung in Salzburg beauftragt. Der Pilotstollen brachte die Erkenntnis, daß sich in der projektierten Tunneltrasse eine ca. 130 m lange Anhydritstrecke sowie ein 110 m langes Teilstück aus komplex aufgebautem Lockergestein befand. Hinzu kam Bergwasser mit hohem, aggressiven Sulfat- und Chloritgehalt.

Im Spätsommer '92 wurde Willich Berg- und Bautechnik von der ausführenden Arge "Umfahrung Unken", gebildet von den Firmen Stettin-Tiefbau-

Phase 2

Um ein weiteres Aufbladdern des Gebirges zu verhindern, wurde eine bewehrte Spritzbetonschale aufgetragen.

In besonders gefährdeten Bereichen mußte in der Sohle ein Ringschluß mit WIBOREX und WILKIT-Injektionen erstellt werden. Nach Beendigung dieser Arbeiten haben sich die Konvergenzenzuwächse dem Nullbereich genähert.

Strecken, die nach dem oben beschriebenen Verfahren aufgefahren oder saniert werden, weisen eine sehr

Wasserabdichtung mit WILKIT-Schaum
Water-sealing with WILKIT-foam

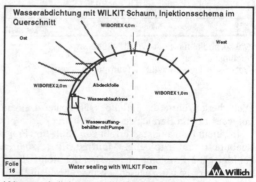

Wasserabdichtung mit WILKIT-Schaum
Water-sealing with WILKIT-foam

Östu, beauftragt einen Wasserzutritt bei Station 1388 bis 1400 zu stoppen. Bis zu 18 l Wasser pro Sekunde traten auf einer Fläche von ca. 3 x 10 m bei ca 11.00 Uhr durch die "Spritzbetonschale" in den Tunnel ein. Es war vorgesehen, im Bereich der Wasseraustritte ein Injektionsraster von 12 x 4,5 m mittels 4 m langen WIBOREX Injektionsbohrankern und dem schnell-reagierenden Injektionsmittel WILKIT-Schaum zu errichten. Danach sollte eine Nachinjektion mit WILKIT zur Abdichtung und zur Verfestigung des Gebirges durchgeführt werden.

Die Rasterinjektionen wurden von Station 1388 aus in Richtung Süden begonnen. Auf den ersten Metern stellte sich sehr rasch ein Abdichtungserfolg ein. Ab Station 1392 zeigte sich, daß die 4 m langen Injektionsanker zu lang warten. Es wurde auf 2 m lange Injektionsanker umgestellt. Die Wasseraustritte wurden weiter Richtung Süden verdrängt. Zur Entlastung wurden von der Arge an den Stationen 1396, 1397, 1398 Stahlrohre mit einem Durchmesser von 20 cm eingesetzt. Durch gezielte Injektionen wurde das Wasser durch diese drei Punkte gefaßt. Anschließend ist durch Injektionen auch hier der Austritt von Wasser gestoppt worden.

Über Nacht staute sich das Wasser an der sanierten Spritzbetonschale auf. Am nächsten Tag zeigte sich eine Anzahl von kleineren Wasseraustrittsstellen, die durch Anbohren mit Injektionsbohrankern und durch Injizieren mit WILKIT-Schaum schnell verschlossen wurden. Das folgende Bild zeigt Ihnen Punkte der Nachinjektion durch WIBOREX-Injektionsbohranker.

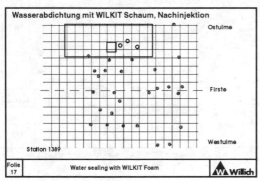

Wasserabdichtung mit WILKIT-Schaum, Nachinjektionen
Water-sealing with WILKIT-foam, re-injection

Die Arbeiten führten ca. ein halbes Jahr später zu einem Folgeauftrag im Bereich der Anhydritstrecke.

In einem Fachaufsatz, verfaßt von Dipl.-Ing. Franz Laabmayr und Dipl.-Ing. Manfred Eder, zu den geologischen Problemen beim Bau der Röhre durch den Achberg ist u.a. zu lesen: "Wie sehr es dem Berg-wasser gelingt, über weite Strecken just an jene Stellen zu gelangen, wo es am wenigsten erwünscht ist, zeigte

sich im südlichen Tunnelabschnitt. Bergwasser drang über 400 Meter in die Kontaktfläche Warfener Strecke/Sohlgewölbe." Es bedrohte die bis dahin trocken geblieben Anhydritstrecke. "Obwohl versucht wurde mit einem Dichtschott samt Ringspaltinjektionen dem Sickerwasser Herr zu werden, drang es in den kritischen Anhydritbereich vor", heißt es weiter in dem Report. Zusätzliche Maßnahmen zum Stop und zum Verdrängen des Wassers wurden notwendig.

Die technische Umsetzung sah Injektionen von WILKIT-Schaum durch Injektionsbohranker vor.

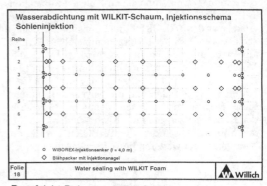

Draufsicht Bohrraster WIBOREX
Plan-view of WIBOREX-drill-holes

Als technische Lösung wurde ein Riegel mit tief-reichenden Injektionen (Länge 4,0 m) vorgesehen. Dies um einen möglichst großen Bereich unterhalb der Bauwerkssohle zu erfassen. Gebohrt wurde direkt mit WIBOREX.

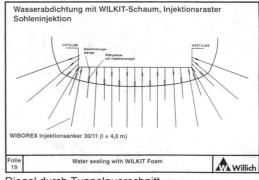

Riegel durch Tunnelquerschnitt
Tunnel with concrete-slab, cross section

Die ca. 1 m tief in das Gebirge reichenden Injektions-bohranker sollen den Nahbereich und die Kontaktfuge zwischen Bauwerk und Fels erfassen.

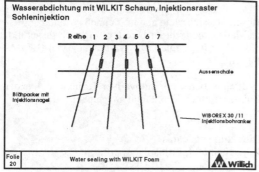

Querschnitt durch Riegel Tunnellängsrichtung
Tunnel with concrete-slab, longitudinal section

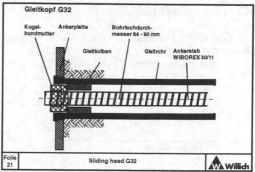

Gleitkopf G32
Sliding head G32

Die Injektionen durch die Injektionsbohranker
WIBOREX, die in diesem Beispiel als Lanzen genutzt
wurden, erfolgte planmäßig. Meßpegel in der Sohle
haben den Injektionserfolg bestätigt.

Diese beiden Beispiele zeigen, wie individuell der
Bohrinjektionsanker WIBOREX technisch sinnvoll
eingesetzt werden kann.

4. ENTWICKLUNGEN

Trotz all dieser vielen Möglichkeiten zeigt sich doch
immer wieder, daß spezielle Anforderungen neue Entwicklungen erforderlich machen.

4.1 Gleitkopfanker

Neben der Konsolidierung der Gebirgsformationen
durch Injektionen in Kombination z.B. mit dem
Injektionsbohranker WIBOREX sind weitere Überlegungen und Entwicklungen im Hause Willich im Gan-
ge, die auch in Verbindung mit der von uns propagierten
Ankertechnik, dem Injektionsbohranker, direkt in Verbindung stehen.

So wurden diese bereits für Streckenbereiche bzw.
Tunnelbereiche, in denen große Konvergenzen zu
erwarten sind, erfolgreich angewandt. Die Gleitkopf-
ankertechnik (s. Folie 21), ist bereits seit Jahren dem
breiten Fachpublikum bekannt. Willich ist es jedoch in
den letzten Monaten gelungen, durch weitere Entwicklungen die bisher für den Tunnelbau relativ unbefrie-
digenden Kennlinien so zu verändern, daß nun auch
Anker mit Gleitköpfen versehen werden können, die
auch unter Vorspannung mit der gewünschten Gebrauchslast eingebracht werden können.

Die einfache Konstruktion und die leichte Hand-
habbarkeit dieses Gleitkopfes ermöglicht sowohl den
Einsatz in Kombination mit Injektionsbohrankern
WIBOREX wie auch den Einsatz mit SN-Ankern (s.
Folie 22).

Wie bereits vorgehend erwähnt, ist es bei der Konstruktion dieses Gleitkopfes möglich, die Trag-fähigkeit

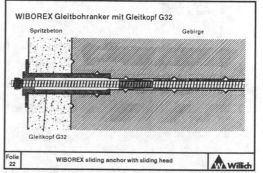

Gleitbohranker mit Gleitkopf G32
Sliding anchor with sliding head G32

von Gleitkopf und Ankerstab so aufeinander abzustim-
men, daß die vorgegebene Tragfähigkeit des Ankersystems ohne weitere Gebirgsschiebungen sofort zur
Verfügung steht. Das bedeutet für den konstruktiven
Ingenieur, daß auch Freispielanker mit Gleitkopf-
konstruktion vorspannbar werden. Als Beispiel sei die
Kennlinie 14 t bei einem Gleitweg von 250 mm gezeigt. Diese Kennlinie, die sich aus theoretischen Überlegungen ergeben hat, wurde durch umfangreiche Versuche beim MPA Dortmund mit einer geringen Streuung bestätigt (s. Folie 23).

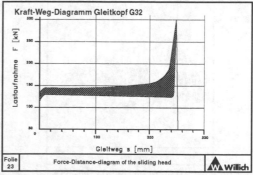

Kraft-Weg-Diagramm Gleitkopf G32
Distance versus force, sliding head G32

4.2 Laststufenindikator

Der Sinn dieser Entwicklung ist sehr einfach erklärt. Durch das Vorspannen des Ankers wird die Vorspannkerbe zusammengedrückt. Man weiß jetzt, daß der Anker mit ca. 60 bis 70 kN vorgespannt ist. Steigt die Last an, zeigen die zweite und dritte Stufe an, welchen Last- und Spannungszustand der Anker erreicht hat.

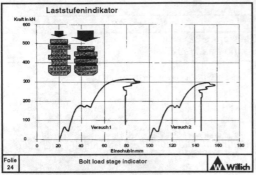

Laststufenindikator
Indicator for load stage

Bei einer Belastung von 180 kN ziehen sich die Arbeitsnuten zusammen und ermöglichen durch einfache visuelle Kontrolle, ob der Anker seine maximale Gebrauchslast erreicht, bzw. überschritten hat. Bei sich verschlechternden Gebirgsverhältnisse bleibt genügend Zeit zur Einleitung geeigneter Zusatzmaßnahmen, wie Nachankern bzw. Injizieren.

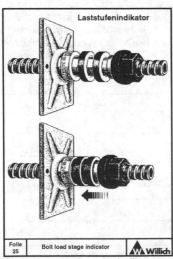

Arbeitsweise des Laststufenindikators
Indicator for load stage, principle of operation

Eine Fülle von Anwendungsfälle sind bei der Böschungssicherung und im Tunnelbau denkbar. Beide Maßnahmen wurden bereits erfolgreich ausgeführt. Auch hier liegt eine LOBA-Zulassung von (LOBA-Nr. 18.24.6-28-4).

Willich ist daher mit innovativer Technik auch in der Zukunft für den Tunnelbau gewappnet.

Literaturverzeichnis
Prof. Steinfeld, Erdbaulaboratorium
   Hamburg, 28.10.1985
Dr. Heinz W. Haag, Laboratorium für Bodenmechanik
   Kornwestheim, 18.11.1986
Prof.Dr.-Ing. R. Floss, TU München,
   Projekt 9941/13, 23.10.1989
Prof.Dr.-Ing. R. Floss, TU München
   Projekt 9941 b/12, 30.10.90
Prof.Dr.-Ing. W. Blümel, TU Hannover
   Fachgebiet Grundbau, 20.06.1991
DIN 4125   DIN 55928 T5
DIN 4128   DIN 50976 (1989)
DIN 4014   DIN 1045
Bundesanstalt für Materialforschung und -prüfung
   (BAM) Berlin, Aktenzeichen 1.3/12279
Dipl.-Ing. Jochen Kasselmann
   F. Willich Berg- und Bauchtechnik GmbH + Co
   "Neue Erkenntnisse zur Beurteilung von Injektionsmitteln im Berg- und Tunnelbau" - Vortrag anlässlich Weiterbildungkurs ETH-Zürich, 1994
Mannesmann Röhrenwerke
   "Mannesmann Corrosion Resistant Alloys for Oil Country Tubular Goods", Edition 1988
Fa. Ischebeck,
   Firmenveröffentlichung "Neue Wege in der Ankertechnik", 10/1994
Fa. Ischebeck,
   Firmenveröffentlichung "Injektionsanker", 04/1994
F. Willich Berg- und Bautechnik GmbH + Co,
   Firmenveröffentlichung "Sanierungsmaßnahmen Unken"
F. Willich Berg- und Bautechnik GmbH + Co,
   Firmenveröffentlichung "Achbergtunnel - Dichtungsmaßnahmen im Anhydritbereich"
Forschung + Praxis, Ausgabe zur STUVA-Tagung
   1993 "Achbergtunnel, Salzburg - Ein Tunnelbauwerk in einer der größten Störungszonen der nördlichen Kalkalpen"

Anchors in Theory and Practice, Widmann (ed.) © 1995 Balkema, Rotterdam. ISBN 90 5410 577 1

Extending the traditional role of rockbolts

Erweiterung der traditionellen Aufgabe des Gebirgsankers

Björn Kåreby
Rock Reinforcement, Atlas Copco Rock Drills AB, Sweden

ABSTRACT: The paper reviews the history of rockbolts and explains why rockbolts form an unique system for rock reinforcement. It is concluded that the variety of rockbolts used today are almost unlimited however they can all be divided into three basic groups depending upon the anchoring mechanism; mechanically -, grouted - or friction anchored rockbolts.

The concept of the ideal rockbolt system is presented and discussed considering the load-displacement characteristics of an installed rockbolt acting across a single joint in tension. Representatives from each of the three groups of rockbolts are tested and the results of these tests are discussed. It is summarised that some bolts better than others resemble the behaviour of the ideal rockbolt.

The development of the Swellex rockbolt system is described. The installation procedure, the different types of bolts and their applications are presented. Examples are given for use of Swellex in some special applications. Further, it is shown that by the introduction of Swellex the use of rockbolts have been extended to include ground conditions which could not previously be reinforced using rockbolts.

ZUSAMMENFASSUNG: Der Vortrag gibt einen Überblick über die Geschichte von Gebirgsankern und erläutert, warum Gebirgsanker ein besonderes System für die Gebirgssicherung darstellen. Er kommt zu dem Schluß, daß die Vielzahl der heute eingesetzten Gebirgsanker nahezu unbegrenzt ist. Allerdings lassen sie sich je nach Ankerungsmechanismus in drei Gruppen einteilen: in mechanische Anker, Injektionsanker und Reibungsanker.

Das Konzept des idealen Gebirgsankersystems wird vorgestellt und hinsichtlich der Lastaufnahme eines belasteten Ankers besprochen. Vertreter aller drei Ankergruppen werden getestet und die Resultate diskutiert. Zusammenfassend wird festgestellt, daß einige Anker dem Verhalten des idealen Gebirgsankers in höherem Maße nahekommen als andere.

Im folgenden wird die Entwicklung des Gebirgsankersystems Swellex beschrieben. Es werden der Einbau, die verschiedenen Ankerarten und ihre Anwendung vorgestellt. Außerdem werden spezielle Einsatzbeispiele von Swellex-Ankern angeführt. Weiterhin wird dargelegt, daß durch die Einführung von Swellex-Ankern die Verwendung von Gebirgsankern auf Gesteinsformationen ausgedehnt worden ist, die früher mit Gebirgsankern nicht gesichert werden konnten.

1. INTRODUCTION

The history of rock bolting dates from at least the 18th century. However, rockbolts were not used extensively until 200 hundred years later. Since then the use of rockbolts has become world-wide and hundreds of millions of bolts are installed annually. In recent years the range of applications for rockbolts have widened due to advances in rock mechanics and the increasing use of rockbolts in underground excavations, as an alternative to more traditional forms of support. Also, the development of new rockbolt concepts has led to the use of rockbolts in non-traditional applications. Figure 1 illustrates the development of rockbolts.

The principal objective in the design of a support system is to help the rock mass to support itself. The rockbolt reinforces and mobilises the inherent strength of the rock mass, it actually form part of the rock mass. Steel set support systems, however, act to restrict movements of the rock mass externally and do not form part of the rock mass.

Rockbolts and systems for external support are often used in combination to obtain the best possible effect on rock mass stabilisation. In fact, many times a combination of rockbolts and e.g. wire mesh and/ or shotcrete to give support to the rock surface between the rockbolts, provides the optimum system for rock mass stabilisation.

There are a number of reasons for the widespread

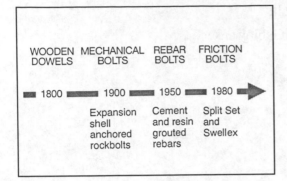

Figure 1. Development of rockbolts.
Abb. 1. Die Entwicklung von Gebrigsankern.

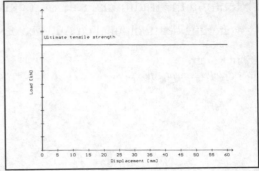

Figure 2. The ideal load-displacement characteristics of an installed rockbolt, independent on ground conditions, after Stillborg (1992).
Abb. 2. Die idealen Last-Weg-eigenschaften eines installierten Gebirgsankers, unabhängig von den Bodenbedingungen, nach Stillborg (1992).

use of rockbolt reinforcement systems. Some of these are as follows;
- versatile, can be used in any excavation geometry
- simple and quick to apply
- relatively inexpensive
- installation can be fully mechanised.

Using rockbolts, the reinforcement density (spacing between bolts and bolt length) can be modified. This is frequently required by local rock mass conditions. Many rockbolt systems offer the advantage of an immediate support action after installation. Another advantage is that rockbolts can easily be combined with e.g. the above mentioned additional external support systems wire mesh, shotcrete or a concrete lining.

A number of different types of rockbolts are used world-wide. Many rockbolt types show however, only minor differences in their design and are basically varieties of the same concept. It is therefore possible to arrange the different types of rockbolts in groups and present representatives from each group.

Classification of the different types of rockbolts are made on the basis of their anchoring mechanism.

The following groups of bolts are considered:
- Mechanically anchored rockbolts
- Grouted rockbolts
- Friction anchored rockbolts

2. ROCKBOLT LOAD - DEFORMATION CHARACTERISTICS

Traditionally, standard pull-out tests are used by engineers to obtain guidance on the load bearing capacity and the load-deformation characteristics of installed rockbolts. There are however, depending on the type of rockbolt tested, a number of disadvantages in using standard pull-out tests as a means of testing and comparing different rockbolt types. One obvious deficiency being that a standard pull-out test simply does not provide the load-deformation char-

acteristics of a rockbolt subjected to loading across a joint in the rock, a situation for which rockbolts are designed. In order to obtain the load-deformation characteristics of rockbolts which realistically resemble the characteristics of the installed rockbolt, and to be able to compare, the general load-deformation characteristics of different types of rockbolts, the block test arrangement was developed, Stillborg (1992).

Reported are test results, employing the block test arrangement, of some commonly used types of rockbolts.

2.1 The concept of the ideal rockbolt

Here it is interesting to consider the ideal load-displacement characteristics of an installed rockbolt acting across a single joint in tension. A relationship that can be given independent of available ground conditions, from swelling/squeezing rock to extremely hard and brittle rock, in low or high stress conditions, Stillborg (1992).

The ideal bolt system, should initially act infinitely stiff in order to attract load and by doing so help to maintain the integrity of the rock mass as much as possible. However, as the load on the bolt gets near its ultimate tensile strength the bolt should have the ability to accommodate large rock deformation, not fail or drop in its load bearing capacity. The bolt behaviour should be rigid/perfectly-plastic.

The ideal load-displacement characteristics of an installed rockbolt is schematically illustrated in Figure 2.

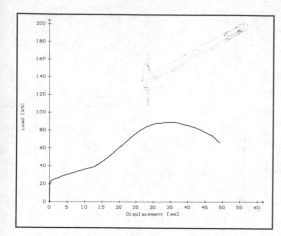

Figure 3. Expansion shell anchored rockbolt, tensile loading across a joint, after Stillborg (1992).
Abb. 3. Installierter Spreizhülsenanker, Zugbelastung über einer Kluft, nach Stillborg (1992).

2.2 Mechanical anchored rockbolts

The expansion shell anchored rockbolt, is the most common form of mechanically anchored rockbolt. A wedge attached to the bolt shank is pulled into a conical expansion shell as the nut on the bolt is rotated. This forces the shell to expand against and into the wall of the borehole. The nut is rotated until a pre-set torque is reached, resulting in the desired bolt tension.

The effectiveness of an expansion shell anchored rockbolt strongly depends basically on two "points", the grip of the shell against the borehole wall and the contact between the rock and the face plate.

The two mechanisms by which the shell is anchored against the borehole wall are; friction and interlock. The second of the two is the most significant in order for the rockbolt to provide optimum support action.

The common mechanical anchored rockbolt is designed for use in moderately hard to hard rock conditions. They are not recommended for use in *very* hard rock, since a very hard rock will prevent the expansion shell from "gripping" the rock, and the anchor will slip under load. The bolt is used in temporary support systems.

Tests of expansion shell anchored (17 mm) rockbolts, gave the following result, Stillborg (1992), see Figure 3; Before the bolt load reaches the level of pre-tensioning 22.5 kN, no rockbolt deformation occurs. At 22.5 kN, the face plate starts to deform. At a load of 30 kN the face plate is deformed 4.5 mm and at the load of 40 kN the deformation of the face plate is 9.5 mm.

The bolt shank has deformed an additional 3.5 mm at a load of 40 kN. This gives a total rockbolt

deformation of approx. 13 mm at the load of 40 kN. At the load of 40 kN the triangular bell plate is completely flat and only the bolt shank deforms. At a load of approx. 80 kN and 25 mm of rockbolt deformation, the expansion shell anchor fails progressively when the wedge, attached to the bolt shank, is pulled through the conical anchor shell. The rockbolt fails completely at a load of approx. 70 kN.

The bolt accommodate a total displacement of approx. 35 mm, combined face plate deformation, bolt shank deformation and "anchor slippage", under an increasing load bearing capacity up to approx. 90 kN.

2.3 Grouted rockbolts

The most commonly used grouted rockbolt is the fully grouted rebar. Cement or resin are used as grouting agents. The rebar used with resin can be used in a system for tensioned rockbolts. Most common is however, the rebar used as untensioned bolts with cement or resin as the grouting agent.

The grouted rockbolt is confined inside the borehole by means of the grout. Anchoring, (bond) between the bolt and the rock is provided along the whole length of the reinforcing element by means of three mechanisms; chemical adhesion, friction and interlock. The second and third mechanisms are by far the most significant.

The effectiveness of a grouted rockbolt strongly depends on two factors; the quality of the grout and the quality of the grouting. Both factors are equally important however difficult to control. Grouted rockbolts are used for temporary as well as permanent support under various rock conditions.

Tests of resin grouted (20 mm) rebar rockbolts, gave the following result, Stillborg (1992), see Figure 4; The rockbolt does not slide but is loaded

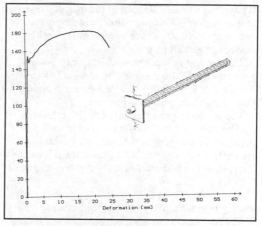

Figure 4. Resin grouted rockbolt - rebar, tensile loading across a joint, after Stillborg (1992).
Abb. 4. Klebeanker - Amierung, Zugbelastung über einer Kluft, nach Stillborg (1992).

up to failure which occurs between the blocks, in the joint, at approx. 180 kN, (18 tons) and 20 mm of rockbolt deformation. The resin is stiffer than the cement and local fracturing as well as bond failure in and near the joint is limited, resulting in comparatively smaller rockbolt deformation. The loading of the rockbolt is concentrated over a short section of the rockbolt. The sudden drop in load which can be seen in the graph at approx. 150 kN reflects the typical characteristics of the hot rolled rebar steel subjected to tensile loading.

Tests of cement grouted (20 mm) rebar rockbolts, gave the following result, Stillborg (1992), see Figure 5; The rockbolt does not slide but is loaded up to failure which occurs between the blocks, in the joint, at approx. 180 kN, (18 tons) and 30 mm of rockbolt deformation.

2.4 *Friction anchored rockbolts*

Friction anchored rockbolts represent the most recent development in rock reinforcement techniques. Two friction anchored rockbolt types are available, the Split Set and the Swellex. For both types of rockbolt system, the frictional resistance to sliding, (for the Swellex combined with interlocking) is generated by a radial force against the borehole wall over the whole length of the bolt. The Split Set rockbolt is forced into a borehole slightly smaller than the bolt whereas the Swellex is expanded inside the borehole into the irregularities of the borehole wall by means of a high pressure water.

Although the two systems are presented under a common heading, they display some major differences. These are related to the anchoring mechanism and the support action, as well as the installation procedure. Friction anchored rockbolts are the only type of bolts where the load of the rock is transferred to the reinforcing element directly, without any necessary auxiliaries such as mechanical locking devices or grouting agents.

The anchoring mechanism of the Split Set will prevent the bolt from sliding up to a load of approximately half that of the ultimate tensile strength of the bolt steel, when the bolt will slide.

The anchoring mechanism of the EXL Swellex will however, as the load level approaches the ultimate tensile strength of the steel, start to slide. This means that for Swellex bolts almost the full tensile strength of the bolt steel is utilised in the support of the rock.

Tests of the Split Set, type SS 39, gave the following result, Stillborg (1992), see Figure 6;

The frictional resistance is overcome and the bolt starts to slide, at approx. 50 kN, (5 tons). The sliding of the bolt is preceded by no measurable rockbolt deformation. The rockbolt maintains however, a constant load bearing capacity for at least the dura-

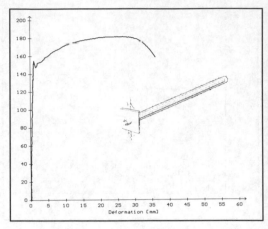

Figure 5. Cement grouted rockbolt - rebar, tensile loading across a joint, after Stillborg (1992
Abb. 5.Mörtelanker, Zugbelastung über einer Kluft, nach Stillborg (1992

tion of the test which was 150 mm of joint opening.

Tests of the Swellex EXL tested in the block test arrangement, gave the following result, Stillborg (1992), see Figure 7; Initially no deformation occurs in the rockbolt up to a load of approx. 50 kN, (5 tons). At 50 kN, the bolt starts to deform locally between the two blocks, in the joint. At the same time "bond failure" occurs near the joint, (some of the frictional and interlock resistance are overcome, partly due to lateral contraction of the bolt). As the load increases the "bond failure" progress and the bolt deforms over a progressively longer "free" length. General "bond failure" reaches the far end of the bolt at approx. 115 kN, (11.5 tons). This corre-

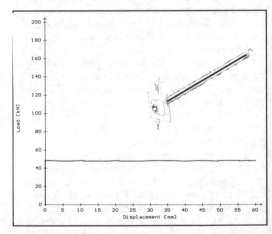

Figure 6. Split Set type SS 39, tensile loading across a joint, after Stillborg (1992).
Abb. 6. Split Set, Typ SS 39, Zugbelastung über einer Kluft, nach Stillborg (1992).

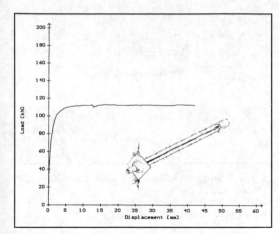

Figure 7. EXL Swellex, tensile loading across a joint, after Stillborg (1992).
Abb. 7. EXL Swellex, Zugbelastung über einer Kluft, nach Stillborg (1992).

sponds to a rockbolt deformation of approx. 10 mm. At general "bond failure" the bolt starts to slide. The rockbolt maintains however, a constant to increasing load bearing capacity for at least the duration of the test which was 150 mm of joint opening, divided into initially 10 mm of rockbolt deformation followed by approx. 140 mm of sliding.

2.5 Discussion

It is clear from the test results given in figures 3 to 7, matched with the technical specifications for the respective bolts tested, that some bolts better than others resemble the ideal load-displacement characteristics of an installed rockbolt, as presented in Figure 2. The information should serve most valuable in the choice of rockbolt system.

The results reported above must however, not lead one to believe that this information alone is the key to a successful rockbolt design. Apart from the load-displacement characteristics of the rockbolt it is important to consider the following rockbolt system (rockbolt and installation equipment) properties;
- versatile, can be used in any excavation geometry,
- installation procedure must be simple and reliable such that the rockbolt can be installed with a high rate of success,
- the rockbolt should give immediate support action after installation and should be able to install in water-filled boreholes,
- the system should be relatively inexpensive.

3. SWELLEX - A VERSATILE SYSTEM FOR ROCK REINFORCEMENT

3.1 General description of the Swellex system

Swellex rock reinforcement system consists of rockbolts, face plates and a portable pump with attached installation arm.

The principle behind Swellex is as simple as it is ingenious. The Swellex bolt is placed in a pre-drilled hole. High-pressure water expands the bolt along its entire length, pressing tightly against the surrounding material and forming itself after the irregularities of the hole. During the expansion the bolt decreases in length, pulling the face plate against the rock with a force of 20 kN. When the proper water pressure is reached, the water is turned off and installation is complete.

Two forces remain which lock the bolt in the rock. One is mechanical - the bolt has formed itself to the irregularities of the rock. The other is friction, which is mobilized as the bolt is subjected to load. At the end of expansion the bolt contracts and pulls the face plate against the rock with a force of 20kN, see Figure 8.

With the aid of a pull-out test equipment, the locking effect of an installed bolt can be tested at any time, even long after the installation has been performed.

Summary of Swellex features
- Immediate full-column rock reinforcement
- Instant full load-bearing capacity
- Swellex accommodates large ground movements
- The bolt adapts to hole irregularities and grips along its entire length
- Standard length up to 8 meters
- Simple installation procedure for all bolt lengths
- Swellex can easily be installed manually or mechanically
- Quality assured with each installation

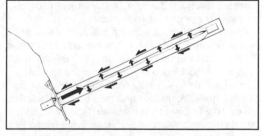

Figure 8. Swellex provides immediate frictional and mechanical anchoring.
Abb. 8. Swellex bietet sofortige Sicherheit durch Reibhaftung und mechanischen Verbund.

339

3.2 Swellex rockbolts

Atlas Copco has developed a family of different Swellex bolts, each designed to meet the customers individual demands. Rockbolts capable of handling large rock formations, high loads and corrosive environments.

All Swellex rockbolts are designed, manufactured and tested to meet ISO 9000 international standards as well as ASTM standards.

EXL Swellex works within the hole range 32 to 39 mm. This 100 kN rockbolt has an original tube diameter of 41 mm with a material thickness of 2 mm.

EXL Swellex is the perfect rockbolt with a load bearing capacity approaching ideal theoretic limits. It can absorb large rock movements without affecting its load bearing capacity. The perfect solution for mining and other applications where large ground movements are to be expected.

Midi Swellex works within the hole range 43 to 52 mm and is a lighter version than the Super Swellex bolt. Its original tube diameter is 54 mm with a material thickness of 2 mm. It is developed to fulfil the demands on a 120 kN rockbolt to be used within the most common hole diameter range utilizing high capacity drill jumbos.

The Midi Swellex incorporates the ideal physical characteristics of a perfect rockbolt. Elongation 20% (A5) means that it does not fail when reaching its tensile strength, but deforms to accommodate large rock movements.

Super Swellex works within the hole range 43 to 52 mm. Its original tube diameter is 54 mm with a material thickness of 3 mm. This 200 kN rockbolt is designed to meet demands for increased speed and safety in mining and construction applications where high breaking loads are at premium.

Yielding Super Swellex is a 160 kN rockbolt that has the same dimensions and work within the same hole range as the ordinary Super Swellex.

The Yielding Super Swellex incorporates the ideal physical characteristics of a perfect rockbolt. Elongation 30% (A5) means that it does not fail when reaching its tensile strength, but deforms to accommodate very large rock movements.

Coated Swellex - Corrosion resistant versions

All common types of rockbolts made from carbon steel are susceptible to corrosion. To overcome this problem, Atlas Copco has developed the special Coated Swellex with a corrosion-protective coating. When expanded, the bolt becomes fully protected as the coating forms a barrier between the rock and the bolt, preventing water and moisture from interacting with the steel.

The coating is factory applied under strict quality control. All the variants of the Swellex rockbolts areavailable in Coated Swellex versions.

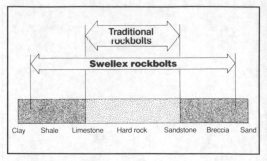

Figure 9. Swellex rock bolting extends the traditional role of rockbolts.
Abb. 9. Sicherung mit Swellex-Ankern erweitert die traditionelle Rolle von Gebirgsankern.

Swellex Face Plate The choice of face plate is crucial for the uniform distribution of the load throughout the surrounding rock. Swellex Face Plate, will secure the rockbolt installation without introducing unfavourable tensile stresses to the bolt head and will ensure that the face plate always lies flat against the rock face. This is facilitated by the elliptical bolt hole in combination with the domed plate.

3.3 Swellex - A versatile system for rock reinforcement

The Swellex system is highly efficient in regular, systematic rock bolting operations in the mining and civil engineering industry. Owing to the unique properties of the system, however, new opportunities for effective and economical rock reinforcement and ground stabilization now exist in situations where rock bolting was not previously considered possible.

This is because the Swellex bolt can be used effectively in almost any ground conditions, ranging from hard rock to sand and clay, see Figure 9.

The effectiveness of Swellex rockbolts in such a broad range of conditions, makes the system a reel problem solver. It has provided new solutions for tunnel front stabilization, enlargement of tunnels, heavily fractured rock, rock reinforcement in connection with TBMs, cable anchoring and micropiling.

The inflation pressure causes large radial forces to be exerted by the Swellex bolt, which in turn compact and consolidate the surrounding rock mass. This is the key to the good performance of Swellex bolts in almost any rock or ground conditions.

Installation of Swellex bolts is cost-effective and adaptable to fully mechanised rock reinforcement using various types of equipment. Faster and easier rock reinforcement means a more efficient use of machinery with minimal downtimes, increased productivity and safety on the worksite. Something of interest to all those involved in mining and construction operations.

The unique characteristics of the Swellex bolt make it perfect for applications which would normally require expensive time-consuming methods in order to achieve efficient rock reinforcement.

Simplicity, speed and immediate reinforcing effect that Swellex offers has shown itself invaluable when minimising downtime due to rock reinforcement work. This is especially important when boring tunnels using TBM for example, where also the space around the machinery is limited. Bolt lengths almost equal the tunnel's diameter can be installed.

3.4 *Swellex in rock burst tunnelling*

Prior to excavation of a tunnel, stresses are in balance. When a volume of rock is removed, for instance by blasting of a tunnel round, the balance is disturbed. After blasting, stresses redistribute and move away from the opening to resume the balance. If stresses in this process exceed the strength of rock closest to the excavation, failure occurs and a zone of fractured rock is formed. The fracture zone deforms until the high stresses are reduced to a level in equilibrium with the strength of the rock, which is lower than that of the undisturbed rock.

Different rock masses behave differently as regards the redistribution of rock stesses. Highly fractured hard rock and soft sedimentary rock both exhibit time-dependent effects with no violent fracturing. Rock that is hard and brittle, with few or no fractures, is however, burst-prone when subjected to high rock stress. It usually behaves elastically, but strain bursting (deformation of the rock beyond it´s elastic limit) may occur for a period of time after the blast. When this period of stress adjustment takes a longer time than that required for the crew to return to the face after the blast, rock bursts become hazardous. The direction of stresses that build up around the tunnel depends on the orientation of the natural ground stresses before the tunnel was excavated. Stress concentrations may very well occur at the face, especially at corners. That is why the face corners of a tunnel passing through stressed rock often develop a rounded shape. This fracturing to a new shape may be violent, with rock fragments being ejected from the face.

The common technique of dealing with rock bursting in this situation is destress blasting to precondition the rock ahead of the face. The blasting creates fractures in the high stress area and movements are induced. This movement relaxes the stress. However, the tuning of a satisfactory blasting arrangement takes time. This is a disadvantage of the destress blasting technique, particularly applied to tunnelling projects, where delays are very costly.

The Kan-Etsu Expressway links the Japan Pacific highways with the coast on the Japanese Sea, and connects the urban Tokyo region with the important cities of Niigata and Naga-Oko. The expressway runs across a mountain range where the first Kan-Etsu Tunnel takes traffic through mountains at an altitude of 1100 m. The tunnel is the longest road tunnel in Japan with a length of 10.925 m. The tunnel passes through hard, volcanic rock. The geological formations consists largely of quartz diorite and hornfels. Rock bursts at the tunnel face was a serious hindrance to tunnelling progress. Construction of the second Kan-Etsu tunnel started in 1986 and it was open for traffic in 1991. When excavation of the second tunnel was due to start, using drill and blast, the contractor Taisei-Nishimatsu-Sato looked for a rock stabilisation system that could prevent sever rock bursts at the face and minimise disruption to the tunnel advance.

Shotcrete combined with pattern bolting proved satisfactory for reinforcement under normal tunnelling conditions. Cement grouted 3 m steel rebars were used for pattern bolting. In sections of the tunnel where rock burst at the face was experienced the contractor tried various bolting techniques to stabilise the burst-prone rock. The first test was made with fibre glass bolts, which were installed in the tunnel face with fast hardening resin following each round. The result was disappointing. Nothing of the bolts remained in the holes after the blast and consequently nothing to support the rock. The rock bursts persisted.

In a second test bolts made of steel rebars were installed with resin. Results were again disappointing. Nothing remained in the holes after the round had been fired, and rock kept on bursting at the face. A test with Swellex bolts was devised. Now things changed. The outer part of the Swellex bolts was cut off by the blast but two thirds of its length stayed in the rock in front of the tunnel face. Rock bursts were reduced to a magnitude that did not disturb tunnelling progress.

As a result of the positive test result, Swellex bolting become the contractor´s standard method for face stabilisation in burst-prone rock. The ease of installation and immediate support action of the bolt also convinced the contractor to switch to Swellex as the standard bolt in the pattern bolting of the tunnel roof and walls. In highly stressed brittle rock, drill hole walls may ´spall´, that is, failure takes place inside the hole. The fractured rock that spalls can make the insertion of the rockbolt difficult, however the small outer diameter of the Swellex, when inserted into the borehole, makes installation very easy.

The application of Swellex bolts for face stabilisation in rock burst prone tunnel sections is

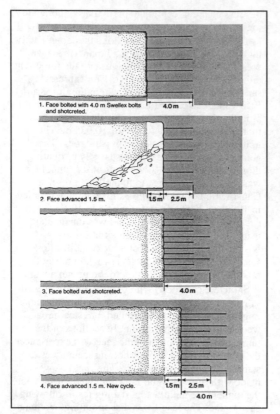

Figure 10. Four stages in the use of Swellex bolts to prevent the effects of rock bursts at the Kan-Etsu tunnel.
Abb. 10. Vier Stufen beim Einsatz von Swellex-Ankern zur Verminderung der Folgen von Gebirgsschlägen beim Bau des Kan-Etsu-Tunnels.

unconventional. In the Kan-Etsu tunnel twenty-two, 4m-long Swellex bolts were installed straight into the face. The round was then drilled and blasted to a depth of 1.5m. The bolts were cut off by the blasting, but the remaining 2.5m length provided support to keep the face intact. This permitted drilling of the

next round without troublesome effects of rock burst, see Figure 10.

Rock bursts induce very high momentary loading on rockbolts and it is therefore important that the bolt can sustain this load and still retain substantial holding force after the burst. In this context most rockbolts are too stiff, but by placing e.g. a wooden block under the face plate, the wooden block act as the fast response absorber of the momentary high strain resulting from the loading the bolt in a rock burst situation.

Stillborg (1992), has given the following general recommendations for levels of reinforcement based on the severity of the conditions.
Level 1. Spalling rock but no violent failures. Shotcreting, (fibre reinforced) and bolting as required. Bolts at face shall be 1.5m longer than the length of the blasted round, to pre-stabilise the rock for the next blast, see Figure 11a.
Level 2. Medium to severe rock bursts. Shotcreting as in Level 1. Bolts in a systematic pattern. Face bolts shall be 1.5m longer than the length of the round length. A timber board fitted between the bolt plate and the rock surface. The board will absorb rapid displacement of rock that might otherwise break the bolt. The board will also bridge over fractured rock between bolts, see Figure 11 b.
Level 3. Severe rock bursts. Thicker layer of fibre reinforced shotcrete and systematic pattern of rockbolts as in Level 2, see Figure 11 c.

The recommendation of reinforcement levels are to be seen as guidelines and should be adopted to conditions in the actual tunnel. The target is to obtain safe working conditions and inhibit the effect of rock bursts that otherwise would slow down tunnelling progress.

3.5 Swellex success in NATM Tunnelling

Austrian tunnellers have found Swellex to be a perfect component in their NATM concept, as it contributes to their high speed and cost-effective

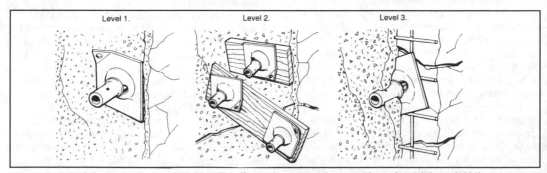

Figure 11a-c. Reinforcement levels to prevent the effects of rock burst in tunnelling, after Stillborg (1992).
Abb. 11a-c. Verschiedene Sicherungsgrade zur Verminderung der Folgen von Gebirgsschlägen, nach Stillborg (1992).

tunnelling.

The essential idea behind the NATM concept is that the initial support system restricts ground movements after excavation. Thereby maintaining the inherent strength of the rock mass. The permanent lining is installed when the rock has reached a state of equilibrium and deformation has ceased.

The main reasons for the perfect compatibility between the Swellex concept and the NATM are as follows:

- rapid installation means that less time is needed to effect rock support
- immediate support provides added safety for tunnel personnel
- immediate support results in less loosening of the rock surrounding the excavation which in turn means that less reinforcement is needed to support this rock
- since Swellex is not sensitive to vibrations, the blasting of subsequent rounds does not   destroy existing reinforcement
- firm contact with the rock along the full length of the bolt effectively binds together highly fissured rock

The introduction of Coated Swellex versions , as well as the Midi Swellex and Super Swellex bolts, have increased still further the applications for this revolutionary concept for rock support.

Neuere Entwicklungen in der DYWIDAG Ankertechnik: Der elektrisch isolierte Daueranker und der ausbaubare Temporäranker

Recent developments in DYWIDAG anchor technology: Electrically isolated permanent anchors and removable temporary anchors

Reinhard Klöckner
DSI München, Germany

ZUSAMMENFASSUNG: Die vorhandene Kunststoffverrohrung eines Dauerankers kann mit vergleichsweise geringem Aufwand vollkommen elektrisch isolierend ausgebildet werden. Wird der Isolationswiderstand des Ankers einschließlich des Ankerkopfes nach dem Spannen, also im endgültigen Gebrauchszustand des Ankers, geprüft und nachgewiesen, kann der Anker als sicher gegen elektrochemische Korrosionsvorgänge am Stahlzugglied und sicher gegen Eindringen korrosiver Medien sowie Zirkulieren von Druckwasser gelten. Im Beitrag werden für den genannten Einsatz konzipierte Verrohrungstypen und ein Verfahren zum Einbau vorgestellt.

Weiterhin wir ein rückbaubarer Litzen-Kurzzeitanker beschrieben, dessen Litzen mit einer Sollbruchstelle versehen sind, an der sie nach dem Gebrauch des Ankers durch Überspannen abgerissen werden. Die jeweiligen Verbundstrecken der Litzen sind einzel oder in Gruppen über die Gesamtverbundstrecke des Ankers gestaffelt angeordnet. Sie sind zugleich durch verbundverbessernde Maßnahmen (Aufspreizungen) so in ihrer Länge minimiert, daß sie im Boden verbleiben können, ohne wesentliche Hindernisse für spätere Arbeiten im durchankerten Bereich darzustellen.

ABSTRACT: The existing plastic sheathing of a permanent rock or soil anchor can relatively easily be conceived as a complete electrical insulation. An anchor, the insulation resistance of which - including the anchor head - is checked and accepted after stressing, i.e. in final condition, can be considered as safe against any electrochemical corrosion process and safe against any ingress of corrosive agents and circulation of pressurised ground water. The presented sheathing systems and an installation procedure were designed to match with the particular conditions of electrically isolated anchors. The anchor head arrangement allows for cased drilling.

Furthermore, a removable temporary strand anchor is described, the strands of which are weakened in a controlled way at the transition point between free length and bond length. The free length will be removed from the ground by loading the strands to a force exceeding the predetermined breaking load. The individual bond lengths of the strands are arranged in a staggered way along the overall bond length of the anchor. These individual bond lengths are minimized by bond enhancing measures (local spreading of the strands) such that they can remain in the ground without representing serious obstacles to following works.

A method particularly appropriate for weakening the strands is electrical high frequency induction using a coil, into which the strands are introduced during the anchor fabrication. The strands can be weakened to a breaking load above the yield load, and   ductile behaviour of the anchor prior to failure can be obtained.

1 DER ELEKTRISCH ISOLIERTE DAUERANKER

1.1 Grundgedanke

Die Idee, sich die elektrisch isolierende Wirkung der ohnehin vorhandenen Kunststoffverrohrung eines Bodenankers oder eines Spanngliedes im Spannbetonbau zum Zweck eines gesicherten, nachprüfbaren Korrosionsschutzes für Zugglieder aus Spannstahl zunutze zu machen, ist schon relativ früh entstanden. Der Grundgedanke ist so einfach wie folgerichtig und überzeugend: Korrosion in Stählen ist stets ein elektrochemischer Vorgang, und in einem mit Zement injizierten Stahlzugglied, das gegen seine Umgebung vollkommen elektrisch abisoliert ist, kann kein korrosionsfördernder Strom fließen. Dies gilt besonders für Daueranker in Bereichen, wo Streuströme auftreten können, wie Tunnel mit elektrischem Bahnbetrieb und Kraftwerkskavernen, oder wo mit der Bildung sog. Makroelemente gerechnet werden muß.

Da zudem eine wirksame elektrische Isolierung von Haus aus vollkommen wasserdicht ist, wird der Ankerinnenraum gleichzeitig gegen das Eindringen oder gar Zirkulieren aggressiver Sickerwässer, insbesondere auch Druckwasser, zuverlässig geschützt. Hier sind auch Seewasser und tausalzhaltige Wässer zu erwähnen.

1.2 Schweizer Richtlinie "Korrosionsschutz für permanente Boden- und Felsanker"

Erstmals öffentlich eingeführt wurde das Konzept eines elektrisch isolierten Daurankers im Jahre 1989 in der Schweiz durch die "Empfehlungen für Projektierung und Ausführung des Korrosionsschutzes von permanenten Boden- und Felsankern" der fünf schweizerischen Spannverfahrensinhaber. Die Empfehlung galt dort zunächst als Richtlinie und wurde später von den Baubehörden zur verbindlichen Vorschrift erklärt.

1.3 Entwurf Euronorm "Verpreßanker"

In nur wenig veränderter Form fand die Schweizer Empfehlung Eingang in den Entwurf des Eurocode pr EN 28802 "Verpreßanker", der im Mai 1994 herauskam, und der gegenwärtig in den zuständigen Fachausschüssen verhandelt wird.

Im wesentlichen besagt die Vorschrift, daß das Zugglied eines elektrisch isolierten Boden- oder Felsankers nach dem Einbau im Bohrloch mit einer Prüfspannung von 500 Volt Gleichstrom zu beaufschlagen ist, und daß der gemessene Isolationswiderstand gegenüber dem Erdreich v o r und n a c h dem Spannen des Ankers nicht unter den Wert von 100 kOhm absinken darf.

Zu beachten ist, daß die Nichterfüllung dieser Prüfbedingung nicht unbedingt die Untauglichkeit des betroffenen Ankers, sondern nur die "Nicht-Nachprüfbarkeit" seines Korrosionsschutzes anzeigt.

1.4 Der DYWIDAG Quetschrohranker

Auf der Suche nach einem robusten, baustellengerechten Anker, der die Bedingungen der Schweizer Richtlinie (EN 28802) zuverlässig erfüllt, und dessen Verrohrung nicht bereits beim Transport auf der Baustelle, im Tunnel etc. beschädigt oder durchschlagen wird, kam ursprünglich von unserem Schweizer Lizenznehmer, der SpannStahl AG, die Idee, den Anker auf seiner ganzen Länge mit einem handelsüblichen HDPE-Glattrohr, Wandstärke 4 bis 5 mm, zu umgeben, das im Verbund-

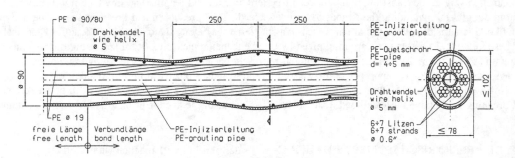

Bild 1 DYWIDAG-Quetschrohranker mit innenliegender Drahtwendel
DYWIDAG DT anchor with internal wire helix

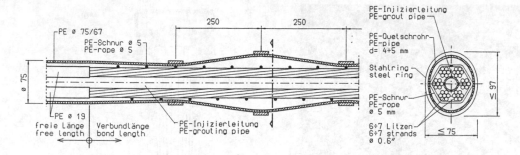

Bild 2 DYWIDAG Quetschrohranker mit außenliegenden Stahlringen
DYWIDAG DT anchor with external steel rings

bereich in regelmäßigen Abständen bleibend ovalisiert wird, wie bereits im Beitrag [1] erwähnt (Bild 1 und 2).

Die durch Druck von außen aufgebrachten Verformungen werden durch mitverformte Stützelemente wie außen übergeschobene Stahlringe oder eine innen eingelegte Drahtwendel stabilisiert. Die Drahtwendel wirkt als rißverteilende Umschnürungsbewehrung zumal im hochbelasteten Bereich am Beginn der Verbundstrecke. Sie verhindert - wie auch die übergeschobenen Stahlringe - das Auftreten von klaffenden Längsrissen im Verpreßkörper und den damit einhergehenden progressiven Verbundbruch längs des Zuggliedes. Wenn der Untergrund dies zuläßt, kann hierdurch die Verbundlänge und somit die Gesamtlänge des Ankers verringert werden. So können Kosten eingespart werden.

Der von der elektrischen Isolierung her kritische Übergang zwischen dem traditionellen Ripprohr der Haftstrecke und dem Glattrohr der Freispielstrecke entfällt. Der Anker ist aufwickelbar.

Bei mehr als ca. 4 Litzen im Anker ist es zweckmäßig, die Litzen innerhalb der Verbundlänge mit Hilfe eines sog. "integrierten Abstandhalters" (Bild 3) gegeneinander und gegen das Hüllrohr zu distanzieren. Es handelt sich hierbei um Aufspreizungen der Litzendrähte auf das etwa 1,5-fache des ursprünglichen Durchmessers und Einfügen eines Distanzstreifens aus PE in regelmäßigen Abständen. Dieses Konzept führt zu einer parallelen Litzenführung in L_v und zum Wegfall der andernfalls erforderlichen Sternabstandhalter und der Zickzackführung der Litzen, ohne daß Zweifel an der vollständigen Umschließung jeder Litze durch Zementmörtel bestehen.

Keiner der bisher (Mai 1995) ausgeführten ca. 140 Stück DYWIDAG Quetschrohranker mit 4 bis 9 Litzen ⌀ 0,6" [2] hat bei der elektrischen Widerstandsprüfung versagt (Bild 4).

1.5 Ankerkopf

Der Ankerkopf unserer elektrisch isolierten Anker ist teilweise oder ganz in die Isolierung einbezogen (Bild 5), und der elektrische Widerstand kann wahlweise bis zur Abnahmeprüfung oder während der Betriebszeit des Ankers nachgemessen werden.

Der Anschluß der Verrohrung an den Ankerkopf beim oben erwähnten Quetschrohranker erlaubt auch das verrohrte Bohren, d.h. die Montage der Ankerkopfkonstruktion am zuvor eingeschobe-

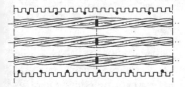

Bild 3:   Integrierter Abstandhalter und Verbundverbesserung der Litzen durch örtliche Aufspreizung
Integrated spacer and bond enhancement by local spreading of the strands

Bild 4:   DYWIDAG-Quetschrohr-Anker
9 ø 0,6" einbaufertig (Projekt Axenstra-
ße, Schweiz)
DYWIDAG DT-Anchors 9 strands 0,6"
ready for installation (Axenstrasse Project, Switzerland)

nen Anker kann nach dem Ziehen der Bohrverroh-
rung erfolgen.

2 DER AUSBAUBARE KURZZEIT-LITZEN-ANKER

2.1 Übersicht über vorhandene Systeme

Die internationale Fachwelt arbeitet seit vielen
Jahren an der Entwicklung von Temporärankern,
die nach Gebrauch möglichst weitgehend wieder
aus dem Untergrund entfernt werden können, so
daß nachfolgende Aushub- oder Gründungsarbei-
ten im durchankerten Bereich möglichst wenig
behindert werden.

Hierbei gibt es Systeme, die vollständig oder
fast vollständig, also einschließlich der Verbundlänge, ausgebaut werden können, und solche, bei
denen die Zugbewehrung im Bereich der Haftstrecke im Boden verbleibt (Bild 6).

Aus der Vielzahl bekannter Methoden seien

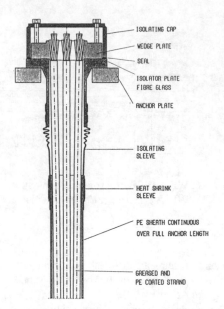

Bild 5 Elektrisch isolierte Bündelanker, Ankerkopfkonstruktion
Electrically isolated multistrand anchor,
anchor head arrangement

hier nur solche erwähnt, die DYWIDAG bereits
angewendet oder untersucht hat:

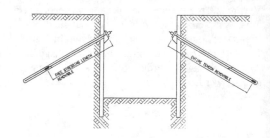

Bild 6:   Rückbaubare Kurzzeitanker, Schema
Removable temporary anchors, schemati-
cal

1. Stabanker, die aus einer Muffe (am oberen
Ende der Haftstrecke) oder einer Druckplatte mit
aufsteigenden, kurzen Bewehrungsstäben (am
unteren Ende der Haftstrecke) zuverlässig her-
ausgeschraubt werden können,
2. Litzenanker, die am Übergang zwischen
Freispiel- und Verbundstrecke entweder mit einem
Brandsatz aus Thermit-Schweißpulver oder mit
einer elektrischen MF-Induktionsspule ausgestattet
sind. Die Trennvorrichtung muß den Einbau, den

Verpreßvorgang und die Betriebszeit des Ankers unbeschadet überstehen. Sie wird nach dem Gebrauch des Ankers in Funktion gesetzt und führt dann zum Abreißen der Litzen meistens unter der Wirkung der eigenen Vorspannung.

3. Litzenanker, die vor dem Einbau im Bohrloch an einer Sollbruchstelle gezielt geschwächt und später durch Überspannen abgerissen werden,

4. Anker aus Spannstahllitzen, die im Bohrlochtiefsten an einem kraftübertragenden Formteil (Sattel) mit engem Radius umgelenkt werden, und die nach dem Lösen der Verankerung an einem der beiden Litzensträge mit noch beherrschbaren Kräften aus dem Bohrloch herausgezogen werden können.

2.2 Merkmale des ausbaubaren Stufenankers System DYWIDAG

Der nachfolgend vorgestellte ausbaubare bzw. rückbaubare Kurzzeitanker weist folgende Merkmale auf:

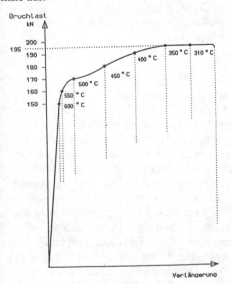

Bild 7 Kraft-Dehnungsverhalten einer örtlich induktiv geschwächten Litze ⌀ 0,52"
Load-elongation behaviour of a ⌀ 0,52" strand, locally weakened by inductive heating

1. Jede Litze ist am Übergang zwischen Freispiel- und Verbundstrecke mit einer Sollbruchstelle versehen, an der sie nach Gebrauch des Ankers

durch Überspannen abgerissen wird. Diese Trennstelle entsteht vor der Herstellung des Ankers durch kontrollierte örtliche Erwärmung und Schwächung der Litze mit Hilfe einer HF-Induktionsspule, in welche die Litze kurzzeitig eingeführt wird.

Versuche haben ergeben, daß die Methode der elektrischen Induktion eine sehr fein gesteuerte Abminderung der Traglast erlaubt, und daß insbesondere eine zuverlässig begrenzte Schwächung bis oberhalb der Streckgrenze möglich ist, so daß noch ein duktiles Verhalten des Ankers vor dem Bruch erreicht werden kann (Bild 7).

2. Die Verbundlänge jeder Einzellitze ist auf ca. 1,5 m beschränkt. Zur Sicherstellung des Haftverbundes zwischen der Litze und dem umgebenden Verpreßkörper über diese geringe Länge wird die Litze in regelmäßigen Abständen mit den bereits in ihrer Funktion als "integrierte Abstandhalter" beschriebenen Aufspreizungen versehen. Versuche haben ergeben, daß eine   e i n z i g e   derartige Aufspreizung in einem vermörtelten Stahlrohr von ca. 300 mm Länge die volle Bruchlast der Litze aufnehmen kann.

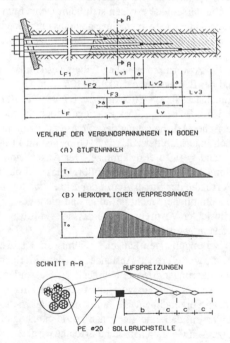

Bild 8 Rückbaubarer Stufenanker System DYWIDAG
Removable staggered strand anchor type DYWIDAG

3. Die Verbundlängen (L_{Vl}) der Einzellitzen (Bild 8) werden gestaffelt über die gesamte Verbundlänge des Ankers verteilt eingebaut. Dieses Prinzip des "Stufenankers" führt zu folgendem doppelten Vorteil:

a) Die kurzen, noch dazu mit Zwischenräumen (a) über die Länge des Verpreßkörpers verteilt angeordneten Litzen-Restlängen stellen keine wesentliche Behinderung späterer Arbeiten im Ankerbereich dar.

b) Der Verlauf der Krafteinleitung vom Verpreßkörper in den Untergrund ist günstiger als beim herkömmlichen Verpreßanker, weil die kritischen Verbundspannungsspitzen am Beginn der Haftstrecke und somit die Gefahr des progressiven Verbundbruches deutlich abgemindert werden. Bei gleicher Gesamtverbundlänge eines Verpreßankers kann so eine höhere Kraft verankert, oder bei gegebener Kraft eine kürzere Verbundstrecke gewählt werden.

Zu beachten sind bei einem Anker mit gestufter Lastabgabe die unterschiedlichen Spannkräfte in den verschieden langen Litzen bzw. Litzengruppen beim Spannen mit der Bündelspannpresse. Diese Kraftunterschiede ergeben sich beim Dehnen ungleich langer Litzen an einer gemeinsamen Spannstelle, wobei die kürzeren, steiferen Litzen höhere Kräfte erhalten als die längeren.

Bei Ankern mit üblichen Längenverhältnissen und bei Bemessung nach DIN 4125, d.h. Gebrauchslast F_w = Streckgrenze/1,75 und Prüfkraft = 1,5 x F_w bzw. 1,25 x F_w, liegt jedoch die rechnerische Mehrbelastung der kürzesten Litzen meist noch unterhalb der zulässigen Grenzlast im Prüfzustand = 0,9 x Streckgrenze. Der Anker kann dann - bei Einstellung der Sollbruchlast auf einen Wert nahe der Streckgrenze der Litze - wie ein herkömmlicher, nicht gestufter und nicht geschwächter Verpreßanker bemessen, geprüft und gespannt werden.

Bei ungünstigen Längenverhältnissen, also vergleichsweise großen Abständen (s) der einzelnen Verankerungshorizonte und geringer Gesamtlänge des Ankers, kann es vorteilhaft sein, die längeren Litzen vorzudehnen. Die Vordehnung läßt sich so bemessen, daß eine gleichmäßige Ausnutzung aller Litzen im Gebrauchszustand erreicht wird.

Wird, wie z.B. in USA und England, die Gebrauchslast eines Ankers auf die Bruchlast der Litze bezogen, muß die Ausnutzung der gewählten Litzen im Maße der vorgenommenen Schwächung an der Sollbruchstelle verringert werden.

Die PE-Einzelverrohrung der Litzen muß am unteren und am oberen Ende sorgfältig gegen das Eindringen von Zementmörtel infolge des Verpreßdrucks abgedichtet werden. Andernfalls würde erhöhte Reibung in der freien Länge den Ausbauerfolg in Frage stellen, d.h. die Litzen neigen dann zum Abreißen am Ankerkopf oder an der Spannpresse statt an der Trennstelle.

Beim Ausbau, also Abreißen der Litzen mittels einer Einzelpresse, können die brechenden Litzen ohne Gefahr in einem PE-Rohr aufgefangen werden, wie es üblicherweise beim Einschieben der Litzen im Spannbetonbau zum Einsatz kommt.

Bild 9 Rückbaubarer Kurzzeitanker - Abreißen der Einzellitzen
Removable temporary anchor - single strand pullout jack

3 AUSBLICKE

3.1 Elektrisch isolierte Anker

Im Bereich der Daueranker mit elektrischer Isolierung arbeiten wir am Konzept eines Ankers, dessen Verrohrung vor dem Einführen einer Zugbewehrung im Bohrloch installiert und mittels elektrischer Widerstandsmessung auf Dichtigkeit geprüft und abgenommen werden kann. Das mit den zuvor beschriebenen integrierten Abstandhaltern versehene Litzenbündel kann dann nachträglich eingebaut werden, und durch eine im Hüllrohr innen anliegende Drahtwendel ist die Forderung nach 5 mm Zementsteinüberdeckung der Litzen (DIN 4125, BS 8081) erfüllt. Auch kann das Prinzip der gestuften Lasteintragung (Abschnitt 2.2.3) in den Untergrund bei einem solchen Anker vorteilhaft angewendet werden.

3.2 Ausbaubare Litzenanker

Beim ausbaubaren Litzen-Temporäranker, der von den Märkten in zunehmendem Maße nachgefragt wird, arbeiten wir an einer speziellen Abreißpresse, um den benötigten Litzenüberstand, der derzeit noch etwa 700 mm beträgt, auf etwa 200 mm zu verkürzen.

4 SCHRIFTTUM

[1]   Jungwirth, Dieter: "Korrosionsschutzsysteme bei der Verwendung hochfester Zugglieder" (Beitrag zu diesem Symposium)

[2]   Müller/Jungwirth/Klöckner/v. Allmen: "EW Obwalden - Felsanker mit neuartiger Verrohrung" (Vorgespannter Beton in der Schweiz; Beitrag zum 12. FIP-Kongreß 1994, Washington DC, USA)

Weiterentwicklung der Ankertechnik im Steinkohlenbergbau für den Einsatz im geklüfteten Fels

Developments of rock bolting in coal mining in layered rocks

W. Müller
DMT-Gesellschaft für Forschung und Prüfung, Germany

Zusammenfassung: Der Aufsatz beschäftigt sich mit den im Steinkohlenbergbau Deutschlands verwendeten Dimensionierungsverfahren für die Ankertechnik. Der Ankerausbau wird in zunehmendem Maße in Strecken mit großen Konvergenzen sowie in anspruchsvollen Bauwerken eingesetzt. Dabei kommen weiterentwickelte Verfahren zur Ausbaubemessung und zur Verformungsprognose zum Einsatz. Diese basieren auf der Verknüpfung physikalischer und numerischer Modelltechniken mit unter Tage Meßergebnissen. Anhand zweier Beispiele wird der Einsatz der gekoppelten Simulationstechnik veranschaulicht.

Abstract: Coal mining in Germany is characterized by increasing mining depths and unfavourite geological conditions. The deposits consists of layered structures with coal seams, sandstones and shales with layers from some centimeters up to 50m. While the average mining depth increases the stress state increases in the some way but the strength of the rock mass remains mostly independent of the depth. This fact leads to unfavourite stress strength conditions with high convergences in many cases. On this basis techniques have been developed which calculate the support, the rock deformations and the rock support interaction. For cable bolting nowadays three rules exist. They describe the deformations and the load bearing conditions the cable support has to fulfill as main support element. Up to now the bolting of roadways has a low importance in German coal mining. The developments during the last years tends to expand the conditions for usage of bolting. The development of highly deformable bolts as Kombi bolts as well the new possibilties of the application of combined physical and numerical modelling to solve geotechnical problems stimulate this tendency.

The publication illustrates the procedure of this combined technique at two examples. The first one investigates the stabiltiy of a roadway face. The roadway is driven as middle roadway in a longwall. The roadway faces have to be supported in dependance of the behaviour of the dams. A bolting of the faces does not yield the expected result. The second example describes the behaviour of Kombi bolts and of stiff bolts which are used in a roadway with double usage. The shear and the axial deformation of each bolt has been investigated in dependance of the development of the convergences and of the fracture processes in the rock.

1 EINFÜHRUNG

Der Steinkohlenbergbau in der Bundesrepublik Deutschland erstreckt sich zur Zeit auf vier Abbauregionen, von denen das Ruhrkarbon mit zur Zeit ca. 40 Mio. Tonnen Jahresförderung ca. 70% der Gesamtförderung stellt. Der Abbau ist hier durch Strebbau von Flözen in unterschiedlichen geologischen Horizonten geprägt. Dabei nimmt die mittlere Gewinnungsteufe durch den Einfall der Lagerstätte nach Norden hin und durch die Erschließung tiefer liegender Flöze seit Jahren stetig zu. Die heutige Abbauteufe erreicht im Schnitt annähernd 1000m Teufe mit Spitzenwerten von 1400m.

Während der Gebirgsdruck im wesentlichen linear mit der Teufe zunimmt ist die Festigkeit der Flöze und der umgebenden Schichten fast unabhängig von der Größe der Überlagerung. Dies führt dazu, daß sich das Verhältnis zwischen Gebirgsdruck und Verbandsfestigkeit ständig verschlechtert. Entsprechend muß bei der Auffahrung von Strecken im Flöz und im umgebenden Gestein als auch während der Nutzungszeit dieser Strecken mit erheblichen Streckenverformungen gerechnet werden. So haben Absolutspannungsmessungen und Verformungsmessungen gezeigt, daß die Druckumlagerungen im Bereich von Strecken ohne Berücksichtigung zusätzlicher Abbauwirkungen ca. 30m weit reichen können. Dabei treten Streckenverformungen auf, die 100% der Streckenhöhe bzw. bei einer zwischenzeitlichen Sanierung der Strecke auch über 100% der Streckenausgangshöhe erreichen können. Bild 1 zeigt den Zustand einer derart verformten Strecke. Aus diesen Erläuterungen wird deutlich, daß trotz einer generell abnehmenden Bergbautätigkeit auf Steinkohle in Deutschland die Gebirgsmechanik von zunehmender Bedeutung wird.

Bild 1: Zustand einer stark verformten Strecke
Situation of a highly deformed roadway

2 AUSBAUDIMENSIONIERUNG

Für die Belange des Steinkohlenbergbaus sind speziell entwickelte Dimensionierungstechniken und -verfahren entwickelt worden. Sie sind für die meisten Abbaustrecken und einen Teil der Gesteinsstrecken auf die Entwicklung und Auslegung eines verformungsfähigen Ausbaus ausgerichtet. Hintergrund dieses Konzeptes ist die Einsicht, daß unter den gegebenen Gebirgsdrücken und -bewegungen Streckenverformungen nicht zu vermeiden sind. Nur in einer Minderzahl von Gesteinsstrecken kann mit starren Ausbaukonzepten gearbeitet werden.

Die Dimensionierungsverfahren basieren auf der Verbindung von systematischen Verformungs- und Spannungsmessungen in situ, mit Modellversuchen an körperlichen Modellen im reduzierten Maßstab, mit numerischen Simulationsverfahren, empirischen Modellvorstellungen und ausbautechnischen Entwicklungen. Beispielhaft sei auf Veröffentlichungen von Götze[1], Jacobi, te Kook, Müller, Stephan, und Zischinsky hingewiesen. Aus diesen Untersuchungen ist das Verformungsverhalten der

Strecken in Abhängigkeit von der Geologie, der Teufe, der Wirkung von Abbautätigkeiten sowie des zu wählenden Ausbaus bekannt.

Für den Ankerausbau ergeben sich aus diesen Betrachtungen festgelegte Anforderungen an seine Auslegung. Dabei spielen im Wesentlichen sicherheitliche Aspekte eine Rolle, da schwere Gesteinsbrüche in gankerten Strecken nach einer Phase der Euphorie der Ankertechnik in den 70´ger Jahren eine veränderte Philosophie erforderten.

Die Ankertechnik wird in verschiedenen Bereichen des untertägigen Bergbaus eingesetzt. Im nachfolgenden möchte ich mich auf den Einsatz des Ankerausbaus in Strecken als wesentliche Ausbaukomponente konzentrieren. In diesen Fällen ist die Ankerdimensionierung zum gegenwärtigen Zeitpunkt durch drei Regeln geprägt.

Die erste wird als Kluftkörpertheorie bezeichnet. Ihr liegt zu Grunde, daß sich im geschichteten Gebirge des Ruhrkarbon typische Kluftkörper im Streckenmantel bilden, deren Größe

[1] siehe Literaturverzeichnis

primär von der Streckengeometrie abhängen. Diese Kluftkörper sind sowohl unter Tage als auch im Labor in körperlichen Modellversuchen vielfach nachgewiesen. Die Zahl und Festigkeit der Anker wird auf das Halten dieser Kluftkörper ausgelegt. Dabei ist das Doppelte des statischen Gewichtes dieser Körper von demjenigen Ankerteil zu tragen, der außerhalb der Kluftkörper im intakten Gebirge verankert ist. Die Auslegung auf das Doppelte des statischen Gewichtes ist auf die dynamische Belastung der Anker bei Aktivieren der Kluftkörper ausgerichtet.

Die zweite Regel bezieht sich auf das Verformungsverhalten der Anker. Aus den für das Gebirge hergeleiteten Konvergenz-Prognoseverfahren läßt sich die Konvergenz für jede Strecke verläßlich vorhersagen. Aus dieser Vorhersage ist der Anker hinsichtlich seiner Verformbarkeit auf die zu erwartenden Ankerdehnungen auszulegen.

Die dritte Regel bezieht sich schließlich auf die konvergenzmindernde Wirkung eines jeden Ausbaus. Für jede Strecke läßt sich in Abhängigkeit von der zu erwartenden Belastung der notwendige Ausbaustützdruck berechnen, der für eine definierte Konvergenzreduzierung notwendig ist. So ist z.B. für eine geregelte Wetterführung ein Mindestquerschnitt offen zu halten. Der Ausbaustützdruck wird zur Zeit aus dem Produkt der Zahl der Anker pro m² und der Streckgrenze der Anker berechnet.

Während die beiden ersten Regeln als Grundvoraussetzung generell anzusetzen sind, greift die dritte Regel je nach notwendigem Ausbaustützdruck.

Die Entwicklung der Ankertechnik im Steinkohlenbergbau ist gegenwärtig durch eine Renaissance dieser Technik geprägt. Dabei spielen Kostengesichtspunkte eine wichtige Rolle. Von der technischen Seite wird diese Tendenz durch den Einsatz von stark verformbaren Ankern in Strecken mit sehr großer Konvergenz, die bessere Beherrschbarkeit der Anker-Verzugwirkung als auch die Weiterentwicklung der Dimensionierungsverfahren gefördert.

Hinsichtlich der leistungsfähigen hoch verformbaren Anker sei auf die Kombianker in unterschiedlichen Bauformen hingewiesen, die im Vortrag von Witthaus näher beschrieben werden. Die Weiterentwicklung der Dimensionie-

rungsverfahren beruht auf der gekoppelten Modelltechnik mittels derer der Einsatz von Ankern als Ausbau nachgebildet und optimiert wird. In dieser Modelltechnik werden Prüfstandsversuche an körperlichen Modellen, die in einem reduzierten Maßstab durchgeführt werden, mit numerischen Berechnungen gekoppelt. Die Eigenschaften der Anker werden aus Prüfstandsversuchen an Originalankern, die unter Zug- und/oder Scherbelastung bis zum Versagen belastet werden, ermittelt. Die Eigenschaften des Gebirges sind aus petrographischen und gesteinsmechanischen Untersuchungen ausreichend bekannt. Dasselbe gilt für den Gebirgsdruck bzw. seine bergbaubedingte Änderung. Diese Größen werden in numerischen Berechnungen verwendet. Die Plausibilität der hierbei ermittelten Ergebnisse kann durch die Versuche an den körperlichen Modellen nachgewiesen werden. Die Einzelheiten dieser Techniken und ihrer Wechselwirkung sind bei Würtele näher beschrieben.

3 FALLBEISPIELE

Ich möchte anhand einiger Beispiele die Wirkungsweise der gekoppelten Modelltechnik aufzeigen. Das erste Beispiel ist ein Sonderfall - es zeigt aber eindrucksvoll den Nutzen der Vorausplanung mittels der Modelltechnik. Es handelt sich hierbei um die Aufgabe, eine sogenannte Mittelstrecke, die mittig in einem Streb geführt ist, hinsichtlich ihres Ausbaus zu optimieren. Konvergenzvorrausberechnungen lieferten ein Prognose von ca. 2.4-2.7m Konvergenz bei einer Streckenhöhe von 4.0m. Da die Strecke u.a. der Bewetterung dienen soll, muß ein Mindestquerschnitt während der gesamten Nutzungsdauer gewährleistet sein. Dies erfordert ein Senken der Sohle nach Erreichen dieses Mindestquerschnittes und eine ausbautechnische Sicherung der neuen eingeschnittenen Stöße. Bild 2 veranschaulicht diese Problematik.

Zur Optimierung des Ausbaus und zur Klärung der gebirgsmechanischen Fragestellungen sind in den körperlichen Modellen zwei Ausbauvarianten und zwar mit einem starren Baustoffbegleitdamm und mit einem nachgiebigen Damm, hier als Holzkasten nachgebildet, untersucht worden. Bild 3 zeigt die Ausgangssituation des Modells mit Baustoffdamm.

Zusätzlich sind numerische Berechnungen

Bild 2: Zustand der Mittelstrecke nach Senkung
mit Holzkästen als Damm
Roadway deformation after subsidence
of the floor with wodden dam

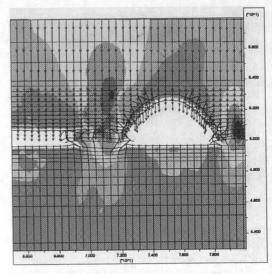

Bild 4: Verformungszustand der Strecke nach
1,8m Senkung
Deformations after 1.8m subsidence

Es zeigt sich aus den Berechnungen, daß eine
Ankerung mit den üblich verwendeten Ankern
nicht zu einer ausreichenden Standsicherheit
führt. Statt dessen ist der Damm durch eine
ausreichende Komprimierbarkeit in die Lage zu
versetzen, die Druckumlagerungen von den
Liegendstößen fernzuhalten. In diesem Falle ist
das Augenmerk auf die Dammgestaltung zu
legen. Wird dieser als starr konzipiert, sind
spätere Sanierungsmaßnahmen unverhältnis-
mäßig.

Bild 3: Mittelstrecke mit Baustoffdamm
Central roadway with cement dam

Das zweite Beispiel bezieht sich auf das Konzept des Kombiausbaus, mit dem zielgerecht
Anker unterschiedlicher Eigenschaften angepaßt an die Geologie und die auftretenden
Gebirgsbewegungen im Streckenmantel angebracht werden. Der Kombiausbau ist ausgelegt
auf Strecken mit sehr hoher Konvergenz, wie
sie etwa in Strecken mit zweifacher Nutzung
auftreten. In den Berechnungen werden die
Konvergenzentwicklung, die Gebirgsbewegungen und die Ausbauverformungen in den Zu-
ständen ohne Abbauwirkung - als Auffahrkonvergenz bezeichnet - , nach erstem Streb-
durchgang und nach zweiter Nutzung ermittelt.
Die Bilder 5 bis 7 beziehen sich auf diese Berechnungen. Es wird in Abhängigkeit von den
Gebirgsbewegungen die Belastung eines jeden
Ankers in axialer und Querrichtung ermittelt.

durchgeführt worden, mit denen die Ergebnisse
der Versuche optimiert wurden. Bild 4 zeigt den
Verformungszustand der Strecke nach Absen-
kung des Gebirges um 2/3 der Flözmächtigkeit.
Der Schwerpunkt der Untersuchungen lag auf
der Dammgestaltung und der Stoßsicherung.
Fragestellungen zur Hangendsicherung bleiben
in der jetzigen Betrachtung unberücksichtigt.
Die Stöße zeigen im Konzept mit starrem
Damm erhebliche Bruchverformungen, so daß
die Standsicherheit nicht gegeben ist. Für diesen Fall ist eine Sicherung der Liegendstöße
nötig.

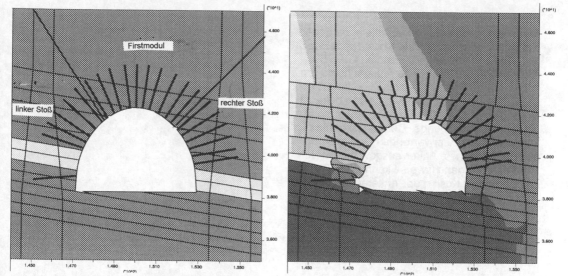

Bild 5: Anordnung der Anker und Lage des Flö-
zes im Streckenquerschnitt
Position of the bolts and the coal seam

Bild 7: Verformungszustand nach 30% Konvergenz
Deformations after 30% convergence

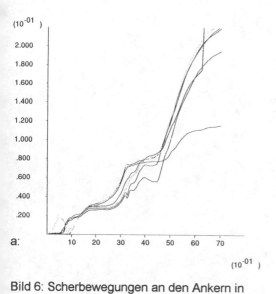

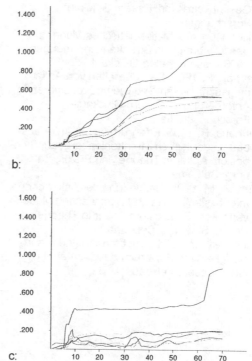

Bild 6: Scherbewegungen an den Ankern in
Abhängigkeit von der Firstabsenkung für
a: Firstmodul, b: rechter Stoß, c: linker
Stoß
Shear displacements of the bolts in dependance of the roof subsidence
a: roof modul, b: right face, c: left face

357

Dabei spielen aufgrund der ausgeprägten Trennflächen im Gebirge vor allem die Scherbewegungen eine zentrale Rolle. Nach Erreichen der Endkonvergenz kann aus den Berechnungen das optimierte Ankerschema ermittelt werden. Für jede Ankerposition sind die Belastungen und Scherbewegungen bekannt. Unter Verwendung der für jeden Ankertyp verschiedenen Spannungs- Dehnungskennkurven können für die zu erwartenden Gebirgsverformungen geeignete Anker ausgewählt werden. Mit dieser Vorgehensweise kann unter Berücksichtigung der Kosten der Ankerausbau in Bereiche hoher Streckenverformungen eingebracht werden. Dabei werden die älteren Anforderungen der Kluftkörpertheorie zwangsläufig mitberücksichtigt.

LITERATUR

Götze, W. 1968. Bruchvorgänge und - verformungen in Strecken im geschichteten Gebirge, Ergebnisse aus Modellversuchen und Untertagebeobachtungen, Berlin: TU

Jacobi, O. 1981. Praxis der Gebirgsbeherrschung, Essen: Verlag Glückauf

te Kook, J. 1986. Konvergenzvorausberechnungen und Ausbaubemessung für Gesteinsstrecken, Essen: Glückauf-Forschungshefte

Müller, W. 1993. Mathematisches Modell zur Beschreibung von Streckenverformungen, 1993, Essen: Verlag Glückauf

Stephan, P. 1985. Das Verhalten von Ankern im geschichteten Gebirge und resultierende Anforderungen, Essen: Glückauf-Forschungshefte

Witthaus, H. 1995. Erfahrungen mit Ankerausbau im deutschen Steinkohlenbergbau, Symposium Anker in Theorie und Praxis, Salzburg: Balkema

Würtele, M. 1995. Darstellung der Ankertechnik mit physikalischen und numerischen Modellverfahren, Symposium Anker in Theorie und Praxis, Salzburg: Balkema

Zischinsky, U. 1984. Bruchformen und Standfestigkeit von Stollen im geschichteten Gebirge, Essen: Felsbau, Nr. 3

Einsatz von neuartigen Ankermörteln mit hoher Frühfestigkeit beim Bauvorhaben Galgenbergtunnel

Application of new types of high early-strength bolt-grout in the Galgenberg tunnel project

Rudolf Röck
Schretter & Cie, Portlandzement- und Kalkwerk, Vils

Peter Schwab
ARGE Tunnelbau Galgenberg, Bauunternehmen Stettin, Leoben

Manfred Blümel
Institut für Felsmechanik und Tunnelbau, TU Graz, Austria

Zusammenfassung
Anhand eines Baustellenversuchs wird die Praxistauglichkeit eines
neuen Ankermörtels auf Basis eines schnellerhärtenden Portlandzements
dargestellt. Bei Einsatz dieses Ankermörtels ist es möglich, bereits
nach einer Stunde teilzuspannen. Weiterführende Laboruntersuchungen
belegen die vorteilhaften Eigenschaften dieses Mörtels bezüglich
Ausziehverhalten unter dynamischen, die Gebirgsbewegung simulierenden
Bedingungen. Darüberhinaus wird der Einfluß der Ankerprofilierung und
jener von Erschütterungen, wie sie beim Sprengvortrieb vorkommen,
untersucht.

Abstract
This paper treats the results of a building site test to determine the
practical suitability of new types of bolt-grout on the basis of a rapid
hardening portland cement. These bolt-grouts permit prestressing already
after one hour. Follow-up laboratory tests demonstrate their advantageous
properties with regard to tensioning behaviour under dynamic conditions
simulating the typical movements of rock strata. A further point of investigation is the influence of varying bolt profiles and as well of vibrations as they occur when driving a tunnel by means of blasting.

Allgemeines

Die äußerst schwierigen Gebirgsverhältnisse, wie sie beim Bau des
Galgenbergtunnels vorgefunden wurden, gaben den Anstoß zur Weiterentwicklung bestehender und Erprobung neuartiger Ankermörtel. Über-
raschenderweise ergab eine kurze
Durchsicht der Standardliteratur
auf dem Gebiet der Ankertechnik
zwar den Bestand zahlreicher Untersuchungen über das Tragverhalten mörtelgebetteter Anker
aber so gut wie keine Hinweise
darauf, wie die Eigenschaften des
Ankermörtels auf die Wirksamkeit
der Anker Einfluß nehmen, ge-
schweige denn darauf, welche Anforderungen überhaupt an den An-

kermörtel zu stellen sind. Auch
die diversen Normenwerke über Anker enthalten diesbezüglich keinerlei Hinweise.Es wird lediglich
vereinzelt Sulfatbeständigkeit gefordert. Man hat sich offenbar damit begnügt, daß der Ankermörtel
irgendwie erhärtet und zwar unge-
stört. In stark verformungsfreudi-
gem Gebirge mit hoher Überlagerung
trifft die Bedingung der ungestörten Erhärtung jedoch nicht zu. Bereits in der Erstarrungsphase kann
es dabei zu nennenswerten Relativ-
verformungen zwischen Ankerstab,
Mörtel und Gebirge kommen, welche
unter Umständen den Abbindeprozeß
und damit das Tragverhalten des
Ankers negativ beeinflussen können.

Die Notwendigkeit, beim Durchör- tern solch schwieriger Gebirgs- formationen am Galgenberg die An kerbelastbarkeit auf einen mög- lichst frühen Zeitpunkt vorzuver- legen, führte dazu, daß am Insti- tut für Felsmechanik und Tunnelbau der Universität Graz die dafür gängigen Vorgangsweisen einer kri- tischen Überprüfung unterzogen wurden. Es zeigte sich dabei, daß mit dem Einsatz von Beschleuniger- patronen unter diesen Verhält- nissen nur unbefriedigende Ergeb- nisse erzielbar waren und die Vor- teile gegenüber dem herkömmlichen Ankermörtel, was den Haftverbund anbelangt, in der Praxis kaum er- faßbar waren.

Aufgrund dessen wurde an uns als Mörtelhersteller der Wunsch heran- getragen, die Frühbelastbarkeit der Anker durch verbesserte Anker- mörtel zu erhöhen bzw. sicherzu- stellen.

Neuentwicklungen

Da die Zeit drängte, zielten die ersten Versuche darauf ab, bei der Herstellung der Mörtel durch Sen- kung des W/Z-Wertes und Verwendung von Zement der Festigkeitsklasse PZ 475 eine Beschleunigung der Bindezeiten und der Früherhärtung zu erwirken. Die Versuchsergebnis- se, die ich hier nicht im Detail wiedergeben möchte, waren zwar vielversprechend und wiesen den Weg in die richtige Richtung, konnten jedoch nicht restlos über- zeugen. Der Beginn der Ankerbe- lastbarkeit konnte zwar um ein bis zwei Stunden vorverlegt werden, entsprach aber noch nicht unseren Zielvorstellungen.
Die Wunschvorstellung der Baustel- le war, daß die starken Gebirgs- verformungen bereits nach wenigen Stunden zu einer Lastaufnahme durch die eingesetzten SN-Anker führen sollten, bzw. eine

Tabelle 1. *Festigkeitsvergleich verschieden eingestellter Ankermörtel auf Basis schnellerhärtendem Zement und Normalzement.*
Table 1. *Comparison of strengths of differently proportioned boltgrouts on the basis of rapid hardening cement and ordinay portlandcement.*

	Schnell - AM			AM
	1	2	3	4
Beginn Minuten	35	70	125	175
Ende Minuten	45	85	155	250
l/kg Wasserbedarf	182,5	180,0	177,5	200,0
ABM cm	26,5	26,0	26,5	26,0
kg/l Rohdichte	2,243	2,236	2,249	2,157

	N/mm2		N/mm2		N/mm2		N/mm2	
	BZF	DF	BZF	DF	BZF	DF	BZF	DF
3 Stunden	1,1	3,8	1,0	3,6	0,6	2,9	0	0
6 Stunden	1,2	4,3	1,2	4,2	1,4	4,9	0	0
12 Stunden	3,0	13,0	2,3	10,1	1,9	7,0	3,6	15,3
24 Stunden	5,0	26,0	5,4	27,1	5,1	27,4	5,0	29,9
3 Tage	8,7	44,3	9,2	43,7	9,1	48,8	7,5	42,3
7 Tage	9,8	53,8	9,3	50,1	9,9	56,1	7,8	54,9
21 Tage	10,1	60,6	9,4	56,3	10,4	64,5	8,2	58,6
28 Tage	9,7	62,1	9,4	59,6	10,5	64,8	8,4	62,3

Teilspannung der Anker bereits nach 5 Stunden möglich sein sollte. Dafür war ein schneller-härtender Ankermörtel notwendig, der auch unter ständiger Gebirgsbewegung zur Kraftübertragung zu möglichst frühem Zeitpunkt in der Lage sein müßte.

Da in unserem Haus bereits seit über einem Jahrzehnt Erfahrung mit Schnellzementen zur Herstellung von Spritzbeton vorliegt, konnten wir auf unsere diesbezüglichen Versuchsergebnisse zurückgreifen und in sehr kurzer Zeit durch geeignete Modifikation des Spritzbeton-Schnellzementes eine passende Variante für den gewünschten Ankermörtel entwickeln.

Dieses spezielle System von schnellabbindendem Portlandzement läßt sich ziemlich exakt in weiten Bereichen auf eine bestimmte Erstarrungszeit einstellen. Wie bei allen Zementen ist die Erstarrungszeit abhängig vom W/Z-Wert

aber auch, wenn auch in eher geringem Maße, von der Verarbeitungstemperatur. Diese Gruppe von Ankermörteln, hergestellt aus besagten Schnellzementen, sind in Tabelle 1 dargestellt. Dabei läßt sich unschwer erkennen, daß diese sich im wesentlichen lediglich im Erstarrungsverhalten unterscheiden. Die übrigen Parameter sind praktisch ident.

Baustellenversuche

Zur Abklärung der Gebrauchstauglichkeit eines schnellen Ankermörtels war es notwendig, die geeignete Erstarrungszeit in einem praxisgerechten Baustellenversuch unter Tunnelbetrieb zu ermitteln. Zweites Versuchsziel war es, die Auszugsfestigkeit der 4 m langen SN-Anker (Durchmesser 21 mm) nach verschiedenen Zeiten zu messen.

Zu diesem Zweck gelangten drei

Tabelle 2. Ergebnisse der Ankerausziehversuche
Table 2. Results of the bolt-tensioning tests

Bohrlochdurchmesser 45 mm			SM - Ankerdurchmesser 21 mm, Ankerlänge 4m			
Bohrloch	Erstarrungsbeginn in Minuten	Alter bei Zugversuch Stunden	Manometerdruck Bar	Kraft kN		Bemerkungen
7	20	1:20	410	239		
8	20	1:20	500	293		Höchstlast, Anker konnte nicht gezogen werden.
9	70	1:35	460	269		Wasser im Bohrloch. Gewinderiß am Anker. Anker konnte nicht gezogen werden.
9*	70	2:00	280	224		Wasser im Bohrloch. Anker wurde durch den Gewinderiß bereits losgeschlagen.
2	20	7:40	485	283		Wasser im Bohrloch.
* Neuerlicher Zugversuch mit 60 Tonnen - Zylinder durchgeführt						

verscheidene Ankermörtel mit unterschiedlichen Erstarrungszeiten auf die Baustelle Galgenbergtunnel. Die Erstarrungszeiten waren mit 35, 70 und 125 Minuten, bestimmt nach Laborbedingungen, eingestellt. Die Laborbedingungen bezogen sich auf ein Ausbreitmaß von 26 cm, d.h. sehr weich bis fließend und 20°C Verarbeitungstemperatur. Wie aus Tabelle 2 zu entnehmen ist, ergaben die Versuche vor Ort zusammengefaßt etwa folgendes Bild:
1. Die Wasserdosierung an der Mörtelpumpe war geringer als im Labor angenommen, was dazu führte, daß die Erstarrungszeit deutlich verkürzt wurde. Möglicherweise hat die hohe Scherbeanspruchung in der Mischschnecke ebenfalls dazu beigetragen, die Erstarrungszeit zu verkürzen.
2. Die Verarbeitungstemperatur von 8°C im Tunnel hat sich offenbar nicht verzögernd gegenüber der Labortemperatur von 20°C ausgewirkt.
3. Die arbeitstechnisch untere Grenze der Verarbeitungszeit des Mörtels unter Baustellenbe-

dingungen soll 45 Minuten nicht unterschreiten.
4. Bei allen Ausziehversuchen konnte bereits 1/2 Stunde nach Erstarrungsende, was etwa 1,5 bis 2 Stunden nach Mischbeginn entspricht, nahezu die maximale Ankerspannung von 290 kN erzielt werden, sodaß die für spätere Termine vorgesehenen Ausziehversuche entfallen konnten.
5. Schräg nach unten gebohrte Ankerlöcher bergen die Gefahr der Wasseransammlung in sich, was zu deutlicher Verminderung des Haftverbundes und zu W/Z-Werterhöhung führt.
6. Bei einem Versuch (Bohrloch 9) gab es bei der Spannkraft von 260 kN einen Gewinderiß. Beim neuerlich durchgeführten Ausziehversuch versagte der Anker schon bei 224 kN. Damit konnte unfreiwillig die negative Wirkung starker Erschütterungen demonstriert werden.
Mit Abschluß dieser Versuche war die Praxistauglichkeit dieses Systems von Ankermörteln basierend auf schnellbindende Portlandzemente erwiesen. Es folgten nun-

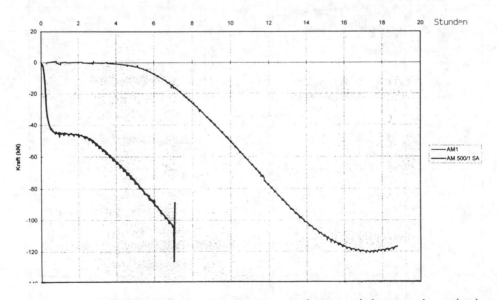

Bild 1. Vergleich der Kraftaufnahme beim Ausziehversuch zwischen schnell erhärtendem und herkömmlichen Mörtel.
Figure 1. Comparison of the taking up of force by rapid hardening and conventional grout during the tensioning test.

mehr umfangreiche Untersuchungen
am Institut für Felsmechanik und
Tunnelbau an der TU Graz über das
Auszugverhalten der neuen Mörtel
unter dynamischen, die Gebirgsbewegung simulierenden Bedingungen.
Das Ziel dieser Untersuchungen war
einerseits das Ermitteln kritischer Deformationsgeschwindigkeiten, unter welchen die Funktion
des Verbundsystems noch sichergestellt ist. Andererseits wurde
versucht das System Mörtel – Ankerstab für verformungsfreudiges
Gebirge hinsichtlich Mörtelzu-
sammensetzung und Ankerprofilierung zu optimieren.

Versuchsaufbau und -durchführung

Ein mit Beton (B300) gefülltes
Stahlrohr(St 37 101,6x2,9;l=500mm)
wird mit einer Kernbohrung
(da=57mm), welche das Ankerloch
nachbilden soll, versehen. Das
Versuchsrohr besitzt am unteren
Ende einen Deckel mit Dicht-
lippendurchführung um ein Austre-
ten des Frischmörtels zu verhindern. Nach dem Abmischen wird der

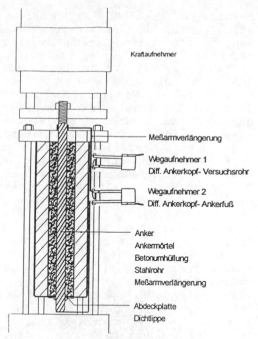

Bild 2. Versuchsaufbau
Figure 2. Testing apparatus

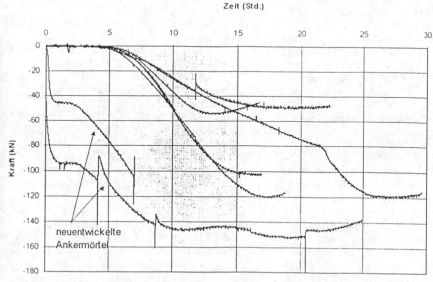

Bild 3. Arbeitslinien mit verschiedenen Ankermörteln und verschiedenen An-
kerprofilen.
Figure 3. Time force-diagramms of bolt-tensioning tests with different
bolt-grouts and bolt-profiles.

Ankermörtel in das Versuchsrohr gefüllt und der Anker versetzt. Danach erfolgt der Einbau in eine servohydraulisch gesteuerte Prüfanlage. Die Vorlaufzeit vom Abmischen bis zum Versuchsstart beträgt ca. 20 Minuten. Die Regelung des Prozeßablaufes wird weggesteuert durchgeführt. Zur Wegerfassung sind zwei elektronische Wegaufnehmer installiert, die die Verformung Ankerkopf-Versuchsrohr bzw. Ankerkopf-Ankerfuß messen. Die Wegrate wird mit Hilfe eines 32 Bit-Prozessors generiert und die Regelung im "Closed Loop" mit einer Abtastrate von 5000 Hz durchgeführt (Bild 1). Die bisherigen Versuche wurden mit einer Ausziehgeschwindigkeit von 0.012 mm/min durchgeführt.

Das entspricht, umgelegt auf einen 4 m Anker, einer radialen Ersttagesverformung von ca. 7 cm, unter der Annahme, daß der differentielle Verformungsbetrag linear über die Ankerlänge ist. Zur weiteren Untersuchung werden die Anker nach Versuchsende überbohrt und geschlitzt .

Im Bild 3 sind einige Arbeitslinien von Ausziehversuchen dargestellt. Bei diesen Versuchen betrug, wie bereits erwähnt, die Ankerausziehgeschwindigkeit konstant 0,012 mm/min. Variiert wurde hierbei der Ankermörtel bzw. der Anker (unterschiedlicher Rippenabstand und Rippenhöhe). Deutlich erkennbar ist die schnelle Kraftaufnahme der beiden Versuche mit dem neuentwickelten Ankermörtel.

Um den physikalischen Mechanismus des Verbundes, welcher im wesentlichen aus Haftung, Reibung und Formschluß bestimmt wird, in Abhängigkeit der Zeit, des Ansteifverhaltens, der Verformungsgeschwindigkeit und Ankerprofilierung richtig interpretieren zu können, bedarf es noch weiterer Versuche. In weiterer Folge sind auch Versuche mit Ausziehbeschleunigung (alle 6 - 8 Stunden) geplant, welche erhöhte Deformationsgeschwindigkeiten bei Abschlägen simulieren.

Literatur

Maidl. Handbuch des Tunnelbaus. Verlag Glückauf. Essen 1994.

Habenicht, R. Anker und Ankerungen zur Stabilisierung des Gebirges. Springer-Verlag 1976.

XXXV. Geomechanikkolloquium. Diskussion zum Thema "Anker im Hohlraumbau". Salzburg. Felsbau 5 (1987) Nr. 1.

Jirovec, P. Untersuchungen zum Tragverhalten von Felsankern. Veröffentlichung des Inst. f. Boden- und Felsmechanik. Uni Karlsruhe. Heft 79. 1979.

Mayer, G. Untersuchungen zum Tragverhalten von Verpressankern in Sand. Veröffentlichung des Grundbauinstitutes der TU Berlin. Heft 12. 1983.

Soretz, S. Einige Bemerkungen zur Berechnung und Ausführung von Stahlverankerungen im Fels. Geologie und Bauwesen. Sonderabdruck Jahrgang 21. Heft 4. 1955.

Schubert P., Das Tragverhalten des mörtelversetzten Ankers unter aufgezwungener Kluftverschiebung. Bericht Nr. 4/94. Institut für Konstruktiven Tiefbau. Montanuniversität Leoben.

Rehm, G. Kriterien zur Beurteilung von Bewehrungsstäben mit hochwertigem Verbund. Stahlbetonbau. S. 79-96 W. Ernst und Sohn. Berlin 1989.

Farmer, I.W. Stress Distribution along a Risin Grouted Anchor. Int. J. Rock Mech. Min. Sci. and Geomech. Abstr.. Vol 12 pp 347-351. Pergamon Press GB.

Coates, D.F. + Yu Y. S. Three-Dimensional Stress Distributions Around a Cylindrical Hole and Anchor. Proc. 2 nd Cong. Int. Soc. Rock Mech. Belgrad 1970/3.

Feder, G. Zur Wirkungsweise und Gestaltung voll eingemörtelter Stabanker. Tunnel Mai 1982. Bertelsmann GmbH.

Feder, G.+ Zitter, H. + Schubert, P. + Kohlbach, K. Verhalten von voll vermörtelten Felsankern in Bereichen vollständiger und unvollständiger

Mörtelumhüllung. Berg- und
Hüttenmännische Monatshefte.
Jahrgang 129. Heft 5. 1984.
Springer-Verlag. Wien.
Fuller, P. G. + Cox, R.H.T.
Mechanics of Load Transfer from
Steel Tendons to Cement Based
Grout. Fifth Australian
Conference on the Mechanics of
Structures and Materials. Proc.
Melbourne 1975 pp, 189-203.
Sprang, K. Beitrag zur rechneri
schen Berücksichtigung voll-
vermörtelter Anker bei der
Sicherung von Felsbauwerken in
geschichtetem oder geklüfteten
Gebirge. These No. 740 (1988).
Ecole Polytechnique Federale de
Lausanne.

Anchors in Theory and Practice, Widmann (ed.)© 1995 Balkema, Rotterdam. ISBN 90 5410 577 1

Sand-anchors, theory and application

Sandanker, Theorie und Anwendung

S. B. Stazhevsky
Institute of Mining, Novosibirsk, Siberian, Division of Russian Academy of Sciences, Russia

D. Kolymbas
Institute of Geotechnics and Tunnelling, University of Innsbruck, Austria

I. Herle
Institute of Soil- and Rock Mechanics, University of Karlsruhe, Germany

Abstract
A new type of anchors is presented based on the wedge action of sand. In contrast to traditional anchors, the force is transmitted to the ground not by means of cement but by means of sand. The new type anchors can be built in very quickly and they can be immediately prestressed. They are applicable in a wide range of soft to hard rock and they can be easily removed for inspection. A simplified calculation of the bearing capacity is also presented.

Zusammenfassung
Ein neuer Ankertyp wird vorgeschlagen, bei dem die Verankerung, d.h. der Kraftschluß zwischen dem Stahlzugglied und dem umliegenden Gebirge nicht durch Zement oder Harz, sondern durch Sand bewerkstelligt wird (siehe Bilder 1 und 4). Sandanker zeichnen sich dadurch aus, daß sie sehr schnell installiert werden können und sofort belastbar sind. Den größten Teil der Installationszeit beansprucht das Bohren. Die anschließende Installation z.B. eines 2,2 m langen Ankers in ein Bohrloch mit dem Durchmesser von 42 mm und die Einbringung des Sandes haben nur 20 Sekunden in Anspruch genommen. Die Vorspannung dieses Ankers wurde innerhalb einer Minute bewerkstelligt. Die Korngröße des Füllmaterials (Sand bzw. Bohrgut) richtet sich nach der Größe des Ringspalts zwischen Stahlzugglied und Bohrloch. Die Sandanker bieten folgende Vorteile: (1) Einfache, zuverlässige und schnelle Installation, (2) Niedrige Kosten, (3) Großer Anwendungsbereich (intakter, verwitterter und geklüfteter Fels, steifer Boden), (4) Sofort nach Installation belastbar, (5) Reversibel, d.h. der Anker kann zur Kontrolle ausgebaut und wiedereingebaut werden, (6) Keine Verwendung von Zement oder chemischen Klebern, daher problemlos bei Sanierung historischer Bauwerke verwendbar. Die Sandanker wurden in zwei Fallbeispielen bereits erfolgreich getestet: Bei der Sanierung einer Felshöhle bei Ivanovo (Bulgarien), in der sich eine Kirche aus dem XIV. Jahrhundert befindet (UNESCO-Denkmal) und bei der Sanierung ähnlicher Höhlenanlagen bei Vardzia (Georgien). Dort hat sich u.a. gezeigt, daß die Vorspannkraft mit der Zeit (Kriechen, Erschütterungen) nicht abgebaut wird. Ist die Bohrlochwand starr, so ist die Ankertragkraft allein durch die Zugfestigkeit des Stahlzuggliedes determiniert. Andernfalls spielt die Nachgiebigkeit (bzw. der Bettungsmodul bzw. die Elastizitätskonstante) des Geomaterials die bestimmende Rolle für die Ankertragkraft, wie eine vereinfachende Analyse zeigt (siehe Gleichung 4).

1 Lay-out of the anchor

Fig. 1 shows the principal lay-out of the anchor. The steel tension rod 1 (tendon) is connected with a steel disk 2 that represents the anchor tip. The annular gap between tension rod and rock is filled with dense sand (3). The anchor top plate 4 is pressed against the rock by means of a screw. The distribution of stresses is shown in Fig. 2. The bearing capacity of the anchor depends on the stiffness of the rock. If the rock is sufficiently rigid, the bearing capacity is only limited by the ten-

sile strength of the tendon. Otherwise the bearing capacity is reduced and the subgrade reaction markedly influences its value.

A technology for the installation of sand anchors has been worked out with laboratory and field tests. It enables to install anchors of up to 10 m length in arbitrary spatial orientation. The duration of installation is mainly determined by the drilling of the borehole. As soon as the borehole is ready, the installation and pre-stressing of the sand anchor can be carried out within a time considerably shorter than for conventional anchors. E.g., the

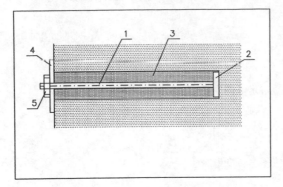

Figure 1: Principal lay-out of the anchor, *Prinzip-skizze. 1: Stahlzugglied, 2: Ankerfußplatte, 3: Sandfüllung, 4: Ankerkopfplatte, 5: Festziehmutter*

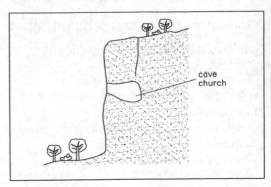

Figure 3: Church near the village Ivanovo (Bulgaria), *Höhlenkirche in der Nähe des Dorfes Ivanovo (Bulgarien)*

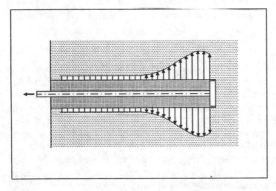

Figure 2: Distribution of stresses, *Spannungsverteilung (ermittelt durch eine spannungsoptische Analyse).*

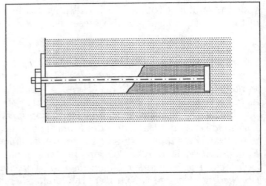

Figure 4: Partfilled borehole, *Teilgefülltes Bohrloch*

installation of an anchor of 2.2 m length within a borehole Ø=42 mm and the subsequent placement of sand took only 20 sec. The duration for the prestressing of the anchor was ca. 1 minute. The diameter of the sand grains should be chosen in accordance to the diameter of the borehole and of the tendon.

2 Case record A

The first application of sand anchors took place in September 1989 in Bulgaria. Near to the village Ivanovo (province of Russe) there is a churche within a cave, see Fig. 3. This churche, built in the XIVth century, has been declared by UNESCO as cultural monument to be conserved. Due to water erosion of the rock an urgent restoration, i.e. strengthening of the rock, has become necessary. The sand anchors installed by the first author have since then been subject to a long term action of humidity, seasonal variation of temperature and seismic activity. The details are as follows:

The unconfined compression strength of the rock was 7.5 - 10 MN/m². The anchors had a length of 1.8 m and the steel tendons had a diameter of 20 mm. The borehole diameter was 42 mm. The pre-stress force was 80 kN. A part of the boreholes has been completely filled with sand (Ø=1 mm), as shown in Fig. 1, but the majority of the boreholes have only been filled with sand up to a length of 30 cm (see Fig. 4). The anchors have been checked a year later (i.e. 1990). No decrease of the pre-stress force was found. This means that humidity, temperature variation and seismicity do not deteriorate the bearing capacity. Based on this results, in August 1990 another 20 anchors have been installed. Since then the restoration works have ceased for economic reasons.

3 Case record B

Similar protection works have been carried out in Georgia: In Upliszikhe, near to the city of Vardzia, there is an ancient town, the houses of which are built within caves. The surrounding rock is partly tuff and partly weathered sandstone. Due to erosion, restoration works became necessary. 14 sand anchors have been installed with lengths varying between 1.5 and 1.8 m. The diameters of the steel tendons and of the boreholes were 20 and 42 mm, respectively. 7 anchors have been prestressed with 60 kN and 7 anchors with 70 kN. No decrease of the prestress-force with time has been detected. Unfortunately, the continuation of the investigations became impossible for economic reasons.

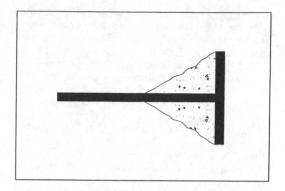

Figure 5: Rigid sand cone after pull-out test, *Verfestigter Sandkegel nach dem Ausziehversuch*

4 Evaluation

According to laboratory tests, the prestress force has not decreased due to vibrations. Protection against corrosion for long term anchors can be provided as usual. Moreover, sand anchors can be easily removed for inspection. If necessary, they can be replaced and then installed and pre-stressed. Thus, the advantages of these new type anchors can be summarized as follows:

1. simple, reliable and quick installation

2. low costs

3. wide range of applicability (i.e. integral, weathered and fissured rock, stiff soil)

4. can be loaded immediately after installation

5. reversible

6. No use of cement or chemical adhesives. Therefore, no hazards to historic monuments

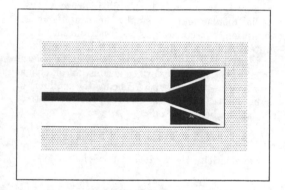

Figure 6: Wedge-mechanism, *Keil-Anker*

5 Fields of applicability

Support in tunneling, mining, excavations, slopes and cuts. Repair and/or replacement of conventional anchors. Applications in cases with prescribed removal of anchors after construction, e.g. in excavations, especially in archaeological ones.

6 Mechanical analysis

Visual inspection of the anchor after its removal from the borehole showed that an almost rigid cone consisting of highly compacted sand was formed on the tip plate, see fig. 5. This fact reveals that the bearing mechanism of the sand anchor is a

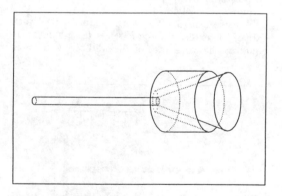

Figure 7: Axisymmetric wedge, *Axialsymmetrischer Keil*

conical wedge. Such wedge-mechanisms are easily conceivable for the plane case and they are in fact applied to anchor rods within boreholes, see fig. 6. However, in the axisymmetric case a solid wedge consisting of a convex and a concave part (see fig. 7) is not feasible, because the cleaving force exerted from the inner (convex) part is carried by the tensile strength of the outer (concave) part such that no lateral force can be exerted upon the borehole. This is, however, not the case for a wedge consisting of sand, because sand has no tensile strength. Based on this concept, the bearing capacity of sand anchor can be analytically estimated as follows. The axial displacement u_z of the inner wedge results in a radial displacement u_r of the outer wedge. The distribution of u_r over r within the outer sand wedge has been investigated with finite elements. To this purpose the code ABAQUS has been used with a hypoplastic constitutive equation [1] to solve the two-dimensional problem of the inflation of a hollow cylinder made from dense sand. At the outer surface of this cylinder a constant subgrade reaction modulus has been assumed. If the sand in the fill is dense and the radial displacement is small, the simplifying assumption of a constant stress distribution on the outer cylindrical surface of the sand fill (see Fig. 8) is reasonable. It implies that the radial displacement of the borehole surface is constant and equal to $u_r = u_z \cdot \tan \alpha$. With K being the modulus of subgrade reaction we obtain thus the radial stress σ_{r0} at the borehole wall:

$$\sigma_{r0} = K \cdot u_r \quad . \tag{1}$$

From elasticity theory it is known that

$$K = \frac{E}{(1+\nu)r} = \frac{E}{(1+\nu)(r_0 + u_r)} \tag{2}$$

with E and ν being Young's modulus and Poisson's ratio, respectively.

Now we consider the work balance. The work rate of the external force F reads $F\dot{u}_z$. It must be equal to the work rate of the wall stress σ_{r0} plus the dissipation rate D. The latter results from the deformation of the outer wedge and the friction at the interface between inner and outer wedge.

$$F\dot{u}_z = \sigma_{r0} \cdot 2\pi(r_0 + u_r)\left(\frac{r_0}{\tan \alpha} - u_z\right)\dot{u}_r + D$$

If we neglect D then we underestimate F (provided $D > 0$), i.e. we obtain a safe estimation of F. Setting thus $D \approx 0$ we obtain

$$F = \sigma_{r0} \cdot 2\pi(r_0 + u_r)\left(\frac{r_0}{\tan \alpha} - u_z\right)\frac{\dot{u}_r}{\dot{u}_z}$$

With $\frac{\dot{u}_r}{\dot{u}_z} = \tan \alpha$ and σ_{r0} according to equations 1 and 2 we obtain

$$\begin{aligned} F &= \frac{2\pi E}{1+\nu} \cdot u_r \left(\frac{r_0}{\tan \alpha} - u_r\right)\tan \alpha \\ &= \frac{2\pi E}{1+\nu} \cdot u_z \left(\frac{r_0}{\tan \alpha} - u_z\right)\tan^2 \alpha \end{aligned} \tag{3}$$

The plot of $F(u_z)$ according to equation 3 is shown in Fig. 9. The maximum force (pull-out force) is obtained from $\frac{dF}{du_z} = 0$ to

$$F_{max} = \frac{\pi E}{2(1+\nu)} \cdot r_0^2 \tag{4}$$

In equation 4 the undetermined angle α does not appear. The bearing capacity after equation 4 was compared with a FEM-calculation using the code ABAQUS with a hypoplastic constitutive equation [1]. Assuming $K = 60\,\mathrm{MN/m^3}$ and $r_0 = 50\,\mathrm{mm}$ the FEM calculation resulted in an overall radial force (i.e. σ_{r0} integrated over the mantle surface) of $R_{max} = 32\,\mathrm{kN}$. Taking a wall friction angle of $42°$ one obtains the pull-out force

$$F_{max} = \tan 42° \cdot 32 = 28.8\,\mathrm{kN}$$

From equation 2 it follows with $\nu = 0.2$:

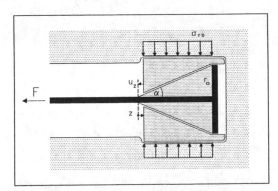

Figure 8: Interaction of wedges (schematically), *Wechselwirkung der Keile (schematisch)*

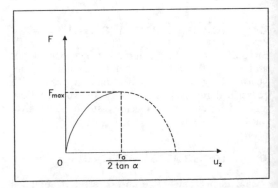

Figure 9: Force F over axial displacement u_z, *Abhängigkeit der Ausziehkraft F von der Verschiebung u_z*

$$E = K(1 + \nu) \cdot r = 60 \cdot 1.2(0.05 + 0.05) =$$
$$= 7.2 \frac{MN}{m^2}$$

and therefrom

$$F_{max} = \frac{\pi \cdot 7.2}{2 \cdot 1.2}(0.05)^2 = 23.6\,\text{kN} \quad,$$

a value which compares well with the FEM result.

References

[1] D. Kolymbas, I. Herle, P. A. von Wolffersdorff: Hypoplastic constitutive equation with back stress. To appear in Int. J. Num. Anal. Meth. Geomechanics

Der wiedergewinnbare Litzenanker System Keller

The Removable Multiple Anchor System Keller

P.Stockhammer
Keller Grundbau Wien, Austria

F.Trummer
Keller Grundbau Graz, Austria

Zusammenfassung:

Durch die Erfindung des wiedergewinnbaren Litzenankers System Keller wurde die
Möglichkeit geschaffen, sämtlichen Spannstahl nach dem temporären Gebrauch der
Anker wiederzugewinnen. Die Anker finden durch die Ausnutzung des Single Bore
multiple Anchor Systems (SBMA) spezielle Anwendung in bindigen Böden. Es können
wesentlich höhere Ankerkräfte in den Boden eingeleitet werden, als dies bei
herkömmlichen Verfahren der Fall ist. Dabei sind Verpreßstrecken bis zu 25 m
möglich.

Abstract:

The Removable Multiple Anchor System allows a complete removal of the steel
tendon form the borehole after use. In utilising the single bore multiple anchor
concept it allows use of very high anchor loads also in cohesive soil with the
efficent use of fixed lengths up to 25 m.
The Removable Multiple Anchor involves the installation of a series of unit
anchorages down the one borehole. The fixied length of the unit anchorages are
located at staggered depths in the borehole and transfer the load from each unit
anchorage in a controlled manner to a discreet length of the borehole. It ensures
a uniform mode of load transfer to the ground and a gross increase in efficiency
in the mobilisation of ground strength. This system also allows the utilisation
of an almost unlimited length of bore over which the load may be transferred from
the multiple of anchorages. The anchors are stressed to develop their full
working capacity.
When the final supports is in position the Removable Multiple Anchor can be de-
stressed and the entire strand tendon removed. Removal is undertaken by a winch
arrangement employing a total pulling force of less than 100 kN to pull a
multiple of strands simultaneously.

1) Einführung

Seit ihrer Erfindung stehen Vorspannanker für die vorübergehende Sicherung
von tiefen Baugruben in Konkurrenz zu
anderen Aussteifungssystemen. Bei Anwendung der Vorspannanker besteht der
große Vorteil insbesondere darin, daß
keine störenden Aussteifungen den Ablauf des Baues behindern und die Baugrube frei zugänglich ist. Im Zusammenhang mit der innerstädtischen Verbauung
und insbesondere der rechtlichen Nutzung von Nachbargrundstücken bestehen
jedoch gewisse Vorbehalte gegen die
Nutzung der temporären Vorspannanker.

Dabei kann meist ein Übereinkommen getroffen werden, wenn durch die Nutzung
des Nachbargrundstückes zur Verankerung
für den Grundeigentümer kein Nachteil
entsteht oder ein allfälliger Nachteil
finanziell abgegolten wird. In der Regel
besteht der Nachteil darin, daß Stahlteile im Boden verbleiben, die bei einer
späteren Bebauung des Nachbargrundstückes Behinderungen für die Baumaßnahme darstellen könnten.

Es wurden daher Systeme entwickelt,
Vorspannanker nach Gebrauch wieder auszubauen, doch sind bisher keine Verfahren von wirtschaftlicher Bedeutung
bekannt, den Spannstahl zu 100 % und
ohne Schadensgefahr wieder aus dem
Boden zu entfernen.

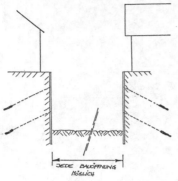

Bild 1 Verankerung für größere Aushubbreiten
Fig. 1 Anchors for large excavations

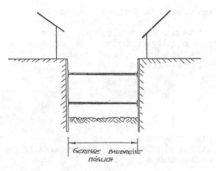

Bild 2 Aussteifung für Aushub geringer Breite
Fig. 2 Bracing of small excavations

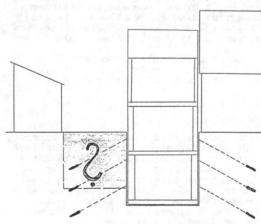

Bild 3 Nachbarschaftsrechte und nachträgliche Verbauung
Fig. 3 Rights of neighbours and later excavations

Die Entwicklung des ''wiedergewinnbaren
Ankers, Litzen – Anker – System Keller''
(''WGL'') ermöglicht die Wiedergewin-
nung des gesamten Spannstahles über die
volle Ankerlänge ohne irgendwelche Er-

schütterungen. Weiters werden die Vorteile des ''SBMA'' (Single bore multiple
anchors) Ankers voll genutzt, womit
auch hohe Einzellasten möglich sind.

2) Das ''WGL'' Anker System

Das SBMA (Single Bore Multiple Anchors)
System beinhaltet den Einbau von
mehreren Ankern in ein Bohrloch. Jeder
Einzelanker besitzt eine eigene Haftstrecke, und eigene Litzendrähte im
Bohrloch und wird mit einer eigenen
Spannpresse simultan mit den anderen im
Bohrloch eingebauten Ankern vorgespannt. Das System erlaubt eine effi-
ziente Anpassung der möglichen Haftspannung des Bodens mit kurzen Haftstrecken (ca. 2,5 m je Anker) des
Einzelankers. Die gesamte Haftstreckenlänge kann den erforderlichen Ankerkräften angepaßt werden.

(Haftstreckenlängen bis 25 m sind schon
ausgeführt worden.)

Beim WGL-Anker, der ausbaubaren Version
des SBMA-Ankers wird der Litzendraht in
einem mit Korrosionsschutzmasse gefüllten Hüllrohr geführt und am unteren
Ende der Verankerungsstrecke um 180 °
umgelenkt.

Die Herstellung dieser vorgefertigten
Umlenkung des Litzendrahtes erfolgt
über einen kleinen Stahlsattel, welcher
die Kraft auf einen Druckstab überträgt, der wiederum die Kraft auf den
Zementstein und weiter in den Boden ableitet. Die Entwicklung und Herstellung
des Sattels und des Druckstabes erfolgte im mehrjährigen Versuchsreihen
im Labor und auf Baustellen, um
wichtige Details zu klären. Die Umlenkung erfolgt mit speziellen Werkzeugen
und wird, ebenso wie die gesamte Fabri-
kation des Einzelankers, unter werk-
mäßigen Bedingungen ausgeführt.

Jeder Einzelanker hat eine definierte
zulässige Ankerkraft der Stahllitze

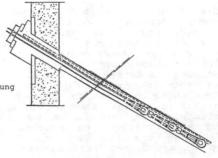

Bild 4 WGL-Anker System
Fig. 4 WGL-Anchor System

entsprechend Durchmesser und Stahlgüte.
Die Versuche an der Umlenkung haben
gezeigt, daß die Bruchgrenze mit 90 %
der maßgeblichen Stahlgüte angesetzt
werden kann.

Bei Testankern sollte die max. Prüflast
80 % der Streckgrenzenlast nicht übersteigen. Für die zulässige Ankerkraft
ist, ausgehend von 95 % der Streck-
grenzlast, der jeweilige Sicherheitsfaktor lt. Ö-Norm anzusetzen. Weiters
muß gewährleistet sein, daß die Ankerkraft auch von der Grenzfläche Zementstein – Bohrlochwand aufgenommen wird.

Die Haftstrecke bewegt sich zwischen
2 m und 5 m Länge.

Von wesentlicher Bedeutung ist der
Einfluß der am Sattel und der Druck-
stange anliegenden Litzen auf die
Kraftübertragung der Einzelhaftstrecke.

Der Einfluß der vorbeilaufenden Stahl-
drähte wurde eingehendst untersucht und
führte zu speziellen Konstruktionen, um
ungünstige Auswirkungen auf den Bereich
der Haftstrecke zu beseitigen.

3) Der Einbau des WGL Ankers

Der Einbau des Ankers erfolgt in verrohrte oder unverrohrte Bohrungen,die
nach allen bekannten Bohrverfahren hergestellt werden. Die üblichen bei der

Ankerherstellung nötigen Bohrdurchmesser sind für den Einbau geeignet.
Die Injektion mit Zement erfolgt über
die Verrohrung während des Ausbaues der
Verrohrung oder über eigens instal-
lierte Injektionsschläuche. WGL-Anker
System Keller sind auch mit Nachinjek-
tionsystem bestellbar. Als Injektionsgut kommt Zementsuspension oder Zementmörtel in Frage, die Mischungsver-
hältnisse entsprechen den üblichen Vorgaben bei der Ankerherstellung.

4) Die Prüfung und Vorspannung des WGL Ankers

Die Prüfung und Vorspannung erfolgt je
nach Wahl der Zementsorte zwischen 3
und 14 Tagen nach Injektion der Anker.
Für die Vorspannung wurde ein eigenes
Spannsystem, welches den unterschiedlichen Dehnlängen Rechnung trägt, entwickelt.

Die Vorspannung erfolgt mit hydrau-
lischen Einzelpressen, welche synchron
und simultan arbeiten. Beim WGL System
werden die zwei Enden der Litzen-
schlaufe eines jeden Einzelankers durch
eine Hohlkolbenpresse geführt. Dadurch
wird sichergestellt, daß es keine Bewegung über die Umlenkung des Sattels
gibt, sondern nur eine Vorspannung erfolgt, wenn beide Teile der Litze über
eine Presse gezogen werden.

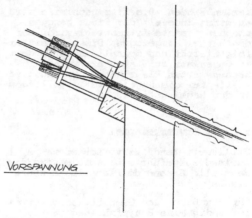

VORSPANNUNG

Bild 5   Vorspannung
Fig. 5   Prestressing

Die Einzelanker werden zyklisch und
simultan gespannt. Die Kraftprüfung
erfolgt nach Vorgabe der Normen.

Es werden die Dehnungen und Kriechmaße
geprüft und aufgezeichnet.

Nach Beendigung der Prüfung werden die
Anker durch Verkeilen der einzelnen
Litzen festgelegt. Der erforderliche
Litzenüberstand bleibt für den Rückbau
erhalten.

5) Der Ausbau des WGL- Ankers

Nach dem Gebrauch der WGL-Anker werden
die Litzen entspannt. Die Entspannung
kann mit jeder geeigneten Hohlkolbenpresse erfolgen. Der Ausbau der Einzellitzen erfolgt mit speziell konstru-

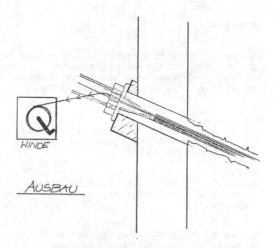

HINDE

AUSBAU

Bild 6   Ausbau WGL-Anker
Fig. 6   Removal of WGL-Anchor

375

ierten Winden. Die Litzenschleife wird
an einem Ende mit der Winde gezogen,
und der gesamte Litzenstahl wird über
den Sattel ausgezogen. Diese Arbeit erfolgt leicht und schnell und erfordert
im Gegensatz zu Stabankern keinen
nennenswerten Platzbedarf. Die Ausziehkräfte für die umgelenkte Litze liegen
in der Regel unter 10 to.

6. Ausführungsbeispiel

Stellvertretend für verschiedene im
Ausland ausgeführte Beispiele wird die
Baustelle Greenside Place in Edinburgh
erwähnt.

Für ein Büro- und Geschäftshaus war es
notwendig eine Bohrpfahlwand in fünf
Ankerebenen mit Vorspannanker für die
erforderliche Baugrube zu sichern.

Die Forderung bestand darin, daß ein
Teil der Anker nach Gebrauch vollständig rückgebaut werden mußte.

Insgesamt wurden 85 SMBA Anker und
25 Stück WGL-Anker System Keller in
bindigem Baugrund eingebaut. Die Gebrauchslasten bewegten sich zwischen
450 - 600 kN, die Ankerabstände betrugen 1,3 - 4,0 m.

Der Spannstahl sämtlicher eingebauter
WGL-Anker wurde zu 100 % wieder gewonnen.

Erste Versuche zur Anwendung wurden
auch in Österreich vorgenommen, welche
die in das Ankersystem gesetzten Erwartungen voll erfüllen. Von wesentlichem
Interesse bei der Anwendung dieses
neuen Ankersystems ist die hohe Belastbarkeit von bis zu 150 to und die verläßliche Rückbaubarkeit des gesamten
Spannstahles.

Literatur:

Ref 1
Barley A. D. 1995. Theory and practice
of the single bore multiple anchor
system. Symposium on ''Anker in Theorie
und Praxis'' Salzburg, October 1995.

Ref 2
Barley A. D. 1993. The capacity of pre-
bend pre-stressing strands around tight
radii. Confidential.

Ref 3
Barley A. D. 1993. The load transfer
capacity of grouted steel saddles.
Confidential.

Ref 4
Barley A. D. 1993. Load transfer
capacity of grouted plastic bars.
Confidentional.

Rockbolt anchors for high convergence or rockburst conditions

Felsanker für große Gebirgsverformungen oder unter Bergschlag-Verhältnissen

Dwayne D.Tannant
Geomechanics Research Centre, Laurentian University, Sudbury, Ont., Canada

ABSTRACT: Pull-out tests were conducted to measure the load-displacement response of fully grouted rebars, cone bolts and smooth-walled bolts and to assess their use in highly converging and burst-prone ground. Well-grouted rebars are stiff and fail at loads of 170 kN before being pulled 20 mm. Cone bolts can tolerate between 100 and 150 mm of displacement in resin grout before failing at loads of 160 kN. The behaviour of smooth bolts is sensitive to the bond length and grout properties.

ZUSAMMENFASSUNG: In den tieferen Bergwerken Kanadas sind die hohen Belastungen Ursache für die Verformungen und ein Versagen des Gebirges. Manchmal tritt das Versagen in Form eines Gebirgsschlages auf. Konventionelle Felsanker funktionieren bei mäßigen Verformungen gut, versagen jedoch bei größeren Verformungen. Bei großen Gebirgsverformungen müssen spezielle Felsanker verwendet werden, die bei Belastung gleiten und sich dehnen können und so das Gebirge stützen und absichern.
In der Creighton Mine der INCO Ltd. wurden Versuche durchgeführt, die Tragkraft von Ankern bei deren Verformungen zu messen.
Es erfolgten Versuche mit Betoneisenstangen, Kegelbolzen und glatten Bolzen. Die Gleit- und Dehnfähigkeit wurde bei verschiedenen Belastungen gemessen.
Voll einzementierte Betoneisenstangen sind starr und brechen bei 170 kN bei maximal 20 mm Dehnung. Kegelbolzen, in Resin einzementiert, gleiten und dehnen sich 100 - 150 mm bevor sie bei 160 kN brechen. Die Tragfähigkeit von glatten Bolzen ist von der Haftlänge und den Hafteigenschaften des Bindemittels abhängig.

1 INTRODUCTION

In the deeper hardrock mines of Canada, the rockmass and stress conditions can lead to extensive rock fracturing and dilatation resulting in high ground convergence around mine openings and drifts. In some cases, the convergence occurs violently and rapidly in the form of rockbursts. As these mines proceed to greater depths, the ground control problems will become worse and new rock support systems must be developed and used.

Conventional rockbolts such as resin-grouted rebars and mechanical, end-anchored rockbolts have proved highly effective under most ground conditions, even for drifts in highly fractured and stressed ground where moderate ground convergence occurs. However, when these rockbolts are used in tunnels or mine drifts that experience large ground convergence or rockbursts, their limited ability to yield often results in their failure. Within economic and practical limits,

ground support cannot be used to prevent the ground movement and rock fracturing. Under these conditions, the primary support design philosophy is not to enhance the stiffness and strength of the rockbolts to resist or prevent the ground movement. Instead, the rockbolts must be designed to yield or slide with the ground movements while simultaneously providing a substantial resistive force, thereby helping to control rock displacements and minimizing damage to the excavation.

The need for yielding tendons or rockbolts for use in highly stressed and rockburst-prone drifts has long been recognized in the deep gold mines of South Africa (Ortlepp 1970; 1983). This has lead to the development of recommended performance requirements for yielding rock tendons (Wojno et al. 1987) and the development and testing of yielding rockbolts (Ortlepp 1970; 1983; 1992; Tannant and Buss 1994). A yielding rockbolt based on a fully grouted deformed bar with a sliding nut near the plate has been developed for use in burst-prone ground (Spann 1987). This sliding nut concept was also adapted for squeezing ground and openings subject to high convergence (Herbst 1990). Tannant and Kaiser (1995) discuss the use of Swellex bolts to provide frictional anchorage of wire rope or cables. If the overlap between the rope and the Swellex bolt is kept below a critical length (about 1.5 to 2 m) then when the rope is loaded it can slide past the Swellex bolt without breaking.

The fracturing and dilatation of the rockmass during tunnel convergence and the ejection of rock during a rockburst generally cause axial loading of the rockbolts. Often these loads are sufficent to fail the bolt. Tests in which the rockbolt is simply pulled out of the ground is used to represent these types of loading conditions.

2 PULL TESTS

In situ pull-out tests were conducted at INCO Limited's Creighton mine over a two year period to measure the load–displacement characteristics of grouted bolts, some of which were designed to yield by anchor slip. These tests were aimed at determining the most appropriate bolt for use in highly converging and burst-prone ground. Five types of bolts were tested and the peak load capacity and the displacement at peak were tabulated.

2.1 Bolt types tested

The following bolt types were tested:
A Conventional resin-grouted rebars to provide a baseline for comparison with the other tests.
B Rebars installed in boreholes containing a fast setting resin at the toe of the hole with the remainder of the bolt grouted in a very slow setting cement.
C Cement-grouted smooth-walled tubular bolts.
D Cement-grouted cone bolts.
E Resin-grouted cone bolts.

All bolts were 1.8 m long and they were installed in 32 mm diameter, 1.8 m long boreholes that were drilled into granitic rock. The rebars were made from 20M or #7 reinforcing bars that had threads machined onto one end. The rebars were made from steel with a yield strength of 400 MPa and an ultimate strength of 550 MPa.

The smooth-walled bolts were made by cutting off the anchor portion of tubular rockbolts. The resulting bolt consists of a high strength, low alloy, tubular shaft was that was roll-threaded on one end. These bolts were made from steel with a minimum tensile strength of 758 MPa resulting in a bolt with a minimum breaking load of 133 kN.

The cone bolt, originating from South Africa,

consists of a 16 mm diameter steel bar with a wax coated smooth shank and a flattened conical flaring forged to the "anchor" end of the bolt. The wax coating is designed to minimize the friction and cohesion resistance along the bolt shank while the cone creates a pull-out resistance as it shears through the grout column (Jager 1992). The cone bolts were 1.8 m long and had a minimum yield load of 90 to 100 kN and a minimum ultimate load of 120 to 135 kN. The yielding mechanism of the cone bolt is based on the shear strength of the grout and the geometry of the cone and as such differs from a grouted smooth bar, which relies on friction.

Cone bolts were designed to be grouted in cement. However, as there is a need for an immediate holding action in many mining situations, cone bolts were also installed in quick setting, resin-grouted holes for comparative purposes.

2.2 Test setup

Figure 1 shows a schematic of the pull-test setup. The pull-out tests were conducted with a hydraulic hand pump and a 300 kN capacity, hollow cylinder hydraulic ram with a long, 150 mm stroke. The loads were measured with a 900 kN capacity, hollow cylinder, load cell and displacements with a rotary potentiometer. Both were logged by a computer-based data aquisition system.

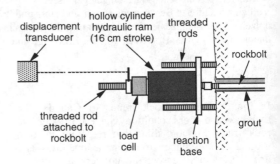

Figure 1 Pull test setup
(Versuchsplan)

3 RESULTS

Plots of the load–displacement results from the pull tests are presented in Figures 2 to 7. The load–displacement curves with displacements exceeding about 100 mm were constructed from multiple pull tests. The load–displacement data from each successive pull were appended together to create these plots. Table 1 summarizes the pull test results.

Table 1 Pull test results

Bolt No.	Age (days)	L_p (kN)	D_p (mm)	D_t (mm)	Result
A1	5	170	15	17	failed
A2	5	165	11	-	stopped
A3	5	169	237	241	failed
B1	5	170	33	37	failed
B2	7	172	21	23	failed
B3	7	159	13	-	stopped
B4	9	163	19	-	stopped
B5	9	165	14	-	stopped
C1	28	28*	54	~200	pulled out
C2	35	120	119	>320	pulled out
C3	35	135	50	53	failed
D1	42	156	250	>253	failed adapter
E1	1	127	94	-	failed adapter
E2	13	156	150	152	failed
E3	13	152	93	-	stopped
E4	15	110	335	>400	stopped
E5	15	157	157	169	failed
E6	42	159	100	106	failed

L_p is the peak load
D_p is the measured displacement at the peak load
D_t is the total measured displacement at failure
* Load measured during sliding stage

3.1 Conventional resin-grouted rebars

Figure 2 shows the test results for two rebars installed with DuPont Fasloc resin grout. The grout is a water-based polyester resin with a catalyst hardener. One 0.45 m long fast-set cartridge was placed at the toe of the hole and two 0.6 m long cartridges (2 to 4 minute gel time) were placed in the remainder of the borehole. The cartridges were 29 mm (1 1/8") in diameter.

Well-grouted rebars typically failed by rupture at the threads before a total displacement (elongation) of less than 20 mm. The rebars had cut threads and the peak load capacity was about 170 kN. One rebar (A3; Figure 7) was poorly grouted into the borehole and it was able to slip a considerable distance (>200 mm) before it eventually broke.

3.2 Rebars installed with cement and resin grout

Five rebars were installed in boreholes containing both resin and cement grouts. One fast-set resin cartridge was placed at the toe of the hole and the remainder of the hole was filled with slow setting Celtite cement cartridges. The resin grout at the borehole toe provided sufficient bonding to achieve an immediate load carrying function. The slow set time of the cement (10 to 14 days) was designed to increase the short-term bolt deformability and to

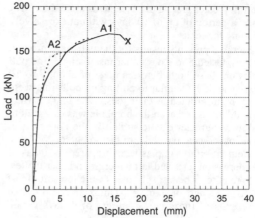

Figure 2 Resin-grouted rebars
(In Resin einzementierte Betoneisenstangen)

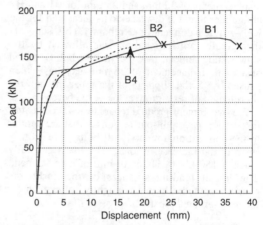

Figure 3 Cement- and resin-grouted rebars
(Resin und Zement einzementierte Betoneisenstangen)

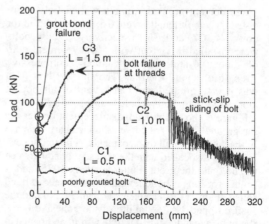

Figure 4 Cement-grouted smooth bars
(Einzementierte glatte Bolzen)

permit greater ground displacements around the drift over a period of up to 10 days without excessively loading the rebar.

Figure 3 presents results for the two-stage grouted rebars, tested 5 to 9 days after installation. The pull tests on these bolts before an age of about 7 to 10 days should be similar to pulling on bolts that were only resin-grouted at their toes. The longer "free-length" of the rebar in the borehole was expected to allow for greater elastic stretch in the bolt when it was loaded near the collar of the borehole.

Cement grout samples were prepared for uniaxial compression tests and left underground near the test site to cure. The test specimens were too soft to test before an age of 7 days. Uniaxial compressive strengths of the grout at an age of 9 days varied from 15 to 19 MPa.

The first two cement-grouted rebars were pulled to failure with peak loads of about 170 kN. The total displacements of 36 mm and 23 mm for B1 and B2 are greater than the displacements measured for the well-grouted conventional rebars. This suggests that greater bolt elongation was allowed by the slow-setting cement. The fact that the bolt B1 failed at a total displacement of 36 mm may have been caused by partial slippage of the rebar through the short length of resin grout at toe of the borehole.

The two cement-grouted rebars tested 9 days after installation gave pull-out responses that are similar to the conventional rebars. All bolts gave a fairly steep initial increase in load. The loads generally reached 150 kN by about 10 mm of displacement. The initial loading stiffness of the early-age, cement-grouted rebars was slightly softer than for the resin-grouted rebars.

3.3 Cement-grouted smooth-walled tubular bolts

Three cement-grouted, smooth-walled bolts were tested. The bolt lengths were 0.5, 1.0, and 1.5 m for bolts C1, C2, and C3, respectively. These bolts were installed in a grout with a water:cement ratio of about 0.4 and a 28 day uniaxial compressive strength of about 40 MPa. The shortest bolt (C1) was poorly grouted and had a void or air pocket along the top of the borehole near the collar.

Figure 4 shows the load–displacement curves for the grouted bolts. All curves show a sharp increase in load at very small displacements. The initial peak in load is associated with the cohesive bond between the bolt and the grout (Fuller and Cox 1975) and is higher for longer bond lengths. The cohesive bond failed at small displacements (< 1 to 2 mm) and the load dropped. After failure of the cohesive bond, the bolts began to slide out of the grout and began to mobilize frictional pull-out resistance. The load caused by the frictional resistance increased as the bolts were pulled.

The maximum pull-out load for these frictionally anchored bolts depended on the length of the bolt in contact with the grout. If the contact length was longer than about one metre, then the bolt broke before it could be pulled from the hole. For shorter lengths, the bolts slid out of the borehole and sometimes displayed stick-slip sliding behaviour.

3.4 Cement-grouted cone bolts

One cone bolt (D1) was installed in a borehole filled with cement grout. The grout was slightly wetter than that used for the smooth bolts and had an estimated water:cement ratio of 0.42.

Figure 5 shows the load–displacement plot for the pull-out test on this bolt. The cone bolt showed an initial peak load of about 120 kN at small displacements. After the initial peak there was drop in load that was probably related to cohesive failure and elongation of the bolt along the debonded section of the shank. The drop in load was followed by an increase in load as the shearing resistance of the conical anchor was mobilized. Some of the load increase at loads above 90 to 100 kN may have been caused by strain hardening in the steel.

Although the pulling adapter failed before the cone bolt itself, the load was almost equivalent to the ultimate load capacity of the cone bolt.

3.5 Resin-grouted cone bolts

Previous tests by Tannant and Buss (1994) showed that cone bolts installed in typical resin grouts could not be pulled more than about 0.1 m before the bolts failed. Therefore, a modified resin was provided by

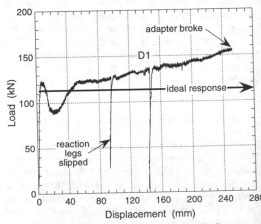

Figure 5 Cement-grouted cone bolt (Einzementierte Kegelbolzen)

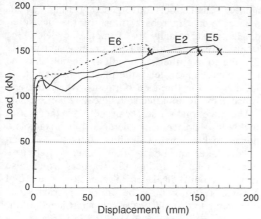

Figure 6 Resin-grouted cone bolts
(In Resin einzementierte Kegelbolzen)

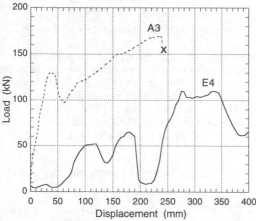

Figure 7 Poorly grouted rebar (A3) and cone bolt (E4)
(Schlecht zementierte Betoneisenstangen, A3
und Kegelbolzen, E4)

DuPont with the intent to permit greater yielding of the cone bolt by anchor shear. Five cone bolts (E1 to E5) were installed with grout with a 60:40 polyester resin:catalyst mixture. For comparison, a sixth cone bolt (E6) was installed with standard fast-set resin cartridges.

Figure 6 shows some of the pull-test results. The pull-out behaviour of the resin-grouted cone bolts was variable. Two cone bolts pulled out of the borehole at very low loads (E1 and E4) whereas others carried significant load over large displacements, until the bolts broke. The well-grouted cone bolts behaved in a similar manner to the cement-grouted cone bolt except that the displacements were smaller. In these bolts, the load rapidly increased to about 120 kN and then remained roughly constant or dropped slightly with

continued pulling. After about 30 to 40 mm displacement, the loads steadily and gradually increased until the bolts broke. The total displacements for the bolts in the modified resin grout were roughly 50 mm more than in the standard resin grout.

Figure 7 presents a pull-out curve for a resin-grouted cone bolt that was believed to be poorly grouted. This bolt gave low and highly variable loads during the pull-out tests. The low loads are thought to result from inadequate mixing between the resin and the catalyst during installation. The observed load oscillations were probably associated with the cone shearing through sections of resin grout with variable strengths.

4 DISCUSSION

The load–displacement curves in Figures 2, 3, 6, and 7 were obtained by smoothing thousands of data points recorded during the pull-out tests. The data were not smoothed for the curves in Figures 4 and 5. These two figures indicate the load fluctuations that occur during the pull-out tests. Usually the loads fluctuate by less than ± 2 kN. However, when stick-slip sliding occurred during the tests on the grouted smooth bolts, large load fluctuations were observed. These fluctuations were accompanied by snapping noises and an easily visible and sudden displacement of the bolt.

4.1 Bolts in highly converging ground

In highly stressed rock, fracturing and inward ground motion can occur. In these situations the displacement capacity of rockbolts is important. Note that the displacement and load capacities of fully grouted bolts can be affected by straining that is localized around joints or rock fractures that separate and by shear displacements at joints or fractures.

Bolts that are designed to yield by anchor slip or shear can result in displacements greater than 10 times larger than those possible with conventional grouted bolts. Cone bolts could accommodate roughly 10 times more displacement than rebars before failing. The peak load in the cone bolts before failure was about 160 kN. This is slightly less than the 170 kN peak load in the rebars but is understandable given the smaller shank diameter on the cone bolts. Cone bolts may work well in areas experiencing significant ground convergence if their pull-out performance could be made more reliable.

There are important quality control issues related to the use of two-part resin grouts with cone bolts. The mixing must be performed by the small conical section of the bolt during installation. The cone bolt does not have the advantage of the ribbed surface texture and the larger diameter of the rebar to aid in mixing the resin. Very little mixing occurs if the cone bolt is pushed into the borehole too quickly or without adequate rotation. In addition, the small shank on the cone bolt does not displace the resin as well as the rebar. The latter problem may be overcome by manufacturing cone bolts with a larger shank diameter.

The cone bolt was originally designed to yield by shearing of the conical section through the cement annulus around the bolt at loads less than the yield load (90 to 100 kN) of the steel shank (Jager 1992). A theoretical ideal load–displacement response for the cone bolt is indicated on Figure 5. The cement grout provided better load–displacement behaviour than the resin grout. However, for the borehole and grout conditions of these tests, the shearing resistance of the cone was too high; the 0.4 water:cement grout contributed to pull-out forces that exceeded both the yield and ultimate strengths of the bolt. The cement grout gave better displacements than the resin grouts, suggesting that further improvement in the design of resin grouts for cone bolts is needed.

A cement-grouted smooth bolt may provide a yielding anchorage through frictional slip. However, to optimize the pull-out response of a grouted smooth bolt, the length of the bolt in contact with the grout must be selected to allow the bolt to slide from the grout column without breaking, while also ensuring that sufficient frictional resistance is mobilized to generate pull-out capacities that are near the yield load of the bolt. The pull-out force and hence, the optimal contact length between a bolt and the grout will depend on the grout properties, borehole conditions (stiffness and stress), and the shape and surface properties of the bolt. Cement grout properties such as compressive and tensile strengths are sensitive to water:cement ratio and curing conditions. Therefore, if cement-grouted smooth bars are used as yielding rockbolts, then very careful quality control of the grout properties is needed to ensure consistent pull-out behaviour. The sensitivity of the pull-out behaviour of the smooth bar to the installation conditions may prohibit the use of grouted smooth bolts in practice.

The use of a two-stage grout for installing rebars temporarily creates a debonded length of bolt between the collar and the resin-grouted section of bolt. Debonding a length of bolt can increase the displacement capacity of the bolt. Unfortunately, the elastic elongation of the steel up to the yield load can only contribute about 2 mm of elongation per metre of free bolt length. Therefore, increasing the free length of the bolt to increase the elastic bolt elongation will only marginally improve the bolt's

ability to handle rock convergence. During the pull-out tests on the rebars, most of the measured displacements actually arise from plastic yield of the steel. However, this plastic strain is not evenly distributed over the free length of the bolt as is the elastic stretch. Instead, the plastic strains tend to localize in the threaded portion of the rebar and hence, they are not affected by the length of the bolt.

The face plate itself can help to increase a rockbolt's displacement tolerance. For example a domed rockbolt plate will allow about 15 to 20 mm of displacement before the dome is flattened. Yielding nuts or collapsible washers would also increase a bolt's displacement capacity. The use of surface fixtures may be more practical than attempting to debond a relatively short length of the bolt. In addition, the debonding is lost after the cement grout has set whereas the yield capabilities of the surface fixtures do not change with time.

4.2 Bolts in rockburst prone ground

In highly stressed rock, fracturing and inward ground motion can occur rapidly in the form of rockbursts. In these situations the displacement capacity and the energy absorption capacity of rockbolts are important. The bolt's energy absorption capacity can be calculated from the work done during the pull-out tests. Table 2 lists the

Table 2 Energy absorption capacity

Bolt No.	D_{150} (mm)	E_{150} (kJ)	E_t (kJ)	Comment
A1	6	0.7	2.5	-
A2	6	0.5	1.5	-
A3	157	17.2	31.2	poorly grouted
B1	14	1.9	5.6	-
B2	9	1.0	3.4	-
B3	11	1.4	1.7	-
B4	12	1.5	2.6	-
B5	8	1.0	2.0	-
C1	-	-	5.3	poorly grouted
C2	-	-	24.4	load <150 kN
C3	-	-	4.1	load <150 kN
D1	234	29.7	32.7	
E1	-	-	9.7	poorly grouted
E2	115	15.0	20.7	-
E3	90	11.8	12.5	-
E4	-	-	>23	poorly grouted
E5	142	17.7	22.2	-
E6	70	9.0	14.7	standard resin

D_{150} and E_{150} are the displacement and energy absorption up to a load of 150 kN
E_t is the total energy absorbed by the bolt during the test

energy absorption capacities of the different bolt types up to a load of 150 kN and at the end of the tests (or until bolt failure).

The energy (or work) was calculated from the area under the load–displacement curves. When the loads in either the cone bolts or the rebars exceed 150 kN, the steel is beyond the yield point and the energy absorption is primarily by plastic deformation of the steel. The energy dissipated in plastic yield is limited to about 1 kJ for both bolt types.

For resin-grouted rebars, a significant portion of the energy absorption comes from the plastic yield of the steel. Only a minor amount of energy was absorbed by shearing or sliding of the rebars through the grout column except for the poorly grouted rebar (A3). Since the volume of steel involved in the plastic straining is approximately constant for all the rebars, the energy absorption capacities up to a load of 150 kN fall within a narrow range between 0.5 and 2 kJ for the well-grouted rebars. Note that at a load of 150 kN, the two-stage, grouted rebars had roughly twice the displacements of the resin-grouted rebars. As a consequence, the two-stage, grouted rebars absorbed about two to three times more energy than the resin-grouted rebars.

An increase in the energy absorption by factor of two to three may, at first, seem significant. But when compared to the energy absorption of the cone bolts or the cement-grouted smooth-walled bolts, the increased energy absorption resulting from the two-stage grouting process is only minor.

The resin-grouted cone bolts consumed energy by shearing and frictional sliding of the conical section through the grout plus plastic strain after the loads exceeded the yield point of the steel. A majority of the energy consumption was caused by the cone bolt pulling through the grout. Energy absorption capacities (up to a load of 150 kN) ranged from a 9 to 18 kJ. The cone bolts that easily pulled from the boreholes were still able to absorb more energy than the rebars although at the expense of much larger displacements. For example, at one extreme, the poorly grouted cone bolt, E4 was pulled 0.43 m while absorbing more than 23 kJ of energy.

On the basis of one test it appears that the cement-grouted cone bolts (D1) are capable of absorbing two to three times more energy than resin-grouted cone bolts. Again, further improvements in the resin grout properties are needed in order to increase the cone bolt's displacement and energy capacities.

The smooth-walled bolts were manufactured from tubular steel with an ultimate load capacity of about 135 kN. Therefore, these bolts were not able to attain a pull-out load of 150 kN. The total energy absorbed by these bolts varied considerably because the pull-out response is highly dependent on the contact length between the bolt and the grout as well as the grout properties. Under optimal conditions, grouted smooth bolts can absorb a signicant amount of energy. For example, bolt C2 absorbed 24 kJ of energy. However, if the contact length is too long (C3) the bolt will break prematurely. Alternatively, if the contact length is too short or if the bolt is poorly grouted (C1), then the bolt will pull out of the borehole at a low load. In either case the energy absorption is much lower.

5 CONCLUSIONS

The use of rockbolts or other forms of tendons to support rock around underground openings that experience large convergence or in excavations that are prone to rockbursts requires a re-evaluation of the support design, particularly with respect to the anchor design. Generally, the rockbolts or tendons cannot stop or prevent the ground movement and they must therefore deform with the ground whilst helping to support and maintain the stability of the opening.

Designing a practical anchor that can slide at a high and predictable load is a greater challenge than designing an anchor that is strong and non-yielding.

A limited number of tests have shown that resin-grouted cone bolts perform better in highly converging ground than grouted rebars, but only if the installation procedure ensures adequate mixing of the resin. A practical and dependable solution to the requirement of a fully grouted bolt that provides both corrosion protection and reliable yieldability still evades us. A fully grouted bolt must provide yieldability by sliding or shearing through the grout column. Plastic yield of the steel does not provide sufficient displacment or energy absorption capacity for bolts used in areas that experience large ground convergence or rockbursts.

A rockbolt with a yielding anchorage, whether it be a grouted smooth bar or a cone bolt, should have the ability to slide substantial distances (100 to 300 mm or more) while maintaining reasonable load bearing capacity. Bolts that slide under high loads can absorb tremendously more energy than conventional rockbolts and this property is very important when they are used to support areas prone to rockbursts.

The following subsections highlight the important conclusions found for each bolt type during this testing programme. However, additional testing should be performed to confirm these findings and to assess the influence of installation procedures on the load–displacement response of grouted bolts.

5.1 Rebars

Resin-grouted rebars that are well grouted are very stiff and can only be pulled about 20 mm before they fail. Using a two-stage cement and resin grout decreases the pull-out stiffness of the bolt slightly and permits slightly larger displacements while the cement is curing. However, this improvement is marginal and may not justify the increased complexity (and cost) of a two-stage grouting system. The 19 mm (3/4") rebars failed at the machined threads after peak loads of about 170 kN. The use of yielding nuts and domed plates should improve the ductility of rebars.

5.2 Grouted smooth bolts

Smooth-walled bolts that are grouted into a borehole can provide a yielding anchorage. The maximum pull-out load for these frictionally anchored bolts depends on the bolt length in contact with the grout. If the contact length was longer than about one metre the bolt broke before it could be pulled from the hole. For shorter lengths, the bolts slid out of the borehole and sometimes displayed stick-slip sliding behaviour.

The pull-out behaviour of these bolts is highly sensitive to the bond length and the grout and borehole properties. If cement-grouted smooth bars are used as yielding rockbolts then careful quality control of the grout properties is needed to ensure consistent pull-out behaviour. The sensitivity of the pull-out behaviour of a smooth bar to the installation conditions may prohibit the use of grouted smooth bolts in practice.

5.3 Cone bolts

Resin-grouted cone bolts can tolerate much larger displacements than grouted rebars. However, ensuring adequate mixing of the resin during installation is difficult resulting in a wide range in load capacities. Well-grouted cone bolts failed after peak loads of about 160 kN. The total pull-out displacements for these bolts ranged from 100 to 150 mm. These displacements depend on the type of resin used to grout the bolts. A modified DuPont resin with a 60:40 resin:catalyst mixture produced roughly 50 mm more displacement than standard fast-set resin cartridges. However, the quality of mixing during installation is probably more dominant than the mix design of the resin.

ACKNOWLEDGEMENTS

This work was partially funded by the Mining Research Directorate as part of the Canadian Rockburst Research Program. I am grateful for the assistance and cooperation of INCO's Creighton mine, in particular the ground control engineers, Chris Langille and Brian Buss. Glenn McDowell from the Geomechanics Research Centre provided field assistance for the first phase of the testing and Dennis Shannon, Executive Coordinator for the Ontario Centre for Ground Control Training helped with student support.

REFERENCES

Fuller P. G. and Cox, R.II.T. 1975. Mechanics of load transfer from steel tendons to cement based grout. 5th Australasian Conf. on Mechanics of Structures and Materials, Melbourne, 189-203.

Herbst, T.F., 1990. Yieldable roof support for mines. Rock Mechanics Contributions and Challenges, 31st U.S. Rock Mechanics Symposium, Golden, 807-814.

Hyett, A.J., Bawden, W.F. and Coulson, A.L. 1992. Physical and mechanical properties of normal Portland cement pertaining to fully grouted cable bolts. *Rock Support in Mining and Underground Construction*, Balkema, 341-348.

Jager, A.J. 1992. Two new support units for the control of rockburst damage. *Rock Support in Mining and Underground Construction*, Balkema, 621-631.

Ortlepp, W.D. 1970. An empirical determination of the effectiveness of rockbolt support under impulse loading. Int. Symp. Large Permanent Underground Openings. Oslo, 197-205.

Ortlepp, W.D. 1983. Considerations in the design of support for deep hard-rock tunnels. 5th Int. Congress on Rock Mechanics, Melbourne, D179-D187.

Ortlepp, W.D. 1992. Impulsive-load testing of tunnel support. *Rock Support in Mining and Underground Construction*, Balkema, 675-682.

Spann H.P. 1987. A Yielding Anchor System to Resist Rockbursts. SANGORM Symposium, Design of Rock Reinforcing: Components and Systems, Johannesburg, 5-7.

Stillborg, B. 1990. Rockbolt and cablebolt tensile testing across a joint. Unpublished report. Lulea, Sweden.

Tannant D.D. and Buss B. 1994. Yielding rockbolt anchors for high convergence or rockburst conditions. 47th Canadian Geotechnical Conference, Halifax, 428-437.

Tannant D.D. and Kaiser P.K. 1995. Friction bolt anchored wire rope for support in rockburst-prone ground. *Canadian Mining and Metallurical Bulletin*, 88(988): 98-104.

Wojno L,, Jager A.J and Roberts M.K.C. 1987. Recommended Performance Requirements for Yielding Rock Tendons. SANGORM Symposium, Design of Rock Reinforcing: Components and Systems, Johannesburg, 71-74.

Anker in Theorie und Praxis, Widmann (Herausgeber) © 1995 Balkema, Rotterdam. ISBN 90 5410 577 1

30 Jahre Ankertechnik in Österreich – Technologische Entwicklung aufgrund praktischer Erfahrungen

30 years of know-how in Austria in the field of anchor technology – Technological development based on practical experience

Walter Tischler
Kontinentale Bau, Wien, Austria

Zusammenfassung

Im vorliegenden Bericht wird auf die Bedeutung der Ankertechnik mit oder ohne Vorspannung hingewiesen. Die Entwicklung aus den Anfängen der Ankertechnik bis zum heutigen vorgespannten Daueranker mit ministerieller Zulassung zeigt ein High-Tech Produkt des Spezialtiefbaues. Damit untrennbar verbunden ist die erfolgreiche Anwendung der Bohr- und Injektionstechnik, sowie das dafür erforderliche hochqualifizierte Personal, welches ausreichende praktische Erfahrung für die ordnungsgemäße Ausführung solcher Arbeiten besitzt.

Abstract

For decades the use of prestressed rock and alluvial anchor systems have represented a recognised and proven method of construction. Because of the prevailing geological and topographical conditions in Austria, development, particularly of transportation systems, would not have been feasible without the availability of suitable anchor technology.

It was back in 1966, that the present classic method "from top to bottom" design was produced for first time for anchoring the castle wall at Innsbruck in the wake of building the highway.

The article shows the development from the initial beginnings of anchor technology through to the present day prestressed permanent anchor, provided with ministry approval, which has become a high-tech product of special civil-engineering structures. Inseparably associated with this is the successful use of drilling and grouting technology, as well as the highly qualified personnel required to carry this out, who possess sufficient practical experience for effective execution of such work.

Civil engineering structures, selected in the example, are discussed in the article with regard to experience acquired in the use of component anchor design, in which the problems of drilling are highlighted, especially in respect of difficult geological conditions in Austria.

The evolved method of grouting and the specified parameters are discussed, together with essential retro-compaction, necessary in the case of increasing load bearing properties of the available foundation substrate, as an improvement so to speak.

A distinctive aspect of anchor technology is the required facility to provide for future potential development when forming encasements and enclosures to the foundations of redundant anchor in urban areas. The example in the report, therefore, deals with implemented building projects.

Finally, the article deals with non-prestressed anchor, the most frequently used form of which is the terrestrial bonding of inclined slopes.

Seit Jahrzehnten stellt die Anwendung vorgespannter Fels- und Alluvialanker eine anerkannte und bewährte Bauweise dar.

Insbesondere der Ausbau des Verkehrswegesystems wäre aufgrund der in Österreich vorherrschenden geologischen und topographischen Verhältnisse ohne entsprechende Ankertechnologie nicht möglich gewesen.

Bereits im Jahre 1966 wurde erstmalig die heute als klassisch angesehene Bauweise "von oben nach unten" bei der Sicherung der Schloßwand Innsbruck im Zuge der Errichtung der Autobahn ausgeführt.

Nicht nur bei Großprojekten, die sich aus Neuanlagen von Verkehrswegen ergeben stehen die planenden Ingenieure sehr oft vor schwierigen Gründungsproblemen. Abgesehen davon, daß zumindest der Ausbau von Autobahnen und Schnellstraßen in der Endausbaustufe ist und künftig der Ausbau der Bahn im Vordergrund stehen wird,

Kombinierte Anker- und Brunnenwand
A 2 Wolfsberg / Kärnten

Combined anchor and well wall
A 2 Wolfsberg / Carinthia

Ankerwand A 12 Innsbruck / Tirol

Anchor wall - A 12 Innsbruck / Tyrol

Baugrubensicherung
Veldidenapark Innsbruck / Tirol

Inset
Veldidenapark Innsbruck / Tyrol

stellen sich bei vielen Projekten die weitläufig dem Gebiet des Tiefbaues zugerechnet werden, ähnliche Probleme. Wenn man weiter bedenkt, daß der Trend zur Verkehrsberuhigung im innerstädtischen Bereich mit dem unterirdischen Ausbau des Schienen- und Straßenverkehrs zur Notwendigkeit von tiefen Baugrubenumschließungen führt, so sind auch diese Bauweisen ohne der Ankertechnik meistens nicht anwendbar.

Weiters erfordert die Kostenexplosion für die Mangelware Bauland eine wirtschaftliche Nutzung der Baugrundstücke mit der Folge, daß die zu errichtenden Gebäude immer mit mehr Untergeschossen ausgebildet werden. Solche Baugrubenumschließungen erfordern sowohl Böschungssicherungen als auch direkte bzw. indirekte Unterfangungen die ihrerseits wieder das Konstruktionselement „Anker" benötigen.

Mit der bewährten Ankertechnik mit oder ohne Vorspannung und im Zusammenwirken mit einer angepaßten Bohr- und Injektionstechnik, lassen sich bei vorhandensein einiger wesentlicher Parameter überaus wirtschaftliche Lösungen entwickeln, die aus einer Kombination von Gründungsmaßnahmen, durch die Ankertechnik zusammengeführt, entstehen.

Flexibilität in der Anwendung läßt dabei selbst bei fortgeschrittenem Baustadium dem

Baugrubenumschließungen

Ankertechnik als Integrationselement
Anchor technology as an integral element

planenden Ingenieur größtmögliche Freiheit.

Ein Blick zurück in die Vergangenheit soll vor Augen führen, welchen Einfluß die technologische Entwicklung auf den „Uranker" Anfang der sechziger Jahre genommen hat. Die Idee war bestechend einfach. Man nehme ein Stahlzugglied, vorzugsweise Spannstangen oder Spanndrähte, verankert diese durch verpressen des künstlich geschaffenen Hohlraumes, nämlich des

389

Modernes Ankerbohrgerät
Modern anchor Drill-Rig

Bohrloches, im gesteinseitigen Ende des Ankers und fixiert durch Vorspannung das luftseitige Ende. Erbrachte das Spannen des Ankers die gewünschte Kraft durch Ablesen am Manometer der Hydraulikpumpe, war der Erfolg sichtbar und gegeben.

Relativ rasch erkannte man die vielseitige Anwendungsmöglichkeit des vorgespannten Ankers und dementsprechend entwickelte sich der Ankermarkt in Österreich speziell in den siebziger Jahren. Eine stürmische technische Entwicklung setzte ein, die nicht nur den Aufbau des Ankers erfaßte, sondern auch die, für eine erfolgreiche Anwendung unabdingbare, angepaßte Bohr- und Injektionstechnik.

Diese Entwicklung führte einerseits zur Tragfähigkeitserhöhung durch Aufweiten des Bohrloches, aufsprengen und nachverpressen in der Haftstrecke und andererseits zur Verbesserung des Korrosionsschutzes. Gleichzeitig wurden entscheidende Verbesserungen am Gerätesektor und der dazugehörigen Ausrüstung vorgenommen, sodaß selbst die Ausführung von Ankerlängen um die 100 m technisch und wirtschaftlich möglich wurde.

Damit diese Entwicklung speziell in Richtung auf den Korrosionsschutz und die Sicherheit der vorgespannten Anker erfolgte, wurde von Fachnormenunterausschuß „Besondere Gründungsverfahren" unter dem Vorsitz von Prof. Dr. Veder die Ö-Norm B 4455 erarbeitet, die im Frühjahr 1978 erschienen ist. Eine wesentliche Erkenntnis, die bis heute unverändert Gültigkeit hat, wurde damals festgeschrieben:

Der Einsatz von zuverlässigen Führungskräften und fachlich geschultem Personal ist die Voraussetzung zur Ausführung dieser Arbeiten.

Bereits im Jahr 1985 erschien eine wesentlich überarbeitete und bis auf eine 1992 vorgenommene Änderung den Spannverlauf betreffend, noch immer aktuelle, dem letzten Stand der Technik Rechnung tragende Ö-Norm B 4455.

Das Produkt dieser Entwicklungsperiode ist ein in jeder Hinsicht entsprechendes, modernes Konstruktionselement, gleichsam ein High-Tech Produkt des Spezialtiefbaues:
- Technisch durch Verwendung hochwertiger Materialien.
- Umweltverträglich durch Einsatz von speziell geeigneter Korrosionsschutzmassen.
- Qualitätssicherung auch durch Anwendung ministerieller Zulassung.

Dieser Anker der nun verfügbar ist, stellt aber nur einen Teil des Gesamten dar, mit dem der Anwender noch wenig anzufangen weiß. Erst durch das Zusammenführen mit der Bohr- und Injektionstechnik entsteht das Fertigprodukt das eine, sowohl in technischer als auch in wirtschaftlicher Hinsicht

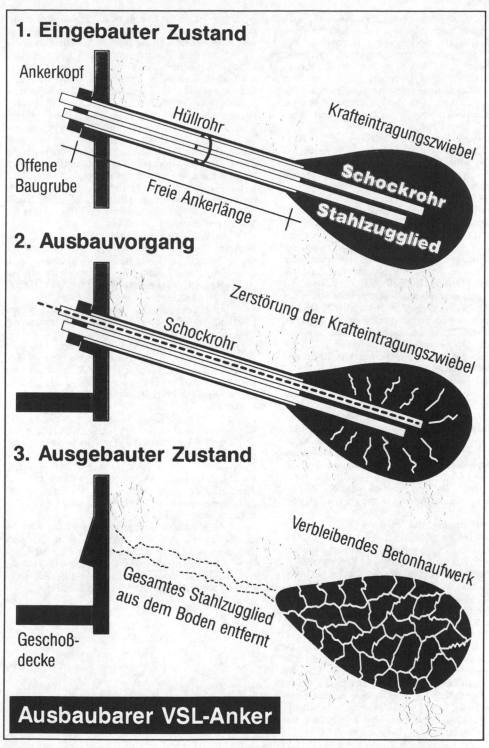

1. Eingebauter Zustand

Ankerkopf

Hüllrohr

Krafteintragungszwiebel

Offene Baugrube

Freie Ankerlänge

Schockrohr

Stahlzugglied

2. Ausbauvorgang

Zerstörung der Krafteintragungszwiebel

Schockrohr

3. Ausgebauter Zustand

Verbleibendes Betonhaufwerk

Gesamtes Stahlzugglied aus dem Boden entfernt

Geschoß-decke

Ausbaubarer VSL-Anker

Voll ausbaubarer VSL-Anker
Fully reusable VSL-Anchor

erfolgreiche Ausführung ermöglicht. Auch ein noch so hochentwickelter Motor im Kraftfahrzeugbau, der durch Einsatz modernster Elektronik in der Fertigung und im Motormanagement optimiert wurde, kann seine Vorzüge nur durch ebenso hochstehende Fahrwerkstechnik und neuester Reifentechnologie umsetzen und die Kraft auf den Boden bringen. Genau so kann der Anker nur seine Kraft in den Boden bringen, wenn Bohr- und Injektionstechnik darauf abgestimmt sind.

Die Herstellung des Bohrloches bei der Ausführung stellt eine Klippe höchsten Schwierigkeitsgrades dar. Müssen hiebei doch nahezu sämtliche Bodenarten von Lehm, Sand, Schotter über groben Blockwerk bis zu Fels in den verschiedensten Härtegraden bewältigt werden. Österreichs Geologie mag für den interessierten Geologen eine wahre Fundgrube für tektonische Vorgänge, Auffaltung von Tälern oder glaziale Abtragungen und erneute Auffüllung mit jüngeren Ablagerungen sein. Dem Bohrmeister hingegen gibt sie manchmal scheinbar unüberwindbare Probleme auf. Sind die zu durchbohrenden Bodenformationen standfest oder muß mit einem Verfall des Bohrloches hinter dem Bohrwerkzeug gerechnet werden? Welche Maßnahmen sind in wasserführenden Schichten zu setzen? Grundsätzliche Entscheidung über das einzusetzende Bohr- und evtl. erforderliche Spülsystem sind zu treffen und setzen viel Erfahrung und Wissen voraus.

Von den angetroffenen geologischen Verhältnissen und deren Beurteilung hat die Wahl der einzelnen Injektionsparameter zu erfolgen. Die Abstimmung von Injektionsdruck und Injektionsmenge, vom Mischungsverhältnis und der Einpresszeit, von der Festlegung der Abbruchkriterien und ob eine Nachinjektionsphase vorgesehen werden muß, sind Voraussetzung für eine erfolgreiche Ausführung. Also eine Vielzahl von Forderungen, ohne die Möglichkeit zu besitzen, den Verlauf und damit den Erfolg des Arbeitsvorganges optisch oder irgendwie zugänglich zu überprüfen.

Die Primärinjektion muß die Krafteinleitung in der Haftstrecke gewährleisten, vergleichbar, wie vorher erwähnt, mit der Reifentechnologie die alle auftretenden Kräfte aufnehmen und ableiten muß.

Der Spannvorgang ist die einzige Prüfung jedes Ankers, wo Erfolg oder Mißerfolg klar erkennbar werden. Bei einem dabei aber auftretenden Mißerfolg ist es nur mehr sehr schwer möglich, Versäumnisse aus den vorhergehenden Arbeitsvorgängen nachzuholen.

Als wirkungsvolle Maßnahme hat sich das Nachverpressen mit dem Aufsprengen des versagendem Haftteiles erwiesen. Ursprünglich also als Mängelbehebung gedacht, hat sich im Laufe der Zeit das Nachverpressen zu einer ausgefeilten Injektionstechnik entwickelt, die heute zur Tragfähigkeitserhöhung, gleichsam als Nachbesserung des zur Verfügung stehenden Baugrundes, dient.

Der Zweck der Nachverpressung besteht darin, die Mantelreibung in der Haftstrecke des Ankers zwischen Verpresskörper und umgebendem Boden zu erhöhen. Die in vielen Fällen entstehenden Sichelschnitte, gefüllt mit Injektionssuspension vergrößern zusätzlich die fiktiven Scherflächen.

Eine dem Anforderungsprofil eines Ankers konträre Eigenschaft muß bei der Anwendung im städtischen Tiefbau verlangt werden. Bei der Ausführung von z.B. verankerten Baugrubenumschließungen verbleiben nach Fertigstellung des Bauvorhabens eine Vielzahl von Ankern, durchwegs aus hochwertigen Spannstählen hergestellt, im umliegenden Boden obwohl ihre Wirksamkeit nicht mehr erforderlich ist. Diese Spannglieder bilden ein unerwünschtes Hindernis für spätere Bauaufgaben wie Kanäle, Fernwärmeleitungen oder auch U-Bahnlinien. Besonders delikat sind die privatrechtlichen Ansprüche Dritter aus dem Nachbarrecht, wo technische Zwänge nicht immer im Vordergrund stehen.

Es war daher notwendig Methoden zu entwickeln, versetzte Anker sobald sie funktionslos waren, möglichst vollständig auszubauen. Auch diese Forderung konnte die Ankertechnik erfüllen und bereits 1976 bei einem Bauvorhaben in Wien in die Praxis umsetzen.

Abschließend soll noch auf die bereits erwähnte Ankertechnik ohne Vorspannung eingegangen werden.

Die häufigste Anwendungsform ist die Bodenvernagelung zur Böschungssicherung. Die Nagelwand ist ein verbundartiger Stützkörper, der aus dem anstehenden Boden den eingebohrten, kraftschlüssigen Nägeln sowie einer Außenhaut aus bewehrtem Spritzbeton besteht. Die Herstellung erfolgt vorzugsweise von oben nach unten, d.h. der Abtrag des Bodens erfolgt etagenweise. Voraussetzung für eine erfolgreiche Anwendung dieser Baumethode ist eine den Erfordernissen angepaßte Bohr- und Ankertechnik, sowie ein optimierter Bauablauf zwischen Aushub - Spritzbeton - und Ankerarbeiten. Dies deshalb, da die Schlußzeiten zwischen Aushub und Anker anspannen möglichst kurz zu halten sind.

Einen Überblick über die vielfältigen Anwendungsmöglichkeiten der Ankertechnik in den vergangenen 30 Jahren in Österreich soll ein Streifzug über Ankerbaustellen bringen, wobei nicht nur auf die Größe, sondern auch auf die speziell technischen Anforderungen hingewiesen wird.

4 Long-term behaviour
Langzeitverhalten

Anchors in Theory and Practice, Widmann (ed.) © 1995 Balkema, Rotterdam. ISBN 90 5410 577 1

Long-term service behaviour of cement grouted anchors in the laboratory and field

Verankerungsverhalten in Gussmörtel auf lange Dauer im Laboratorium und an der Baustelle

Brahim Benmokrane, Mohamed Chekired & Haixue Xu
Department of Civil Engineering, Université de Sherbrooke, Que., Canada

ABSTRACT

Long-term monitoring of the behaviour of ground anchors constitutes an important aspect of the stability of structures reinforced by such structural members. This paper describes full-scale tests performed on two instrumented cement grouted anchors installed in the laboratory and field, respectively. Due to the long-term reliability, stability and accuracy, two types of surface-mounted vibrating wire gauges were employed to instrument the anchors. The performance of these gauges, load distribution, and long-term performance of the anchor were investigated in the laboratory. A practical application and test results of an instrumented anchor at Jeffrey Mine in Asbestos, Canada, are described. The results obtained show that this instrumentation technique of grouted anchors is efficient for long-term monitoring.

ZUSAMMENFASSUNG

Die Überwachung von Verankerungsverhalten in Gussmörtel mit verstärkten Elementen stellt einen wichtigen Aspekt der Stabilität im Bauwesen dar. Diese Veröffentlichung beschreibt die Versuche von zwei vollkommen kontrollierten Verankerungen im Laboratorium und an der Baustelle. Wegen der Verlässlichkeit auf lange Dauer und vom Gesichtspunkt der Stabilität und Genauigkeit wurden zwei verschiedene an der Oberfläche angebrachte Drahtschwingmeßgeräte benützt. Die Leistung dieser Meßgeräte, die Lastverteilung und die Wirksamkeit der Anker wurden im Laboratorium untersucht. Eine praktische Verwendung und die Versuchsergebnisse eines Ankers bei Jeffrey Mine in Asbestos, Canada, sind hier beschrieben. Die Ergebnisse zeigen, daß die Technik der Meßgeräte der Verankerung für die Überwachung auf lange Dauer wirksam ist.

1. INTRODUCTION

Ground anchors are widely used in civil engineering as structural members to ensure the stability of retained structures, such as slopes, retaining walls, dam foundations, underground excavations, and bridge abutment. Ground anchors are also used for the strengthening and rehabilitation of existing concrete dams (Bruce 1993; Xu and Benmokrane 1995).

Ground anchors, which can be temporary or permanent structural members, are often in service with the reinforced structure. Their reliable performance during construction and operation is needed to ensure the service life of the whole structure. Therefore, the instrumentation of ground anchors is vital for monitoring the performance of the structure.

Three quantities are of direct interest for ground anchors:
1) the applied load;
2) the anchor head displacement; and
3) the load distribution over the anchored length.

The load at the anchor head can be measured during the service of the anchor using a load cell or by applying a lift-off load by a jack (Littlejohn and Bruce 1979; Benmokrane and Ballivy 1991a; Littljohn and Xu 1993). The anchor head displacement can be monitored using remote survey technique (Franklin and Denton 1973; Stirling et al. 1992). The above two monitoring methods are common in practice. However, monitoring of the load distribution over the anchored length is less common in practice, although it can provide useful a data particularly on anchor-stratum interaction, load transfer, stress level, creep or displacement of the anchored length. The information obtained from the gauges over the anchored length can also be used as a back analysis for numerical modeling.

Traditionally, electrical resistance strain gauges have been widely used to instrument grouted anchors in the laboratory test as well as in the field practice. In various cases of field application, it has been reported that a number of strain gauges (11 - 75 %) are damaged during anchor installation and service (Shields et al. 1978; Freeman 1978; Björnfot and Stephansson 1983; Myrvang et al. 1992; Xu 1993). Generally, it appears that the performance of electrical resistance strain gauges is reasonable good when instrumentation is developed for short-term purposes. However, for long-term monitoring, the reliability and stability of the instruments are the essential consideration when selecting an instrumentation system.

Successful field measurement can only be made if the instruments are reliable and adequate. Thus, essential features of any instrument are reliability, stability, accuracy, simplicity and ease of calibration, installation, and use. All instruments must be durable in the long-term and not prone to damage during or after installation. The vibrating wire strain gauges fulfill these general requirements. In addition, the induced signals are immune to electrical noises, and long cables can be used without affecting the signals.

This paper describes full-scale tests performed on two instrumented cement-grouted anchors installed in a concrete cylinder in the laboratory and in a rock mass in the field. The primary objective of this study was to investigate the performance of this instrumentation technique of grouted anchors. The performance of the vibrating wire gauges, load distribution, and long-term performance of the anchor were investigated in the laboratory test. A practical application and test results of an instrumented anchor at the Jeffrey Mine in Asbestos, Canada, are discussed.

2. LABORATORY TESTS

2.1 Experimental program

Two types of surface-mounted vibrating wire strain gauges of IRAD and SINCO were selected to instrument grouted anchors. The characteristics of the two types of gauges are described elsewhere (Benmokrane et al. 1995). The two types of vibrating wire strain gauges have different plucking and reading systems: IRAD gauges use the plucking and reading system,whereas, SINCO gauges use the continuous excitation system. The principle of the vibrating wire gauges has been described elsewhere (Hanna 1985; Dunnicliff 1988; Benmokrane et al. 1995).

One anchor was employed to perform the laboratory test. The anchor tendon was composed of a 36 mm in diameter continuously threaded Dywidag steel bar with an elastic limit of 860 kN and ultimate load of 1030 kN. In order to examine the load transfer mechanism, seven gauges were micro-welded on the surface of the bar at intervals between 15 to 25 cm. One gauge was also micro-welded on the free length of the anchor to provide an additional load measurement to the load cell on the anchor head.

The bar surfaces were prepared by removing rust and scale using a sander, with progressively finer grits to expose the bare mental and remove pits. The gauges were micro-welded on the flanges of the bars. The welding operation was carried out using a discharge spot welder with a capacity of 40 watts of power. After welding the gauges, the coil/gauge covers were installed on the gauges as shown in Figure 1(a). Figure 1 (b) shows part of an instrumented bar with a vibrating wire strain gauge covered with a

(a)

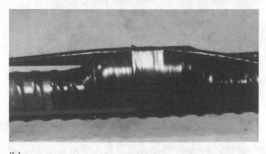

(b)

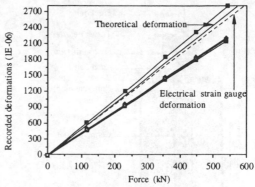

Figure 2   Calibration curves of an
            instrumented anchor
Bild 2   Kalibrierte Bewertungs-
            kurve von einem
            kontrollierten Anker

Figure 1   Vibrating wire strain gauge:
            (a) covered with electrical
            coil and gauge cover
            (b) part of an instrumented
            anchor
Bild 1   Drahtschwingenmeßgerät
            (a) mit elektrischer Spule
            und Meßgerätabdeckung
            (b) Teil einer kontrollierten
            Verankerung

mastic compound for waterproofing.
Prior the installation, the instrumented
steel bar was calibrated in a special frame
in oder to check the functioning of gauges
and eliminate the effect of possible
variations in steel cross-section properties
(Chekired 1993). Several repeated
loadings were applied to allow seating of
the gauges prior to the calibration, then
three calibration runs were conducted on
the instrumented bar. The reponse of the
gauges was measured for the applied
loads. The calibration lines of vibrating
wire gauges were linear, and the
repeatability was very consistent as
shown in Figure 2.

A concrete cylinder of 600 mm in diameter
and 1500 mm in height was cast in a steel
barrel as a medium for installing the
instrumented anchor. The compressive
strength and modulus of elasticity of the

concrete were 36 MPa and 32 GPa,
respectively. A practical hole size of 76
mm was found adequate to accommodate
the instrumented anchor, but a large hole
of 127 mm in diameter was adopted in the
model test. The hole was centrally drilled
in the concrete cylinder by a percussive
and rotary machine. The instrumented
anchor was then inserted into the hole
and the hole was injected with cement
grout (water/cement ratio of 0.4). Type I
Portland cement was used with an
expansive agent (aluminium powder) at a
ratio of 0.005 % of cement by weight.

No problem was observed in the vibrating
wire gauges during the installation. No
changes in gauge readings were
observed during anchor installation and
grouting.

A hollow hydraulic jack was used to apply
load to the anchor. The applied load was
measured using a hollow load cell
installed at the anchor head. Readouts
were taken at each loading cycle using a
Sens-Log data-acquisition system. The
test was conducted 14 days after the
grouting of the anchor when the grout had
developed a compressive strength and
modulus of elasticity of 48 MPa and 11
GPa, respectively. A schematic
representation of an instrumented anchor
subjected to a loading test is shown in
Figure 3. The test setup for anchor
stressing, instrumentation, and data
acquisition is shown in Figure 4.

397

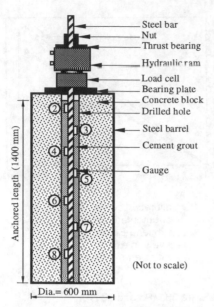

Figure 3   An instrumented anchor in
            the laboratory test
Bild 3   Ein kontrollierter Anker im
            Labortest

Figure 4   Test setup for anchor
            stressing, instrumentation,
            and data acquisition
Bild 4   Aufstellung für den Test zur
            Ankerbelastung, -kontrolle
            und -datengewinnung

Prior to performing the tests, a pull out test on an uninstrumented bar with an anchored length of 400 mm was conducted. Based on this result, it was determined that the anchor had a load

holding capacity equal to the ultimate load of the bar (1030 kN).

2.2   Long-term monitoring of the anchor

The anchor was initially loaded to a load of 620 kN, and then was unloaded and locked off at 440 kN. It is known that when steel is subjected to loads less than 55 % of its elastic limit, the load losses due to relaxation are negligible (Antill 1965; Mihajlov 1968). The lock-off load of 440 kN recorded on the load cell corresponds to 52 % of the elastic limit load. Therefore, the load losses due to steel relaxation can be considered negligible.

Figure 5 shows the load evolution recorded by the data acquisition system at the anchor head and along the anchored length of the anchor over a period of 80 weeks. The long-term behavior of the prestressed anchor occurred in two distinct phases in terms of load changes or of rate of prestress loss at the anchor head and the proximal end of the anchor (situated nearest to the anchor head, gauges 2-5). Phase I occurred within approximately 8 weeks, where rapid losses of prestress were recorded.

Phase II represented a small and uniform rate of prestress loss. This behavior is similar to that of the long-term behaviour of prestressed rock anchors during

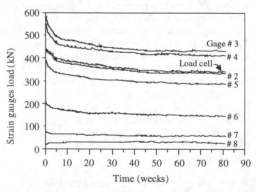

Figure 5   Load variation with time of
            anchor
Bild 5   Zeitliche Änderung der
            Lastverveilung

previous studies (Littlejohn and Bruce 1979; Benmokrane and Ballivy 1991a; Littlejohn and Xu 1993).

The load variation at the anchor head and along the anchored length can be inferred to be a result of the creep of the anchored length. At the proximal end of the anchor (gauges 2 - 5), creep was created under the sustained load and the strain or load in the anchor was reduced. On the other hand, the distal end of the anchor (situated furthest from the anchor head, gauge 8) tended to resist the creep, and the strain or load in the anchor was gained. With time, the load variations stabilized after a period of time (25 weeks) after locking off the applied prestress load. The magnitude of load decrease at the proximal end was much greater than that of load increase at the distal end. The maximum load loss recorded on the load cell was 20 % of the lock-off load after a period of 80 weeks. This load loss could be attributed to the creep of the anchor system and the concrete medium.

The load distribution at different time intervals after locking-off is shown in Figure 6. This indicates a different load distribution profile to that measured during loading (Benmokrane et al. 1995). Similar results were observed by Shields et al. (1978), Indraratna and Kaiser (1990), Benmokrane and Ballivy (1991b) during previous studies. This phenomenon was caused by the loading process. The initial loading to 620 kN created debonding and permanent residual load in the anchor. When unloading an anchor to the seating load, the creep or plastic displacement in the anchor system will create a permanent residual load in the anchor. In the debonding zone, the debonded grout attempts to prevent or limit anchor contraction and creates friction which is opposite to the direction of the anchor contraction. This friction will consequently restrain the anchor from returning to its initial position corresponding to the seating at the beginning of the test. Therefore, the load at the anchored length can be higher than that at the anchor head. Beyond the debonding zone, the shear stress is in the same direction of anchor contraction, and the load distribution reverts to its exponential profile. As a result, the load at the proximal end of the anchor was higher than that on the anchor head.

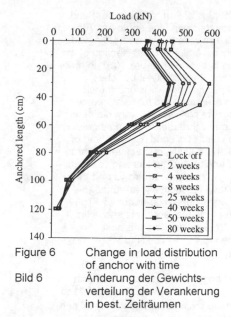

Figure 6   Change in load distribution of anchor with time

Bild 6   Änderung der Gewichts-verteilung der Verankerung in best. Zeiträumen

3.   FIELD APPLICATION OF AN INSTRUMENTED ANCHOR

3.1   Test site and location of a tested anchor

Following the laboratory testing, a field test of an instrumented anchor was conducted at Jeffrey Mine in Asbestos, Canada. The north wall of the open pit mine is 315 m high in low grade slates. The slates dip at approximately 55° into the open pit and are composed of black phyllite, impure quartzite, and quartz-sericite-chlorite schist. The black phyllite is thin-bedded and crumpled. Foliation is almost parallel to bedding. The bedding is characterized by large scale dip variations of between 40° and 75° with an average bedding dip of 55°. The jointing is generally subordinate to the bedding structures and no significant joints at unfavorable attitudes are observed. The details of the site geology and mechanical properties of the rock have been described elsewhere (Sharp et al. 1987).

The north wall consists of benches which are divided into two sub-benches. The main purpose of the benches is to retain any loosened materials that may fall from the upper levels. The benches are 24.4 m

399

in height (2 x 12.2 m sub-benches) and are excavated with face angles of 70° to give a 9.1 m wide bench and 1.8 m wide sub-bench. A typical cross-section of the benches is shown in Figure 7(a). In order to maintain the stability of benches, cement grouted post-tensioning rock anchors are employed for reinforcement. The anchors consist of 32 mm diameter continuously threaded Dywidag steel bars with an elastic limit and ultimate load of 690 kN and 830 kN, respectively. Three rows of anchors with a spacing of 4.5 m and 9 m at the upper two rows and the bottom row, respectively, are used to reinforce the upper sub-benches. The location of the test anchor is shown in Figure 7(a). All anchors are post tensioned to 450 - 470 kN (55 % of the ultimate load).

The test anchor was instrumented with six vibrating wire strain gauges marked #1 to #6 from the proximal end to the distal end of the anchored length, as shown in Figure 7(b). The test anchor was calibrated in the laboratory using the method described above and was then transported to the site. After drilling a hole of 76 mm diameter, the instrumented bar was inserted and grouted to an anchored length of 3.65 m using cement grout (water/cement ratio of 0.35, type III Portland cement, and superplasticizer of 1 % of cement by weight). The grout was allowed two days to harden. The free anchor length was then grouted, and a tension load was applied and locked off into the anchor thereafter.

During the transportation, installation, grouting, and tensioning of the anchor, no damage to the gauges was observed, and all gauges were functioning well. The installation work was easily performed without any difficulties.

3.2 Change in load distribution with time

The load distribution over the anchored length was monitored for the 18 weeks, (and is continuing to be recorded). Figure 8 shows the load profiles recorded by the gauges before locking-off, after locking-off, and after one and 18 weeks. There was an instantaneous load loss due to the lock-off operation, which was approximately 4.9 % of the initial load at

gauge 1, where gauge 1 was located within 20 cm of the proximal end of the anchored length as shown in Figure 7(b). The load loss then occurred gradually with time, and was 5.4 % of the locked-off load at gauge 1 after 18 weeks. The load over the anchored length was redistributed with time, and tended to decrease at the proximal end and increase towards the distal end of the anchor. This result is very similar to the laboratory test result. The load transfer length was approximately 2.0 m, 55 % of the anchored length. At the proximal end (less than 0.70 m of anchored length), there was partial debonding. The bond at the grout/tendon interface behaved

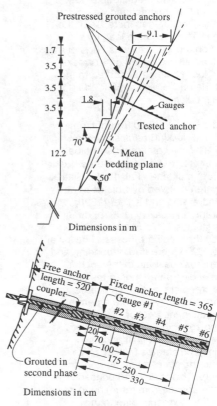

Fig. 7   A tested anchor in the field
(a) typical cross section of benches and location; and
(b) dimension of the instrumented anchor

Bild 7   Ein getesteter Anker auf der Baustelle
(a) typ. Querschnitt; und
(b) Dimension der kontrollierten Verankerung

400

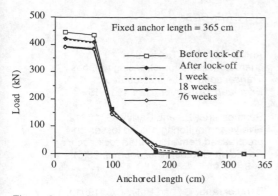

Figure 8   Change in load distribution
            over fixed length with time
            in the field
Bild 8   Wechsel der Gewichts-
            verteilung über eine best.
            Strecke, in einem best.
            Zeitraum, gemessen an der
            Baustelle

elastically between the anchored length of
0.7 m to 2.5 m, and the load distribution
was exponential. No load was transferred
beyond the anchored length of 2.5 m.

3.3   Implication of tests results to the practice of grouted anchors

The field and laboratory tests confirmed
the results determined in previous
experiments and theoretical studies.
Although the bond stress distribution over
the anchored length is non-uniform in
medium to strong rock masses, the design
approach currently used to calculate the
anchored dimensions has seen little
change since the publication of the state-
of-the-art report (Littlejohn and Bruce
1977).

By assuming a uniform bond stress
distribution along the anchored length,
knowing the ultimate rock-grout bond
stress (τ_{ult}), the ultimate load (T_f) and
fixed anchor diameter (D), the anchored
length (L) can be calculated using the
following formula:

$$L = \frac{T_f}{\pi D \tau_{ult}} \qquad (1)$$

In the massive rock, the ultimate bond
stress is often taken as 10 % the uniaxial
compressive strength (UCS), but is not
assumed to exceed 4 MPa for UCS
values exceeding 40 MPa. In the field,
the UCS of the rock is typically 50 MPa,
with lower values of 15 to 20 MPa. The
ultimate bond stress of 2.0 MPa is
obtained by assuming a UCS of 20 MPa
for the rock. The working bond stress (τ_w)
of 0.8 MPa is obtained by assuming a
safety factor of 2.5 at the rock/grout
interface. For a design load of 460 kN
and a fixed anchor diameter of 76 mm,
the anchored length is 2.4 m.

As shown in Figure 8, the load transfer
length is approximately 2.0 m, the design
bond length is close to the load transfer
length. In practice, the anchored length
should not be more less than 3 m in order
to avoid the effect of sudden change in
rock quality along the anchored length. In
this case, the anchor length was chosen
to be 3.6 m, which provided an additional
safety factor of 1.5.

As shown in Figure 8, the top of the
anchored length has suffered debonding.
During prestressing, cracks at the
grout/tendon interface are created and
propagate radically towards the rock/grout
interface with increasing load. This
phenomenon has been observed along
the anchored length, especially at
debonding zone in the laboratory tests
(Natau and Wullschlager 1983;
Weerasinghe 1993). As a result, cement
grout cannot protect the steel tendon from
corrosion over debonding zone.
Corrosion protection system must be
defined for permanent anchors in
practice.

4.   CONCLUSION

Vibrating wire strain gauges performed
satisfactorily during the laboratory and
field tests. No damage to the gauges was
observed during installation,
transportation, tensioning, and long-term
monitoring. Vibrating wire strain gauges
meet the general requirements of
monitoring, particularly in regard of the
long-term reliability and stability over a
long period of time.

Monitoring of the load distribution of ground anchors is a very important aspect. The collected data can provide a complete image of anchor loading and reevaluate the original design and factors of safety.

The long-term load loss can be divided into two phases: rapid load loss and slow rate load loss. The rapid load loss occurred within 8 weeks, thereafter, the rate of load loss became stable.

Load redistribution after lock-off occures in the anchored length both observed in the laboratory and field tests. The load tends to decreases with time at the proximal end, which has a behavior similar to that recorded at the anchor head; whereas, tends to pick up with time at the distal end. The load transfers further from the proximal end with time.

It is recommended that this instrumentation technique be considered where long-term monitoring, no accessibility for repair and recalibration, and long cable runs and immunity to electrical noise are required. This instrumentation technique can also be applied to the post-tensioning steel bars used for prestressed concrete and for retrofitting applications.

ACKNOWLEDGMENTS

This study was made possible by financial support from the Natural Science and Engineering Research Council of Canada (NSERC) and from the Ministère de l´Education du Québec (Fonds pour la formation de chercheurs et l´aide à la recherche, FCAR). The authors wish to acknowledge the assistance of Mr. James L. Deacon and Mr. Jocelyn Frechette of Jeffrey Mine, Asbestos Inc. during the field test, and the participants from the following parties: Dywidag DSI Canada Ltd. (St. Bruno, Quebec), Geostructures Instruments Inc. (La Praitie, Quebec), Rocktest Ltd. (St-Lambert, Quebec), and Slope Indicator Canada Ltd. (Richmond, British Columbia).

REFERENCES

Antill, J.M. 1965. Relaxation characteristics of prestressing tendons. *Transactions of the Institution of Engineers*, Australia, 7(2): 151 - 159.

Benmokrane, B., and Ballivy, G. 1991a. Five year monitoring of load losses on prestressed cement grouted rock abchors. *Canadian Geotechnical Journal*, 28(5): 668 - 677.

Benmokrane, B., and Ballivy, G. 1991b. Investigation of stress distribution along cement routed rock anchors under uplifting loads (in French). *CIM Bulletin*, 84(951): 45 - 52.

Benmokrane, B., Chekired, M., and Xu, H. 1995. Monitoring behavior of grouted anchors using vibrating-wire gauges. Journal *of Geotechnical Engineering*, 121(6): 466 - 475.

Björnfot, F., and Stephansson, O. 1983. Interaction of grouted rock bolts and hard rock masses at variable loading in a test drift of the Kiirunavaara Mine, Sweden. *Proc. of the Int. Symp. on Rock Bolting*, Abisko, Sweden, 377 - 396. Balkema, Rotterdam.

Bruce, D.A. 1993. The stabilization of concrete dams by tensioned rock anchorages: The State of American practice. *Geotechnical Special Publication No. 35, Geotechnical Practice in Dam Rehabilitation*, Raleigh, NC, 320 - 332. ASCE, New York,.

Chekired, M. 1993. The use of vibrating wire strain gauges in the instrumentation of injected anchors (in French), Master thesis, Université de Sherbrooke, Québec, Canada, 164 p.

Dunnicliff, J. 1988. *Geotechnical instrumentation for monitoring field performance*. A Wiley-Interscience Publication, New York, 577 p.

Franklin, F.A., and Denton, P.E. 1973. The monitoring rock slopes. *Quarterly Journal of Engineering Geology*, 6(3 & 4): 259 - 286

Freeman, B.A. 1978. The behavior of fully bonded rock bolts in the Kielder Experimental Tunnel. *Tunnels and Tunneling*, 10(5): 37 - 40.

Hanna, T.H. 1985. *Field instrumentation in geotechnical engineering*. Trans Tech Publications, Clausthall-Zellerfield, 843 p.

Indraratna, B., and Kaiser, P.K. 1990. Design for grouted rock bolts based on the convergence control method. *Int. J. Rock Mech. Min. Sci. and Geomech. Abstr.*, 27(4): 269 - 281.

Littlejohn, G.S., and Bruce, D.A. 1977. Rock anchors: State-of-the-art. Foundation Publication Ltd., Brentwood, Essex, England, 20 p.

Littlejohn, G.S., and Bruce, D.A. 1979. Long-term performance of high capacity rock anchors at Devonport. *Ground Engineering*, 12(7): 25 - 33.

Littlejohn, G.S., and Xu, H. 1993. The service behavior of ground anchors at Pen y Clip Tunnel (1990 - 1992). TRRL *Ground Anchor Monitoring Project Report*, University of Bradford, Bradford, England, 21 p.

Mihajlov, K.V. 1968. Stress relaxation of high tensile steel. *Proc. of Fédération International de la Precontrainte, Symp on Steel for Prestressing*, Madrid, Spain, FIP, Paris, 57 - 78.

Myrvang, A., Stjern, G., and Morseth, B. 1992. Performance of fully grouted rebar anchors and shotcrete in a 61 m span. *Proc.Int.Symp. on Rock Support in Mining and Underground Construction*, Sudbury, Ontario, Canada, Edited by Kaiser, P.K., and McCreath, D.R., 117 - 122. Balkema, Rotterdam.

Natau, O.P., and Wullschlager, D.H. 1983. Theoretical and experimental studies in bearing behavior and corrosion protection of rock anchors up to a load limit of about 4900 kN. *Proc. 5th Cong., Int. Soc. Rock Mech.*, Melbourne, Australia, A 59 - A 64.

Sharp, J.C., Lemay, C., and Neville, B. 1987. Excavation, reinforcement and monitoring of a 300 m high rock face in slates. *Proc. 6th Int. Cong. on Rock Mechanics*, Montreal, Canada, 1: 533 - 540. Balkema, Rotterdam.

Shields, D.R., Schnabel, H.Jr., and Weatherby, D.E. 1978. Load transfer in pressure injected anchors." *J. of the Geotech. Eng. Division*, ASCE, GT9, 1183 - 1196.

Stirling, D.M., Chandler, J.H., and Clark, J.S. 1992. Monitoring one of Europe's largest retaining walls using oblique aerial photograph. *Int. Soc. for Archives Photogrammetry and Remote Sensing XXIX*, Washington DC, B5, 701 - 708.

Weerasinghe, R.B. 1993. The behavior of anchorages in weak mudstone, Ph.D. thesis, University of Bradford, Bradford, England, 313 p.

Xu, H. 1993. The dynamic and static behavior of resin bonded rock bolts in tunneling, Ph.D. thesis, University of Bradford, Bradford, England, 249 p.

Xu, H., and Benmokrane, B. 1995. Strengthening existing concrete dams using post-tensioned anchors: a state-of-the-art-review. Paper submitted to *Canadien J. of Civil Engineering*.

Tension fatigue failure of short anchor bolts in concrete

Ermüdungsbruch von kurzen, in Beton versetzten Ankerbolzen

E.Cadoni
European Commission, Joint Research Centre, Safety Technology Institute, Ispra, Italy

ABSTRACT: Fatigue failure of short anchor bolts has been studied by means of pull-out tests carried out on concrete slabs.
Three types of anchors have been tested by applying sinusoidal shaped loading cycles. Analysis was performed with fatigue behaviour of anchorage according to the maximum cyclic load. Damage propagation was studied as a function of the number of loading cycles. A comparison among energy dissipated of cycles, compliance and displacement of anchorage variations has been made during the fatigue life in order to verify which is the most adequate tool for the fatigue process control.
Experimental results show that a relationship between the displacement in static tests and the displacment in dynamic tests exists. Therefore by means of the static pull-out test, it is possible to point out the fundamental features that govern the cyclic behaviour (maximum displacement, failure displacement etc.).

ZUSAMMENFASSUNG: Der Ermüdungsbruch von Verankerungen mit kleiner Versenkung ist mittels Ausziehtests auf Betonplatten untersucht worden.
Es sind drei Verankerungstypen unter Anwendung einer zyklisch sinusförmigen Belastung getestet worden. Es ist ein Vergleich angestellt worden zwischen den Größenvariationen von abgegebener Energie pro Zyklus, Nachgiebigkeit, Verschiebung der Verankerung über die ganze Ermüdungsdauer hinweg, um festzustellen, welche die geeignetste für die Kontrolle des Ermüdungsprozesses ist.
Die Versuchsergebnisse haben eine Beziehung zwischen der Verschiebung in einem statischen Test und jener in einem zyklischen Test gezeigt. Es ist also möglich, über einen statischen Ausziehtest die grundsätzlichen Grenzen zu bestimmen, die das zyklische Verhalten lenken (max. Verschiebung bei Betrieb, Verschiebung bei Bruch usw.).

1 INTRODUCTION

Anchor bolts embedded in concrete are found in many kinds of structures, often they are subjected to cyclic loading. A short anchor bolt is usually defined as one whose embedded length is insufficient to develop tensile yield in the bolt.

In this paper, an experimental research about short anchor bolts in concrete carried out at the Fracture Mechanics and Non Destuctive Tests Laboratory - Structural Engineering Department of Politecnico di Torino (Cadoni,1994) is reported.

The fatigue failure has been investigated by means of pull-out tests carried out on concrete slabs by applying sinusoidal shaped loading cycles to three types of anchor bolts previously embedded in the casting.

The different anchor bolts have been chosen in order to emphasize the different types of concrete brittle failure, with crack propagation, concrete failure due to diffused damage and bond failure.

2 TESTING PROGRAMME

The tests have been carried out on 40 concrete slabs 50x50x15 cm in size, with a short anchor in the middle. The tests were performed by means of an MTS with maximum load of 250 kN. A large-diameter contrast ring (50 cm) was used in order not to affect the cracking surface, in both static and dynamic tests.

Compressive strength, R=24.7 MPa, was evaluated on cubes (with 160 mm-long sides). The determination of the secant modulus, E=20380 MPa, and of fracture energy, G_F=62 N/m, were performed on a 160x160x500 mm prism and 100x100x840 mm notched prisms, respectively.

It should be pointed out that before the tests, the slabs and test pieces were kept for 30 days at a temperature of about 20°C and at a relative humidity of approx. 65%.

Static pull-out tests were performed by means of an MTS with maximum load of 250 kN by imposing a constant velocity of the load application point of $d\eta/dt=5\cdot10^{-6}$ m/s. Instantaneous displacement, η, was calculated as the arithmetical mean of η_1 and η_2 values measured by two inductive transducers placed in a diametrically opposed position with respect to the anchor bolt, see Fig.1 (Layout of testing equipment: 1. Specimen; 2. Extractor; 3. LVDT; 4. Contrast ring; 5. Load cell; 6. MTS).

The measuring points of the transducers on the surface of the slab and on the testing machine were chosen so as to minimize possible displacement errors due to play in the mechanical connection or elastic strains in the materials.

Dynamic tests were carried out by applying a sinusoidal loading cycle with a frequency of 1Hz. Maximum load, P_{max}, was kept constant throughout the test. Force-displacement diagrams (P,η) were recorded according to the following procedure:
- recording the first loading/unloading cycle by keeping the load increase rate constant;
- recording the (F,η) loading/unloading diagram after N_0, N_1, N_2, ..., N_i fatigue cycles.

The three types of anchorage are shown in Fig. 2. The anchor bolts were embedded at 40 mm while the

Fig. 2 Anchorage geometries chosen

Abb. 2 Anker

rod and ribbed bars were embedded at 10 cm in depth. All three types had a nominal diameter of 16 mm .

3 FAILURE MODE OF SHORT ANCHOR BOLTS

The anchor bolt behaviour is influenced by many factors which must be considered in order to develop a correct design (i.e. the steel-concrete interaction, the presence of rib-in-the-rod or one head at its end).

Among the parameters which determine the behaviour of anchor bolts, some are reported as follows: matrix material; matrix condition (i.e. cracked concrete); matrix resistance; rod material resistance; load direction; embedded length; distance between the anchors and between them and the free edges; type of load; presence of reinforced bar near the anchors; environmental condition (corrosion, fire exposition, thermic variations, etc.).

The failure characteristics of the component material, on their interaction as well as on loading mode, makes different failure modes possible. These failure modes are classified as follows (Fig. 3):

a) rod failure:
it is the mode present in the case of very strong matrix. This is the failure mode recommended by the norms, like ACI 349. In fact, the current design philosophy is based on the hypothesis of yielding of the rod before its slipping or concrete cone failure;

b) concrete cone failure:
it is the typical failure of short anchors but it is also possible to observe it when the rod is made in high strength material. This failure occurs even when the anchors are near the edges or when the distance between each other is too small. Normally the crack enucleates from the end of the bolt head, or from the expansion mechanism, and develops in the matrix with an angle that depends on the strength of concrete and on the embedded length;

c) rod slipping:
This failure can occur due to the lack of mechanical

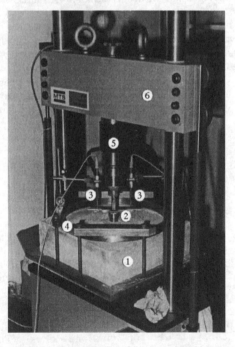

Fig. 1 Testing set-up

Abb.1 Testvorichtung

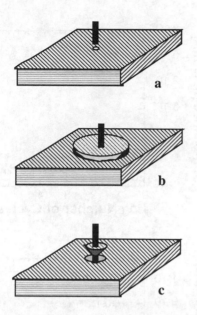

Figure 3 Failure mode of anchor bolts

Abb. 3 Buch Anzeigen

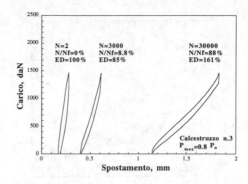

Fig. 4 Variation of cycle shape in fatigue test.

Abb. 4 Variationen von Fatigue Zyklen

response, like in the case of expansion anchors, or even when the chemical bond was broken.

4 DAMAGE EVOLUTION DURING THE FATIGUE LIFE

As the number of cycles increases the anchorage subjected to fatigue loading shows progressive damage. The fatigue damage is a consequence of increasing internal cracking in concrete. In Fig. 4 the variation of the cycle shape during the fatigue test of anchor bolt is shown.

The anchorage behaviour is strongly influenced by its geometry. The substantial difference is located in the damage types produced during the anchorage life. Consequently, the anchorage life is strictly connected to damage developed. According to the brittle or ductile failure of anchorage its geometry must be strictly examined.

Energy dissipated (E_D) of cycles, compliance (C) and displacement of anchorage (η_{max}) have been chosen in order to study anchorage behaviour.
In the case of anchor bolts the energy dissipated decreases after the beginning cycles, and increases steadily thereafter, up to failure. This does not occur for the rod and ribbed bars, where the energy dissipated continues to decrease, as shown in Fig. 5. The same behaviour was observed for the compliance of anchorage (Fig. 6).

Strain evolution and stiffness degradation with the number of cycles in anchor bolts were found to be similar to those of normal concrete.
Fig. 7 shows that all the three anchorage geometries

have a similar displacement behaviour.

In the case of anchorage the displacement of anchor bolt appears in suitable quantity so as to check a fatigue process (Shah, 1984).

Strain evolution and stiffness degradation with the number of cycles in anchor bolts were found to be similar to those of normal concrete.
Fig. 7 shows that all the three anchorage geometries have a similar displacement behaviour.

In the case of anchorage the displacement of anchor bolt appears in suitable quantity so as to check a fatigue process (Shah, 1984).

These three parameters clearly indicate that the increase of damage with cyclic loading is highly nonlinear; therefore, the Miner's hypothesis is not valid for the structural element examined.

The anchorage fatigue life may be predicted more effectively through a relationship based on the increase in the displacement of the load application point, η_{max}, as a function of the number of cycles rather than through a relationship based on the crack propagation velocity as a function of the number of cycles as in metals (Bocca et al., 1992).

5 DISCUSSION ON FATIGUE FAILURE HYPOTHESIS

In order to check the presence of fatigue process the displacement of anchors is the most suited feature.

The same conclusion as Hordijk's (1991) -reached through a theoric-experimental study that described a local approach to fatigue concrete- was reached by this paper.

The displacement, more generally the deformation, offers the possibility to understand if the fatigue process is in a stable or an unstable zone.

Therefore, the capital problem is to find if a link between the static and dynamic deformation exists.

Hordijk, with his experimental investigation, intended to find out which is this relation and if the descending branch of static test was also the limit for the cyclic test.

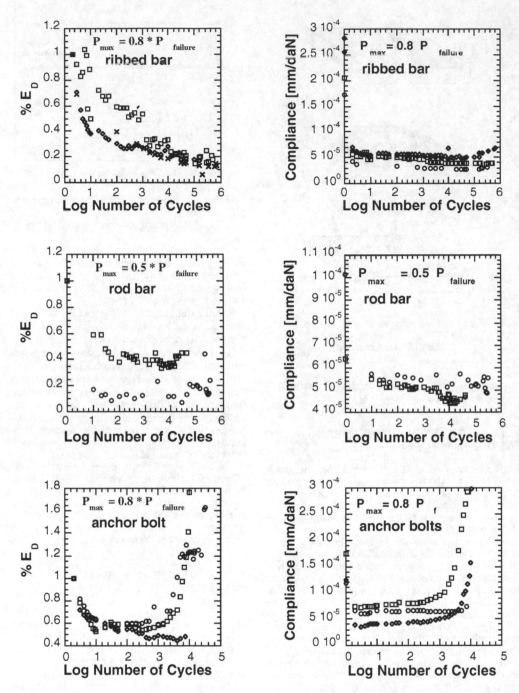

Fig. 5. Energy dissipated per cycles referred to 2°
cycle vs number of cycles.

Abb. 5 Größenvariationen von abgegebener Energie
pro Zyklus

Fig. 6. Compliance vs. number of cycles

Abb 6. Nachgiebigkeit von Zylen

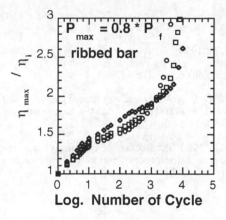

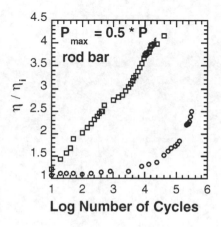

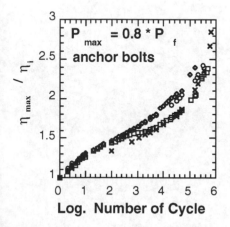

Fig. 7. Displacement referred to first cycle as a function of number of cycles

Abb. 7. Verschiebung der Anker von Zyklen

The hypothesis also discussed this paper, is that the descending branch of pull-out test's load-displacement curve is the boundary for the displacement in fatigue tests.

As it has been reported before, the load-displacement curve during the cycles shows damage that consists of a decrease of slope with respect to the displacement axis hence, increase of compliance for the whole system.

The failure occurs when the cyclic load-displacement intersects the descending branch of the static pull-out curve.

It is necessary to make some comments before verifying this hypothesis.

It is difficult to know the exact pull-out curve for each type of anchor. In fact each anchor possesses its own curve, in the sense that this depends on numerous factors as the concrete composition, the disposition of particular aggregate near the anchor, the modality of extraction and so on.

It must be considered that the deduction and the hypothesis are referred to on a mean behaviour;

Point B in Fig. 8 represent the start of linear growth of displacement; point C represents the end of the such growth and the non linear growth begins up to reach the point D where the failure occurs.

In this paper the hypothesis that point C is the displacement that corresponds to the displacement of static pull-out load has been considered. The same hypothesis for the description of τ-slip law of bars embedded in concrete (Balazs, 1986) was also used.

The tests on anchor bolts have shown a behaviour similar to Hordijk's observations.

In tab 1 the displacement recorded at the stable growth end and the static value for some anchors are reported.

Therefore, it is possible to define a failure criterion based on displacement because a relationship between the static and dynamic deformation (or displacement) recorded in static and cyclic tests respectively exists.

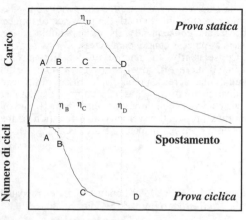

Fig. 8 Hypothesis scheme

Abb. 8 Hypothese

Tab. 1 Experimental results on failure hypothesis

Tabb. 1 Versuchs werte

Type	ηstatic [mm]	ηcyclic [mm]	% σ/σ_u	N_f
anchor bolt	0.310	0.365	60	15000
	0.310	0.327	68	40000
	0.310	0.561	78	2500
	0.310	0.496	89	4000
	0.310	0.320	68	12250
	0.310	0.312	84	5000
ribbed bar	1.093	0.748	80	150000
	1.093	0.914	72	150000
	1.093	1.061	84	5000

As a result of this hypothesis, it is possible to point out the fundamental features that govern the cyclic behaviour by means of the static pull-out test. The maximum displacement is represented by the intersection of the line at predetermined percentage of load and the descending branch of the static pull-out test. However, in the fatigue life the of anchor it is better not to exceed the displacement at static failure because in this case the process of fatigue is in the unstable zone.

6 CONCLUSION

Experimental results have shown that the concrete damage has a non linear growth during the fatigue life. The damage is influenced by anchor geometries thus one needs to consider this aspect in anchor type choice, especially if subjected to cyclic load .

By comparing the load-displacement curves it is observed that a relationship between displacement in static and dynamic tests exists. Therefore, by means of the static pull-out test, it is possible to point out the fundamental features that govern the cyclic behaviour.

The selected variable trends show how only the displacement is able to mark the presence of fatigue process; and consequently, the others should not be chosen as process control parameters.

Consequently, the anchor fatigue life may be predicted more effectively through a relationship based on the increase in the displacement of the load application point, η_{max}, as a function of the number of cycles rather than through a relationship based on the crack propagation velocity as a function of the number of cycles as in metals (Bocca et al., 1992).

REFERENCE

Balazs, G.L. 1986. Bond behaviour under repeated loads. *Studi e Ricerche*. 8: 395-430.

Bocca, P., Cadoni, E., Valente, S. 1992. On concrete fatigue fracture in pull-out tests, in H.P. Rossmanith (ed), *Fracture and Damage of Concrete and Rock*:: 637-646. London: Chapman & Hall.

Cadoni, E. 1994. *Sul comportamento a fatica degli ancoraggi nel calcestruzzo*. Doctoral Thesis, Politecnico di Torino: Torino.

Hordijk, D.A 1991. *Local approach to fatigue of concrete*. Doctoral Thesis, Technische Universiteit Delft: Delft.

Shah, S.P. 1984. Predictions of cumulative damage for concrete and reinforced concrete. *Material and Structure*. 17: 65-68.

Anker in Theorie und Praxis, Widmann (Herausgeber) © 1995 Balkema, Rotterdam. ISBN 90 5410 577 1

Besondere technische Aspekte bei der Überwachung und meßtechnischen Begleitung der Felsankerarbeiten an der Edertalsperre

Notable technical aspects of the supervision and supportive measurement of the rock anchor project at the Eder dam

H.Gaitzsch
Stump Spezialtiefbau GmbH, Hannover, Germany

ZUSAMMENFASSUNG: Gegenstand des Beitrages sind projektspezifische Sondermaßnahmen bei der Planung, Ausführung und Überwachung der komplizierten Bohr-, Injektions- und Ankerarbeiten, die sich aus der ungewöhnlichen Dimension der Ankerkräfte von 4500 kN/Anker, der geforderten Bohrgenauigkeit von < 1 % und den Korrosionsschutzmaßnahmen ergeben. Bei der Bauausführung kam ein umfangreiches Qualitätssicherungssystem zur Anwendung, das an ausgewählten Beispielen erläutert wird.

ABSTRACT: The subject of the presentation centers on specific project requirements in the planning, execution and supervision of the complicated tasks of drilling, injecting and anchoring. These challenges arise through the abnormal dimensions of the anchor force of 4500 kN/anchor, the required drilling precision of less than 1 % and the actions required to prevent corrosion. In the construction process, an extensive system of quality control was implemented, which will be explained in detail through selected examples.

1. Qualitätssicherung durch Eigen- und Fremdüberwachung bei den komplizierten Bohr- und Ankerarbeiten

Im Beitrag von SCHWARZ (1995) wurde über wesentliche kennzeichnende Anforderungen und Ausführungsgrundsätze beim Einbau von 104 Dauerfelsankern System STUMP-SUSPA, Typ 6-34, in die Ederstaumauer berichtet.

Gegenstand dieses Beitrages sind weiterführende Betrachtungen zur Sicherung des erforderlichen hohen technisch-technologischen Qualitätsstandards bei der Planung, Ausführung und Überwachung der einzelnen Arbeitsschritte.

Qualität wird dabei als ein Maß der Übereinstimmung von technisch-technologischen Vorschriften mit der tatsächlichen Ausführung vor Ort betrachtet. Die besonderen Anforderungen an die Qualität der Ausführung ergeben sich aus der ungewöhnlichen Dimension der Ankerkräfte mit 4500 kN/Anker, der geforderten Bohrgenauigkeit von < 1 % im Mauerbereich und den Forderungen an die volle Gebrauchsfähigkeit der Staumauer für weitere 80 - 100 Jahre. Hierzu waren besondere Korrosionsschutzmaßnahmen notwendig.

Die Ausführungsqualität wurde durch ein System der Eigen- und Fremdüberwachung gewährleistet, deren Arbeitsweise an ausgewählten Beispielen dokumentiert werden soll.

2. Die Genauigkeit der Ankerbohrungen - Ausführung, Meßverfahren und Dokumentation

Die abwechselnd 68 und 73 m tiefen Bohrungen mußten mit einer Neigung von exakt 3,2 ° und einem maximal zulässigen Krümmungsradius von 500 m durch das Grauwacke-Bruchsteinmauerwerk der Staumauer und den aus Grauwacke und Tonstein bestehenden Untergrund abgeteuft werden. Der Enddurchmesser der Bohrungen war vom gewählten Ankersystem abhängig und betrug 273 mm. Die zulässigen Bohrtoleranzen von maximal 1 % Bohrabweichung in Höhe des Kontrollganges und 2 - 3 % bei Endteufe der Bohrungen ergaben sich aus konstruktiven Gegebenheiten der Staumauer. So betrug der Abstand der Bohrung zur wasserseitigen Dichtung der Mauer und zum Kontrollgang nur ca. 1,25 m (vgl. Abb. 1). Erschwerend kam hinzu, daß der exakte Verlauf der Außendichtung der Mauer infolge fehlender Vermessungsunterlagen nicht bekannt war, so daß mit einer zusätzlichen Minimierung des zur Verfügung stehenden Bohrbereiches

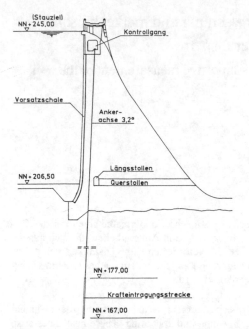

Abb. 1: Staumauerquerschnitt mit Lage der Ankerbohrung und dem unteren Kontrollgang.
Cross-section of the dam, including the position of the anchor drilling and the lower control tunnel

gerechnet werden mußte.

Ähnliches gilt für den unteren Kontrollgang selbst. Er wurde nach dem Staumauerbau mit bergmännischen Methoden aufgefahren und verzeichnete an der wasserseitigen Wandung Toleranzen von mehreren Dezimetern.

Der Abstand der Anker und damit auch der 10 m langen Krafteinleitungsstrecken betrug im Minimum nur 1,85 m (i. M. 2,25 m).

Aufgrund dieser beengten geometrischen Verhältnisse waren Ersatzbohrungen grundsätzlich ausgeschlossen. Die Möglichkeit von Kurskorrekturen während des Abteufens der Bohrungen wurden im Vorfeld eingehend untersucht, infolge der technologischen Unzulänglichkeiten derartiger Bohrverfahren aber verworfen. Daraus ergab sich die Forderung, daß alle 104 Bohrungen, d.h. exakt 7332 Bohrmeter, ohne Fehlversuch und mit der geforderten Genauigkeit abgeteuft werden mußten.

Das Grauwacke-Bruchsteinmauerwerk der Staumauer war durch extreme Druckfestigkeitsunterschiede der Grauwacke einerseits und des z. T. entfestigten Mörtels gekennzeichnet. Die daraus resultierenden Querkräfte auf die Bohrkrone verstärkten

laterale Abweichungstendenzen.

Risiken für den geradlinigen Verlauf der Bohrungen ergaben sich im Untergrund der Staumauer aus einer mit 60 - 70 ° einfallenden rheinisch streichenden Störungszone mit einer Vielzahl eggisch und rheinisch streichenden Klüften. Eine teilweise Tonfüllung mit geringeren Gesteinsfestigkeiten war zu erwarten.

Ausgehend von den Erfahrungen bei der Herstellung der Probeanker im Tosbecken der Staumauer (SCHWARZ, 1995) ergaben sich bei der Herstellung der Bohrungen folgende Überwachungsschwerpunkte für das ausführende Bohrpersonal:

• Exaktes Einrichten des Bohrgerätes am Bohrpunkt und des Bohrstranges am Bohransatzpunkt, der sich 6,6 m unterhalb im Kontrollgang befand. Die Einmessung des Gerätes erfolgte durch einen unabhängigen Vermesser mit Theodolit. Erst

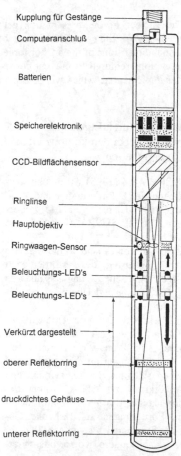

Abb. 2: Aufbau der MAXIBOR-Meßsonde
Construction of the MAXIBOR measurement probe

nach Vorlage der Ergebnisse wurde der Bohrbeginn durch den verantwortlichen Bohringenieur freigegeben.

• Ständige Überprüfungen der Maßhaltigkeit und Funktionsfähigkeit der für das richtungsgenaue Bohren wesentlichen Gerätekomponenten wie z. B. der Spezialhydrauliksysteme zur Neigungseinstellung auf 3,2 °, der Präzisionslafette des Bohrkopfes und des aus hochfesten Prismenschienen bestehenden Verschiebetisches des Bohrgerätes.

• Akribische Untersuchung des Bohrstranges auf Risse, Unwuchten, Abnutzungsgrad der Krone, Zustand der Räumer und Stabilisatoren.

• Durchsetzung eines Bohrregimes, das der Zielgenauigkeit und Sicherheit gegen eine Bohrlochhavarie äußerste Priorität auferlegte.

Die Umsetzung dieser Grundsätze, die ständige Motivation der Mitarbeiter und die Vermeidung von Nachlässigkeiten und Routineerscheinungen war eine zentrale Aufgabe der verantwortlichen leitenden Mitarbeiter. Zur Unterstützung der Zielstellung wurde ein umfangreiches internes Dokumentationssystem eingeführt, das teilweise erheblich über die bauseitigen Forderungen hinausging.

Eine zweite wichtige Komponente der Kontrolle des Bohrverlaufes bestand in der permanenten begleitenden Bohrlochvermessung mit dem REFLEX MAXIBOR System.

Die 3 m lange MAXIBOR-Sonde (vgl. Abb. 2) wird für den Meßeinsatz mit einem Stahlseil in das Bohrloch abgelassen, nachdem sie mit dem trag-baren Computer am Bohrlochmund geeicht wurde. Über Präzisionsstabilisatoren wird die exakte Fixierung im Bohrloch oder im Bohrgestänge gewährleistet.

Die Sonde beinhaltet ein optoelektronisches Meßsystem. Kernstück ist ein CCD-Bildflächensensor, der auftreffende Lichtsignale registriert und in digitale Signale umwandelt, die als Koordinaten in einer Auswerte- und Speicherelektronikeinheit abgespeichert werden. Der Auftreffpunkt des Lichtstrahles ändert sich durch eine minimale Verbiegung der Sonde, entsprechend dem Bohrlochverlauf. In die Ebene der Reflektorringe ist eine Libelle integriert, die die Senkrechte anzeigt.

Registriert werden x-, y- und z-Koordinaten, Neigungs- und Richtungswinkel. Während des Meßvorganges besteht keine Datenverbindung zur Erdoberfläche. Die Werte werden nach Beendigung des Meßdurchganges aus der in der Sonde integrierten Speichereinheit in einen Handcomputer eingelesen und sofort vor Ort ausgewertet und auf dem Bildschirm angezeigt. Damit können die notwendigen Informationen unmittelbar an das Bohrpersonal gegeben werden.

Eine umfangreiche Auswertesoftware gestattet eine zeichnerische Darstellung der

• xy-, yz- und zx-Flächen
• Neigungsdifferenz gegen die Tiefe
• Richtungsdifferenz gegen die Tiefe.

Abb. 3 zeigt beispielhaft ein Meßergebnis mit Soll- und Ist-Linie bei einer Bohrlochteufe von - 57 m.

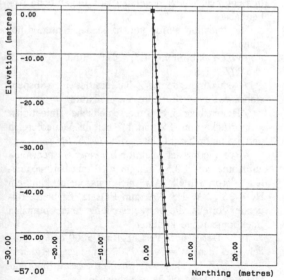
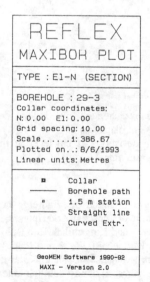

Abb. 3: Grafische Darstellung des Bohrlochverlaufs
Graphic representation of the drill hole path

413

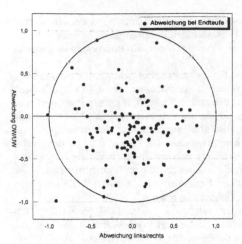

Abb. 4: Bohrlochabweichungen bei Endteufe der Bohrungen
Drill hole deviations at the final drilling depth

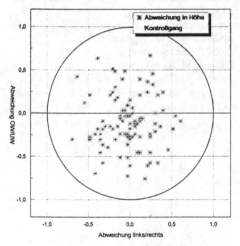

Abb. 5: Bohrlochabweichungen in Höhe des unteren Kontrollganges
Drill hole deviations at the level of the lower control tunnel

Das Ergebnis der Messungen bei Endteufe der Bohrungen und in Höhe des unteren Kontrollganges ist in den Abbildungen 4 und 5 ersichtlich. Die Bohrtoleranzen liegen im Mittel bei 0,45 % und 0,36 %. Damit wurden die Forderungen von 1 % in Kontrollgangshöhe und 2 - 3 % bei Endteufe deutlich unterschritten. Der Koordinatenvergleich der aus dem Bohransatzpunkt berechneten Sollbohrlinie mit Werten, die mit einem zweiten, vollkommen unabhängi-

gen, induktiven Peilverfahren aus dem unteren Kontrollgang gewonnen wurden, ergab nur Abweichungen von wenigen Zentimetern. Damit konnte die Zuverlässigkeit des MAXIBOR-Systems nachgewiesen werden.

Eine zusätzliche Sicherheit erbrachten Zielkernbohrungen in Richtung Ankerbohrloch vom unteren Kontrollgang aus. Der Bohransatzpunkt resultierte aus dem Meßergebnis der induktiv arbeitenden Sonde. Die Ankerbohrung wurde in jedem Fall exakt getroffen.

Damit konnte das mit bemerkenswerter Präzision funktionierende komplexe System der Qualitätskontrolle bei der Herstellung der 104 Ankerbohrlöcher eindrucksvoll bestätigt werden.

3. Ausführungsgrundsätze und Überwachung der Zementinjektionsarbeiten

Nach Abteufen der Kernbohrungen wurden die Bohrlöcher mit Zementsuspension verpreßt. Die Injektion hatte folgende Zielsetzung:

1. Wiederherstellung der vollen Funktionsfähigkeit des durchbohrten alten Injektionsschleiers.

2. Erhöhung der Gefügefestigkeit des Gebirges im Krafteinleitungsbereich der Anker.

3. Optimale Abdichtung des bohrlochnahen Bereichs der Krafteinleitungsstrecke der Anker. Der WD-Wert des abschließenden Wasserdurchlässigkeitstests mußte < 1 l/min $\times$ m bei 10 bar (= 1 Lugeon) betragen.

4. Vermeidung von Umläufigkeiten zwischen den eng beieinander liegenden Bohrungen beim Ankereinbau.

Die Injektion wurde von folgenden Kriterien bestimmt:

• W/Z-Faktoren von 1,5; 1,0; 0,8 und 0,6; Beginn mit W/Z = 1,5.

• Reduzierung des W/Z-Faktors bei einer Suspensionsaufnahme > 400 l pro Verpreßstufe.

• Teufenabhängige unterschiedliche Injektionsgrenzdrücke von 0,5 und 1,5 bar im Mauerbereich und 3,5 und 6 bar im Untergrund (Abb. 6).

• Beendigung der Injektion bei einer Suspensionsaufnahme von < 0,5 l/min über 10 min bei maximalem Verpreßdruck. Als Injektionszement kam ein HOZ 35L NW HS NA zum Einsatz. Er bietet folgende Vorteile, die der Zielstellung einer optimalen Injektion entgegenkommen:

1. Der Blaine-Wert liegt mit 5000 cm²/g deutlich über dem Wert von 3700 für einen vergleichbaren HOZ 35. Damit ist eine verbesserte Eindringfähigkeit in feine Klüfte gewährleistet.

2. Der Korngrößenanteil von 0,05 mm liegt mit 97 % über dem Wert von ca. 90 % für normalen HOZ.

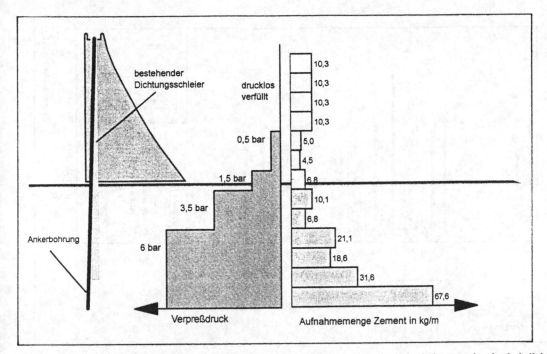

Abb. 6: Querschnitt der Staumauer mit altem Injektionsschleier, neuer Ankerbohrung, durchschnittlicher Zementaufnahme und Injektionsdruck.
Cross-section of the dam with old grout curtain, new anchor drill hole, average volume of cement and injection pressure

3.   Infolge des höheren Schlackenanteiles ist eine verbesserte Beständigkeit gegen aggressive Einflüsse des Grundwassers gegeben. Daraus ergibt sich eine verbesserte Langzeitbeständigkeit des Zementsteines.

4.   Durch den Zusatz eines Plastifikators mit 0,5 Gew.-% des Zementes (Tricosal 188 EH) wurden Sedimentationsstabilität und Fließfähigkeit erhöht.

Die Injektionsarbeiten erfolgten im durchgehenden Tag- und Nachtschichtbetrieb. Sie bildeten einen Schwerpunkt der Überwachungstätigkeit durch das bauüberwachende Ingenieurbüro (WBI) und der Ausführungsfirma. Folgende Werte wurden registriert:

• Injektionsdruck unmittelbar am Bohrloch und Verpreßmenge durch Injektionsschreiber.
• Injektionsrate
• Mischungsverhältnis der Zementsuspension
• Viskosität und Dichte der Suspension
• Druckfestigkeit von Rückstellproben.

Mitarbeiter der Bauüberwachung waren ständig vor Ort, um gegebenenfalls im Einvernehmen mit der Ausführungsfirma in das Injektionsregime eingreifen zu können. Dadurch konnten die Vorgaben wechselnden geologischen Gegebenheiten sofort angepaßt werden. Ein "starres" Injektionsregime wurde dadurch konsequent vermieden. Insgesamt wurden in mehr als 2000 Injektionsbetriebsstunden 100 t Zement verpreßt. Die Zementaufnahmen der einzelnen Bohrlöcher zeigt Abbildung 7.

Unterhalb des in Abbildung 6 dargestellten alten Injektionsschleiers ergab sich eine Zementaufnahme von 44,5 kg/m. Innerhalb des Schleiers lag dieser Wert bei 14 kg/m.

Das Ergebnis der Injektion ist aus Abbildung 8 ersichtlich. Der mittlere ermittelte WD-Wert liegt bei 0,21 Lugeon. Damit wurde die Zielstellung von 1,0 Lugeon deutlich unterschritten und das wesentliche Kriterium für die Freigabe des Bohrloches zum Ankereinbau erfüllt.

Lediglich bei 8 der insgesamt 104 Bohrungen wurde zum Erreichen des geforderten Dichtekriteriums eine zweite Injektion erforderlich.

4.   Qualitätssicherung bei der Fertigung, dem Transport und dem Einbau der 4500 kN Dauerfelsanker

Die 104 Dauerfelsanker sind die tragenden Elemente

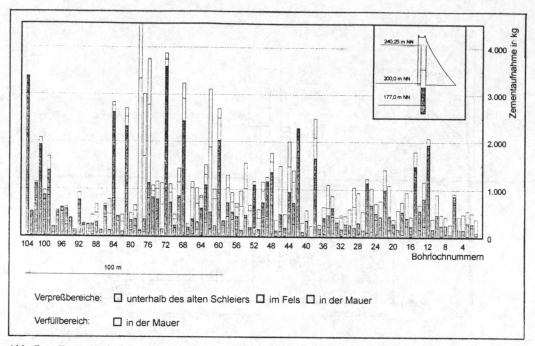

Abb. 7: Zementaufnahme in den Bohrlöchern in kg in verschiedenen Tiefenbereichen
Cement volume in the drill holes in kg at diverse depth levels

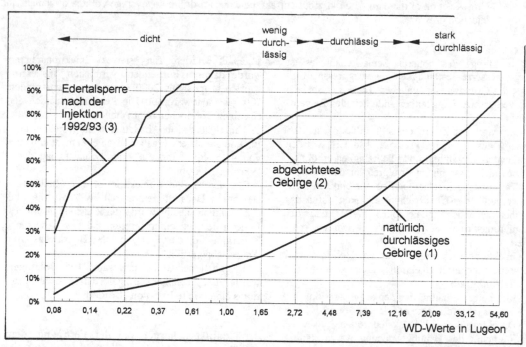

Abb. 8: Summenkurven von WD-Werten im Gebirge in natürlichem (1) und abgedichtetem Zustand (2) nach SCHADE (1977); ergänzt mit der Summenkurve der Ederstaumauer (3).
Summary curves of water permeability test values in rock in natural (1) and sealed states (2), following SCHADE (1977); suplemented by summary curve of the Eder dam (3) (in lugeon).

der Instandsetzung der Edertalsperre. Gemäß DIN 4125 ist bei Dauerankern grundsätzlich der Korrosionsschutz für das Stahlzugglied und alle Einzelteile des Ankers unter werksmäßigen Bedingungen herzustellen. Für die Firma STUMP BOHR GmbH werden deshalb alle Anker in den SUSPA-Werken Königsbrunn und Langenfeld gefertigt und dann auf die Baustellen transportiert.

Für die Daueranker der Ederstaumauer gelten in bezug auf die Fertigung besondere Maßstäbe, die aus ihren Dimensionen ersichtlich werden:
• Länge der Anker im Mittel 71 m
• 10 m lange vorverpreßte Haftstrecke, umhüllt mit einem Ripprohr.
• Durchmesser des PE-Glattrohres im Bereich der freien Ankerlänge 200 mm, Wandstärke der Rohre 11,4 mm, maximaler Biegeradius 2,50 m
• Hohes Gesamtgewicht von ca. 3,5 t/Anker.

Aus diesen Angaben wird ersichtlich, daß ein Antransport auf der Straße nicht in Frage kam. Neben den logistischen Gesichtspunkten war dabei auch die Gefahr von mechanischen Beschädigungen beim Transport, dem mehrfachen Umladen mit Kraneinsatz und der Zwischenlagerung maßgebend für die Entscheidung, die Dauerfelsanker unmittelbar neben der Sperrmauer unter werksmäßigen Bedingungen in einem Baustellenwerk der Firma SUSPA Spannbeton GmbH zu fertigen.

Das ermöglichte eine schnelle Rückkopplung zwischen Einbau und Produktion, eine größere Lagerhaltung konnte dadurch vermieden werden.

Die Fertigung gliedert sich wie folgt:
• Anlieferung der im SUSPA-Werk Königsbrunn vorbereiteten Bestandteile des Ankers im Baustellenwerk.
• Zusammenschweißen der PE-Glattrohre durch Heizspiegelstumpfschweißung.
• Einstoßen der Litzen in das PE-Glattrohr nach vorheriger Kontrolle des freien Durchganges mit einem Kaliberstopfen. An beiden Hüllrohrenden sind Schablonen mit exakter Kennzeichnung der Litzenanordnung vorhanden.
• Aufziehen des PE-Ripprohres nach vorherigem Anbringen der Distanzhalter und Stahlbänder für eine optimale Mörtelausbreitung im Haftstreckenbereich.
• Verbindung von Ripp- und Glattrohr mit einer Stahlmuffe, Abdichtung mit einem aufgeschrumpften Kunststoffschlauch.
• Auflegen des Ankers auf die geneigte Fertigungsbühne.
• Verpressen der Krafteintragungsstrecke mit einer Zementsuspension, W/Z-Wert 0,36 - 0,40, Einpreßhilfe Tricosal H 181 mit 0,5 % des Zementgewichtes.

• Exakte Kontrolle der im Kontraktorverfahren durchgeführten Verpressung über eine später verschließbare Entlüftungsöffnung.
• Abheben des Ankers frühestens nach 3 Tagen von der Fertigungsbühne und Ablage auf ein spezielles Rollenlager.

Alle Arbeitsschritte wurden gemäß DIN 4125 durch ein Eigenüberwachungssystem entsprechend den Vorgaben der Allgemeinen Bauaufsichtlichen Zulassung und der von der Wasser- und Schiffahrtsverwaltung in Zusammenarbeit mit dem Institut für Bautechnik Berlin erteilten Zustimmung im Einzelfall überwacht. In dieser Zulassung sind die Anforderungen an die Ankerkomponenten und alle notwendigen Arbeitsschritte detailliert beschrieben. Beginnend mit dem Eingang der einzelnen Bauteile auf der Baustelle bis zum Ende der Fertigung zeichneten die Mitarbeiter alle Prüfergebnisse in Protokollen ab. Eine dieser Checklisten der SUSPA Fertigungskontrolle und Eigenüberwachung zeigt Abbildung 9.

SUSPA Werk Edertalsperre				FFE-1
Fertigung der Felsanker 6–34 für die Edertalsperre				Schü/Fri
SUSPA–Fertigungskontrolle und Eigenüberwachung				10.11.92

Anker–Fertigungs–Nr.		TypA: lang	TypB: kurz
PE–Glattrohr		DATUM	NAME
Markieren mit Fertigungs–Nr.			
Schweißstelle A			
Schweißstelle B			
Schweißstelle C			
Gesamtlänge	L= m		
Freier Durchgang			
PE–Montageverlängerung			
Sortierscheibe/–büchse			
Herstellen des Zuggliedes			
Anordnen der Stahlmuffe			
Anordnen der 10 Distanzhalter			
Litzen–Trommel Nr.	Nr.:		
Einziehen der 34 Litzen			
Einziehen der 2 Verfülleitungen			
Haftstrecke rostfrei			
Flugrost			
fettfrei			
Länge 10 m +/– 0,2 m			
Abstd. zur Stahlmuffe 0,3–0,5m			
Bündelung u. Schutzkappe			
Aufziehen des PE–Ripprohrs			
Visuelle Kontrolle des Ripprohrs			
Aufziehen bis über Stahlmuffe			
Länge in Ordnung			
Endkappe			
Verbindung Glattrohr/Ripprohr			
Befestigung des Glattrohrs an der Stahlmuffe			
Schrumpfschlauch			
Verpreßebene			
Auflegen des Ankers			
Lagesicherung			
Hochheben der Verpreßebene			
Bohren der Entlüftungsbohrung			
Verpressen mit Zementeinpreßmörtel			
Güteprüfung in Ordnung			
Rückstellproben für Erhärtungsprüfung			
Protokoll 12 erstellt			
Protokoll 13 erstellt			
Einpreßvorgang in Ordnung			
Abheben von der Verpreßebene			
Schließen der Entlüftungsbohrung			
Abheben u. seitlich Lagern			
Bemerkungen			
Freigabe Eigenüberwachung SUSPA:		Datum	Name

Abb. 9: Beispiel einer Checkliste der Ankerfertigung
Sample checklist for anchor production

Transport und Einbau:
Ankernummer: A/B _____
Bohrlochnummer:_____

ℊump BOHR GmbH

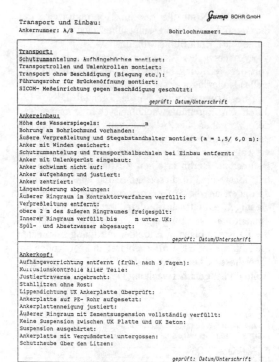

Transport:
Schutzummantelung, Aufhängehöhse montiert:
Transportrollen und Umlenkrollen montiert:
Transport ohne Beschädigung (Biegung etc.):
Führungsrohr für Brückenöffnung montiert:
SICOM- Meßeinrichtung gegen Beschädigung geschützt:

geprüft: Datum/Unterschrift

Ankereinbau:
Höhe des Wasserspiegels: _____ m
Bohrung am Bohrlochmund vorhanden:
Äußere Verpreßleitung und Stegabstandhalter montiert (a = 1,5/ 6,0 m):
Anker mit Winden gesichert:
Schutzummantelung und Transporthalbschalen bei Einbau entfernt:
Anker mit Umlenkgerüst eingebaut:
Anker schwimmt nicht auf:
Anker aufgehängt und justiert:
Anker zentriert:
Längenänderung abgeklungen:
Äußerer Ringraum im Kontraktorverfahren verfüllt:
Verpreßleitung entfernt:
obere 2 m des äußeren Ringraumes freigespült:
Innerer Ringraum verfüllt bis   m unter UK:
Spül- und Absetzwasser abgesaugt:

geprüft: Datum/Unterschrift

Ankerkopf:
Aufhängevorrichtung entfernt (früh. nach 5 Tagen):
Korrosionskontrolle aller Teile:
Justiertraverse angebracht:
Stahllitzen ohne Rost:
Lippendichtung UK Ankerplatte überprüft:
Ankerplatte auf PE- Rohr aufgesetzt:
Ankerplattenneigung justiert:
Äußerer Ringraum mit Zementsuspension vollständig verfüllt:
Keine Suspension zwischen UK Platte und OK Beton:
Suspension ausgehärtet:
Ankerplatte mit Vergußmörtel untergossen:
Schutzhaube über den Litzen:

geprüft: Datum/Unterschrift

Abb. 10: Checkliste Transport und Ankereinbau
Checklist for transportation and anchor
production

Die Fremdüberwachung, das Materialrüfungsamt für Bauwesen der TU München, überprüft das Eigenüberwachungssystem und kontrolliert stichprobenartig die Montagearbeiten und Materialien.

Bedingt durch die ungewöhnliche Dimension der Anker und die hohen Anforderungen vor allem an das sichere Funktionieren des Korrosionsschutzsystems wurde dieses Kontrollsystem durch eine weitere Fremdüberwachung ergänzt. Das Ingenieurbüro Professor Dr.-Ing. W. Wittke Beratende Ingenieure für Grundbau und Felsbau GmbH begleitete die Fertigung der Anker im Auftrage des Bauherrn und gewährleistete eine lückenlose Kontrolle, die sich auch auf den Transport und den Einbau der Anker erstreckte.

Das beschriebene Qualitätssssicherungssystem setzte sich bei den Einbauarbeiten der Anker fort. Besonderes Augenmerk galt auch hier der Vermeidung von mechanischen Beschädigungen. Deshalb wurden alle Transport- und Einbauzustände über die in den Richtlinien der Zulassung im Einzelfall getroffenen Festlegungen hinaus in Arbeitsanweisungen beschrieben und deren Einhaltung durch die Mitarbeiter der Ausführungsfirma protokolliert (Abb. 10).

In den Abbildungen 11 und 12 sind die Arbeitszustände "Aufheben des Ankers auf die Einbaukonstruktion" und "Anheben der Einbaukonstruktion und Einbau des Ankers" dokumentiert. Die Anker wurden durch Umhüllungen aus PE-Material, Stahlhalbschalen und Führungselemente zusätzlich ge-

Phase 2: Aufheben des Ankers auf die Einbaukonstruktion

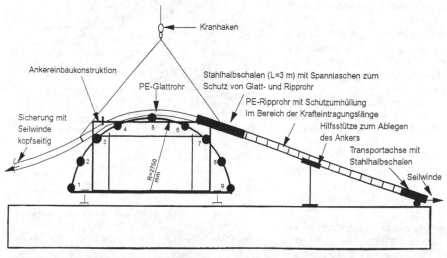

Abb. 11: Arbeitsanweisung zum Aufheben des Ankers auf die Einbaukonstruktion
Work instructions for the lifting of the anchor onto the installation apparatus

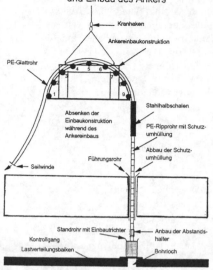

Phase 3: Anheben der Einbaukonstruktion
und Einbau des Ankers

Kranhaken

Ankereinbaukonstruktion

PE-Glattrohr

Absenken der
Einbaukonstruktion
während des
Ankereinbaus

Stahlhalbschalen

PE-Ripprohr mit Schutz-
umhüllung

Abbau der Schutz-
umhüllung

Seilwinde

Führungsrohr

Standrohr mit Einbautrichter
Kontrollgang
Lastverteilungsbalken

Anbau der Abstands-
halter

Bohrloch

Abb. 12:   Prinzip des Ankereinbaus
Principle of the anchor installation

schützt. Eine Kalibrierung der Bohrlöcher garantier-
te einen zwängungsfreien Einbau. Nicht zuletzt der
hohen Fachkompetenz und Sorgfalt der mit dem
Einbau betrauten Mitarbeiter ist es zu danken, daß
auch dieser letzte Schritt des Ankereinbaus rei-
bungslos verlief.

Beim Ablassen der Anker in das Bohrloch wurde
der Hohlraum zwischen den PE-ummantelten Litzen
und dem Glattrohr im Bereich der freien Ankerlänge
mit Zementsuspension verfüllt. Dadurch sollte ein
Einbeulen des Hüllrohres durch den äußeren Wasserdruck vermieden werden. Abschließend erfolgte
die Verfüllung des Ringraumes zwischen Bohrlochwand und Anker über Verpreßleitungen im Kontraktorverfahren, bis die Zementsuspension satt aus dem
Bohrlochmund austrat. Die Anker wurden hängend
eingebaut, die entsprechenden Vorrichtungen sind
frühestens nach 5 Tagen entfernt worden, um so eine
ausreichende Erhärtung des Zementes zu gewährleisten.

Die ca. 300 kg schweren Ankerplatten mit dem
Schutzrohr und der innenliegenden Lippendichtung
wurden im Kontrollgang mit einer speziellen Ver-
fahr- und Hebevorrichtung gesetzt und auf 3,2 °
Neigung justiert. Nach dem Verpressen des äußeren
Ringraumes über eine externe Verfüllleitung erfolgte
das Ausfüllen des Kopfbereiches mit einer Korrosionsschutzmasse.

Mitarbeiter der Bundesanstalt für Wasserbau,

Karlsruhe, die den Ankereinbau und insbesondere
das Spannen der Anker überwachten, untersuchten
vorher den Kopfbereich der Anker durch Endosko-
pieren auf eventuell anhaftendes Kondenswasser.
Spezialgebläse sorgten bei Bedarf für eine zuverlässige Trocknung. Das Setzen der Ankerhauben
bildete nach dem Spannvorgang den Abschluß der
Arbeiten zur Ertüchtigung der Ederstaumauer.

Zur täglichen Kontrolle der Ankerkräfte sind 10
Stück Kraftmeßdosen eingebaut worden. Der Langzeitüberwachung dienen außerdem Lichtwellenleitermessungen, mit denen verfolgt werden kann, ob
und ggf. wie es langfristig zu Kraftumlagerungen
entlang der Verankerungslänge des Stahlzuggliedes
kommt. Die Ankerbüchsen sind so konstruiert, daß
ein Aufsetzen einer Spannpresse jederzeit möglich
ist. Dadurch ist eine Überprüfung der Ankerkraft
durch Abhebeversuche garantiert.

Die zuletzt geschilderten Maßnahmen der Langzeitüberwachung sind der abschließende Baustein
eines umfangreichen Qualitätssicherungs- und Über-
wachungssystems, das der Staumauer für die näch-
ste Generation die geforderte Funktion garantieren
wird.

5.   Qualitätssicherung als Einheit von technisch-technologischen und bauüberwa-
chenden Maßnahmen - zusammenfassende
Betrachtung

Die anspruchsvolle Bauaufgabe zur Verankerung
der Ederstaumauer konnte mit ihren technischen und
technologischen Anforderungen nur durch die
Durchsetzung eines konsequenten Qualitätsmanage-
mentes bewältigt werden.

Die Maßnahmen zur Sicherung einer hohen Qualität der Bauausführung führen u. a. zu einer Minimierung des Ausführungsrisikos und damit für die
Baufirma zu kalkulierbaren Kosten. Für den Bauherrn, der als Auftraggeber der Sanierungsmaßnahme für die Gewährleistung der öffentlichen Sicherheit eines derartig sensiblen Bauwerkes verantwort-
lich ist, ergibt sich im Gegenzug die notwendige
Gewähr für die Betriebsfähigkeit des Staubauwerkes
in den nächsten 80 - 100 Jahren.

Es konnte eindrucksvoll gezeigt werden, daß
Anker einer derartigen Größenordnung in hoher
Qualität herstellbar sind.

Bauausführung und Bauüberwachung müssen bei
derartig komplizierten Bauvorhaben eine untrenn-
bare Einheit bilden.

419

LITERATURHINWEISE:

Schwarz, H. 1995. Die Sicherung der Edertalsperre
- Bericht über eine außergewöhnliche Felsanker-
Anwendung. Tagungsband Int. Symposium -
Anker in Theorie und Praxis - Rotterdam:
Balkema

Schade, D. C. 1977. WD-Versuche in verschiedenen
geologischen Formationen - Ber. 1. Nat. Tagung
Ing.-geol.

Electrical testing of ground anchors

Elektrische Prüfung von Ankern

Marcel Grimm
VSL Ltd, Switzerland

Abstract Permanent ground anchors have to meet stringent corrosion protection requirements because vital parts can not be inspected or replaced. Durability is a sensitive topic, corrosion performance demands continue to increase and due to some bad experiences confidence in permanent installations must be regained.

New concepts of corrosion protection have been developed. Additional barriers, such as double sheating, coated steel and new grouting compounds, have been introduced. While the search for a reliable test method to check the quality of the steel member continues, a procedure to check the integrity of the corrosion protection system has been developed. For the construction of the railway station Stadelhofen in Zürich, Switzerland, ground anchors with a working life exceeding 100 years were required. The anchors had to be electrically insulated to protect them against strong stray currents. To check the integrity of the insulation, a test procedure involving electrical resistance measurements was introduced.

Initial problems with this simple, but sensitive method were overcome and since 1989 electrically insulated anchors have become a standard application in Switzerland.

At first, only the free length and the bond length were "fully" electrically insulated. It was difficult to insulate the stressing anchorage to the same degree. Although a "partial" insulation was achieved to prevent metallic contact between the anchor head and the structure, it was not possible to reinspect the tightness of the sheath after stressing.

In the course of 1992 solutions to these technical problems were found and it is now possible to check the integrity of the encapsulation after stressing, under load and at any time throughout the life of an anchor.

Some important details of the "Recommendations for the Design and Execution of Corrosion Protection of Permanent Soil and Rock Anchors" as produced by the leading prestressing companies in Switzerland including the working and testing procedure are included in the paper.

There is considerable merit in ensuring that the primary corrosion protection barrier is intact and can be kept under surveillance. Perhaps the greatest advantage of electrical testing is that it will lead to improved care and quality of construction.

Zusammenfassung Die Qualität des Korrosionsschutzes ist bei Ankern von großer Bedeutung. Neue Konzepte wie Doppel-Hüllröhre, beschichteter Stahl, etc. sind eingeführt. Noch wird nach einer zuverlässigen Methode für die Stahlprüfung gesucht, anderseits ist das Verfahren der elektrischen Widerstandsmessung zur Integritätsprüfung des Kunststoffhüllröhres voll entwickelt.

Seit 1989 sind in der Schweiz elektrisch isolierte Anker Standard. Zuerst konnte man nur die Haft- und freie Länge prüfen, seit 1992 gibt es jedoch technische Lösungen, die es erlauben, die Integrität des ganzen Hüllrohres inkl. Spannverankerung jederzeit auch nach dem Spannen zu prüfen.

Der Beitrag enthält Einzelheiten der "Empfehlungen für Projektierung und Ausführung des Korrosionsschutzes von permanenten Boden- und Felsankern".

Es ist sicher vorteilhaft, wenn der Korrosionsschutz überprüft werden kann. Der größte Vorteil ist vielleicht die Tatsache, daß bei elektrisch isolierten Ankern mit erhöhter Sorgfalt und Qualität gearbeitet werden muß.

Introduction

General

Permanent ground anchors have to meet stringent corrosion protection requirements because vital parts cannot be inspected or replaced. Following some bad experiences, an increased awareness of the importance of durability has led to demands for improved performance in withstanding corrosion. New concepts of corrosion protection have been developed. Additional barriers, such as double sheathing, coated steel and new grouting compounds have been indroduced. While the search for a reliable test method to check the quality of the steel tendons continues, a procedure to check the integrity of the corrosion protection system has been developed - the Electrical-Resistance-Measurement(ERM) method.

Anchor Construction

In the late 1970s, corrugated plastic sheathing was introduced to protect the bond length, as shown in Fig. 1.

This system remained the standard concept for the corrosion protection of permanent anchors for the following 10 years. The corrosion protection system illustrated in Fig. 1 depicts the state-of-the-practice

in Switzerland in 1985 at the time that development of the electrically isolated anchor described in this paper began.

Corrosion Hazards and Mechanisms

For a description of possible corrosion hazards and mechanisms of corrosion, reference may be made to the FIP state-of-the-art report "Corrosion and corrosion protection of prestressed ground anchorages" published in 1986[8].

Corrosion Protecton Requirements

In principle, the protective system should aim to exclude a moist gaseous atmosphere around the metal by totally enclosing it in an impervious covering or sheath. The American PTI recommendations[5] specify encapsulation over the full length only for anchors placed in an aggressive environment. The German code DIN 4125[6] requires protection over the full length in all cases. The British Standard BS 8081[7] and FIP[4] require at least one physical barrier against corrosion. Finally, the Swiss Standard SIA 191[9], being the oldest of the compared regulations, keeps the requirement very general indeed and implies a minimum grout cover of 20mm. Despite the lack of rigorous official specifications, the standard

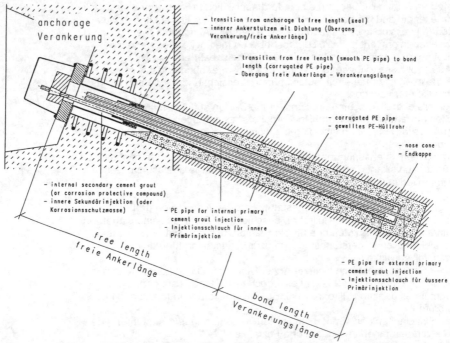

Fig. 1. State-of-the-practice in Switzerland in 1985
Abb. 1. Standard-Ausführung der Anker in der Schweiz 1985

actually practiced in Switzerland is very high, as will be demonstrated here; these higher standards are to be reflected in the 1994 issue of the anchor standard.

Encapsulation with polyethylene pipes is the most widely employed system worldwide. In Switzerland this type of protection has in the past been referred to as double corrosion protection, the PE sheathing provides the first barrier and the grout forms the second barrier. According to FIP[4] and BS 8081[7], the grout injected insitu to bond the tendon does not constitute a part of the corrosion protection system. Accordingly, this arrangement would be classified as single protection. The German code DIN 4125[6], recognises the grout as a corrosion barrier provided that grout is encapsulated in a pipe of sufficient integrity. Apparently, most if not all engineers are required by the prevailing code to supply fully encapsulated anchors. It becomes less relevant whether the system is referred to as "double protection" or "single protection", provided that there is at least one sound barrier and its integrity can be verified.

Development of ERM

Partially Electrically Isolated Anchors

For the construction of the railway station Stadelhofen in Zurich, starting in 1985, 945 permanent soil anchors with a working life of over 100 years were required to support a piled wall. Since the station is located in the middle of the city,

close to one of the city's most powerful rectifiers, it was apparent that stray electric currents would penetrate the retaining wall and the railway tracks. Since stray electric currents are known to produce corrosion it was necessary to provide the anchors with suitable protection.

Electrical Isolation

Suitable protection means electrical isolation provided by an electrically non-conducting and watertight encapsulation as provided by a PE sheath. Due to the exceptional importance of the electrical isolation, a group of specialists[2][3] initiated a proposal to check the integrity of the encapsulation by electrical resistance measurements (ERM). A high resistance confirms that the encapsulation is intact; a low resistance, however, indicates damage to the plastic sheathing. It was determined that the anchor free length and bond length had to provide a minimum electrical resistance of 0.1MOhm.

It was further decided that anchors which failed to meet this requirement would have to be fitted with electrically isolated anchor heads to avoid contact between the anchor and the reinforcement of the retained structure [2]. For measuring techniques and instruments used, refer to the summary of recommendations[1] below as well as to Figures 2,3 and 4.

Bond and Free Length

In principle, the existing concept should have met the specification, except that the grout pipe used to

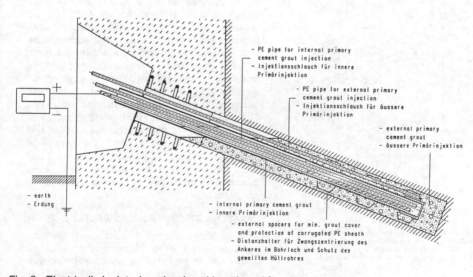

Fig. 2. Electrically isolated anchor bond length and free length

Abb. 2. Elektrisch isolierte Haftstrecke und Freispielstrecke eines Ankers

inject the outer primary grout had to be placed on the outside of the PE encapsulation (Fig. 2). However, the first measurements were very disappointing as significant deficiencies were detected. In addition to the structural deficiencies, it was found that, in the absence of quality control measures, anchors were handled rather carelessly, leading to damage of the encapsulation. It is not surprising that the transition from the smooth pipe to the corrugated pipe, as well as the nose cone at the bottom of the anchor were among the weak spots. Unfortunately, the transition from free length to the bond length is also the location where the grout is subject to cracking due to the strains associated with load transfer. As is well known, two pieces of PE can be bonded to each other by hot-mirror, extrusion or rod welding. However, joining of the PE is more difficult if the two mating pieces are of different diameter and/or wall thickness. Joining with shrink hose provides the best alternative. However, the ERM-method is so sensitive, that even the smallest leaks, humid joints or connections will not escape detection. The corrugated PE sheathing proved to be a "weak spot"

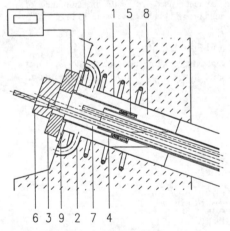

1 – Outer trumpet, flansh and spiral
   – Äusserer Ankerstutzen mit Flansch und Wendel
2 – Pipe for injection of 8
   – Injektionsanschluss für sekundäre Ausseninjektion
3 – Bearing plate
   – Ankerplatte
4 – Inner trumpet
   – Innerer Ankerstutzen
5 – Seal
   – Dichtung
6 – Anchor head
   – Ankerkopf
7 – Internal secondary cement grout (or corrosion protective compound)
   – Innere Sekundärinjektion (oder Korrosionsschutzmasse)
8 – Outer secondary cement grout, injected after ERM
   – Äussere Sekundärinjektion, injiziert nach el. Widerstandsmessung
9 – Insulation plate
   – Isolationsplatte

Fig. 3. Partially electrically isolated anchorage

Abb. 3. Elektrisch teilisolierte Verankerung

in itself. Spacers could not be fitted because of limited drilling diameter, leading to damage during homing of the anchor and withdrawal of the casing. Low temperatures and high pressures during grouting created additional problems. It was further found that 50% of the anchors failing the load acceptence testing, i.e., with large displacement of the bond length, lost their isolating quality. After the introduction of various structural modifications and improved awareness in handling and grouting of the anchors, 90% of the anchors did eventually meet the electrical resistance requirements.

Anchor Head

The remaining 10% that did not meet the specifications regarding isolation were equipped with an isolation plate placed between the anchor bearing plate and the outer trumpet, as shown in Fig. 3. An electrical resistance measurement was performed to check that there was no metallic contact. The minimum resistance required was set at 10Ohm.

Recommendations 1989[1]

After evaluation of the information and experience gathered in the course of the construction of the railway station Stadelhofen [2], it was found that while the minimum isolation requirement of 0.1MOhm for the anchor bond and free length was appropriate, but the criterion for the anchor head had to be raised to 100Ohm in order to eliminate serious defects. The requirement for a complete encapsulation of the anchor is of utmost importance, irrespective of the presence of stray electric currents as other factors may lead to corrosion and endanger the durability of the anchor[8]. Since electrical resistance measurements provide a simple and reliable method, other projects using this new quality control procedure followed. In 1989 the leading prestressing companies in Switzerland, in co-operation with a group of specialists[2][3], produced and published "Recommendations for the Design and Execution of Corrosion Protection of Permanent Soil and Rock Anchors". These recommendations eventually became the accepted standard for permanent anchors in Switzerland (see summary below). Since then, several thousand anchors have been manufactured and tested in accordance with these recommendations.

Fully Electrically Isolated Anchors

With the above concept it was impossible to re-check the integrity of the PE encapsulation once the anchor was stressed. The use of metal for the inner trumpet and reliance on the mechanical seal together with the ever present humidity in the anchorage zone made it impossible to repeat measurement I. The strong desire for a system that

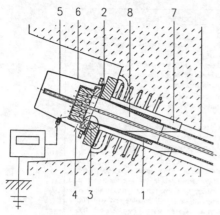

1 – PE trumpet welded to PE pipe
   – innerer PE-Ankerstutzen, verschweisst mit PE-Hüllrohr
2 – Insulation plate
   – Isolationsplatte
3 – Connection detail (patent pending), shown in principle only
   – Anschlussdetail (nur im Prinzip gezeichnet)
4 – Drainage/ventilation hole
   – Entlüftungsloch
5 – Thick coating with corrosion protective compound
   – Anstrich mit Korrosionsschutzmasse
6 – Electrically non-conducting coating
   – elektrisch isolierender Anstrich
7 – External secondary cement grout (optional)
   – äussere sekundär Injektion
8 – Internal secondary cement grout
   – innere Sekundärinjektion

Fig.4. Fully electrically isolated anchorage

Abb. 4. Elektrisch vollisolierte Verankerung

would allow the permanent surveillance of the integrity of the encapsulation was not satisfied until 1992 when the Swiss National Highways Department decided to test the market by calling for tenders for fully electrically isolated anchors for the reinforcement of the northern portal of the Seelisberg tunnel, a major section of the North-South transit route. Subsequently, VSL developed such an anchor head concept and was awarded the contract for the supply of 100 anchors each of 3300kN capacity and up to 40m in length. Details of the new anchor head are shown in Fig. 4.

In the new anchor head the steel trumpet is replaced by a PE trumpet which is shop-welded to the smooth sheath of the free length. Continuity of isolation is achieved by placing an isolation plate between the anchor head and the anchor plate. A non-conductive coating on the protection cap completes the electrical encapsulation. The anchor free length and bond length remained unchanged. The encapsulation of such anchors can be verified both during and after stressing, or at any time after installation (see Fig. 5). At Seelisberg, all the anchors were wired up to two easily accessible control boxes for convenient surveillance after completion of construction. The Swiss Anchor

Committee is at present considering the introduction of requirements for full electrical isolation of all future permanent anchors.

Recommendations 1989[1]

The recommendations first give a short description of the corrosion hazard created by electrical currents, particularly due to formation of galvanic cells, followed by general requirements on how electrical separation is to be achieved (as described above). The second part gives instructions on the technique of electrical resistance measurements and hardware to be applied:

Measurement I (see Fig. 2):
to examine the PE-encapsulation, taken after grouting of the unstressed anchor; voltage applied between tensile member and earth: 500V DC; instrument with a range >ca. 10kOhm; positive terminal connected to the anchor, negative terminal connected to the earth; minimum required resistance: $R_I = 0.1$MOhm.

Measurement II (see Fig. 3):
to examine the isolation of the anchorage, taken on the stressed anchor before grouting of the anchorage; voltage applied between anchorage and earth: approx. 40V AC; instrument with a range <ca. 200kOhm; polarity of the testing circuit makes no difference; minimum required resistance: $R_{II} = 100$Ohm.

The recommendations have been written for partially electrically isolated anchors, therefore two different measurements, I and II, are specified. For fully electrically isolated anchors measurement II is not required; instead measurement I is made at all construction stages.

Tolerances

Of utmost importance is the specification of tolerances. The recommendations allow a maximum of 10% of all measured anchors to exhibit measurements below the specified values. These inadequate anchors should however be distributed evenly. Accordingly, provisions must be made by the engineer to allow the positioning of additional or replacement anchors.

Responsibilities

Responsibilities are defined in Chapter 3 of the Recommendations. Until stressing of the anchor, the contractor is responsible for the quality of the corrosion protection. If the encapsulation is damaged during test loading due to factors beyond his control, the contractor's responsibility may be waived. Such factors include 1) deformations at the abutment leading to damage of the seal between anchorage and free length or of the smooth PE pipe and 2) displacements of the bond length leading to disjointing or damage of the transition from smooth to corrugated PE pipes or damage(due to

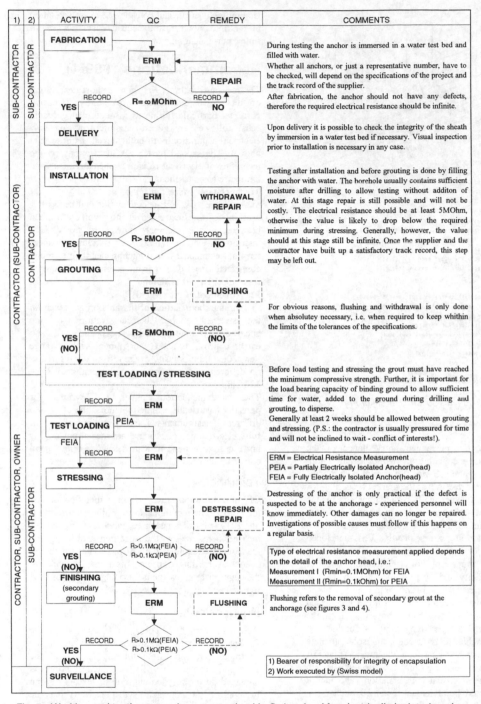

Fig. 5. Working and testing procedure as practiced in Switzerland for electrically isolated anchors
Abb. 5. Arbeits- und Prüfablauf, wie sie in der Schweiz für elektrisch isolierte Anker angewendet werden.

discontinuous deformations along the bond length) of the corrugated PE sheath itself . The contractor will, however, remain responsible if deformations leading to damages are a result of his own negligence. In Switzerland, post-tensioning contractors(sub-contractor) do not generally carry out drilling and grouting work. Instead, subcontract agreements are entered into with drilling contractors (contractor) for supply and stressing of the anchors, as well as completion work of the anchorages only. Typically, the anchors are manufactured at the work-shop of the sub-contractor and then delivered by truck to the construction site. The contractor is then responsible for the installation and injection of primary grout. Acceptance testing and stressing are then carried out by the sub-contractor. Since the contractor, and not the sub-contractor, is responsible for the load capacity of the anchor, the testing by the supplier attains the status of third party surveillance. To clearly define the respective responsibilities of the contractor and the sub-contractor, supplementary conditions have been written with regard to electrically isolated anchors constructed under the terms of the above mentioned Recommendations. These conditions stipulate requirements regarding the delivery, unloading, storage, installation and grouting of the anchors, together with definitions of responsibilities. Fig. 5 shows an example of the typical working and testing procedure practiced in Switzerland for electrically isolated anchors. Chapters 4,5 and 6 of the Recommendations give further guidelines for test anchors, long term monitoring and documentation.

Other state-of- the-art Features

Once electrical isolation of an anchor has been achieved most of the known weak spots, namely the critical interfaces at the anchorage/free length and free length/bond length as well as the end cap at the bottom the anchor will have been initiated. A tough and sensible proof-loading procedure (acceptance testing) is essential for the verification of the load bearing capacity of individual anchors. While early failure of an anchor due to corrosion must be avoided, it is equally important to regularly check the actual load in the anchor.

Load Monitoring

It has become routine for all anchors on most major works to be equipped with anchorages that allow periodic checking of the actual load as well as proof-loading at a later stage. At least 10% of the anchors are usually equipped with electrical load cells for continuous monitoring of the load (surveillance anchors). In the past, hydraulic load cells have been used for this purpose. Today, however, electrical load cells are equally reliable and have the added advantage of being easily linked to a remote computer monitoring system. The possibility to replace a load cell without the need for destressing is given with the range of VSL load cells and anchorages.

Protection Cap

The above mentioned anchorages have to remain accessible and are therefore fitted with protection caps. Instead of using coated steel, these caps may be manufactured from plastic. Durability with regard to UV-radiation will have to be checked. If coated steel is used, special attention must be paid to the choice of a system which can be easily repaired and renewed. The type and thickness of coating must always be chosen with due regard to the prevailing climatic conditions and ecological aspects. If the anchorages are to be wired up for remote control of the electrical isolation it is necessary to use an electrically non-conductive coating. Until recently, such protection caps have been filled with a corrosion protective compound. This measure is no longer ecologically acceptable or necessary. Instead, a thick coating of the steel parts, and the use of a steel cap with a ventilation hole placed at the low point to avoid accumulation of water due to condensation, are adequate.

Future

To make the construction of fully encapsulated anchors simpler and more reliable, thick walled corrugated PE-pipes, such as the PT-PLUS[10] plastic duct, with smooth sections at the ends for butt-welding to adjoining sections, are needed. The minimum wall thickness specified in codes, standards and recommendations reflects the limited availability of such pipes; however a minimum requirement of 0.8mm(FIP[4] or 1mm(PTI[5], FIP[8], BS[7]) is not stringent enough. It is therefore not surprising that some authorities recommend a second sheath to protect the first one. It is not known if a sheath of 0.8 mm or 1.0 mm wall thickness will withstand stressing to proof load without failing. The ERM-method provides a reliable means for verifying sheath integrity.

Related Applications for PT

The same quality control method has already been applied on 3 bridge construction projects in Switzerland using electrically isolated tendons (EIT) together with the new VSL PT-PLUS[10] plastic duct. VSL has also developed a new anchorage, the VSL Composite System CS, which among other advantages allows the designer to select from among the following configurations; STANDARD(corrugated steel ducting), PLUS(PT-PLUS ducting) and SUPER(PT-PLUS ducting and electrically isolated anchor head). This range of new products will help to improve the quality of any post-tensioned structure.

Conclusion

With the details discussed above, considerable progress has been made towards truly monitorable, inspectable permanent anchors. The electrical isolation also allows the use of other inspection techniques such as reflectrometric impulse measurements. It is believed that these techniques will soon become the state-of-the-practice in Central Europe and elsewhere. In fact, quality control by electrical resistance measurements has also found its way into the draft of the FIP recommendations[4] with reference to the simple and convenient nature of the method. The net result is that electrical testing promotes more careful construction practices and improved quality.

References

[1] AVT, Freyssinet, Spannstahl, Stahlton, VSL International: Recommendations for the Design and Construction of Permanent Soil and Rock Anchors 1989.

[2] F. Hunkeler, F. Stalder: Streustromschutz-massnahmen bei Boden und Felsankern. Schweizer Ingenieur und Architekt 33-34/87, page 978.

[3] U. von Matt: Auf dem Weg zu langfristig sicheren Boden- und Felsankern. SIA Dokumentation D 031: Korrosion und Korrosionsschutz, Teil 4, Anker und Spannkabel, 1989. Swiss Society of Architects and Engineers, Zurich, Switzerland.

[4] FIP: Recommendations for the Design and Construction of Prestressed Ground Anchorages (Draft September 1991). FIP-Publications, The Institution of Structural Engineers, London, England, 1991.

[5] American code PTI : Recommendations for Prestressed Rock and Soil Anchors. Post-Tensioning Institute, Phoenix, USA, 1986.

[6] German Code DIN 4125: Kurzzeitanker und Daueranker, Bemessung, Ausführung und Prüfung, November 1990. Beuth Verlag GmbH, Berlin, Deutschland.

[7] British Standard BS 8081: Ground Anchorages.

[8] FIP: Corrosion and Corrosion Protection of Prestressed Ground Anchorages (State of the art report). FIP-Publications, The Institution of Structural Engineers, London, England, 1986.

[9] Swiss Code SIA 191: Ground Anchorages 1977. Swiss Society of Architects and Engineers, Zurich, Switzerland.

[10] PT-PLUS Plastic Duct System, VSL International Ltd., Report No. 241 e, Berne, Switzerland, October 1992.

Anker in Theorie und Praxis, Widmann (Herausgeber) © 1995 Balkema, Rotterdam. ISBN 90 5410 577 1

Korrosionsschutzsysteme bei der Verwendung hochfester Zugglieder in der Geotechnik

Corrosion protection systems for high-strength tendons in geotechnics

Dieter Jungwirth
DSI München, Germany

ZUSAMMENFASSUNG: Hochwertige Stähle sind im Preis-/Leistungsverhältnis gegenüber den herkömmlichen Beton- bzw. Baustählen sowie Alternativwerkstoffen überlegen. Sie werden daher bevorzugt zur Aufnahme hoher Zugkräfte in der Geotechnik verwendet. Damit sie ähnlich robust und dauerhaft wie die weniger korrosionsempfindlichen, niederwertigen Stähle eingesetzt werden können, bedarf es eines besonderen, möglichst redundanten Korrosionsschutzes. Dabei ist in der Geotechnik als typisch zu bewerten, daß eine Überwachung und Nachbesserung der nicht zugänglichen Zugglieder in der Regel nicht möglich ist. Der Beitrag bringt eine Vielzahl an Beispielen unter Einbezug von Qualitätsmanagementsystemen.

ABSTRACT: High-strength steels are superior to conventional reinforcing and construction steels as well as to alternative materials with regard to their price/performance ratio. Thus they are preferably used for taking up high tensile forces in geotechnics. In order to be able to use them similarly robust and durable as the less corrosion-sensitive low strength steels, a special, if possible redundant, corrosion protection is necessary. It has to be considered as typical for geotechnics, that supervision and repair of the not accessible tendons generally is not possible.

Thus this report states different corrosion protection measures, basing on the existing regulations, distinguishing between
different steels,
differently aggressive effects and
different periods of subjection.

The corrosion protection measures reach from different coatings to cement paste, multi-layer systems to special steel and substitutes as glass-, plastic- or carbon fibre composite materials.

The sensible application of monitoring systems (sensory analysis) will be discussed. The reliability and durability of the measures depends on the quality of execution. Quality management systems thus will also be included.

1 EINFÜHRUNG

In den Industrieländern liegt der jährliche, ökonomische Korrosionsschaden vom Auto bis zum Stahlbetonbau bei 3 bis 4 % des Bruttosozialproduktes. Sind Korrosionsschäden bei kurzlebigen Gütern akzeptierbar, muß dies bei langlebigen Investitionen wie Bauwerke verhindert werden, zumal hier Fragen der dauerhaften Sicherheit für Leib und Leben berührt werden.

Ist der Korrosionsschutz nicht vorhanden oder zerstört, kann Korrosion auftreten, wenn ausreichend Sauerstoff und Feuchtigkeit zur Verfügung stehen. Der Korrosionsprozeß ist ein elektrochemischer Vorgang, der in seiner Grundform in einen anodischen und in einen kathodischen Teilprozeß aufgeteilt werden kann. An der Anode gehen positiv geladene Eisenionen in Lösung, an der Kathode werden mit Sauerstoff und Wasser Hydroxylionen gebildet (Bild 1), wobei die im Stahl vom anodischen Prozeß entstandenen Elektronen aufgenommen werden. Eisenionen und Hydroxylionen gehen im Elektrolyten in Lösung. Die Gesamtreaktion führt zu $Fe(OH)_2$, das zu Rost oxidiert, einer Mischung zwischen verschiedenen Eisenoxiden und Eisenhydroxiden. Eine Volumenvergrößerung ist damit verbunden. Neben dieser Grundform der Korrosion spielen Sauerstoffarmut, pH-Wert, Promotoren (Chloride), Mikro- oder Makroelementbildung, Depassivierung oder Potentialunterschiede und Kristallgitterform des Stahles sowie Spannungen eine Rolle, um weitere Korrosionsarten wie Lochfraßkorrosion, Spaltkorrosion, Kontaktkorrosion, Spannungsrißkorrosion, Wasserstoffversprödung, Reibkorrosion, Schwingungsrißkorrosion zu deuten; mehr hierzu siehe z.B. [1 bis 3].

Gegenüber üblichen Bauteilen, die gewartet und nachgebessert werden können, ist der Anker in der Regel nicht zugänglich, nur im meist nicht relevanten Außenbereich. Auch lassen sich die Korrosionsschutzmechanismen im Boden nicht immer eindeutig definieren, so daß eher die jeweils ungünstigeren Verhältnisse zugrunde zu legen sind.

Es bedarf daher der Unterscheidung folgender Parameter:

1. Empfindlichkeit des Stahles bezüglich Stahlfläche und Stahlgüte. Hochwertige Spannstähle sind entsprechend empfindlicher als niederwertige bezüglich gewisser Korrosionsarten, besonders wenn sie hoch ausgenutzt werden.

2. Grad der Einwirkung wie aggressiv bzw. nicht aggressiv, mit der eher ungünstigeren Annahme.

3. Grad des Korrosionsschutzes in Hinblick auf Lebensdauer. Kurzzeitanker werden üblich mit 2 Jahren begrenzt, Daueranker mit 100 Jahren. Ob eine Zwischenstufe, z.B. 2 - 20 Jahre sinnvoll ist, muß noch diskutiert werden.

4. Bei Dauerankern werden zwei Korrosionshüllen (redundant) gefordert, oder eine Hülle, die kontrollierbar ist.

Dieser Philosophie schließt sich auch der Europäische Normenentwurf EN 00 288 02 an. Darüberhinaus machen Regelwerke (z.B. DIN 4125 und ÖNORM B 4455 Verpreßanker, DIN 4128 Verpreßpfahl) oder Zulassungsstellen gewisse Anforderungen an Mindestabmessungen, die in der nachfolgenden Ausarbeitung mit eingehalten sind.

2 WERKSTOFFE

2.1 Baustahl, Betonstahl

Diese meist unterhalb einer Bruchfestigkeit von 600 N/mm² befindlichen Stähle mit hohem Verformungsverhalten, sind robuste Baustoffe, vor allem wenn sie der Abrostung große Querschnitte entgegensetzen.

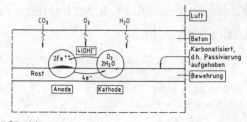

1) Teilreaktionen
Anode: $Fe \longrightarrow Fe^{++} + 2e^-$ (Eisenauflösung)
Kathode: $O_2 + 2H_2O + 4e^- \longrightarrow 4\ (OH)^-$ (Sauerstoffreduktion)
2) Rostbildung
1. Stufe: $Fe^{++} \cdot 2(OH)^- \longrightarrow Fe\ (OH)_2$
2. Stufe: $2Fe(OH)_2 + \frac{1}{2}O_2 \longrightarrow$ z.B. $Fe_2 O_3 \cdot 2H_2 O$ (Eisen-(III)-oxidhydrat)

Bild 1   Modellvorstellung über die Karbonatisierungskorrosion von Bewehrung im Beton
Model on carbonization corrosion of reinforcement in concrete

2.2 Spannstahl

Die Spannstähle erhalten die hohen Festigkeiten durch geeignete Legierung, Kaltverformung oder Vergütung. Entsprechend dominieren am Markt (siehe Euronormentwurf EN 10138):

1. Naturharte, ggf. gereckte und angelassene Stäbe ⌀ 15 bis 50 mm, St 1230 bis 1030 N/mm²

2. Gezogene Drähte und verseilte Litzen, ⌀ 4 bis 10 mm St 1850 bis 1550 N/mm² bzw. ⌀ 5,2 bis 18,0 mm, St 2060 bis 1700 N/mm² (Litzen mit 3 bzw. 7 Drähten)

3. Vergütete Drähte, ⌀ 6 bis 14 mm, St 1600, über Zulassung

Das Verformungs- und Ermüdungsverhalten ist geringer als vergleichsweise bei niederfesten Stählen. Die Kerb- und Querdruckempfindlichkeit ist entsprechend höher [4]. Meist ist Schweißeignung nicht gegeben.

Die Korrosionsempfindlichkeit von Spannstählen ist ebenfalls höher einzustufen als bei normalen Bau-/Betonstählen. Sie nimmt vom naturharten, gezogenen zum vergüteten Spannstahl zu. Es tritt - wenngleich in noch akzeptabler Häufigkeit [2] [3] - hauptsächlich wasserstoffinduzierte Spannungsrißkorrosion auf, wobei häufig Vorschädigung durch fehlenden temporären Korrosionsschutz die Ursache ist. Die Korrosionsempfindlichkeit läßt sich nach Stolte [5] vereinfacht durch die Standzeit $L = C \times 1/\sigma^3 \times R^9$ unter Spannung und korrosiver Bedingung ausdrücken. Neben dem schon erwähnten Einfluß durch das Herstellverfahren C geht die Höhe der Dauerspannung σ in der dritten Potenz und der Einfluß von Überfestigkeit R in der neunten Potenz ein.

Die in jüngster Zeit spektakulären Schäden durch gebrochene Spannstähle [6] sind analysiert und lassen sich im wesentlichen zurückführen auf
1. Planungsfehler,
2. Ausführungsfehler,
3. Überempfindliche (vorgeschädigte, überfeste), nicht mehr zugelassene, vergütete Stähle,
4. mangelnde Bauwerksüberwachung.

Ausreichend technische Regeln, diese Fehler zu vermeiden, sind vorhanden, z.B. in einer Serie von FIP-Papieren (z.B. [3], [4], [7] bis [9]) oder aktualisierten Normen. Mit Hilfe von Qualitätsmanagementsystemen gilt es, diese Informationen über Schulung an die erforderliche Stellen zu bringen.

Folgende Erkenntnisse bei Verwendung von Spannstählen lassen sich damit zusammenfassen:
1. Verwendung robuster Spannstähle, ohne Überfestigkeiten
2. Durchführung eines Korrosionsempfindlichkeitstests, z.B. nach FIP-Richtlinie [10]
3. Durchführung des "Deflected tensile test" für Litzen nach FIP-Richtlinie [11]
4. Sorgfältige Handhabung bei Lagerung und Transport, um Vorschädigung zu vermeiden
5. Verwendung geeigneter Verankerungen [7]
6. Dauerhafter Korrosionsschutz [3]
7. Geeigneter Entwurf. Folgen überlegen, wenn ein Spannstahl ausfällt (Systemreserve, Bruchankündigung [6]
8. Vorhandenes Fachwissen z.B. [8], [9] und QMS nutzen
9. Bauwerksüberwachung

2.3 Nichtrostende Stähle, sog. Edelstähle

Bei klar definierter Einwirkung lassen sich geeignete Sorten finden, gar mit Festigkeiten um St 1000. Leider sind, wie schon erwähnt, die Einwirkungen im Bauwesen nicht immer voraussagbar. Die Gefahr der Spaltkorrosion - bei nichtrostenden Stählen ein besonderes Problem - ist im Verankerungsbereich besonders groß (z.B. Mutter-Platte-Schraubverbindung), so daß von einem alleinigen Einsatz ohne zusätzlichen Schutz aus Sicherheitsgründen abgeraten wird.

2.4 Stahlersatz

Umweltfreundlich, mit einem Energieaufwand von nur 2600 kWh/Tonne in die Wiederverwertung rückführbar, wird Stahl langfristig im Preis-

Bild 2   Glasfaseranker
         Glass fiber anchorage

/Leistungsverhältnis zu Alternativstoffen wie Glas, Kunststoff und Kohlefaser wettbewerbsfähig bleiben. Trotzdem gibt es gewisse Randbedingungen, die anderen Werkstoffen Vorteile einräumen (z.B. nicht magnetisch, nicht strahlend). Sind über Vollschnittmaschinen abbaubare Anker gefordert, eignet sich Glasfaser vorzüglich.

DYWIDAG hat verschiedene Gewindeformen für Glasfaserstäbe entwickelt, wobei sich das eingeschliffene Grobgewinde als wirtschaftlichste Form herausstellte. Durch besondere, bewehrte Muttern und Platten lassen sich hohe Verankerungskräfte erzielen (Bild 2).

2.5 Preis-/Leistungsverhältnis

Spannstähle müssen hohe Festigkeit, damit hohe Dehnfähigkeit beim Vorspannen besitzen, um die Kriech- und Schwindverluste aus zeitabhängiger Betonverformung und Relaxation gering zu halten. Da Herstellkosten nicht proportional zur Festigkeit wachsen, ist das Preis-/Leistungsverhältnis (Kosten in Pfennigen, um eine Tonne Bruchlast 1 m zu tragen) gegenüber anderen Materialien wirtschaftlich interessant. In Tabelle 1 sind verschiedene Materialien/Systeme gegenübergestellt.

3 KORROSIONSSCHUTZSYSTEME, INSBESONDERE IM ANKERBEREICH

Als temporärer Schutz kommen üblicherweise emulgierbare Fette, trockene und gereinigte Luft oder Schutzgase in Frage. Die im Ankerbereich vorgesehene Kunststoffverrohrung mit Endabdichtung läßt Zweifel der Dichtigkeit aufkommen. Ein

Tabelle 1 Preis-/Leistungsverhältnis für verschiedene Zugglieder in Pfg./to,m
Price/performance ratio for different
tendons in Pfg./to,m

PREIS-/LEISTUNGSVERHÄLTNIS	Pfg./to,m
BAU-/BETONSTAHL St 500	9
GEWI-Stahl Ø 50 und 63,5 mm, St 500/550	14
Abrostrate 8 % (1 mm), DIN 4125 Verpreßanker temporär	15
Abrostrate 16 % nach 20 Jahren (0,1 mm/J) EAU 1990	16
3,5 cm Mörteldeckung, DIN 4128 Verpreßpfahl	17
Verzinkt, Abtragrate 0,005 mm/J	20
LITZE St 1570/1770	6
mit PE-Rohr t = 2 mm als Temporäranker	9
Verzinkte Litze	8
Monostrand (Fett, PE 1,5 mm)	8
Epoxy-coated Strand	10
Bündelspannglied mit Anker	19
Externes Spannglied mit Anker	36
Schrägseil mit Anker zementverpeßt	50
SPANNSTAB Ø 36 mm, St 1080/1230	13
Temporäranker, PE 2 mm	16
Daueranker mit Anker	40
EDELSTAHL, St 900	65
GLASFASER	10 - 25
KOHLEFASER	50 - 100

Tabelle 2 Korrosionsschutzsysteme
Corrosion protection systems

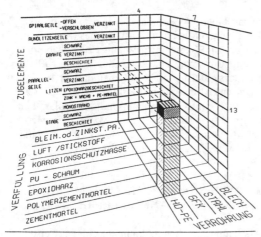

KOMPONENTEN VON KORROSIONSSCHUTZSYSTEMEN
(4 x 7 x 13 KOMBINATIONEN)

zusätzlicher Schutz wie Beschichtung, Verfüllung ist vorzuziehen.

Einen grundsätzlichen Überblick der verschiedene Korrosionsschutzsysteme und deren Kombination zeigt Tab. 2.

Bester aktiver, preiswerter und dauerhafter Schutz bietet Zementleim. Karbonatisierung und Chlorideinwirkung ist bei Ankern zweitrangig. Unzulässige Rißbreite > 0,1 mm müssen vermieden werden. Eine Mindestdeckung zwischen Stahl und Hüllrohr von 5 mm wird verlangt.

Die Einpreßtechnik [9] hat heute hohen Stand. Es steht die Nachpreßtechnik, das Druckverpressen und die Vakuumtechnik zur Verfügung.

Korrosionsschutzanstriche in mehreren Lagen sind dauerhaft (DIN 55928), lassen sich jedoch außer durch Tauchen auf Rundmaterial nicht zuverlässig aufbringen. Dagegen eignen sich Schrumpfschläuche als Korrosionsschutzschicht.

Metallische Beschichtung, wie industrielles Verzinken mit ca. 200 bis 300 gr/m² bietet ausreichend Grundschutz, vor allem im Freien. Der zeitabhängigen Abtragung oder mechanischen Verletzung muß durch eine zweite Barriere Sorge geleistet werden (Duplex- oder Triplex-Systeme). Zink leistet wenig Widerstand gegenüber Chlorideinwirkung aus Tausalz oder Meerwasser. Im Verankerungsbereich sind Folgen aus Schichtverletzungen in Verbindung mit Zementleim oder anderen Elektrolyten zu beobachten [12].

Korrosionsschutzmassen wie Fette oder Wachse bieten zwar keinen Verbund, jedoch preiswerten Schutz, wenn Sie in Form der Monostrand mit einem extrudierten PE-Rohr umgeben werden. An das Fett sind Anforderungen zu richten wie (siehe auch [3] und EN 00 288 02):

- Wasseraufnahme < 0,15 M-%
- Verseifungszahl < 5 mg KOH/g
- Keine S, Cl, NO_2, M_eSCN (Spannstahlgifte)
- Ausölen auf Filterpapier bei 50° < 50 mm
                     bei 30° < 5 mm
- Tropfpunkt > 55°
- Ausdehnungskoeffizient 10^{-4} bis 10^{-3}/Grad
- Elektrischer Widerstand $10^9 \Omega m$

Im Krafteintragungsbereich kann heute durch Unterbrechung der Fettzufuhr die erforderliche Kraftübertragung sichergestellt werden. Auch kann durch entsprechendes Verfahren vorhandenes Fett umweltfreundlich entsorgt werden.

Schließlich bietet Epoxidharzbeschichtung eine interessante Alternative. Industriell aufgebracht, sind Anforderungen an Harz und Verbund -

Stahl/Epoxi und Epoxi/Bauwerk - zu richten [13]. Die Mängel bei der Beschichtung von normalem Betonstahl in den USA sind erkannt und behoben. Sonderkeile verquetschen die Beschichtung aber verletzen sie nicht im Ankerbereich. Es entstehen hochschwingfeste Verankerungen (Bild 3).

Mehrschichtsysteme (redundant) sind je nach Einwirkung aufzubauen (siehe Abschnitt 5).

Schließlich gilt es den konstruktiven Korrosionsschutz zu beachten, der Wasser möglichst fern hält bzw. Feuchtigkeit rasch ableitet (Bild 4).

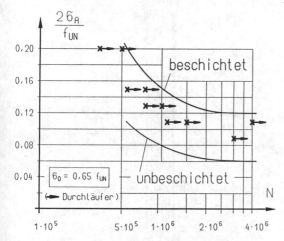

Bild 3   Hochschwingfeste Keilverankerung einer epoxidharzbeschichteten Litze, $\sigma_0 = 1000$ N/mm²
Highly vibration resistant wedge anchorage of an epoxy coated strand, $\sigma_0 = 1000$ N/mm²

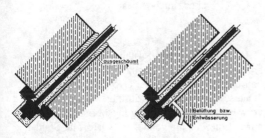

Spannglied ohne Verbund

Wassersäcke durch z.B. Ausschäumen, Kondenswasserbildung durch Belüftung vermeiden.

Bild 4   Konstruktiver Korrosionsschutz, Beispiel
Example for corrosion protection by design measures

4 KONTROLLIERBARE (MONITORING), REPARIERBARE UND AUSWECHSELBARE SYSTEME

Bauwerke müssen zugänglich sein, um Schäden rechtzeitig erkennen zu können (DIN 1076, SIA 169, [4]). Damit lassen sich die Erhaltungs-/Behebungskosten minimieren. Ein logischer Wunsch der Bauherren, die Systeme dann auch reparaturfreundlich, ggf. auswechselbar zu gestalten, ist nur sinnvoll. Für Umweltkatastrophen ist dies gar das richtige Sicherheitskonzept.

Bei Ankern ist dies nicht ohne weiteres möglich. Die Alternative heißt Monitoring. Die Beobachtung kann durch Monitore mit entsprechenden Sensoren erleichtert werden, die Verformung, Temperatur, Feuchte, Korrosionspotentiale usw. erkennen lassen und elektronisch auswerten (Bild 5).

Die einfachste Lösung ist, den elektrischen Widerstand zwischen Ankerstahl und Boden zu beobachten. Ein elektrisch isolierter Anker läßt in der Regel keine Korrosion zu. Der Anker muß entsprechend ausgebildet sein. Mehr dazu in [14].

Automatisierte Bauwerksüberwachung mit Sensoren

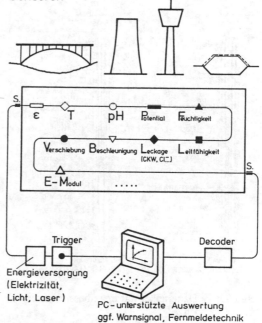

Bild 5   Monitoring
Monitoring

5 ANWENDUNGSBEISPIELE

Den klassischen Korrosionsschutz für die Daueranker mit Gewindestäben zeigt Bild 6. Durch die
kontrollierte Rißbildung (w < 0,1 mm) ist dieser
Anker als doppelt korrosionsgeschützt einzustufen,
mehr siehe [15, 16].

Der epoxidharzbeschichtete Litzenanker (Bild
7) ist ebenfalls als Daueranker einzustufen.

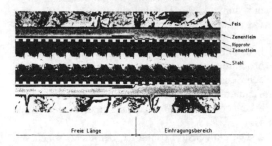

Bild 6   Doppelt korrosionsgeschützte Erd- und
          Felsanker mit DYWIDAG-Gewinde-
          stahl
          Double corrosion protected soil- and
          rock bolts with DYWIDAG-threadbars

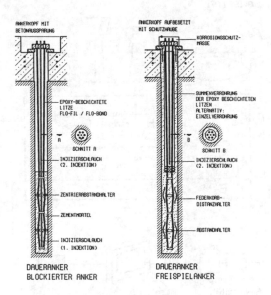

Bild 7   Daueranker mit epoxibeschichteter
          Litze
          Permanent anchorages with epoxy-
          coated strand

In Bild 8.1 wird eine elektrisch isolierte Bün-
delankerversion vorgestellt.

DYWIDAG STRAND ANCHOR WITH ELECTRICAL ISOLATION

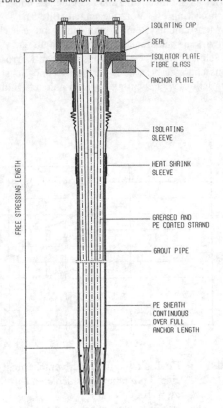

Bild 8.1   Elektrisch isolierter Quetschrohranker,
            Kopfausbildung
            Electrically isoltated indented tube an-
            chor, anchor head

Elektrisch isolierte Systeme lassen sich umso
leichter erreichen, je weniger Übergang/Wechsel
von Glatt- zu Ripprohren erfolgen. So wird beim
"Quetschrohranker" ein durchgehend glattes Rohr
verwendet, das im Eintragungsbereich onduliert
wird. Eine innenliegende, mitverformte Stahl-
spirale z.B. gewährleistet die Form des ver-
quetschten Glattrohres (Bild 8.2) und ermöglicht
so einwandfreie Kraftübertragung in der Verankerungslänge. Der Verbund der Litze mit Einpreßmörtel kann über verbundverbessernde Maßnahmen (Bild 9) angehoben werden, mehr siehe
[17].

434

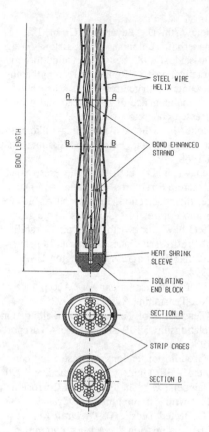

Bild 8.2   Kraftübertragungsstrecke,
            Fußausbildung
            Bond length, anchor foot

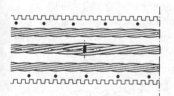

Bild 9   Verbundverbesserung
          Improvement of bond

Dichte Bebauung erfordert immer mehr die Ausbaubarkeit der Anker nach Gebrauch (Bild 10). DYWIDAG hat neben der erprobten Stablösung einen zuverlässigen Abbrennsatz entwickelt. Eine Induktionsspule oder ein Thermitsatz erhitzen die Litzen nach Gebrauch an der Eintragungslänge über 400 °C und lassen sich so leicht abreißen. Weitere Varianten sind in Erprobung, so wird die Litze bereits vor dem Einbau induktiv wärmebehandelt, wodurch die Festigkeit zwischen 10 bis 20 % verringert wird bei Beibehaltung ausreichender Zähigkeit. Nach dem Gebrauch werden die Litzen einzeln abgerissen. In Kombination mit Bild 9 verbleiben nur sehr kurze Verbundstrecken im Boden [17].

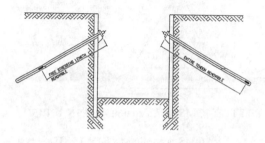

Bild 10   Ausbaubare Temporäranker
           Recoveravble temporary anchor

6   QUALITÄTSMANAGEMENTSYSTEME (QMS)

Das Wissen ist vorhanden; es muß an die richtige Stelle gebracht werden; dazu dienen QMS, wie schon unter Abschnitt 2.2 vermerkt. Außerdem verlangt der europäische Binnenmarkt werkseigene Produktionskontrollen, ein Organisationsversprechen bzw. einen Konformitätsnachweis, was mit zertifizierten QMS (Bild 11) erbracht werden kann. Das Bemühen bei allen am Bau Beteiligten, dieses Ziel zu erreichen, ist groß. Ohne auf Einzelheiten des QMS nach DIN EN ISO 9000 ff einzugehen, sollen nachfolgend nur die wesentlichen Ziele herausgestellt werden [18].

Die Geschäftsleitung muß qualitätsbewußtes Denken in die Köpfe der Beteiligten bringen, sie motivieren. Klare Organisation mit Verantwortlichkeit und Information müssen für die ausgebildeten Mitarbeiter vorliegen. Auf Schwachstellen muß mittels Checklisten aufmerksam gemacht werden. Qualität muß produziert werden, ingenieurmäßiges Vorausdenken ist gefragt. Das System muß schlank bleiben und gelebt werden. Kein überflüssiges Papier erzeugen.

Schließlich ist eine Vernetzung dieses Systems mit vorhandenen technischen Regeln erforderlich. Der Dialog zwischen Regelwerk und QMS hilft, das vorhandene Wissen umzusetzen.

Entwurf, Herstellung und Einbauen von Ankern erfordern hohen Sachverstand. Die Arbeiten dürfen daher nur durch qualifizierte Spezialfirmen ausgeführt werden, die in der Regel nach QMS arbeiten.

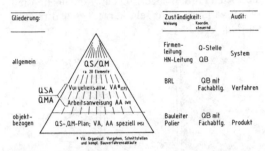

Bild 11   QMS in Anlehnung an DIN EN ISO 9001
QMS according to DIN EN ISO 9001

7 ZUSAMMENFASSUNG

Eine Vielzahl von Korrosionsschutzsystemen stehen zur Verfügung. Die Anwendungsbedingungen, wie Schutzdauer, Umweltbedingungen sind klar zu definieren. Der Konstrukteur, die Ausführung stellen qualitätsgesichert das Wissen und die Ausführungstechnik zur Verfügung. Der Bauherr will zunehmend kontrollierbare, ggf. nachrüstbare, auswechselbare Systeme. Ist dies nicht möglich, sind redundante, doppelte Korrosionsschutzsysteme erforderlich.

Neben Stahl stehen alternative Werkstoffe für Sonderfälle zur Verfügung. Ihr Preis-/Leistungsverhältnis im Vergleich zu Stahl wird sich verbessern.

8 LITERATUR

[1] CEB-Bulletin 148: Durability of Concrete Sturctures - State of the Art Report; 1982

[2] Nürnberger, U., et al: Korrosion von Stahl im Beton - einschließlich Spannbeton; DAfStb Heft 393

[3] FIP State of the Art Report: Corrosion Protection of Prestressing Steel; 1995

[4] Jungwirth, D., Beyer, E., Grübl, P.: Dauerhafte Betonbauwerke; Beton-Verlag Düsseldorf 1986 bzw. FIP Guide to good practice, 1991, "Repair and strengthening of concrete structures" bzw. "Inspection and maintenance of reinforced and prestressed concrete structures", 1986

[5] Stolte, E.: Über die Spannungsrißkorrosion an Spannstählen; Beton- und Stahlbetonbau 5/1968

[6] Jungwirth, D., et al: Durable Prestressed Concrete Structures. Problems and Damage, their Detection, Repair and Avoidance in the Future; FIP Congress Kyoto Japan 1993 bzw. Jungwirth, d, et al: Durability of Prestressed Concrete and its Improvement; FIP-Congress Washington USA, 1994

[7] FIP Recommendations for "Acceptance of Post-Tensioning Systems"; 1993

[8] "Tensioning of Tendons: Force-elongation relationship"; FIP State of the Art Report 1986

[9] "Grouting of tendons in prestressed concrete", FIP Guide to good practice; 1990

[10] Hampejs, G.: The FIP Stress Corrosion Test with Ammonium Thio-Cyanat; 1988

[11] "Deflected Tensile Test", Draft Report of FIP Commission 2 - Ad hoc committee "Multiaxial behaviour"

[12] Hampejs, G., et al: Galvanisation of prestressing steels. Commission report, Prestressing Materials & Systems; FIP News 1991/4

[13] Jungwirth, D., et al: Coating of Tendons; Proceedings FIP Congress Washington, 1994

[14] Anhang A zum prEN 00 288 02: 1994

[15] DIN 4125 und FIP-Recommendations for the Design and Construction of Prestressed Ground Anchorages; Final Draft 1993

[16] Herbst, Th.: "Principles of Corrosion Protection for Soil and rock Anchors", Proceedings Int. Symposium on Prestressed Rock and Soil Anchors, Oct. 1984; Des Plaines Illinois USA organized by PTI Post-Tensioning Institute USA

[17] Klöckner, R.: Neuere Entwicklungen in der DYWIDAG-Ankertechnik; Symposium Anker in Theorie und Praxis; Salzburg 1995

[18] Jungwirth, D. u. A.: Qualitätsmanagement im Bauwesen; VDI-Verlag Düsseldorf, 1994

Anker in Theorie und Praxis, Widmann (Herausgeber) © 1995 Balkema, Rotterdam. ISBN 90 5410 577 1

Die Europäische Norm über Verpreßanker prEN 1537, Harmonisierung durch Vielfalt

European standard on ground anchors prEN 1537, harmonizing by great variety

T. F. Herbst
Allspann GmbH, Puchheim, München, Germany

U. von Matt
Dr. Vollenweider AG, Beratende Ingenieure ETH/SIA, Zürich, Switzerland

L. V. Martak
MA 29-Fachbereich Grundbau, Wien, Austria

ZUSAMMENFASSUNG: Mit einer Bearbeitungszeit von weniger als zwei Jahren wurde der nun zur endgültigen Fertigstellung vorliegende Normenentwurf einer europäischen Norm für Verpreßanker in Locker- und Festgestein von Geotechnikfachleuten aus 10 Staaten Europas eingebracht. Fußend auf den erprobten Normenwerken Englands, Frankreichs und der deutschsprachigen Staaten wurde im Verein mit den mannigfaltigen baupraktischen Erfahrungen der südeuropäischen Länder ein allseitig abgestimmter Vorschlag geschaffen, der auch grundsätzlich neues Gedankengut enthält. Dies gilt insbesondere für die zahlreichen Methoden des Korrosionsschutzes bei Daueranker und ihrer Prüfung und für die Einführung der Teilsicherheitsbeiwerte des probabilistischen Sicherheitskonzeptes, wie es in allen anderen Euronormen eingeführt werden soll.

ABSTRACT: Within of two years the now existing draft of an European Standard on Ground Anchors was delivered by a group a geotechnical engineers coming from 10 European States. The prEN 1537 shall be valid for anchors in rock and soil for temporary and permanent use and is based on the existing standards of United Kingdom, France, Germany, Swiss and Austria. Beyound these the farspread experience on numerous case histories of the southeuropean countries was used to get to a harmonized draft, containing some principal new ideas too. Among these are the many methods on corrosion protection and their control for permanent anchors and the introduction of the probabilistic safety concept on the design to be used in all Euro-Standards in future. Especially the realistic demands on geotechnical design for anchors are missing in Eurocode 7, Part 1. They will now be codified in the normative annex of this anchor standard, possibly taken up into Eurocode 7, Part 4 in the future.

Observing the european regulations to unify technologies in all the memberstates it is impossible to ignore the different existing high sophisticated methods and anchor production and anchor testing all over Europe. The determination of requirements, recommendations, possibilities etc. had to respect these experiences on anchors arised from many years work. Certainly this standardization group has to face the different responsibilities of the geotechnical consulting in United Kingdom f.i. and elsewhere and those of the contractors in Central Europe. A harmonization of the play of roles and of the used technologies was only possible by accepting the equal quality of the existing standards and production methods. The standard on anchors will include therefor at least 4 different systems on corrosion protection for permanent anchors and 3 different testing and monitoring methods respectively. The three autors will give their comments on geotechnical design, on corrosion protection and on testing the anchors to different philosophies agreed in this standard.

1 EINLEITUNG

Mit dem Eurocode (EC) 1 Teil 1 Basis of Design, der demnächst als ENV 1991-1 erscheint, wird für das gesamte Bauingenieurwesen ein neues Bemessungskonzept mit den Nachweisen von Grenzzuständen (GZ) der Tragwerke eingeführt. Der Nachweis des GZ 1 (Tragfähigkeit) ist mit Teilsicherheitsfaktoren zu führen, welche die Einwirkungen erhöhen und die Widerstände reduzieren. Der Nachweis des GZ 2 (Gebrauchstauglichkeit) besteht in der Regel aus einem Nachweis der Tragwerksverformungen. Er wird mit charakteristischen Werten für die Einwirkungen und die Tragwerkseigenschaften geführt.

Für Stahlbeton- und Stahltragwerke ist dieses neue Bemessungskonzept einfach anzuwenden, weil bei diesen Tragwerken Lasten und Widerstände klar trennbar sind. Der EC 2, Teil 1, Planung von Stahlbeton- und Spannbetontragwerken ist denn auch bereits 1992 als europäische Vornorm ENV 1992-1-1 erschienen.

Im Grundbau hingegen ist die Trennung von Last und Widerstand nicht immer ganz einfach. Die Erarbeitung des EC 7 Entwurf, Berechnung und Bemessung in der Geotechnik Teil 1, Allgemeine Regeln, erwies sich deshalb als bedeutend schwieriger und langwieriger. Im Mai 1993 vom CEN angenommen, erschien die deutsche Fassung erst im Oktober 1994, die Veröffentlichung als ENV 1997-1 erfolgte erst im Juli 1995.

1992 wurde vom CEN das TC 288 gebildet, mit dem Auftrag, Ausführungsnormen für besondere geotechnische Bauverfahren zu erarbeiten. Als erstes wurden drei Working Groups (WG) für Pfähle, Anker und Schlitzwände eingesetzt. Die WG 2 (Anker) stellte fest, daß Anker im EC 7 recht stiefmütterlich behandelt sind. Während den Pfählen ein ausführliches Kapitel 7, Pfahlgründungen (14 Seiten) gewidmet ist, werden Anker nur im Unterabschnitt 8.8 des Kapitels 8 Stützkonstruktionen auf drei Seiten behandelt. Ohne dem Umstand Rechnung zu tragen, daß jeder Anker einer Abnahmeprüfung unterzogen wird, wird die Ermittlung des charakteristischen Ankerwiderstandes R_{ak} unverändert aus dem Kapitel Pfahlgründungen übernommen. Ein klares Bemessungskonzept für Verankerungen, welches der Wirkung der Vorspannung Rechnung trägt, fehlt.

Diese unbefriedigende Situation bewog die WG 2, ein ausführliches Kapitel für die Bemessung von Ankern in die Ausführungsnorm 1537 aufzunehmen, wohl wissend, daß dieses Kapitel an sich in den EC 7 gehört. Nach der im März 1995 abgeschlossenen Einspruchsfrist wurde die WG 2 denn auch vom CEN angehalten, das Bemessungskapitel aus der Ausführungsnorm zu entfernen, immerhin mit der Erlaubnis, das Bemessungskonzept, das in der Einspruchsfrist inhaltlich nicht auf Widerstand gestoßen ist, in einem informativen, das heißt unverbindlichen Anhang der Ausführungsnorm EN 1537 beizufügen.

Nun bleibt nur die Hoffnung, daß dieses Bemessungskonzept bei der Umwandlung des EC 7 in eine definitive europäische Norm (EN) übernommen werden wird.

2. DAS NEUE BEMESSUNGSKONZEPT FÜR VERANKERUNGEN

Bei der Bemessung von Verankerungen mit vorgespannten Ankern ist zu unterscheiden zwischen der Bemessung des verankerten Tragwerkes und der Bemessung der Verankerung sowie der einzelnen Anker.

2.1 Bemessung des verankerten Tragwerkes

Verankerte Tragwerke werden nach den Grundsätzen der EC 1, 2, 3 und 7 bemessen. Obwohl Anker Widerstandselemente sind, werden die Ankerkräfte beim Nachweis der Grenzzustände eines verankerten Tragwerkes in der Regel als (günstig wirkende) Einwirkung betrachtet. Dabei ist zu beachten, daß die Größe des Bemessungswertes der Ankerkraft, der beim Nachweis von Grenzzuständen der Tragfähigkeit anzusetzen ist, von der Art der Beanspruchung der Anker im betrachteten GZ abhängt. Wird der Anker nur durch Zug beansprucht, darf der Bemessungswert des Zugwiderstandes des Ankers angesetzt werden ($R_d = R_k / \gamma_R$). Beispiele für solche GZ sind Auftriebssicherungen oder das Umkippen von verankerten Stützwänden (Bilder 1 und 2).

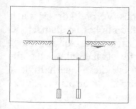

Bild 1: Auftriebssicherung, die Anker sind nur durch Zug beansprucht

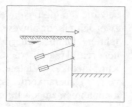

Bild 2: Erd- und Wasserdruck, die Anker sind vorwiegend durch Zug beansprucht.

Wird der Anker im betrachteten GZ nicht nur durch Zug, sondern auch durch Querkräfte beansprucht, kann der Bruch des Tragwerkes eintreten, bevor der Anker seinen vollen Zugwiderstand mobilisieren kann. Beim Nachweis solcher GZ darf als Bemessungswert nur die effektiv beim Eintritt des GZ wirksame Ankerkraft angesetzt werden ($R_d = \gamma_q \cdot P_o$). P_o ist die Festsetzkraft des Ankers, der Ankerkraftbeiwert γ_q berücksichtigt die Ankerkraftänderung zwischen dem Festsetzen des Ankers und dem Eintreten des GZ infolge Relaxation des Zuggliedes, Kriechen des Verpreßkörpers, Bauwerksverschiebungen im Bereich des Ankerkopfes (oft Kraftverluste infolge Konsolidierungssetzungen) und Verschiebungen des Bauwerkes als steifer Körper bis zum Eintreten des GZ (Kraftzunahme). γ_q variiert in der Regel zwischen 0.8 und 1.1. Beispiele für solche GZ sind Instabilitäten in Form von Gleitkreisen oder der Bruch von $\pm$ hangparallel geschichteten Felsanschnitten (Bilder 3 und 4).

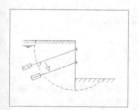

Bild 3: Geländebruch, die Verschiebung erfolgt $\pm$ rechtwinklig zur Ankerachse.

Bild 4: Hang-/Felssicherung, die Verschiebung erfolgt sehr schief zur Ankerachse.

Bei der ersten Gruppe von GZ (Bilder 1 und 2) werden Anker als reine Widerstandselemente analog der Bewehrung in Stahlbeton betrachtet. Bei der zweiten Gruppe hingegen (Bilder 3 und 4) wird die effektive Vorspannkraft maßgebend für die Tragfähigkeit des Tragwerkes.

Der Nachweis der Gebrauchstauglichkeit (GZ 2) wird mit charakteristischen Werten für die Einwir-

kungen und die Tragwerkseigenschaften geführt. Beim Ansatz der Ankerkraft ist häufig die minimale, das heisst die nach Abzug aller Kraftverluste während der Nutzungsdauer sicher noch vorhandene Vorspannkraft maßgebend. Zu beachten ist, daß in bestimmten Fällen ein Überschreiten dieser minimal vorhandenen Ankerkraft beträchtliche Tragwerksverschiebungen zur Folge haben kann, z.B. bei Auftriebsankern oder bei der Verankerung eines Mastes.

Aus diesen Ausführungen ergibt sich, daß je nach Fall die effektiv vorhandene Anker(vorspann)kraft für den Nachweis der Gebrauchstauglichkeit (GZ 2) oder aber für den Nachweis der Tragfähigkeit (GZ 1) massgebend sein kann. Dies wirkt auf den ersten Blick unlogisch, wird deshalb oft nicht verstanden und könnte geradezu als Paradoxon der Ankerbemessung bezeichnet werden.

2.2 Bemessung der Verankerung

Aus der Bemessung des verankerten Tragwerkes, das heißt, aus den verschiedenen dafür erforderlichen Nachweisen von Grenzzuständen ergeben sich die konkreten Anforderungen an die Verankerung, nämlich:

- Zugwiderstand R_k in kN (bei einer konzentrierten Ankergruppe), in kN/m^1 (bei der Ankerlage einer Stützwand) oder in kN/m^2 (bei einer Auftriebssicherung)
- die während der Nutzungsdauer des Tragwerkes minimal erforderliche Ankerkraft P_{min}
- evtl. die während der Nutzungsdauer maximal zulässige Ankerkraft P_{max}
- die erforderliche freie Ankerlänge.

2.3 Bemessung der einzelnen Anker

Die Bemessung der einzelnen Anker und die Festlegung der erforderlichen Ankerzahl erfolgen aufgrund der oben beschriebenen Anforderungen an die Verankerung unter Berücksichtigung der Baugrundverhältnisse, der Tragwerkseigenschaften und der geometrischen Gegebenheiten. Daraus ergeben sich folgende wesentliche Kenngrößen der einzelnen Anker:

- der charakteristische innere Ankerwiderstand R_{ik}

- der charakteristische Herausziehwiderstand R_{ak}
- die Festlegekraft P_o
- die erforderliche freie Ankerlänge L_{fr}

Für diese Kenngrößen gelten folgende Anforderungen und Empfehlungen:

Der charakteristische innere Ankerwiderstand ist gleich der charakteristischen Bruchlast des Ankerzuggliedes

$$R_{ik} = P_{tk} = A_t \cdot f_{pk}$$

Darin sind A_t = Querschnittsfläche des Zuggliedes
f_{pk} = charakteristische Zugfestigkeit des Zuggliedes (gemäß EC 2)

Daraus folgt, daß rechnerisch oder durch Versuche nachzuweisen ist, daß die Tragfähigkeit des Ankerkopfes und der erdseitigen Verankerung des Zuggliedes (Zugglied-Verpreßgut und, wo vorhanden Verpreßgut-Korrosionsschutzumhüllung) gleich oder größer als P_{tk} sind.

Der Herausziehwiderstand des Ankers R_a ist der Herausziehwiderstand des Verpreßkörpers aus dem Baugrund. R_a entspricht jener Kraft, bei der die Verschiebung des Verpreßkörpers nicht mehr abklingt, sondern nach einer gewissen Zeit der Bruch zwischen Verpreßkörper und Baugrund eintritt. Aus praktischen Gründen wird R_a als jene Kraft definiert, die ein bestimmtes Kriechmaß bewirkt. Der charakteristische Herausziehwiderstand R_{ak} eines Ankers ist aus den Ergebnissen von Belastungsversuchen abzuleiten. Für die Bemessung des Verpreßkörpers wird empfohlen, daß R_{ak} gleich oder größer als R_{ik} sein soll.

Diese Festlegungen beinhalten wesentliche Neuerungen gegenüber der bisherigen Praxis. Sie sind aufgrund folgender Erkenntnisse und Zielsetzungen entstanden:

· Die Korrosion von Spannstählen kann sehr rasch vor sich gehen (Spannungsrißkorrosion, Wasserstoffversprödung). Die bisher verbreitete Praxis, die Lebensdauer von Ankern durch eine Vergrößerung des Stahlquerschnittes zu verlängern, welche auf der Vorstellung von jährlichen Abrostungsraten wie im Stahlbau basierte, ist materialtechnisch falsch.

· Das Kriechverhalten des Baugrundes beruht auf einer logarithmischen Zeitfunktion. Daraus folgt,

daß die Tragfähigkeit eines Verpreßkörpers nach zwei Jahren nur unwesentlich höher ist als nach 50 Jahren. Damit erübrigt sich der Ansatz einer höheren Tragsicherheit für Daueranker als für Kurzzeitanker. Vielmehr ist die Dauerhaftigkeit von Dauerankern durch Korrosionsschutzmaßnahmen zu gewährleisten.

· Es ist anzustreben, daß das Versagen eines verankerten Tragwerkes durch Verformungen angekündigt wird (kein spröder Bruch). Deshalb wird das Zugglied, welches in absoluten Zahlen das weitaus größte Verformungsvermögen des Ankersystemes aufweist, als „schwächstes Glied" des Ankers gewählt. Das Verschiebungspotential eines Verpreßkörpers zwischen Gebrauchs- und Bruchzustand liegt bei 10 bis 15 mm, während die Verlängerung eines Litzenankers mit $L_{fr} = 20$ m zwischen Gebrauchs- und Bruchzustand etwa 75 mm beträgt.

· Weil der Stahl beim vorgespannten Anker frei dehnbar ist, besteht kein Grund, für die Bruchsicherheit wie beim Stahlbeton von der Fließgrenze auszugehen (keine örtliche Einschnürung). Dies und die Prüfung jedes Ankers bei mindestens 1.25-facher Festsetzkraft mit vorsichtigen Kriechkriterien (Abnahmeprüfung) ermöglicht die Einführung eines einheitlichen Teilsicherheitsfaktors γ_R für die innere und die äußere Ankertragfähigkeit. Vorgeschlagen wird $\gamma_R = 1.35$ für Kurzzeit- und für Daueranker.

Die maximale Ankerkraft während der Nutzungsdauer wird auf $P \leq 0.65\ P_{tk}$ festgelegt, womit das Risiko von Spannungsrißkorrosion auf ein vertretbares Maß beschränkt bleibt.

Die maximale Festsetzkraft P_o wird auf $P_o \leq 0.60\ P_{tk}$ begrenzt. Dies erlaubt eine Prüfkraft bei der Abnahmeprüfung von mindestens $1.25 \cdot P_o$, ohne den Ankerstahl höher als $0.75\ f_{pk}$ zu belasten (EC 2). Die Begrenzung von P_o erlaubt eine bescheidene Kraftzunahme während der Nutzungsdauer von max. 7.7 %. Sind größere Kraftzunahmen zu erwarten, z.B. bei der Sicherung von Kriechhängen, ist die Festsetzkraft entsprechend tiefer anzusetzen.

Daß die wirksame freie Ankerlänge innerhalb der verlangten Grenzwerte liegt, ist für jeden Anker bei der Abnahmeprüfung nachzuweisen.

3 KORROSIONSSCHUTZ FÜR DAUERANKER

Kapitel 6 des Euronormenvorschlages trägt die Überschrift „Ankerkomponenten und Korrosionsschutz" und hat einen Umfang von 11 Seiten. Es ist das zweitlängste und möglicherweise das schwierigste Kapitel. Es behandelt auf den ersten drei Seiten die Ankerkomponenten, auf den restlichen 8 Seiten wurde der Versuch unternommen, ein einheitliches Konzept für den Korrosionsschutz zu entwickeln.

3.1 Barrieren gegen den Korrosionsangriff

Da in den in der Arbeitsgruppe vertretenen Ländern hinsichtlich des dauerhaften Schutzes vor Korrosion der Stahlzugglieder und Stahlteile große Unterschiede in der Korrosionsschutz-Philosophie und in den Praktiken bestanden, versuchte man, die FIP-Richtlinie für Verpreßanker als Ausgangsbasis zu nehmen. Diese Richtlinie lag bereits in ihrer zweiten überarbeiteten Fassung vor und konnte damit den Anspruch erheben, in der Fachwelt eine Akzeptanz gefunden zu haben, da sie ebenfalls von einem internationalen Experten-Team erarbeitet worden war. Angesichts der beschränkten Zeit von zwei Jahren für die Ausarbeitung des Norm-Entwurfes war es ein glücklicher Umstand, sich auf dieses Dokument abstützen zu können. Hier war ein klar nachvollziehbares Konzept vorhanden, das in der Praxis allerdings nur den Rang einer Empfehlung hat. Die Klassifizierung des Korrosionsschutzes wird dort von der Anzahl der Korrosionsschutzhüllen abhängig gemacht. Einfacher und doppelter Korrosionsschutz sind die Merkmale für Kurzzeit- und Daueranker. Bei der europäischen Norm war ein ähnliches Konzept zunächst konsensfähig, wobei man von einer oder zwei Barrieren gegen den Korrosionsangriff ausging. Als Anforderung wurde definiert, daß alle unzugänglichen Stahlteile eines Dauerankers ausgerüstet sein müssen mit:

a) entweder zwei Korrosionsschutzhüllen, bei denen die zweite unversehrt bleibt, wenn die erste beim Ankereinbau oder Spannen beschädigt wird,

b) oder einer Korrosionsschutzhülle, deren Unversehrtheit an jedem eingebauten Anker überprüft werden muß.

Aufbauend auf diesem Konzept mußte nun definiert werden, was als vollwertige Korrosionsschutzhülle und welche Kombinationen als zwei wirksame Korrosionsschutzhüllen gelten können.

Der Abschnitt Korrosionsschutz gliedert sich in folgende Punkte:

a) Grundsätzliche „Definitionen und Anforderungen" mit 2 Tabellen, die Beispiele für übliche Korrosionsschutzsysteme bei Kurzzeit- und Dauerankern auflisten.

b) „Übliche Komponenten und Materialien für den Korrosionsschutz", wo deren Eignung für die verschiedenen Bauteile eines Ankers festgelegt sind.

c) „Aufbringung des Korrosionsschutzes", in dem wichtige Anforderungen bei den Schutzmaßnahmen für freie Stahllänge und Verankerungslänge des Zuggliedes und des Ankerkopfes aufgeführt sind.

d) „Prüfung des Korrosionsschutzes". Hier werden Apriori-Prüfungen und solche an Bauwerksankern angegeben.

3.2 Vergleichbarkeit der Korrosionsschutzmaßnahmen

Die Ausführlichkeit der Behandlung des Korrosionsschutzes hat zum Hintergrund, daß die Maßnahmen für die Dauerhaftigkeit eines Verpreßankers innerhalb des Gültigkeitsbereiches der europäischen Norm auf vergleichbares Niveau gebracht werden sollen. In manchen Ländern besteht diesbezüglich Informations- und Handlungsbedarf. Hier waren nun von den Vertretern der verschiedenen Länder die Erfahrungen hinsichtlich der zu erwartenden Lebensdauer der Korrosionsschutzhüllen unter den gegebenen Spannungs- und Verformungsbedingungen, bei den angenommenen aggressiven Substanzen in Boden und Grundwasser und unter Berücksichtigung der Alterungserscheinungen der Schutzhülle einzubringen.

Die Aufmerksamkeit konzentrierte sich vor allem auf die Verbundstrecke des Zuggliedes, wo in besonderem Maße unterschiedliche Konstruktionen anzutreffen sind. Dies ist verständlich, da die Korrosionsschutzhülle dort auch die Kraftübertragung in verschiedenartigen Baugrund von bindigem Boden über rollige Böden bis zu Fels mit übernehmen muß. Sowohl der Verbund entlang des Zuggliedes mit einer typischen Rißbildung im Zementmörtel, der als Verfüll- und Verpreßmaterial verwendet wird, als auch die Verpreßtechnik beeinflussen die Wirkungsweise und Dauerhaftigkeit einer Korrosionsschutzhülle. Es wird aber auch die Beurteilung der Wahrscheinlichkeit des Korrosionsangriffes in der Krafteinleitungsstrecke mit eingebracht und allgemein als gering eingestuft, da sie am entferntesten vom Sauerstoffzutritt ist. Ferner wird der Baugrund so weit wie möglich mit dem Verpreßgut durchsetzt und abgedichtet.

Bisher ist es weitgehend üblich, Kunststoff-Ripprohre in Kombination mit Zementmörtel zu verwenden. Wird dabei für den Zementmörtel durch Versuche eine Rißbreitenbegrenzung von 0,1 mm unter Belastung nachgewiesen, ein Wert der aus den Erfahrungen über den Korrosionsschutz im Stahlbetonbau abgeleitet ist, so wird diese Kombination als dauerhafter Schutz angesehen.

Anderenfalls wurde als neue Möglichkeit die Substitution der einen Korrosionsschutzhülle, bestehend aus Zementstein, durch eine Integritätsprüfung der verbleibenden anderen Korrosionsschutzhülle eingeführt. Diese Vorgehensweise stammt aus der Schweiz, wo heutzutage zwingend für alle Daueranker mit einer Korrosionsschutzhülle aus Kunststoffhüllrohren eine elektrische Widerstandsmessung vorgeschrieben ist. Diese in Anhang A der Norm beschriebene Methode sieht die Widerstandsmessung zu verschiedenen Zeitpunkten der Ankerherstellung vor und erlaubt des weiteren eine Ersatzprüfung am gespannten Anker mit veränderten Versuchsbedingungen.

Hinter dem Gedanken der zwei Barrieren steht immer die Überlegung, daß bei Transport, Lagerung und Einbau die äußere Schutzhülle ungewollt beschädigt werden kann und unentdeckt, d.h. ohne Reparatur eingebaut wird. Der seltenere Fall dürfte die zufällige Beschädigung der Korrosions-schutzhülle durch das Anspannen sein. In beiden Fällen soll immer noch eine vollwertige Korrosionsschutzhülle übrig bleiben.

In Großbritannien ist es Praxis, daß die zwei Korrosionsschutzhüllen als doppelte, konzentrische Kunststoff-Ripprohre angeordnet werden, die in Zementmörtel eingebettet sind.

3.3 Stahlrohre als Korrosionsschutzhülle

Die große Anzahl der Einsprüche aus der nationalen Einspruchsperiode zeigt die divergierenden Entwicklungen und Erfahrungen in den verschiedenen Ländern. Es gilt nun, bewährte Konstruktionen, die erst im Einspruchsverfahren vorgelegt wurden, so weit wie möglich zu integrieren, ohne das Grundkonzept des Kapitels in Frage zu stellen. Hier sind besonders Stahlrohre als Schutzhülle in der Verankerungslänge des Stahlzuggliedes erwähnenswert. Sie können wie in Frankreich mit dem Zugglied in der Haftstrecke über Zementmörtel solidarisiert werden und arbeiten dann unter Zugspannung. Sie können aber auch als Druckrohre ausgebildet werden, in die die Kraft vom unteren Ende voll eingeleitet werden kann. Äußere Haftrippen sind dort üblich. Aufgabe ist es nun, die Langlebigkeit der Stahlrohre bzw. deren lokale Beschädigung oder deren planmäßige Durchtrittsöffnungen, die mit Manschetten abgedeckt sind, in korrosionstechnischer Hinsicht zu bewerten.

Die Substitution einer Korrosionsschutzhülle durch ein Prüfverfahren in dem vorliegenden Norm-Entwurf deutet bereits einen gangbaren Kompromiß an, wie der Begriff „Korrosionsschutzhülle" erweitert ausgelegt werden kann. Als Grundsatz könnte festgelegt werden, daß das substituierende Verfahren bzw. die substituierende Konstruktion einer zweiten Korrosionsschutzhülle gegen mechanische Beschädigungen gleichsetzbar ist. Damit könnte z.B. bisher nicht berücksichtigten, in Deutschland aber zugelassenen Ankern, die damit zwangsläufig einer gründlichen korrosions-technischen Prüfung standgehalten haben, die Konformität mit der Norm ermöglicht werden.

3.4 Möglichkeit eines europäischen Konsenses

Es war unvermeidbar und zu erwarten, daß sich in den verschiedenen Ländern durch unterschiedliche Strukturen im Bauwesen, durch andersartige baurechtliche Regelungen und durch firmenspezifische Entwicklungen auch auf dem Gebiet der geotechnischen Arbeiten Fachgebiete unterschiedlicher Ausrichtungen entwickelt haben. Diese zu harmonisieren, ist das Ziel der europäischen Normung. Dies geschieht im wesentlichen dadurch, daß ein hohes Anforderungsniveau, das in dem einen oder anderen Land vorhanden ist, abgesenkt wird, um einen Konsens zu erreichen. Der planende Ingenieur sollte sich dieser Randbedingung bei der Anwendung einer, bzw. dieser speziellen europäischen Norm bewußt sein und seinen Spielraum unter Einbeziehung eines Experten nutzen. Hierdurch kann für einen gewissen Anwendungsfall durchaus eine Auswahl aus den genormten Systemen getroffen werden, die unter dem Gesichtspunkt der Qualität, der bisherigen Erfahrungen und der Bausicherheit am geeignetsten erscheinen.

Voraussetzung für den Erfolg einer europäischen Normungsarbeit ist der Wille zum Konsens und zur Zusammenarbeit in solchen europäischen Gremien. Dieser wurde von den deutschsprachigen Ländern bereits recht gut gepflegt, und wir sind sicher, daß diese schwierige Ausführungsnorm unter Berücksichtigung der nationalen Eigenheiten eine gemeinsame Basis und eine breite Akzeptanz finden wird.

4 ANKERSPANNPRÜFUNG IN EUROPA

Seit Verpreßanker gebaut wurden, und die ersten wurden in den Fünfzigerjahren im Festgestein gebaut, haben verantwortliche Ingenieure diese einer Spannprüfung unterzogen. Sämtliche nationalen europäischen Normen und Richtlinien besitzen solche detailierten Spannvorschriften, die sich an Grundsatzuntersuchungen und an Spannerfahrungen anlehnen. Dabei stehen zwei Prüfungskriterien im Vordergrund :

a) das Tragverhalten des Ankers hinsichtlich zulässiger Vorspannung und langzeitlicher Dauer zu erkennen,

b) Nachweis der im erdstatischen Entwurf ausgewiesenen Festlegekraft und Krafteintragungslänge durch den Spannvorgang

Bei der weit verbreiteten Anwendung von Verpreßankern darf es nicht verwundern, daß sehr unterschiedliche Spannprüfungen und Auswertungen in den europäischen Staaten entstanden sind und einen dementsprechend vielfältigen Erfahrungsschatz darstellen. Bei der Erstellung des europäischen Normenentwurfes wäre es unklug gewesen, auf diese in jahrzehntelanger Spannpraxis gewachsene Kenntnisse zu verzichten zugunsten eines bestimmten Verfahrens. In gesonderten Sitzungen bei Dr. Ostermayer an der TU München war es möglich, die etwas unterschiedlichen Spannvorschriften in Deutschland, in der Schweiz und in Österreich zu einem akzeptablen Prüfverfahren zusammenzufassen, um dieses gleichwertig mit dem englischen und französischen Vorspannverfahren in den Vorschlag der Euronorm aufzunehmen. Der derzeitige Normenvorschlag, der

für eine Endfassung vorbereitet wird, sieht somit drei Prüfverfahren vor, und zwar:

Prüfverfahren 1, das im wesentlichen die Spannvorschriften der deutschsprachigen Staaten wiedergibt,

Prüfverfahren 2, das dem British Standard 8081 von 1989 und der FIP Richtlinie von 1991 folgt und

Prüfverfahren 3, das weitgehend der französischen Ankernorm entspricht.

4.1 Arten der Ankerversuche

Wie die Auswertung der nationalen Einsprüche zum Normenentwurf zeigte, erbrachte diese Art der Harmonisierung nur sehr wenig kritische Stimmen. Es wurde nicht sosehr die mögliche Vielfalt bemängelt, als vielmehr die z.T. noch nicht ausreichend vorhandenen Anweisungen für die unterschiedlichen Arten der Ankerversuche. In Anlehnung an die meisten nationalen Normen wurde unterschieden zwischen:
* Untersuchungsprüfung (investigation test), die an zuggliedmäßig verstärkten Ankern durchgeführt wird oder an Ankern, die gegebenenfalls eine verkürzte Krafteintragungslänge aufweisen dürfen, wenn eine Spannsystemveränderung gegenüber den Bauwerksankern unumgänglich werden würde;
* Eignungsprüfung (suitability test), die an wenigstens drei Bauwerksankern durchzuführen ist und sich an die Ergebnisse der Untersuchungsprüfung anlehnt;
* Abnahmeprüfung (acceptance test), der jeder Anker unterzogen werden muß, die aber möglichst einfach und kurz gehalten werden soll;
* Nachprüfung (monitoring during design life), über deren Meßintervalle und Anzahl der Anker der Entwurfsverfasser zu entscheiden hat.

Ankerprüfungen werden seitens der Baustellen immer wieder ihrer erforderlichen Zeitdauer und ihres Meßaufwandes wegen kritisiert. Dabei wird übersehen, daß die Ergebnisse der Spannprüfungen während der Einsatzdauer des Ankers oft die einzige Information über das boden- oder felsmechanische Zusammenwirken mit dem Untergrund darstellen. Die Verfasser der Euronorm haben sich daher für die geforderte Meßgenauigkeit von der Vorstellung leiten lassen, daß diese in Relation zur maximalen Versuchslast gesehen werden kann und begrenzten die Genauigkeit der Meßeinrichtung der Spannlast auf 2% der maximalen Prüflast. Wenn es um die Erfassung der Spannlastveränderung über die

Spannzeit geht, also um die Messung des Kriechens des Ankers oder seines Vorspannverlust, wurden anspruchsvolle Meßgeräte vorgeschrieben, wie sie auf modernen Ankerbaustellen zu finden sind. Was die Spannzeiten betrifft, wurden für die drei Prüfmöglichkeiten gemäß der Handhabung auf der Baustelle Beobachtungszeiten vorgeschrieben, die sich entweder an der Verschiebungsgeschwindigkeit des Verpreßkörpers (zeitliche Zunahme oder Abnahme des Kriechmaßes) oder am zeitlichen Verlauf des Lastabfalles orientieren und somit untergrundabhängig unterschiedlich lang sein können.

4.2 Gemeinsamkeiten und Unterschiede der drei Ankerprüfverfahren

Im Laufe der jahrzehntelangen Spannpraxis haben sich bei Untersuchungs- und Eignungsprüfungen in Europa grundsätzlich zwei unterschiedliche Prüfphilosophien für die Ermittlung der Traglast der Anker herausgebildet. Die französischen Ankerspezialisten führen die Vorspannung des Ankers in einem einzigen stufenweise unterbrochenen Belastungsvorgang durch und entlasten den Anker erst nach Erreichen der Prüflast. In England und im deutschsprachigen Raum nähert man sich der Prüflast mit Be- und Entlastungsschleifen, die auf ähnlichen Laststufen wie in Frankreich die Prüflast erreichen.

In England stellt man die Zeitbeständigkeit der eingebrachten Vorspannung über den Vorspannverlust nach einer bestimmten Anzahl von Stunden oder Tagen bei der jeweiligen Belastungsstufe fest. In Deutschland, Frankreich, Österreich und in der Schweiz wird der zeitliche Verschiebungszuwachs des Ankerkopfes bei konstant gehaltener Vorspannung über eine bestimmte Anzahl von Minuten oder Stunden ermittelt (Kriechmaßermittlung). Dabei ergeben sich beim französischen Verfahren ohne Entlastungsvorgänge konsequenterweise andere Kriechmaße als beim Lastschleifenverfahren. Im deutschsprachigen Raum wird das maximale Kriechmaß in Millimeter pro Zeitdekade, das der Dimension nach einer Geschwindigkeit gleich kommt, je nach Höhe der Vorbelastung (Prüfkraft, Festlegekraft etc.) zahlenmäßig vorgeschrieben. Bei Erreichen dieser Kriechmaße ist die jeweilige Grenzkraft gegeben.

Die französische Ankernorm sieht eine halbgraphische Ermittlung des Näherungswertes P'_c

für die kritische Kriechlast P_c vor, die sich aus dem Übergang von einer linearen Zunahme des Kriechmaßes zu einer nicht immer leicht zu erfassenden exponentiellen Zunahme desselben bei der weiteren Laststeigerung ergibt (Bild 5). Es gilt dann $P_c = 0.9 . P'_c$

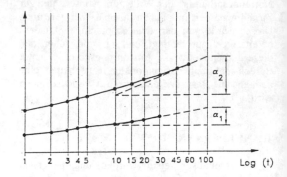

Zeit—Kriechverschiebungs Diagramm mit Kriechmaßen α_n

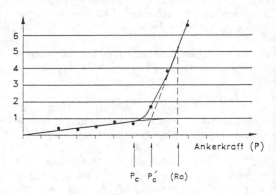

Kriechmaß — Ankerkraft Diagramm

Bild 5: Prüfverfahren 3, Bestimmung des Näherungswertes für die kritische Kriechlast Pc

Viel Diskussion wird es noch bei der endgültigen Festlegung der Länge der Beobachtungzeiten zur Ermittlung des Kriechmaßes bzw. zu den bisher vorgeschlagenen Werten geben, wie die eingesandten Einsprüche erkennen lassen. Ähnlich wie in der DIN 4125 wird versucht werden, zumindest für die Eignungs- und Abnahmeprüfungen eine Reduktion der Mindestbeobachtungszeiten zu erreichen.

Während zwischen dem französischen Beurteilungsverfahren und der deutschsprachigen

Version, was die Tragfähigkeit und die Kriechbeurteilung betrifft bei den zulässigen Grenzlasten nur geringe Unterschiede bestehen, ist eine Vergleichbarkeit mit dem englischen Verfahren des Spannkraftverlustes grundsätzlich nicht gegeben. In den meßbaren Abfall der Vorspannung (load loss) geht die Hysteresis der Lastschleife zwischen Belastung und teilweiser Entlastung ein, die als Systemreibung bis zu 5 % der Prüflast betragen darf und den Vorspannverlust verkleinert. Außerdem bestehen für den zulässigen Spannkraftabfall keinerlei zugehörige Längenangaben der maximalen freien Ankerlänge. Bei besonders langen Ankern können daher aus der elastischen Stahldehnung rückgerechnete Kriechverformungen und damit Kriechmaße entstehen, die die kontinentaleuropäischen Vorschriften um das Mehrfache übersteigen. Zusätzlich können noch unterschiedliche Stahlzugglieder (Stangen, Drähte, Litzen) und Stahlqualitäten unterschiedliche Spannkraftverluste ergeben.

Das englische Prüfverfahren macht für sich den Vorteil der langen Beobachtungsperiode (im Extremfall bis zu 10 Tage) geltend, im Vergleich zu den verhältnismäßig kurzen Meßzeiten der anderen Verfahren. Einen Kompromiß in der Vergleichbarkeit der zulässigen Kriechgrenzkräfte kann es unserer Meinung nach nur dann geben, wenn im englischen Verfahren die freie Ankerlänge L_{free} auf etwa 20 m begrenzt wird. Ob es dazu kommen wird ist ungewiß. Andererseits ist bekannt, daß die englische Spannprüfung durch ihren hohen Zeitaufwand kostenintensiv ist, und es ist anzunehmen, daß durch den nun vorliegenden Normenvorschlag englische Ankerhersteller sich mit der Kriechmaßbestimmung befassen werden. Vielleicht wird sich dann diese Methode einheitlich durchsetzen.

4.3 Protokollierung

Der Euronormenvorschlag pr EN 1537:1994 enthält detaillierte Angaben zur Verfassung des Ankereinbauplans, zum verwendeten Ankersystem, zur Ankerherstellung und zum Spannvorgang. Im Zuge der Behandlung der Einsprüche werden nun die Vorschreibungen zur Ankerprüfung und zur Protokollierung in einem normativen oder informativen Annex zusammengefaßt, da die vorgeschlagene Vielfalt und Detaillierung nach jüngster Mitteilung vom CEN aus Brüssel nur bedingt in das Euronormenschema paßt. Durch die

vorgegebene starre Gliederung der Euronormen war es nicht möglich die drei verschiedenen Ankerprüfmethoden getrennt zu beschreiben, was zu Lasten der Übersichtlichkeit der Norm ging. Ob diesem Übel, das auch in mehreren Einwendungen angesprochen wurde im nunmehr vorgesehenen Annex durch eine Neugestaltung des Kapitels „Ankerprüfung, Überwachung und Nachprüfung" abgeholfen werden kann, hängt vom Konsenswillen des CEN ab.

Um die Bemessung, die Dimensionierung und die Prüfung der Anker zu erleichtern wurde der Versuch unternommen im Zusammenhang der Teilsicherheitsbeiwerte einerseits und die normmäßige Stahlzuggliedauslastung, sowie die Kriechkriterien andererseits in einer Tabelle darzustellen (Tabelle 1). Sie beinhaltet die wichtigsten Entwurfs- und Spannkriterien, erhebt aber keinen Anspruch auf Vollständigkeit. Sie ist auch nicht Teil des Normenvorschlages, sondern nur ein mehrfach überarbeiteter Behelf zum schnellen Vergleich der Ankerprüfungen und stellt lediglich den gegenwärtigen Stand der Vornorm dar.

5 SCHLUSSBEMERKUNG

Als die europäische Normenarbeit für besondere geotechnische Bauverfahren 1992 begann, meinten alle nationalen Normeninstitute, daß die Euronormen nach Fertigstellung für die Staaten der Europäischen Union einen verbindlichen Charakter erhalten würden. Ausgelöst durch diese Vermutung sahen sich die nationalen Spiegelausschüsse veranlaßt, Delegierte in die Arbeitsgruppen zu entsenden um das in jahrelanger Zusammenarbeit mit zahlreichen Tiefbaufachleuten aus Bauwirtschaft, Universitäten und Verwaltung gewonnene Qualitätsniveau möglichst auch in einem enger zusammengeschlossenen Europa zu erhalten. Mit dem nun vorliegenden Normenvorschlag der pr EN 1537 scheint das auch weitgehend erreicht zu sein.

Wenn allerdings das CEN Komitee nunmehr feststellen muß, daß die kommenden Euronormen keinerlei nationale Verbindlichkeiten besitzen werden, sondern es den Auftraggebern, also dem Auftragsmarkt nach wie vor freigestellt bleibt die technischen Vertragsgrundlagen selbst zu wählen, dann stellt sich die Frage nach dem Nutzen der Euronormen. Bei den in vielen europäischen Staaten bereits vorhandenen Normenwerken zum gleichen

Thema wird eine Euronorm wenig Anklang finden, da sie schließlich nur einen mehr oder weniger geglückten Kompromiß in der vorhandenen Meinungsvielfalt darstellen kann. Es bleibt also nur zu hoffen, daß die politischen Kräfte Europas den Willen zur Einheitlichkeit unter Bedachtnahme auf begründete Vielfalt in naher Zukunft bekräftigen, soll nicht die gesamte europäische Normenarbeit einem auseinanderstrebenden Zeitgeist geopfert werden.

Tabelle 1
Bemessungskonzept und Ankerprüfung der Vornorm pr EN 1537

	RESISTANCE		DESIGN		TESTING		CREEPING		LOAD LOSS
	Steel	Ground	Ultimate Limit State	Servicebility Limit State	Investigation	Suitability / Acceptance	D, A,CH Method 1 k_s	F Method 3 α	UK Method 2 k_l
S T R E S S	R_i ($R_{i,k}$) R_a ($R_{a,k}$)					$\leq 1,0 \, P_{t,k}$			
	$R_k \leq R_{i,k}$; $R_{a,k}$				$P_p \geq R_k$	$P_p \leq 1,0 P_{t,o.1k}$	$\leq 2mm$		3 days, 7 %
					$\leq 0,80 \, P_{t,k}$				
					$P_{p\,max} \leq 0,95 \, P_{t,o.1k}$	$P_{p\,max} \leq 0,90 \, P_{t,o.1k}$			
	$R_d = R_k / \gamma_R$		$R_d \geq S_d$		$P_{p\,min} \geq S_d$	$P_{p\,min} \geq 1,50 \, P_o$			3 days, 7 %
L E V E L						$\geq 1,25 \, P_o$	$\leq 0,80$ mm $\leq 1,0$ mm^x	$< 1,2$ mm $\leq 1,5$ mm^{xx} $\leq 1,8$ mm^{xxx}	
	$P \leq 0,65 \, P_{t,k}$						$\leq 0,8$ mm $\leq 1,0$ mm^{xx} $\leq 1,2$ mm^{xxx}		
	$P_{o\,max} \leq 0,60 \, P_{t,k}$		$R_d > C_d$			$P_{o\,max} = \gamma_q \, P_o$	$\leq 0,50$ mm		
	$P_{o\,max} \leq 0,50 \, P_{t,k}$					$P_{o\,max} = \gamma_q \, P_o$			50 ' (3%)

P	Ankerkraft	$P_{t,k}$	Charakterischtische Bruchkraft des Zuggliedes
P_p	Prüfkraft	$P_{t, o.1\,k}$	Tragkraft an der charakteristischen Spannung des
P_o	Festlegekraft		Stahlzuggliedes bei 0,1 % bleibender Dehnung

γ_q Ankerkraftbeiwert $0,80 \leq \gamma_q \leq 1,10$

γ_R Teilsicherheitsbeiwert des Ankerwiderstandes R_k $\gamma_R = 1,35$

x   Werte bei vorliegenden Investigation Tests

xx   Werte bei vorliegenden Investigation Tests für Daueranker

xxx   Werte bei vorliegenden Investigation Tests für Kurzzeitanker

6 WEITERFÜHRENDES SCHRIFTTUM

DIN 4125, Verpreßanker Kurzzeitanker und Daueranker, Bemessung, Ausführung und Prüfung, Nov. 1990.

prEN 1537:1994, Ausführungen von besonderen geotechnischen Bauverfahren, Verpreßanker, deutsche Ausgabe, Sept. 1994.

Federation Internationale de la Precontrainte (FIP) „Recommendations for the design and construction of prestressed ground anchorages", Thomas Telford Ltd, London 1991.

Norme Francaise NF P 94-153, Sols: reconnaissance et essais, Essai statique de tirant d'ancrage, Nov. 1991.

Eidgenössisches Verkehrs- und Energiewirtschaftsdepartment, Richtlinien für permanente Boden- und Felsanker, Bundesamt für Straßenbau, Bern 1992.

British Standard Code of practice for Ground Anchorages, BS 8081:1989.

Ö Norm B 4455, Erd- und Grundbau, Vorgespannte Anker für Festgestein und Lockergestein, Aug. 1992

Spullersee: Sperrenerhöhung

Lake Spuller: Increasing the height of the barrages

M. Schmitter
Elektrobauleitung Innsbruck, Österreichische Bundesbahnen, Austria

Zusammenfassung

Unmittelbar nach dem ersten Weltkrieg wurde die schrittweise Elektrifizierung der österreichischen Bahnstrecken beschlossen. Voraussetzung dafür war die Errichtung von bahneigenen Wasserkraftanlagen. Für die Elektrifizierung der Arlbergbahn wurde in den Jahren 1919 bis 1926 das Kraftwerk Spullersee erbaut. Der Aufstau des in den Lechtaler Alpen gelegenen Spullersees erfolgte durch zwei betonierte Talsperren. In einer weiteren Ausbaustufe kam es in den Jahren 1962 bis 1965 zur Erhöhung der bestehenden Sperren um insgesamt 4.0 m.

Aus statischen und wirtschaftlichen Überlegungen wurde eine Bauweise mit vorgespannten Stahlankern gewählt, wobei mit Zementbrei verpreßte Vieldraht - Anker nach dem System BBRV zum Einbau gelangten. Zweifel an der Alterungsbeständigkeit der Spannglieder machten zerstörungsfreie Untersuchungen an den Ankern erforderlich, die jedoch keine konkreten Aussagen zuließen.

1981 ermöglichte eine Erweiterung des Sperrenmeßsystems unter anderem eine Überwachung der Ankerfunktion. Zusätzlich wurden Drainagierungen des Sperrenaufstandsbereiches mit einer umfangreichen Überwachung des Sohlwasserdruckes vorgenommen. Die Funktionstüchtigkeit der Anker ist durch die zusätzlichen Drainagierungen nur mehr in untergeordnetem Umfang notwendig. Aus heutiger Sicht ist in den nächsten Jahren ein Ersatz der Ankerfunktion geplant.

The decision on the gradual electrification of Austria's railway lines was adopted shortly after World War I. The prerequisite to implement this decicion was the construction of railway-owned hydropower plants. For electrification of the Arlberg railway line the Spullersee power plant was built in the period between 1919 and 1926. The lake, which is located in the Lechtal Alps, has been dammed up by two concrete barrages. In the period between 1962 and 1965, the height of the two existing barrages was increased by a total of 4,0 metres to maximize energy output. Due to static and economic considerations prestressed steel bolts, viz. multi-wire BBRV system bolts with bolt drill holes filled up with cement had to be used. Doubts about the ageing stability of the prestressing elements called for nondestructive testing of the bolts, on the basis of which, however, no concrete statements could be made.

In 1981, the barrage measuring system was enlarged to include, amongst other things, the monitoring of the bolt performance. Moreover, drainages at the barrage contact surface as well as a sophisticated system to control ground water pressure were provided. Due to the additional drainage system bolt efficiency is required only at a subordinate level. Complete replacement of the bolts has been planned to be carried out in the years to come.

1 ALLGEMEINES

Bereits 1891 wurden von der damaligen österreichischen Staatseisenbahnverwaltung Studien mit dem Ziel, Kohle durch die Nutzung der Wasserkraft zu ersetzen, in Auftrag gegeben. Es war naheliegend, Bahnstrecken mit schwer lüftbaren Tunnel - wie etwa auf der Arlbergbahn - als erste für die Einführung der elektrischen Zugförderung auszuwählen. Im weiteren Verlauf der Studien plante man entlang der Bahnstrecke am Inn

Bild 1: Spullersee Südsperre
Fig. 1: Spullersee south barrage

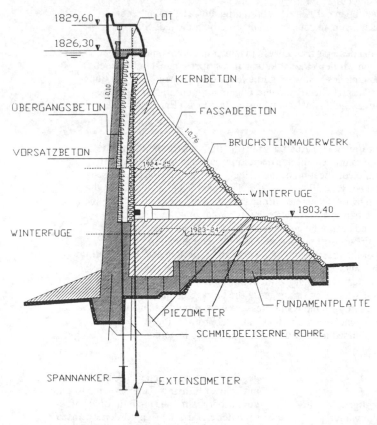

Bild 2: Querschnitt durch die Südsperre
Fig. 2: South barrage cross section

und am Oberlauf des Illflusses bahneigene Wasserkraftwerke. Zu diesen Projekten gesellte sich schließlich auch eines, das die Nutzung des Spullersees vorsah. Der erste Weltkrieg führte dazu, daß diese Projekte vorerst nicht weiter vorangetrieben wurden. Durch den Krieg verlor das neue Österreich nahezu alle kohlenhältigen Landstriche, wodurch die Nutzung der Wasserkräfte an Priorität gewann.

1920 wurde die schrittweise Elektrifizierung der Bahnstrecken beschlossen und auf Grund der Studien konkrete Standorte für den Bau von Wasserkraftanlagen festgelegt. Einerseits sollte das die Mittenwaldbahn speisende Ruetzkraftwerk vergrößert und zur Grundlasterzeugung herangezogen werden, andererseits das Speicherkraftwerk Spullersee zur Spitzenabdeckung ausgebaut werden.

Die Errichtung der Spullerseekraftwerksanlagen vollzog sich in 3 Stufen:
* Bau der Kraftwerksanlage in den Jahren 1919 bis 1926
* Erweiterung der Anlage in den Jahren 1930, 1938 und 1939 durch Beileitungen.

* Sperrenerhöhung in den Jahren 1963 bis 1965

2 ERSTE AUSBAUSTUFE

Der Spullersee befindet sich im Land Vorarlberg in den Lechtaler Alpen nördlich der Arlbergbahn. Der Aufstau des Sees erfolgt mittels zweier Sperren, welche auf erhöhten Felsschwellen gegründet sind. Die Spullerseemauern gelten als die ersten Talsperren, die damals nach 10 jähriger, kriegsbedingter Unterbrechung in Österreich zur Ausführung gelangten. Bis zu diesem Zeitpunkt war es im europäischen Raum üblich, große Staumauern aus Bruchsteinmauerwerk herzustellen. Warum bei diesem Sperrenbau erstmals der Baustoff Beton Verwendung fand, ist unter anderem darauf zurückzuführen, daß es an erfahrenen Steinmetzen mangelte.

2.1 PLANUNG

Zur Ausführung kamen zwei Gewichts - Staumauern mit annähernd dreieckigem Re-

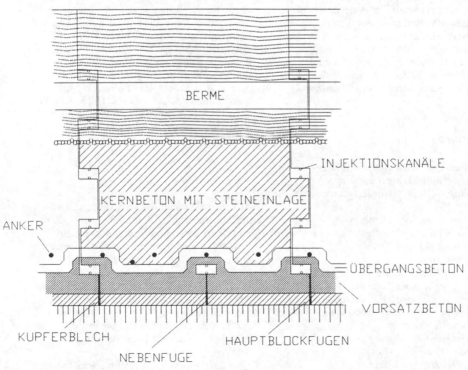

Bild 3: Grundriß eines Abschnittes der südlichen Staumauer
Fig. 3: Part of south barrage, ground plan

gelquerschnitt und aufgesetzter Fahrbahn an der Sperrenkrone. Die Spitze dieses Dreieckes lag in Höhe des höchsten Hochwasserstandes (1825.50 m). Die Mauerneigung ist wasserseitig 1 : 0.10 und luftseitig 1 : 0.76, womit sich die Sohl-breite b = 0.86 h ergibt. Bei der Südsperre sitzt das Regelprofil im mittleren Teil der Sperre auf einem verbreiterten Sohlblock auf, der an der Luftseite eine Berme bildet.

Im Grundriß sind beide Staumauern gekrümmt; bei der Südsperre beträgt der Bogenhalbmesser 5000 m, bei der Nordsperre 400 m. Die gekrümmte Anordnung wurde von den gemauerten Talsperren übernommen, wo sie unter anderem wegen ihres vorteilhaften Verhaltens bei stark schwankenden Temperaturen und wegen des günstigen Einflusses der Gewölbewirkung auf die Dichtheit der Querfugen Verwendung fand.
  Unter Zugrundelegung des tatsächlichen Mauergewichtes und eines Sohlwasserdruckes, der an der Wasserseite 100 % beträgt und zur Luftseite linear auf 0 % abnimmt, verläuft die Resultierende der Belastung bei vollem Speicher durch den luftseitigen Drittelpunkt. Der Querschnitt bleibt also unter diesen Belastungsannahmen gerade noch zugspannungsfrei. Außer diesem Lastfall Vollstau wurden damals keine weiteren Belastungen, wie Überstau, Eislast und Erdbeben in die Berechnung aufgenommen. Die Wirksamkeit der Bogenform wurde aus statischer Sicht ebenfalls nicht berücksichtigt.

2.2 AUSFÜHRUNG

In den einzelnen Querschnittsteilen werden die besonderen Anforderungen durch spezielle Betonsorten erfüllt. Gefordert wurde dabei:
* *Wasserdichtheit der Sohle*: Einbringung einer wasserdichten Sohlplatte und Verpressen des Sohlbereiches mit Zementsuspension.
* *Dichtheit und Frostbeständigkeit der Wasserseite*: Einbau einer Zone aus Vorsatzbeton. Als Besonderheit erhielt die Nordsperre eine wasserseitige Verkleidung aus Betonformsteinen.
* *Frostbeständigkeit der Krone und der Luftseite*: Anordnung einer Zone aus Fassadenbeton bzw. Bruchsteinen.
Damals nahm man von einer Entwässerung der Gründungssohle und des wasserseiti-

gen Betonbereiches Abstand, da günstige Gründungsverhältnisse und ein kräftiger Mauerquerschnitt vorhanden waren.

Der Felsuntergrund besteht aus dichtem Fleckenmergel und Abneter Kalk, dessen Oberfläche nur geringe Verwitterungsspuren zeigt.
  Für die Spullerseemauern kam ein durch Stampfen gut verdichtbarer, plastischer Beton zur Anwendung, wobei sowohl Übergangsbeton, als auch der Kernbeton Steineinlagen bis zu 12 % Massenanteil erhielten. Die Blocklängen betragen 16 bis 22 m . In der Mitte zwischen je zwei Hauptfugen ist eine Nebenfuge eingeschaltet, die nur die wasserseitige Vorsatzbetonschichte durchtrennt. Risse, hervorgerufen durch Schwinden und Temperatureinfluß, können damit unterbunden werden. Die Hauptfugen sind im Grundriß verzahnt angeordnet, dadurch kann ein gewisser Querverband erzielt werden.

3 SPERRENERHÖHUNG

Die Entwicklung des Transportvolumens zeigte für den Raum Tirol und Vorarlberg ein seit dem Jahre 1948 ständig steigendes Energiedefizit, dem man unter anderem mit der Vergrößerung des Spullersee - Speicherraumes Rechnung trug.
  In den Jahren 1962 bis 1965 wurden die beiden Staumauern des Spullersees aufgestockt und das Stauziel von Kote 1825.00 auf 1829.60 erhöht. Dadurch konnte ein Energiezuwachs von rund 57.6 GWh erzielt werden.

3.1 PLANUNG

Aus statischen und wirtschaftlichen Erwägungen wurde diese Aufhöhung von 4.00 m mittels vorgespannter Anker geplant. So konnte mit einem geringen Betonaufwand von nur 17 m3 je Längenmeter Sperrenkrone das nutzbare Speichervolumen um rund. 20 % erhöht werden. Diese Bauweise kam erstmals 1935 bei der im Jahre 1882 erbauten Staumauer Cheurfas (Algerien) zur Anwendung. In Europa gab es bis zum damaligen Zeitpunkt keine derartige Ausführung.
  Bei der Ausarbeitung des Projektes tauchten einige Fragen auf, die mittels Modell - und Großversuch erforscht und geklärt wer-

den mußten. Im Großversuch wurden 5 Anker mit einer Spannkraft von 640 kN bzw. 1380 kN unter verschiedenen Bedingungen in den beiden Sperren erprobt. Während der dreijährigen Beobachtungszeit wurden die Spannglieder zum Teil bis zum Bruch belastet. Parallel dazu wurden spannungsoptische Modellversuche sowie Laboruntersuchungen an den Ankerstählen durchgeführt.

Folgende Punkte wurden untersucht:

* *Die Einleitung der Ankerkraft, besonders in der unteren Haftstrecke.*

Die spannungsoptischen Untersuchungen an mehreren Modellen mit verschieden langen Verankerungsstrecken zeigen folgendes Spannungsbild:

Bei voll wirkender Haftung stellt sich am Kontaktbeginn eine ausgesprochene Schubspannungsspitze mit raschem, konkavem Abfall auf Null ein. Nach Überschreiten der Haftfestigkeit im Bereich der Spannungsspitze tritt, verbunden mit einem begrenzen Schlupf, eine Verflachung des Spannungsverlaufes ein.

Ist die Einbindungslänge kleiner als die Kraftübertragungslänge, dann verteilt sich die Haftspannung mehr oder weniger gleichmäßig auf die ganze Einbindungslänge.

Aus diesen Versuchen ergibt sich eine Mindestlänge der Haftstrecke von 80 cm. Bei einem angenommenen Sicherheitsfaktor von 3 beträgt die Verankerungslänge somit rund 2.50 m.

* *Spaltkräfte im Bereich der Verankerung.*

Zugspannungen im Verankerungsbereich konnten bei den spannungsoptischen Versuchen nicht nachgewiesen werden. Zur Kontrolle der Ankerfunktion sind in der Süd- und Nordsperre insgesamt drei Zweifachextensometer installiert. Ausgehend vom Kontrollgang sind sie parallel zu den Ankern angeordnet. Ein Extensometer reicht bis zur Haftstrecke der Anker, das zweite 10 m darunter. Bei Ausschaltung der Temperaturkomponente zeigen die Meßwerte bei Vollstau nur eine geringfügige Dehnung des Felsbereiches unterhalb der Haftstrecke der Anker. Es ist jedoch nicht auszuschließen, daß der erhöhte Kluftwasserdruck bei Vollstau oder die staubedingte Spannungsänderung die Ursachen dafür sind.

* *Das Korrosionsverhalten*

Untersucht wurde vor allem das Korrosionsverhalten der ungeschützten Anker für den Zeitraum vom Einbau bis zur Verfüllung mit Zementbrei. Als Langzeitschutz für die Anker hat man sowohl die Zementverfüllung, als auch den relativ hohen pH - Wert des gestauten Wassers und das damit

verbundene basische Klima als ausreichend angesehen. Die Möglichkeit einer Funktionskontrolle zu einem späteren Zeitpunkt ist in der Planung nicht berücksichtigt worden.

Es wurden drei Anker in der Südsperre eingebaut; davon zwei mit Inertol II behandelt und einer ohne jeglichen Korrosionsschutz belassen. Begleitend dazu wurden Laborversuche an verschiedenen Ankerstählen durchgeführt. Bei den Streckversuchen am Anker mit Korrosionseinwirkung zeigte sich, daß zu Beginn des normalen Streckbereiches der Dehnung ein erhöhter Widerstand entgegengesetzt wird, der bei weiterem Lastanstieg in ein rasches Nachgeben der spezifischen Spannung übergeht.

Auch Stabanker wurden in Erwägung gezogen. Die Verankerung im Untergrund und die erschwerte Einbringung der Anker in das Bohrloch waren einige der Gründe, die ein Vieldraht - Ankersystem zweckmäßig und sicherer erscheinen ließen. Man wählte schließlich das System BBRV - benannt nach Birkenmayer, Brandestini, Ros und Vogt, die dieses System entwickelt haben - für die Bauausführung. Der Anker besteht aus mehreren Drähten der Güte Delta 100 A, mit einem Durchmesser von 7 mm. Die Drähte wurden an beiden Enden nach genauer Ablängung durch die Ankerköpfe gesteckt und mittels maschinell verpreßter Stauchköpfe gehalten. Der untere Ankerkopf ist als Kegelstumpf ausgebildet, um ein Verkeilen im Auspreßmörtel zu bewirken. Der obere Ankerkopf besteht aus einem Stahlring mit Außen - und Innengewinde, Stellring und Stützplatte. Die Dichtungsmanschette am oberen Ende der Haftstrecke hat die Aufgabe, ein vollständiges Verpressen der Anker - Haftstrecke zu gewährleisten.

Durch den Einbau der Spannanker wird in den Sperrenquerschnitt eine Druckbelastung eingebracht, die wegen ihrer Exzentrizität zur Ankerachse insbesondere an der Wasserseite zum Tragen kommt und dem Kippmoment aus der Staubelastung entgegenwirkt. Es läßt sich so auf sparsamste Weise erreichen, daß die Standsicherheit der Sperren auch dem Katastrophensicherheitsnachweis gerecht wird.

3.2 AUSFÜHRUNG

Die exponierte Lage der Baustelle, vor allem die extreme Lawinengefahr ließ nur eine Bauzeit von 5 Monaten im Sommer zu.

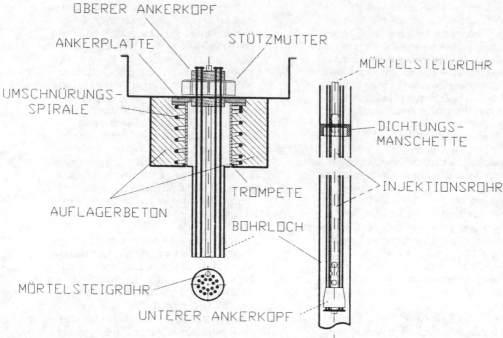

Bild 4: Stahldrahtanker nach System BBRV
Fig. 4: BBRV system steel wire bolt

Die Bauarbeiten mußten daher zwangsläufig auf 3 Jahre aufgeteilt werden.
AUFBETON
  Der oberste Sperrenbereich wurde schonend abgetragen. Vor der Betonierung der Sperrenerhöhung versah man den Übergang zum alten Sperrenkörper mit einer Haftbrücke. Die Aufbetonierung erfolgte zweilagig mit Beton der Güte B 225.
BOHREN DER ANKERKANÄLE
  Die Ankerkanäle wurden als lotrechte Bohrungen von der neuen Sperrenkrone aus abgeteuft und reichen bis zu 4.0 m unter die Sohlfuge. Der kritische Bereich liegt einerseits in der überschlanken Zone des Sperrenquerschnittes, andererseits in der Gründungsfuge. Unterhalb dieses Bereiches wird die Ankerkraft aus der unteren Verankerung voll wirksam in den Sperrenquerschnitt eingeleitet. Eine Ausnahme bildet der Mittelteil der Südsperre; hier weist die Mauer unterhalb der luftseitigen Berme eine Überbreite auf. An dieser Stelle wurden nur die Injekionsbohrungen tiefer abgeteuft, um in den Untergrund einen Dichtungsschleier aus Zementsuspension einpressen zu können. Die Anker reichen hier jedoch nicht in den Felsuntergrund.

  Die Ankerkanäle wurden als Rotationsbohrungen abgeteuft. Mittels Multi - shot - Gerät erfolgte die Prüfung der Bohrlöcher, wobei die Abweichungen vom Lot kleiner 2 % waren. Das Ergebnis der anschließend vorgenommenen Wasserabpreßversuche in den für die Übertragung der Ankerkräfte bestimmten untersten Abschnitten der Bohrlöcher machte es vereinzelt notwendig, Ankerkanäle mit Zementinjektionsgut auszupressen und wiederaufzubohren.
VERSETZEN DER ANKER
  Nach Versetzen der Vieldraht - Anker wurden die unteren Haftstrecken bis zu den Dichtungsmanschetten mit Zementbrei verfüllt. Das Spannen der Anker erfolgte nach 28 Erhärtungstagen auf die volle Last. Ein Nachspannen nach einer Woche zeigte keine nennenswerte Relaxation. Unmittelbar anschließend an die Nachspannung wurden die Ankerkanäle mit Zementsuspension von unten nach oben mit Hilfe der eingebauten Injektionsrohre gefüllt. Bis zu 5 Tage nach erfolgter Zementverfüllung konnten Nachsetzungen festgestellt werden, die drucklos nachgefüllt werden mußten. In der Krone der beiden Sperren waren entsprechend den Spannankerabständen Spanngruben ausge-

DATEN DER VERANKERUNG

		NORDSPERRE	SÜDSPERRE
Erf. Ankerkraft / Sperrenlänge	kN/m	150.0	400.0
Spannlast pro Anker	kN	630.0	1320.0
Ankerabstand	m	4.4	3.3
Ankeranzahl	Stk	43	75
Ankerlänge minimal	m	12.9	13.3
mittel	m	25.4	31.0
maximal	m	33.4	41.3
gesamt	m	1 092.0	2 320.0
Drahtanzahl	Stk	17	35
Drahtdurchmesser	mm	7.0	7.0
Stahldrahtquerschnitt gesamt	mm2	655.0	1 348.0
Bruchfestigkeit	N/mm2	1740.0	
Streckgrenze (0.2 %)	N/mm2	1550.0	
Elastizitätsgrenze (0.01 %)	N/mm2	1250.0	
Aufgebrachte Vorspannung	N/mm2	980.0	
Elastizitätsmodul	N/mm2	20 220.0	
Ankerkanaldurchmesser	mm	76	101
Bohrlochgesamtlänge	m	1 100	2 315

spart worden, die man nach dem Versetzen der Anker ausbetonierte.

4 WEITERE MASSNAHMEN

Etwa 10 Jahre nach der Herstellung der Verankerung kamen auf Grund von Versagens - Mechanismen an Ingenieurbauten - unter anderem an vorgespannten Konstruktionen - Zweifel an der Alterungsbeständigkeit der Spannglieder bei den Spullerseesperren auf. Die behördlichen Sachverständigen forderten einen Funktionsnachweis der Anker.

Geprüft wurden verschiedene Möglichkeiten der Untersuchung:

Zerstörungsfreie Untersuchungen

Ein Anker - Spannversuch bringt keine Aussage über dessen Zustand, da dieser fest in Zementstein eingebettet ist. Auch eine Kontrolle der Verformung der Mauer läßt keinen direkten Rückschluß auf die Funktion der Spannanker zu. Eine visuelle Untersuchung kann sich nur auf den obersten Ankerbereich beziehen. Sie läßt keine verläßliche Aussage über den gesamten Ankerbereich zu. Elektrische Prüfverfahren können nur bedingt eingesetzt werden. Sie können keine eindeutige Aussage über den Zustand der Spannglieder abgeben.

Entnahme und Untersuchung von Ankern

Ein Herausbohren von Ankern schied aus Kostengründen und vor allem wegen des sehr hohen Ausführungsrisikos aus.

SPANNANKERKONTROLLE

1977 wurde der Spannanker Nr. 73/Südsperre auf eine Länge von 0.7 m unterhalb des Spannkopfes freigelegt und kontrolliert. Der behördliche Sachverständige stellte fest, daß eine Rostgefährdung auszuschließen ist.

1982 wurden zur Kontrolle des Sperrenverhaltens parallel zur Ankerebene Extensometer versetzt. Aus den Meßwerten läßt sich kein außergewöhnliches Verhalten feststellen.

1986 hat Prof. G. Feder durch indirekte Meßverfahren die Anker 58 und 59 geprüft. Zur Anwendung kamen folgende Prüfverfahren:

* Elektrisches Impulsechoverfahren
* Elektrische Widerstandsmessung
* Messung des elektrochemischen Ruhepotentiales
* Elektrochemischer Kurzschlußstrom zwischen den Ankern
* Messung des elektrochemischen Polarisationswiderstandes der Anker
* Besondere Korrosionsmechanismen

Die Messungen ergaben, daß beim Anker 58 ein Korrosionvorgang abläuft. Bei Anker 59 ist er nicht auszuschließen, beschränkt sich jedoch auf den unteren Bereich. Der Korrosionsvorgang läuft sehr langsam ab und ist nicht örtlich konzentriert. Ein Drahtriß wurde nicht festgestellt.

SOHLWASSERDRÜCKE

Die Größe des Sohlwasserdruckes hat erfahrungsgemäß einen hohen Einfluß auf die Standsicherheitsbetrachtungen der Sperren.

Parallel zur Funktionskontrolle der Spannglieder wurde eine Reduktion des Sohlwasserdruckes vorgenommen, um die Ankerfunktion auf einen statisch untergeordneten Bereich zu beschränken.

Die bei der Sperrenerhöhung versetzten 3 Piezometer ließen keine Aussage über die vorhandenen Auftriebsverhältnisse im gesamten Sperrenaufstandsbereich zu. Daher war die Installation zusätzlicher Piezometermeßeinrichtungen erforderlich. Um die Drainagewirkung zu verbessern, wurden die Bohrungen vom luftseitigen Fuß der Sperre aus abgeteuft. Die Bohrungen reichen bis 5.0 m unter die Gründungssohle bzw. bis zu 4.0 m an die durch die Anker errichtete Dichtfront heran. Durch die schräge Lage der Bohrungen konnten besonders bei den Drainagebohrungen die lotrechten, nahezu achsparallelen Schieferungs- und Trennflächen ideal aufgefädelt werden. Der Einbau erfolgte in zwei Phasen:

In der ersten Phase wurden in einem Abstand von 25 m Druckmeßquerschnitte eingerichtet. Jeder Meßquerschnitt besteht aus zwei Piezometerbohrungen, die den Sohlwasserdruck in den Drittelpunkten erfassen. In einer zweiten Phase wurde der Sohlwasserdruck auf Grund der gewonnenen Meßergebnisse gezielt durch Drainagefächer abgesenkt. Zur ständigen Kontrolle der Sohlwasserdrücke sind in den Piezometerrohren elektrische Druckaufnehmer mit einer Grenzwerterfassung installiert.

STATISCHE ÜBERPRÜFUNG

Die Gleit - und Kippsicherheit der Blöcke ist bei allen in Betracht zu ziehenden Lastfällen auch ohne Ankerwirkung gegeben. Im Sohlbereich der niederen Randblöcke und im überschlanken Betonquerschnitt treten jedoch wasserseitige Zugspannungen auf. Beim Lieckfeldtnachweis mit den Sonderlastfällen kann jedoch auf die Funktion der Anker nicht verzichtet werden. Weist man der Bogenwirkung in einer groben Abschätzung 10 % der Staubelastung zu, so treten im Querschnitt der Nordsperre keine Zugspannungen mehr auf. Der Ansatz einer geringen Bogenwirkung ist auf Grund der speziellen Fugenausbildung und des Bogenhalbmessers von 400 m gerechtfertigt. Die Meßergebnisse der langjährigen Fugenüberwachung bei den Staumauern bestätigen dieses monolitische Verhalten.

5 RESÜMEE

Aus all den Untersuchungen läßt sich die Alterungsbeständigkeit nicht uneingeschränkt nachweisen. Die Notwendigkeit der Anker ist aus statischen Gesichtspunkten nur mehr in einigen wenigen Bereichen der Südsperre erforderlich. Für diese Bereiche werden spezielle Ankerersatzmaßnahmen geplant und in den nächsten Jahren ausgeführt. Dabei werden mehrere Varianten untersucht, wie:

* Versetzen von prüfbaren Ankern im oberen überschlanken Betonbereich und im Aufstandsbereich der niederen Randblöcke.
* Ersatz der Ankerkraft durch Betonauflast an der Sperrenkrone.
* Luftseitige Mauerverstärkung.
* Luftseitige Mauerverstärkung im Zuge einer weiteren Sperrenerhöhung.

LITERATUR:

BIENER E., 1983
Zur Sanierung älterer Gewichtsmauern.
Mitteilung Inst. f. Wasserbau und Wasserwirtschaft, TH Aachen.

GESSMANN H., 1981
Großerneuerung Kraftwerk Spullersee: Erster Abschnitt fertiggestellt.
ÖBB spezial, 1981

ILLWERKE A.G., 1992
Machbarkeitsstudie, Sperrenerhöhung.
Bericht 1992.

MÜHLHOFER L., 1933
Die Staumauern des Spullerseewerkes.
Die Wasserwirtschaft, Nr. 17 - 19 - 1933.

RAETSCH W., 1980
Über die Kapazitätserweiterung von Talsperren.
Mitteilung Inst. f. Wasserbau und Wasserwirtschaft, TH Aachen.

RUTTNER A., 1964
Großversuche mit Spannankern an Talsperren der Österreichischen Bundesbahnen und die Anwendung der Vorspannbauweise auf den Talsperrenbau.
Österr. Bau - Zeitung, Sonderdruck Nr. 38, 1964.

RUTTNER A., 1968
Die Erhöhung der Spullersee - Talsperren.
Schweizer Bau - Zeitung, Heft 50, 1968.

ÖSTERREICHISCHE BUNDESBAHNEN
Sperrenbuch, Bestandsunterlagen.

Kritik an der Ankertechnik anhand von Beispielen aus der Vergangenheit

Critique on the anchor technique shown by history cases

H. Stäuble

Tauernkraftwerke AG / Tauernplan Consulting GmbH, Salzburg, Austria

ZUSAMMENFASSUNG: Obwohl die Ankertechnik in den letzten beiden Jahrzehnten große Fortschritte gemacht hat, kam es immer wieder zu Schadensfällen. An Beispielen aus der Vergangenheit wird dies aufgezeigt. Die Zweckmäßigkeit des Einsatzes von Daueranker ist daher bei jedem Bauwerk gewissenhaft zu prüfen. Empfehlungen für die Anwendung von Daueranker werden gegeben.

ABSTRACT: Although the development of the anchor technique has made great process in the past two decades defects again and again occur. In most cases defects of permanent anchors can lead to additional expensive repairs. History cases show the defects that can occur. The first case is an example for an application of unsuitable permanent anchor by the existing ground conditions. The second case is an example for poor protection against corrosion. The third case is an example for the use of unsuitable material. Concerning all structures it should be carefully investigated if an application of a permanent anchor is the best solution for the whole life of the structure. For application of permanent prestressed anchors some advice is given on design, installation and monitoring.

1 EINLEITUNG

Anker kommen im Fels- und Grundbau immer mehr zur Anwendung. Mit diesen hoch beanspruchten Konstruktionsteilen können große Kräfte zwischen den Bauwerken und dem Untergrund übertragen werden. Ihre Anwendung erweist sich in vielen Fällen als die wirtschaftlichste Baumethode.

In den letzten beiden Jahrzehnten wurde die Konstruktion der Anker hinsichtlich ihres Einbaues, des Korrosionsschutzes und der Kontrollmöglichkeit immer mehr verbessert. Normen und Vorschriften über Anker, die in vielen Ländern erlassen wurden, trugen dazu nicht unwesentlich bei.

Trotzdem sollte bei jeder Bauausführung sorgfältig geprüft werden, ob der Einsatz von Ankern die beste Methode ist.

Bei Ankern, die nur für temporäre Bauzustände verwendet werden, ergeben sich bei sorgfältiger Prüfung beim Vorspannen meist keine Probleme.

Schwierig ist die Entscheidung bei Ankern, die auf Dauer zur Standsicherheit eines Bauwerkes notwendig sind. Kann die Tragfähigkeit von so hoch beanspruchten Konstruktionsteilen auf Dauer sichergestellt werden? Wie erfolgt die Überprüfung? Was tun, wenn später Schäden auftreten?

An einigen Beispielen aus der Vergangenheit wird gezeigt, wie schwierig es ist, Schäden an Daueranker mit Sicherheit zu vermeiden.

2 SCHADEN INFOLGE DES EINBAUES UNGEEIGNETER ANKER

Bei einer Bergstraße mußte eine Steinschlaggalerie errichtet werden. Eine Fundierung an der talseitigen Straßenseite war nicht möglich. Zur Ausführung kam eine Stützmauer mit einer 4,5 m langen, aus der Stützmauer auskragenden Kragplatte. Die Stahlbetonkonstruktion wurde mit Vorspannanker in dem bis zu 35° steilen Gelände, bestehend aus gemischtkörnigem Hangschuttmaterial, verankert (s. Abb. 1).

Die Galerie wurde im Winter 1974/75 errichtet. Zur Ausführung kamen 68 Litzenanker mit einer Länge von je 22 m, davon 7 m Haftstrecke. Die Vorspannkraft betrug 70 to. Zur Injektion der Haftstrecke wurde Zementmörtel, zum Teil mit Sand

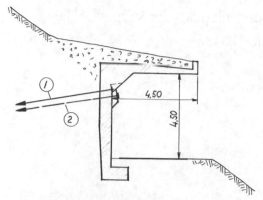

Abb. 1: Querschnitt durch die Steinschlaggalerie.
1 ... ursprüngliche Anker
2 ... zusätzliche Anker
Cross section through the protection
structure against rock fall.
1 ... first anchor
2 ... additional anchor

vermischt, verwendet. Die Injektionsgutaufnahme
betrug i.M. 720 kg pro Anker. Schon beim
Vorspannen der Anker ergab sich eine relativ hohe
Ausfallsquote. 11 Anker wurden ausgezogen und
mußten durch neue ersetzt werden.

Aufgrund dessen und der großen Bedeutung der
Anker für die Standsicherheit des Bauwerkes wurden
alle Jahre ca. 25 % der Anker auf ihre Vorspannkraft
geprüft. Im Jahre 1983 betrug die Vorspannkraft der
Anker i.M. nurmehr 67 % der ursprünglichen.

Abb. 2: Ansicht des Schutzbauwerkes mit den alten
und neuen Ankerköpfen.
View of the protection structure with the old an new
anchorheads.

Genaue Untersuchungen ergaben, daß der Abfall
der Vorspannkraft im Versagen der
Verankerungsstrecke lag. Es stellte sich heraus, daß
Anker eingebaut wurden, deren Litzen in der
Verankerungsstrecke freilagen. Die Injektion der
Verankerungsstrecke erfolgte mit einem PE-Schlauch
mit nur einer Öffnung am Ende der
Verankerungsstrecke. Die Anker waren für den
Einbau in Felsbohrungen geeignet, aber nicht für
Alluvionen. Die durch die Injektion gebildeten
Verankerungskörper bildeten sich großteils nur am
Ende der Anker. Am Beginn der
Verankerungsstrecke lagen die Stahllitzen zum Teil
frei. Trotz getrennter Bearbeitung von Planung,
Ausführung und Bauaufsicht kam es zum Einbau
nicht geeigneter Anker!

Aufgrund dieser Feststellung wurden im Jahre 1984
Ersatzanker eingebaut. Zur Ausführung kamen
Litzenanker mit doppeltem Korrosionsschutz mit
einer Länge von 30 m, davon 8 m
Verankerungsstrecke. Bemerkenswert ist, daß beim
Einbau dieser Anker die Injektionsgutaufnahme i.M.
7040 kg Zement pro Anker betrug gegenüber 720 kg
beim Einbau der ersten Anker.

3 SCHADEN INFOLGE KORROSION

Im Jahre 1966 wurde auf 2560 m Seehöhe eine hohe
Seilbahnstütze errichtet. Die Stütze hatte ein
Mittelrohr und 4 Streben. Um Beton zu sparen,
erfolgte die Fundierung der Stütze mit
Vorspannankern. Die Gebrauchslast der Anker
betrug 65 to bzw. 144 to. Verwendet wurden Anker
mit 20 bzw. 44 Spanndrähten der Stahlgüte 155/175
mit einem Durchmesser von 6 mm. Die Länge der
Anker betrug 22 bis 27 m.

Die Anker mußten in einer Zone mit Permafrost
eingebaut werden. Vor dem Einbau wurden die
Bohrungen mit Warmwasser längere Zeit gespült.
Die Injektion der Verankerungsstrecke erfolgte mit
Zementmörtel (W/Z 0,4). Der Korrosionsschutz der
Freispielstrecke erfolgte damals mit einer
Bentonitsuspension. Nach Meinung aller Beteiligten
war dies unter den gegebenen Verhältnissen die beste
Methode, um die Anker auch später kontrollieren zu
können.

Eine Kontrolle der Anker im Dezember 1967 ergab,
daß einzelne Spanndrähte abgerissen waren. Die
Untersuchung zeigte, daß die Litzen 0,6 bis 1,2 m
unter dem Ankerkopf gerissen sind. Bis auf diese
Höhe ist die Bentonitsuspension abgesessen und
ermöglichte eine Korrosion der Spanndrähte.

Metallurgische Untersuchungen ergaben, daß die
Risse durch Spannungsrißkorrosion entstanden sind.
Kleine Verletzungen der Drahtoberfläche in

Abb. 3: Stützenfundament mit der ursprünglichen und zusätzlichen Ankerkonstruktion.
Foundation of the cableway mast with the first and additional anchor construction.

Abb. 4: Seilbahnstütze mit alten verankerten Fundamenten und neuen Gewichtsfundamenten.
Cableway mast with the old anchor foundation and the new gravity foundation.

Verbindung mit hohen Zugspannungen führen zu dieser Korrosionsform. Erwähnenswert für diese Korrosionsform ist, daß der Zeitraum vom Beginn der Korrosion bis zur Zerstörung nur wenige Monate beträgt.

Um den Betrieb der Seilbahn sicherzustellen, mußte im Dezember 1966 bei Temperaturen bis - 26°C unter großem Zeitdruck eine zusätzliche, notdürftige Verankerung hergestellt werden (s. Abb. 3).

Im Jahre 1980 erfolgte ein Umbau der Seilbahnanlage, um ihre Kapazität zu erhöhen. Dafür mußte auch die Seilbahnstütze erhöht und mit 4 zusätzlichen Streben verstärkt werden. Aufgrund der

Schäden im Jahre 1966 entschloß sich der Bauherr, auf weitere Vorspannanker zu verzichten und Gewichtsfundamente auszuführen (s. Abb. 4).

4 SCHADEN INFOLGE FEHLENDER KONTROLLMÖGLICHKEIT DER ANKER

In der Bergstation einer Seilbahn wurde das Kuppengerüst und die Tragseiltrommel in Fels verankert. Da sich hinter der Seilbahnstation ein Stollen befand, erfolgte die Verankerung durch Bohrungen bis in diesen Stollen. Die Anker einschließlich der für damals nach Ö-Norm B 4455 vorgeschriebenen Hüllrohre wurden durch die Bohrungen geschoben und vorgespannt. Der Einbau erfolgte im Jahre 1965.

Verwendet wurden Anker mit je 32 Spanndrähten, Durchmesser 6 mm, Stahlgüte 155/175. Die Vorspannkraft betrug 105 to je Anker.

Aus konstruktiven Gründen erfolgte die Vorspannung der Anker für das Kuppengerüst vom Kuppengerüst aus am unteren Ende der Anker. Für die Tragseiltrommel erfolgte die Vorspannung vom Stollen aus ebenfalls am unteren Ende der Bohrung (s. Abb. 5).

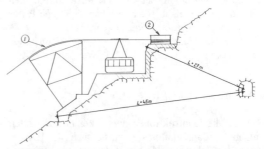

Abb. 5: Schemaquerschnitt durch die Bergstation mit der Verankerung der Einfahrtsstütze (1) und der Tragseiltrommel (2).
Schematic illustration of the mountain station with the anchoring of the entrance mast (1) and cable drum (2).

Nach dem Vorspannen wurden die Anker als Schutz gegen Korrosion mit Zementmörtel verfüllt.

Bei einer Kontrolle im Jahre 1977 wurde festgestellt, daß bei den Ankern für das Kuppengerüst 2 Spanndrähte gerissen sind. Bis zum Jahre 1979 erhöhte sich diese Zahl auf 5 Stück. Genauere Kontrollen ergaben, daß sich das Injektionsgut abgesetzt hat und im oberen Teil der Anker kein ausreichender Korrosionsschutz gegeben war.

457

Da durch die Zementinjektion keine Möglichkeit mehr bestand, die Anker zu prüfen, wurden alle Anker durch neue ersetzt.

Die neuen Anker wurden so ausgeführt, daß eine Prüfung der Anker von beiden Enden aus möglich ist und die Anker jederzeit ausgetauscht werden können. Um einen Austausch auch während des Betriebes der Seilbahn durchführen zu können, wurde je ein Reserveanker eingebaut.

5 SCHADEN INFOLGE UNGEEIGNETEN MATERIALS

Bei einer Druckrohrleitung wurden, um Beton zu sparen, zwei Fixpunkte verankert. Verwendet wurden doppelt korrosionsgeschützte Felsanker aus Spanndrähten der Stahlgüte ST 135/150. Die Anker wurden 1974 eingebaut.

Bei Kontrollen wurde festgestellt, daß die Vorspannkraft der Anker abgenommen hat und mehr als 25 % der Spanndrähte gerissen sind.

Untersuchungen ergaben, daß der Stahl sehr empfindlich gegen Korrosion ist. Die Spanndrähte wurden zwischen Fertigung und Einbau an der Baustelle verschmutzt. Nach dem Einbau kam es zu einer wasserstoffinduzierten Spannungsrißkorrosion. Verstärkt wurde dieser Effekt noch durch den Umstand, daß nach dem Einbau kein perfekter Korrosionsschutz der Anker gegeben war.

Vom Ankerhersteller wurde der Stahl damals aus dem Verkehr gezogen.

6 SCHLUSSFOLGERUNGEN

Obwohl die Kontruktionen der Anker in den letzten beiden Jahrzehnten hinsichtlich Einbau, Korrosionsschutz und Kontrollmöglichkeit immer wieder verbessert wurden, sind auch in Zukunft Schäden an Dauerankern nicht ausgeschlossen. Der Einsatz von Dauerankern bei Bauwerken, deren Standsicherheit bei Versagen der Anker in Frage gestellt ist, soll daher gewissenhaft geprüft werden. Bei Vergleich mit anderen Alternativen ist zu berücksichtigen, daß die Kontrollen der Anker später oft nicht unerhebliche Betriebskosten verursachen.

Bei einem Einsatz von Dauerankern ist jedenfalls folgendes zu beachten:
- Die Ankerkonstruktion ist sorgfältig, abgestimmt auf den jeweiligen Anwendungsfall auszuwählen. Vorsicht bei Sonderkonstruktionen, für die noch keine Erfahrungen vorliegen!
- Der Einbau, die Injektion, die Vorspannung und der Korrosionsschutz ist streng zu überwachen.

Oft wird sowohl von den ausführenden Firmen als auch von den kontrollierenden Stellen Personal eingesetzt, das nicht genügend Erfahrung besitzt. Die Anwesenheit eines Ingenieurs beim Vorspannen allein genügt nicht!
- In den Terminplänen sind genügend Zeitreserven für Zusatzmaßnahmen einzuplanen. Dies ist besonders bei wasserführendem Untergrund wichtig. Mehrmaliges Nachinjizieren und Wiederaufbohren kann für einen ordnungsgemäßen Einbau der Anker notwendig werden. Auch genügend Abbindezeit für das Injektionsgut ist einzuplanen.
- Anker mit geringerer Stahlausnützung sind vorzuziehen.
- Bei Ankern, die für den Bestand eines Bauwerkes entscheidend sind, sollte immer eine Kontrollmöglichkeit gegeben sein.
- Außerdem soll immer die Möglichkeit bestehen, erforderlichenfalls später Ersatzanker einbauen zu können.
- Bei Bauwerken, bei denen bei Versagen von Dauerankern die Standsicherheit unter S = 1 sinkt, sollte immer geprüft werden, ob nicht durch andere, wenn auch teurere Lösungen auf so hoch beanspruchte Konstruktionsteile, wie es Vorspannanker sind, verzichtet werden kann. Dies ist die persönliche Meinung des Berichtsverfassers aufgrund von vielen Beispielen aus der Vergangenheit.

Anchors in Theory and Practice, Widmann (ed.) © 1995 Balkema, Rotterdam. ISBN 90 5410 577 1

Author index
Autorenverzeichnis